Clinical Neuroanatomy
and Neuroscience | FIFTH
EDITION

Commissioning Editor: Inta Ozols
Development Editor: Joanne Scott
Project Manager: Bryan Potter
Design Manager: Jayne Jones
Illustrator: Richard Tibbitts
Marketing Manager(s) (UK/USA): Jeremy Bowes/John Gore

Clinical Neuroanatomy and Neuroscience | FIFTH EDITION

M J Turlough FitzGerald, MD, PhD, DSc, MRIA

Emeritus Professor
Department of Anatomy
University College
Galway, Ireland

Gregory Gruener, MD, MBA

Professor of Neurology, Vice-Chairman of Neurology,
Associate Dean of Educational Affairs
Stritch School of Medicine
Loyola University Medical Center
Chicago, IL, USA

Estomih Mtui, MD

Associate Professor of Clinical Anatomy in Neurology and
Neuroscience, Director, Program in Anatomy and Body
Visualization
Cornell University
Weill Medical College
New York, NY, USA

ELSEVIER
SAUNDERS

ELSEVIER
SAUNDERS

An imprint of Elsevier Limited

First edition 1985
Second edition 1992
Third edition 1996
Fourth edition 2002
Fifth edition 2007

Main Edition
ISBN: 978-1-4160-3445-2
 Reprinted 2007

International Edition
ISBN: 978-0-8089-2376-3
 Reprinted 2008

British Library Cataloguing in Publication Data
A catalogue record for this book is available from the British Library

Library of Congress Cataloging in Publication Data
A catalog record for this book is available from the Library of Congress

Notice
Medical knowledge is constantly changing. Standard safety precautions must be followed, but as new research and clinical experience broaden our knowledge, changes in treatment and drug therapy may become necessary or appropriate. Readers are advised to check the most current product information provided by the manufacturer of each drug to be administered to verify the recommended dose, the method and duration of administration, and contraindications. It is the responsibility of the practitioner, relying on experience and knowledge of the patient, to determine dosages and the best treatment for each individual patient. Neither the Publisher nor the author assume any liability for any injury and/or damage to persons or property arising from this publication.
The Publisher

Printed in China

Contents

CONTENTS

The table below lists the main clinical perspectives covered in the book

CLINICAL PERSPECTIVES

Chapters	Perspectives
1 Embryology	(Explanatory layout)
2 Cerebral topography	(Explanatory layout)
3 Midbrain, hindbrain, spinal cord	(Explanatory layout)
4 Meninges	Extradural hematoma. Subdural hematoma. Hydrocephalus. Meningitis. Spinal tap. Epidural analgesia. Caudal analgesia.
5 Blood supply of the brain	Blood-brain barrier pathology.
6 Neurons and neuroglia	Brain tumors. Multiple sclerosis. Neuronal transport disorders.
7 Electrical events	(Explanatory layout)
8 Transmitters and receptors	Some general clinical applications concerning malfunctions and pharmacology.
9 Peripheral nerves	Degeneration and regeneration.
10 Innervation of muscles and joints	(Explanatory layout)
11 Innervation of skin	Neurogenic inflammation. Leprosy.
12 Electrodiagnostic examination	Peripheral neuropathies including entapment syndromes. Myasthenia gravis.
13 Autonomic nervous system and visceral afferents	Horner's syndrome. Raynaud syndrome. Stellate block. Lumbar sympathectomy. Visceral pain. Drug actions on the sympathetic and parasympathetic systems.
14 Nerve roots	Spina bifida. Cervical spondylosis. Prolapsed intervertebral disc.
15 Spinal cord: ascending pathways	Syringomyelia
16 Spinal cord: descending pathways	Upper motor neuron disease. Lower motor neuron disease. Spinal cord injury.
17 Brainstem	(Explanatory layout)
18 The lowest four cranial nerves	Supranuclear, nuclear, infranuclear lesions.
19 Vestibular nerve	Vestibular disorders. Lateral medullary syndrome.
20 Cochlear nerve	Conduction deafness. Sensorineural deafness.
21 Trigeminal nerve	Trigeminal neuralgia. Referred pain in diseases of the head and neck.
22 Facial nerve	Lesions of the facial nerve.
23 Ocular motor nerves	Several well-known ocular palsies.
24 Reticular formation	Cardiovascular, respiratory, urinary, locomotor controls. Spinal and supraspinal antinociception.
25 Cerebellum	Characteristic clinical pictures associated with lesions of vermis, of anterior lobe and of neocerebellum. Cerebellar cognitive affective syndrome.
26 Hypothalamus	Hypothalamic disorders including major depression.
27 Thalamus, epithalamus	(Explanatory layout)
28 Visual pathways	Detection of lesions of the visual pathways, segment by segment.
29 Cerebral cortex	(Explanatory layout)
30 Electroencephalography	Narcolepsy. Seizures of several kinds and their EEG detection.
31 Evoked potentials	Use of visual, auditory, somatosensory and motor evoked potentials in disease detection. Clinical neurophysiology in relation to acupuncture.
32 Hemispheric asymmetries	The aphasias. Developmental dyslexia. Frontal lobe dysfunction. Parietal lobe dysfunction.
33 Basal ganglia	Parkinson's disease. Cerebral palsy. Huntington's disease. Hemiballism.
34 Olfactory and limbic systems	Alzheimer's disease. Schizophrenia. Drug addiction.
35 Cerebrovascular disease	Eight Clinical Panels about Strokes of various kinds.

Preface

This new edition has the same *Structure-Function-Malfunction* organization as the fourth. Throughout, descriptions of *Structure* are followed immediately by descriptions of *Function* in the form of everyday applications. The primary aim of this book is then expressed, in the description of *Malfunction* in the context of structural and/or functional derangements. The outcome is in keeping with the established principle of *vertical integration*, as expressed at undergraduate level. This approach has enabled 200 neurological disorders to be touched upon, as indicated in the Clinical Perspectives list.

Every effort has been made to deliver information in a graded manner. The basic assumption is that the reader is new to Neuroscience in general and to Clinical Neuroscience in particular. Any attempt to make the nervous system appear simple would be quite inappropriate, and the style of presentation is rigorous. But a helping hand *is* offered, in the Study Guidelines provided for each chapter.

The material covered has increased substantially for this edition:

- The molecular biology of neurotransmitters and their receptors is introduced in a new Chapter 8, providing a basic foundation for the Clinical Neuropharmacology required by all medical graduates.
- Clinical electrophysiology appears for the first time in this edition. New Chapters 7 and 12 are primarily concerned with normal and deranged electrical responses in the peripheral nervous system. New Chapters 30 and 31 address the most frequent derangements in the brain, and their causation.

The large number of new topics required both text and illustrations to increase in amount by one-third. Events involving sequential activation of neural pathways are explained in *flow diagrams*, where numbered arrows guide the viewer through the normal sequence of events within the maze between input and output. Examples include: second messenger pathways in the context of neurotransmission; sequences involved in automatic execution of learned movements; operation of brainstem controls of circulation and respiration; intestinal, urinary and genital reflex responses to excitation; transfer of novel sensory experiences into memory; and reading something out loud.

It is a pleasure to acknowledge the honorary Panel of Consultants, nine of whom were members for the fourth edition and all of whom have been helpful within their specialist areas.

Particularly helpful in identifying items requiring amendment in the first print of this edition has been Cairo-based Prof. Kamal Asaad.

Also acknowledged, within several information Boxes/Clinical Panels, are specialists who kindly volunteered expert advice.

At NUI, Galway, most consistent supporters over the years have been Professors Michael Kane (Physiology), Brian Leonard (Pharmacology) and Peter McCarthy (Radiology). Another most helpful colleague has been Professor Hugh Staunton, Department of Clinical Neurological Sciences, Royal College of Surgeons in Ireland.

It has been a particular privilege to have had the continued artistic support of Richard Tibbitts and his team at Antbits Illustration, Saffron Walden, UK. Their illustrations are a joy to behold.

At Elsevier London, editors Inta Ozols and Joanne Scott provided constant support and advice.

TF
GG
EM

Panel of Consultants

Kamal Asaad,
Department of Anatomy,
Faculty of Medicine,
Ain Shams University Cairo,
Egypt

J. Andrew Armour,
Centre de recherche,
Hôpital du Sacré-Coeur de Montréal,
Canada

N.E. Bharucha,
Chief, Department of Neuroepidemiology,
Medical Research Centre,
Bombay Hospital,
Mumbai,
India

J.S. Chopra,
Department of Anatomy,
Chandigarh University School of Medicine,
Chandigarh,
India

Timothy Counihan,
Department of Neurology,
University College Hospital,
Galway,
Ireland

J. Paul Finn,
Director,
Magnetic Resonance Research,
Department of Radiology,
David Geffen School of Medicine at UCLA,
California,
USA

Chizuka Ide,
Department of Anatomy and Neurobiology,
Kyoto University Graduate School of Medicine,
Kyoto,
Japan

Brian E. Leonard,
Department of Pharmacology,
National University of Ireland at Galway,
Ireland

Hans-Joachim Kretschmann,
Department of Neuroanatomy,
Medical University of Hannover,
Germany

Pearse Morris,
Director, Interventional Neuroradiology,
Wake Forest University School of Medicine,
Winston-Salem,
N. Carolina,
USA

Masao Norita,
Division of Neurobiology and Anatomy,
Department of Sensory and Integrative Medicine,
Niigata University Graduate School of Medical and
    Dental Sciences,
Japan

Wei Yi Ong,
Department of Anatomy,
Yong Loo Lin School of Medicine,
National University of Singapore,
Singapore

Tetsuo Sugimoto,
Department of Neuropsychiatry,
Kansai Medical University,
Osaka,
Japan

Shen Yucun,
Department of Psychiatry,
The Chinese University of Engineering,
Beijing,
China

Li Shwei,
Department of Anatomy,
Shandong University faculty of Medicine,
Jilin,
Shandong,
China

Hugh Staunton,
Department of Clinical Neurological Sciences,
Royal College of Surgeons in Ireland,
Dublin,
Ireland

B. Ulfhake,
Department of Neural Transmission,
Karolinska Institutet,
Stockholm,
Sweden

Mario Rende,
Section of Anatomy,
Department of Experimental Medicine and
    Biochemical Sciences,
University of Perugia School of Medicine,
Perugia,
Italy

J.R. Sanudo,
Department of Anatomy and Human Embryology,
Faculty of Medicine,
Complutense University of Madrid,
Spain

David Yew,
Department of Anatomy,
The Chinese University of Hong Kong,
Hong Kong S.A.R.

Dedication

For Ellen Green, who realized the fourth edition

Embryology

STUDY GUIDELINES
This chapter aims to give you sufficient insight into development to account for the arrangement of structures in the mature nervous system. If not already familiar with adult brain anatomy, we suggest you read this chapter again following study of Chapters 2 and 3.

For descriptive purposes, the embryo is in the prone (face-down) position, whereby the terms *ventral* and *dorsal* correspond to the adult *anterior* and *posterior*, and *rostral* and *caudal* correspond to *superior* and *inferior*.

SPINAL CORD

Neurulation

The entire nervous system originates from the **neural plate**, an ectodermal thickening in the floor of the amniotic sac (Figure 1.1). During the third week after fertilization, the plate forms paired **neural folds**, which unite to create the **neural tube** and **neural canal**. Union of the folds commences in the future neck region of the embryo and proceeds rostrally and caudally from there. The open ends of the tube, the **neuropores**, are closed off before the end of the fourth week. The process of formation of the neural tube from the ectoderm is known as *neurulation*.

Cells at the edge of each neural fold escape from the line of union and form the **neural crest** alongside the tube. Cell types derived from the neural crest include spinal and autonomic ganglion cells and the Schwann cells of peripheral nerves.

Spinal nerves

The dorsal part of the neural tube is called the **alar plate**; the ventral part is the **basal plate** (Figure 1.2). Neurons developing in the alar plate are predominantly sensory in function and receive dorsal nerve roots growing in from the spinal ganglia. Neurons in the basal plate are predominantly motor and give rise to ventral nerve roots. At appropriate levels of the spinal cord, the ventral roots also contain autonomic fibers. The dorsal and ventral roots unite to form the spinal nerves, which emerge from the vertebral canal in the interval between the neural arches being formed by the mesenchymal vertebrae.

The cells of the spinal (dorsal root) ganglia are initially bipolar. They become unipolar by the coalescence of their two processes at one side of the parent cells.

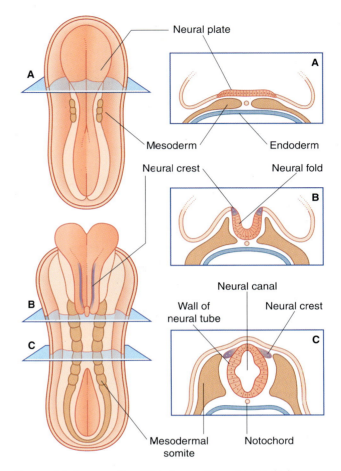

Figure 1.1 Cross-section **(A)** is from a 3-somite (20-day) embryo. Cross-sections **(B)** and **(C)** are from an 8-somite (22-day) embryo.

BRAIN

Brain parts

Late in the fourth week, the rostral part of the neural tube undergoes flexion at the level of the future midbrain (Figure 1.3A). This region is the **mesencephalon**; slight constrictions mark its junction with the **prosencephalon** (future forebrain) and **rhombencephalon** (future hindbrain).

The alar plate of the prosencephalon expands on each side (Figure 1.3A) to form the **telencephalon** (cerebral hemispheres). The basal plate remains in place here as the **diencephalon**. Finally, an **optic outgrowth** from the diencephalon is the forerunner of the retina and optic nerve.

The diencephalon, mesencephalon, and rhombencephalon constitute the embryonic brainstem.

The brainstem buckles as development proceeds. As a result, the mesencephalon is carried to the summit of the

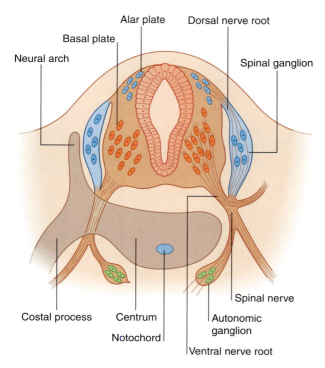

Figure 1.2 Neural tube, spinal nerve, and mesenchymal vertebra of an embryo at 6 weeks.

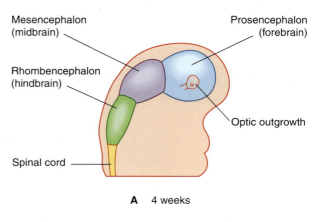

A 4 weeks

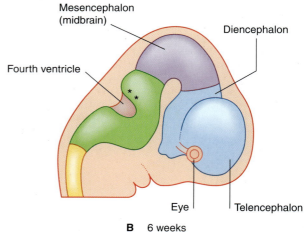

B 6 weeks

Figure 1.3 (A and B) Brain vesicles, seen from the right side. Asterisks indicate the site of initial development of the cerebellum.

brain. The rhombencephalon folds on itself, causing the alar plates to flare and creating the rhomboid (diamond-shaped) fourth ventricle of the brain. The rostral part of the rhombencephalon gives rise to the pons and cerebellum. The caudal part gives rise to the medulla oblongata (Figure 1.4).

Ventricular system and choroid plexuses

The neural canal dilates within the cerebral hemispheres, forming the lateral ventricles; these communicate with the third ventricle contained within the diencephalon. The third and fourth ventricles communicate through the aqueduct of the midbrain (Figure 1.5).

The thin roofs of the forebrain and hindbrain are invaginated by tufts of capillaries, which form the choroid plexuses of the four ventricles. The choroid plexuses secrete cerebrospinal fluid (CSF), which flows through the ventricular system. The fluid leaves the fourth ventricle through three apertures in its roof (Figure 1.6).

Cranial nerves

Figure 1.7 illustrates the state of development of the cranial nerves during the sixth week after fertilization.

- The olfactory nerve (I) forms from bipolar neurons developing in the epithelium lining the olfactory pit.
- The optic nerve (II) is growing centrally from the retina.
- The oculomotor (III) and trochlear (IV) nerves arise from the midbrain, and the abducens (VI) nerve arises from the pons; all three will supply extrinsic muscles of the eye.

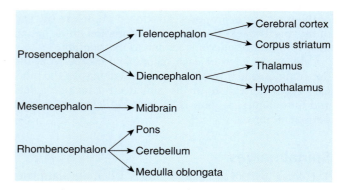

Figure 1.4 Some derivatives of the brain vesicles.

- The three divisions of the trigeminal (V) nerve will be sensory to the skin of the face and scalp, to the mucous membranes of the oronasal cavity, and to the teeth. A motor root will supply the muscles of mastication (chewing).
- The facial (VII) nerve will supply the muscles of facial expression. The vestibulocochlear (VIII) nerve will supply the organs of hearing and balance, which develop from the otocyst.

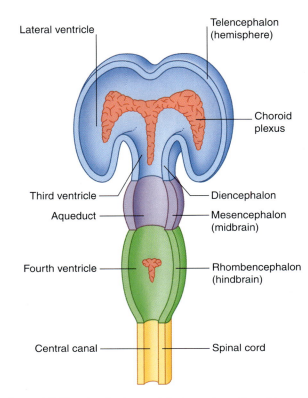

Figure 1.5 The developing ventricular system. Choroid plexuses are shown in red.

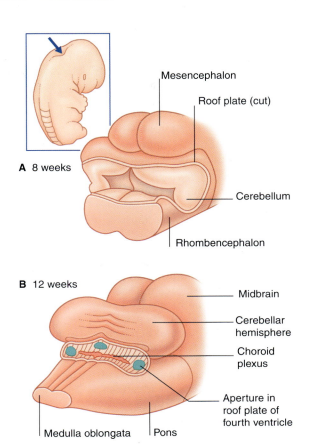

Figure 1.6 Dorsal views of the developing hindbrain (see arrow in inset). **(A)** At 8 weeks, the cerebellum is emerging from the fourth ventricle. **(B)** At 12 weeks, the ventricle is becoming hidden by the cerebellum, and three apertures have appeared in the roof plate.

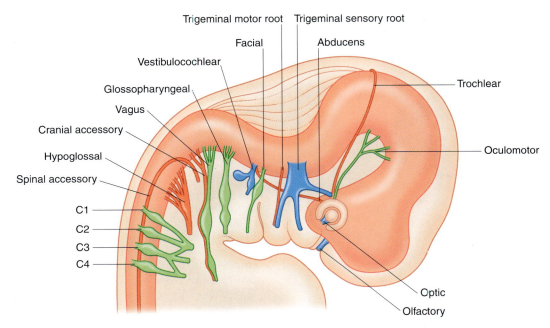

Figure 1.7 Cranial nerves of a 6-week-old embryo. (After Bossy et al. 1990, with permission of Springer-Verlag.)

- The glossopharyngeal (IX) nerve is composite. Most of its fibers will be sensory to the oropharynx.
- The vagus (X) nerve too is composite. It contains a large sensory element for the supply of the mucous membranes of the digestive system, and a large motor (parasympathetic) element for the supply of the heart, lungs, and gastrointestinal tract.
- The cranial accessory (XIc) nerve will be distributed by the vagus to the muscles of the larynx and pharynx.
- The spinal accessory (XIs) nerve will supply the sternomastoid and trapezius muscles. The hypoglossal (XII) nerve will supply the muscles of the tongue.

Cerebral hemispheres

In the telencephalon, mitotic activity takes place in the **ventricular zone**, just outside the lateral ventricle. Daughter cells migrate to the outer surface of the expanding hemisphere and form the cerebral cortex.

Expansion of the cerebral hemispheres is not uniform. A region on the lateral surface, the insula (*L*. 'island'), is relatively quiescent and forms a pivot around which the expanding hemisphere rotates. Frontal, parietal, occipital, and temporal lobes can be identified at 14 weeks' gestational age (Figure 1.8).

On the medial surface of the hemisphere, a patch of cerebral cortex, the hippocampus, belongs to a fifth, limbic lobe of the brain. The hippocampus is drawn into the temporal lobe, leaving in its wake a strand of fibers called the fornix. Within the concavity of this arc is the choroid fissure, through which the choroid plexus invaginates into the lateral ventricle (Figure 1.9).

The anterior commissure develops as a connection linking olfactory (smell) regions of the left and right sides. Above this, a much larger commissure, the corpus callosum,

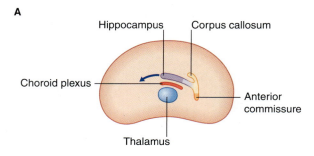

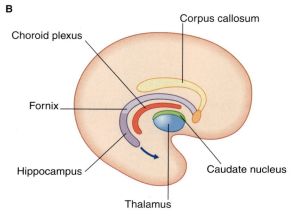

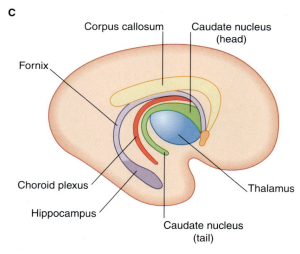

Figure 1.9 Medial aspect of the developing left hemisphere. The hippocampus, initially dorsal to the thalamus, migrates into the temporal lobe (arrows in **A** and **B**), leaving the fornix in its wake. The concavity of the arch so formed contains the choroid fissure (the line of insertion of the choroid plexus into the lateral ventricle) and the tail of the caudate nucleus.

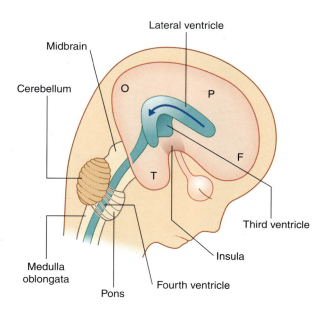

Figure 1.8 Fetal brain at 14 weeks. The arrow indicates the C-shaped growth of the hemisphere around the insula. F, P, O, T; frontal, parietal, occipital, temporal lobes.

links matching areas of the cerebral cortex of the two sides. It extends backward above the fornix.

Coronal sections of the telencephalon reveal a mass of gray matter in the base of each hemisphere, which is the forerunner of the corpus striatum. Beside the third ventricle, the diencephalon gives rise to the thalamus and hypothalamus (Figure 1.10).

The expanding cerebral hemispheres come into contact with the diencephalon, and they fuse with it (see 'site of fusion' in (Figure 1.10A). One consequence is that the term 'brainstem' is restricted thereafter to the remaining,

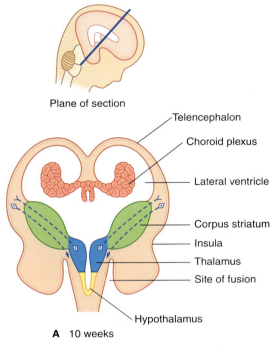

Plane of section

Telencephalon

Choroid plexus

Lateral ventricle

Corpus striatum

Insula

Thalamus

Site of fusion

Hypothalamus

A   10 weeks

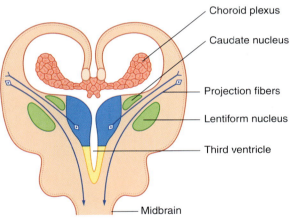

Choroid plexus

Caudate nucleus

Projection fibers

Lentiform nucleus

Third ventricle

Midbrain

B   17 weeks

Figure 1.10 Coronal sections of the developing cerebrum. In **A**, the corpus striatum is traversed by fibers projecting from thalamus to cerebral cortex and from cerebral cortex to spinal cord. In **B**, the corpus striatum has been divided to form the caudate and lentiform nuclei (fusion persists at the anterior end, not shown here).

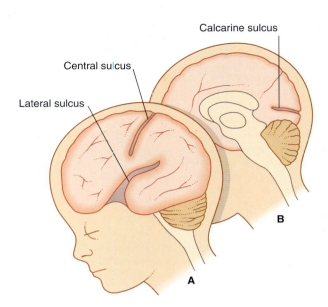

Calcarine sulcus

Central sulcus

Lateral sulcus

B

A

Figure 1.11 Three major cortical sulci in a fetus of 28 weeks. **(A)** Lateral surface of left hemisphere; **(B)** medial surface.

free parts: midbrain, pons, and medulla oblongata. A second consequence is that the cerebral cortex is able to project fibers direct to the brainstem. Together with fibers projecting from thalamus to cortex, they split the corpus striatum into caudate and lentiform nuclei (Figure 1.10B).

By the 28th week of development, several sulci (fissures) have appeared on the surface of the brain, notably the lateral, central, and calcarine sulci (Figure 1.11).

Core Information

The nervous system takes the initial form of a cellular neural tube derived from the ectoderm and enclosing a neural canal. A ribbon of cells escapes along each side of the tube to form the neural crest. The more caudal part of the tube forms the spinal cord. The neural crest forms spinal ganglion cells that send dorsal nerve roots into the sensory, alar plate of the cord. The basal plate of the cord contains motor neurons that emit ventral roots to complete the spinal nerves by joining the dorsal roots.

The more rostral part of the tube forms three brain vesicles. Of these, the prosencephalon (forebrain) gives rise to the cerebral hemispheres (telencephalon) dorsally and the diencephalon ventrally; the mesencephalon becomes the midbrain, and the rhombencephalon becomes the hindbrain (pons, medulla oblongata, and cerebellum).

The neural tube expands rostrally to create the ventricular system of the brain. CSF is secreted by a choroid capillary plexus that invaginates the roof plates of the ventricles.

The cerebral hemispheres develop frontal, parietal, temporal, occipital, and limbic lobes. The hemispheres are cross-linked by the corpus callosum and anterior commissure. Gray matter in the base of each hemisphere is the forerunner of the corpus striatum. The hemispheres fuse with the side walls of the diencephalon, whereupon the mesencephalon and rhombencephalon are all that remain of the embryonic brainstem.

REFERENCES

Bossy J, O'Rahilly R, Müller F. Ontogenese du systeme nerveux. In: Bossy J, ed. Anatomie clinique: neuroanatomie. Paris: Springer-Verlag;1990:357–388.

Cabana T. Development of the nervous system. In: Cohen M, ed. Neuroscience for rehabilitation. Philadelphia: Lippincott; 1993:357–387.

FitzGerald MJT, FitzGerald M. Human embryology. London: Baillière Tindall; 1994.

Larsen WJ. Human embryology. New York: Churchill Livingstone; 1993.

Muller F, O'Rahilly R. Embryonic development of the central nervous system. In: Paxinos G, Mai JK, eds. The human nervous system. Amsterdam: Academic Press; 2004:22–48.

O'Rahilly R, Gardner E. The initial development of the human brain. Acta Anat 1979; 104:123–133.

Cerebral topography

STUDY GUIDELINES

1 The most important objective is for you to get able to recite **all** the central nervous system items identified in the MRI pictures without looking at the labels.

2 Try to get the nomenclature of the component parts of the basal ganglia into long-term memory. Not easily done!

3 Because of its clinical importance, you **must** get able to pop up a mental image of the position and named parts of the internal capsule, and to appreciate continuity of corona radiata, internal capsule, crus cerebri.

SURFACE FEATURES

Lobes

The surfaces of the two cerebral hemispheres are furrowed by **sulci**, the intervening ridges being called **gyri**. Most of the cerebral cortex is concealed from view in the walls of the sulci. Although the patterns of the various sulci vary from brain to brain, some are sufficiently constant to serve as descriptive landmarks. Deepest sulci are the **lateral sulcus** (*Sylvian fissure*) and the **central sulcus** (*Rolandic fissure*) (Figure 2.1A). These two serve to divide the hemisphere (side view) into four **lobes**, with the aid of two imaginary lines, one extending back from the lateral sulcus, the other reaching from the upper end of the **parieto-occipital sulcus** (Figure 2.1B) to a blunt **preoccipital notch** at the lower border of the hemisphere (the sulcus and notch are labeled in Figure 2.3). The lobes are called **frontal**, **parietal**, **occipital**, and **temporal**.

The blunt tips of the frontal, occipital, and temporal lobes are the respective **poles** of the hemispheres.

The **opercula** (lips) of the lateral sulcus can be pulled apart to expose the **insula** (Figure 2.2). The insula was mentioned in Chapter 1 as being relatively quiescent during prenatal expansion of the telencephalon.

The medial surface of the hemisphere is exposed by cutting the **corpus callosum**, a massive band of white matter connecting matching areas of the cortex of the two hemispheres. The corpus callosum consists of a main part or **trunk**, a posterior end or **splenium**, an anterior end or **genu** ('knee'), and a narrow **rostrum** reaching from the

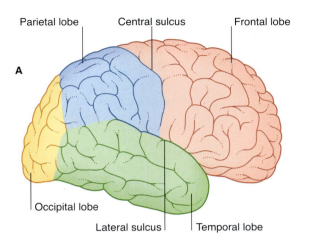

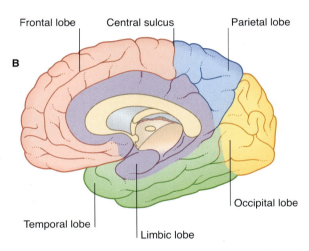

Figure 2.1 The five lobes of the brain. **(A)** Lateral surface of right cerebral hemisphere. **(B)** Medial surface of right hemisphere.

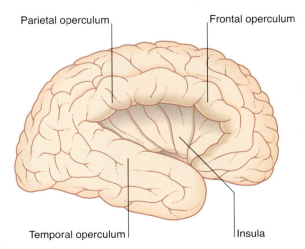

Figure 2.2 Insula, seen on retraction of the opercula.

genu to the **anterior commissure** (Figure 2.3B). The frontal lobe lies anterior to a line drawn from the upper end of the central sulcus to the trunk of the corpus callosum (Figure 2.3B). The parietal lobe lies behind this line, and it is separated from the occipital lobe by the parieto-occipital sulcus. The temporal lobe lies in front of a line drawn from the preoccipital notch to the splenium.

Figures 2.3–2.6 should be consulted along with the following description of surface features of the lobes of the brain.

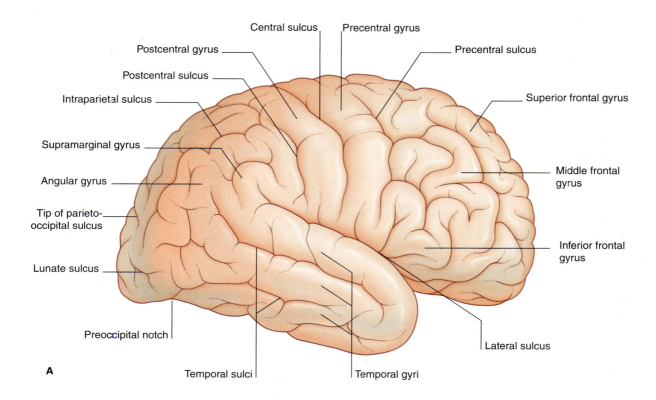

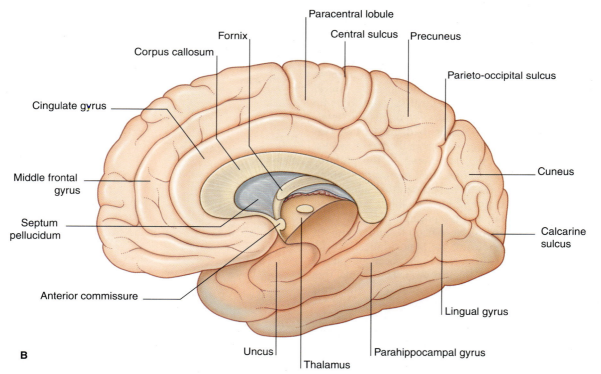

Figure 2.3 (A) Lateral and **(B)** medial views of the right cerebral hemisphere, depicting the main gyri and sulci.

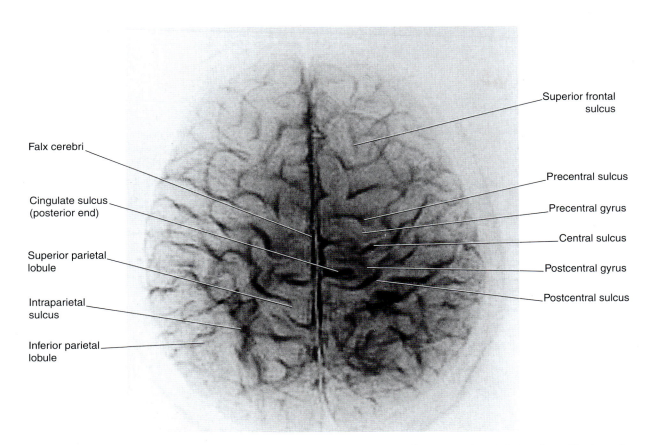

Superior frontal sulcus

Falx cerebri

Precentral sulcus

Cingulate sulcus (posterior end)

Precentral gyrus

Central sulcus

Superior parietal lobule

Postcentral gyrus

Postcentral sulcus

Intraparietal sulcus

Inferior parietal lobule

Figure 2.4 'Thick slice' surface anatomy brain MRI scan from a healthy volunteer. (From Katada K. MR Imaging of brain surface structure: surface anatomy scanning (SAS). Neuroradiology 1990; 32(5): 439–448).

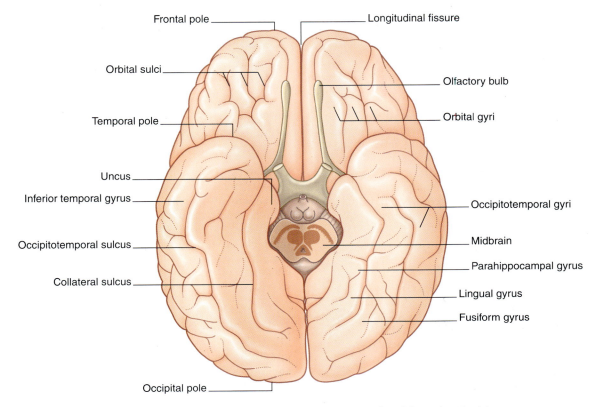

Frontal pole

Longitudinal fissure

Orbital sulci

Olfactory bulb

Temporal pole

Orbital gyri

Uncus

Inferior temporal gyrus

Occipitotemporal gyri

Midbrain

Occipitotemporal sulcus

Parahippocampal gyrus

Lingual gyrus

Collateral sulcus

Fusiform gyrus

Occipital pole

Figure 2.5 Cerebrum viewed from below, depicting the main gyri and sulci.

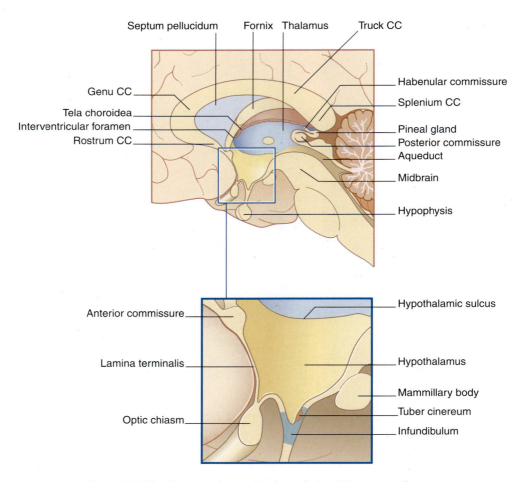

Figure 2.6 The diencephalon and its boundaries. CC, corpus callosum.

Frontal lobe

The lateral surface of the **frontal lobe** contains the **precentral gyrus** bounded in front by the **precentral sulcus**. Further forward, **superior**, **middle**, and **inferior frontal gyri** are separated by **superior** and **inferior frontal sulci**. On the medial surface, the superior frontal gyrus is separated from the **cingulate gyrus** by the **cingulate sulcus**. The inferior or orbital surface is marked by several **orbital gyri**. In contact with this surface are the **olfactory bulb** and **olfactory tract**.

Parietal lobe

The anterior part of the parietal lobe contains the **postcentral gyrus** bounded behind by the **postcentral sulcus**. The posterior parietal lobe is divided into **superior** and **inferior parietal lobules** by an **intraparietal sulcus**. The inferior parietal lobule shows a **supramarginal gyrus**, capping the upturned end of the lateral sulcus, and an **angular gyrus** capping the superior temporal sulcus. The medial surface contains the posterior part of the **paracentral lobule** and, behind this, the **precuneus**. The paracentral lobule (partly contained in the frontal lobe)

is so called because of its relationship to the central sulcus.

Occipital lobe

The lateral surface of the occipital lobe is marked by several **lateral occipital gyri**. The medial surface contains the **cuneus** ('wedge') between the parietooccipital sulcus and the important **calcarine sulcus**. The inferior surface shows three gyri and three sulci. The **lateral** and **medial occipitotemporal gyri** are separated by the **occipitotemporal sulcus**. The **lingual gyrus** lies between the collateral sulcus and the anterior end of the calcarine sulcus.

Temporal lobe

The lateral surface of the temporal lobe displays **superior**, **middle**, **and inferior temporal gyri** separated by **superior and inferior temporal sulci**. The inferior surface shows the anterior parts of the occipitotemporal gyri. The lingual gyrus continues forward as the **parahippocampal gyrus**, which ends in a blunt medial projection, the **uncus**. As will be seen later in views of the sectioned brain, the parahippocampal gyrus underlies a rolled-in part of the cortex, the **hippocampus**.

Limbic lobe

A fifth, **limbic lobe** of the brain surrounds the medial margin of the hemisphere. Surface contributors to the limbic lobe include the cingulate and parahippocampal gyri. It is more usual to speak of the *limbic system*, which includes the hippocampus, fornix, amygdala, and other elements (Ch. 34).

Diencephalon

The largest components of the diencephalon are the **thalamus** and the **hypothalamus** (Figures 2.6 and 2.7). These nuclear groups form the side walls of the third ventricle. Between them is a shallow **hypothalamic sulcus**, which represents the rostral limit of the embryonic sulcus limitans.

Midline sagittal views of the brain

Figure 2.8 is taken from a midline sagittal section of the head of a cadaver, displaying the brain in relation to its surroundings.

INTERNAL ANATOMY OF THE CEREBRUM

The arrangement of the following structures will now be described: thalamus, caudate, and lentiform nuclei, internal capsule; hippocampus and fornix; association and commissural fibers; lateral and third ventricles.

Thalamus, caudate and lentiform nuclei, internal capsule

The two thalami face one another across the slot-like third ventricle. More often than not, they kiss, creating an **interthalamic adhesion** (Figure 2.9). In Figure 2.10, the thalamus and related structures are assembled in a mediolateral sequence. In contact with the upper surface of the thalamus are the **head** and **body** of the **caudate nucleus**. The **tail** of the caudate nucleus passes forward below the thalamus but not in contact with it.

The thalamus is separated from the lentiform nucleus by the **internal capsule**, which is the most common site for a *stroke* resulting from local arterial embolism (blockage) or hemorrhage. The internal capsule contains fibers running from thalamus to cortex and from cortex to thalamus, brainstem, and spinal cord. In the interval between cortex and internal capsule, these ascending and descending fibers form the **corona radiata**. Below the internal capsule, the **crus** of the midbrain receives descending fibers continuing into the brainstem.

The lens-shaped **lentiform nucleus** is composed of two parts: **putamen** and **globus pallidus**. The putamen and caudate nucleus are of similar structure, and their anterior ends are fused. Behind this, they are linked by strands of

gray matter that traverse the internal capsule, hence the term **corpus striatum** (or, simply, **striatum**) used to include the putamen and caudate nucleus. The term **pallidum** refers to the globus pallidus.

The caudate and lentiform nuclei belong to the **basal ganglia**, a term originally applied to a half-dozen masses of gray matter located near the base of the hemisphere. In current usage, the term designates four nuclei known to be involved in motor control: the caudate and lentiform nuclei, the subthalamic nucleus in the diencephalon, and the substantia nigra in the midbrain (Figure 2.11).

In horizontal section, the internal capsule has a dog-leg shape (see photograph of a fixed-brain section in Figure 2.12, and living-brain magnetic resonance imaging [MRI] 'slice' in Figure 2.13). The internal capsule has four named parts in horizontal sections:

1 **anterior limb**, between the lentiform nucleus and the head of the caudate nucleus;

2 **genu**;

3 **posterior limb**, between the lentiform nucleus and the thalamus;

4 **retrolentiform part**, behind the lentiform nucleus and lateral to the thalamus.

The **corticospinal tract** (CST) descends in the posterior limb of the internal capsule. It is also called the **pyramidal tract**, a *tract* being a bundle of fibers serving a common function. The CST originates mainly from the *motor cortex* within the precentral gyrus. It descends through the corona radiata, internal capsule, and crus of midbrain and continues to the lower end of the brainstem before crossing to the opposite side of the spinal cord.

From a clinical standpoint, *the CST is the most important pathway in the entire central nervous system (CNS)*, for two reasons. First, it mediates voluntary movements of all kinds, and interruption of the tract leads to motor weakness (called *paresis*) or motor paralysis. Second, it extends the entire vertical length of the CNS, rendering it vulnerable to disease or trauma in the cerebral hemisphere or brainstem on one side, and to spinal cord disease or trauma on the other side.

A coronal section through the anterior limb is represented in Figure 2.14; a corresponding MRI view is shown in Figure 2.15. A coronal section through the posterior limb from a fixed brain is shown in Figure 2.16; a corresponding MRI slice is shown in Figure 2.17.

Lateral to the lentiform nucleus are the **external capsule**, **claustrum**, and **extreme capsule**.

Hippocampus and fornix

During embryonic life, the **hippocampus** (crucial for memory formation) is first seen above the corpus callosum. The bulk of it remains in that position in lower mammals, including rodents. In primates, it retreats into the temporal lobe as this lobe develops, leaving a tract of white matter, the **fornix**, in its wake. The mature hippocampus stretches the full length of the floor of the inferior (temporal) horn of the lateral ventricle (Figures 2.18 and 2.19). The mature fornix comprises a **body** beneath the trunk of the corpus callosum, a **crus**, which enters it from each hippocampus,

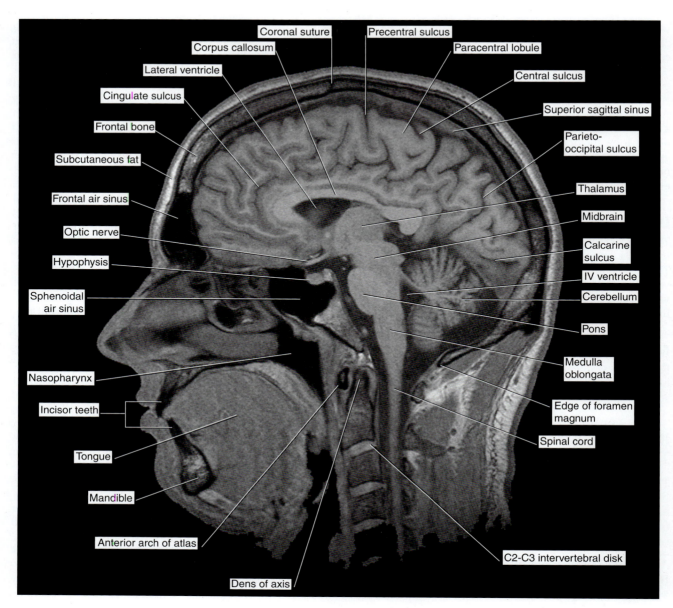

Figure 2.7 Sagittal MRI 'slice' of the living brain. (From a series kindly provided by Professor J. Paul Finn, Director, Magnetic Resonance Research, Department of Radiology, David Geffen School of Medicine at UCLA, California, USA.)

and two **pillars** (**columns**), which leave it to enter the diencephalon. Intimately related to the crus and body is the **choroid fissure**, through which the choroid plexus is inserted into the lateral ventricle.

Association and commissural fibers

Fibers leaving the cerebral cortex fall into three groups:

1 **association fibers**, which pass from one part of a single hemisphere to another;

2 **commissural fibers**, which link matching areas of the two hemispheres;

3 **projection fibers**, which run to subcortical nuclei in the cerebral hemisphere, brainstem, and spinal cord.

Association fibers   (Figure 2.20)

Short association fibers pass from one gyrus to another within a lobe.

Long association fibers link one lobe with another. Bundles of long association fibers include:

- the **superior longitudinal fasciculus**, linking the frontal and occipital lobes;
- the **inferior longitudinal fasciculus**, linking the occipital and temporal lobes;
- the **arcuate fasciculus**, linking the frontal lobe with the occipitotemporal cortex;
- the **uncinate fasciculus**, linking the frontal and anterior temporal lobes;
- the **cingulum**, underlying the cortex of the cingulate gyrus.

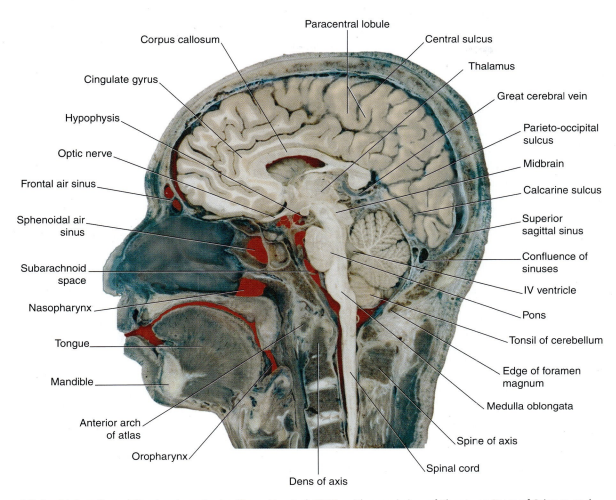

Figure 2.8 Sagittal section of fixed cadaver brain. (From Liu et al. 2003, with permission of Shantung Press of Science and Technology.)

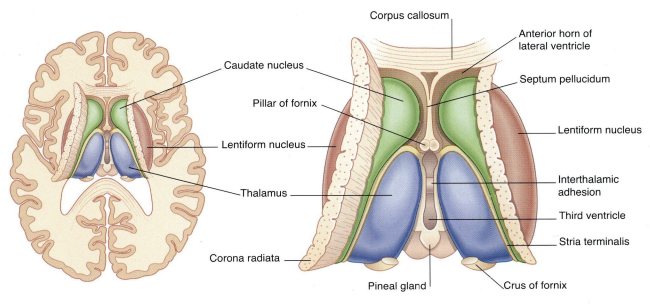

Figure 2.9 Thalamus and corpus striatum, seen on removal of the trunk of the corpus callosum and the trunk of the fornix.

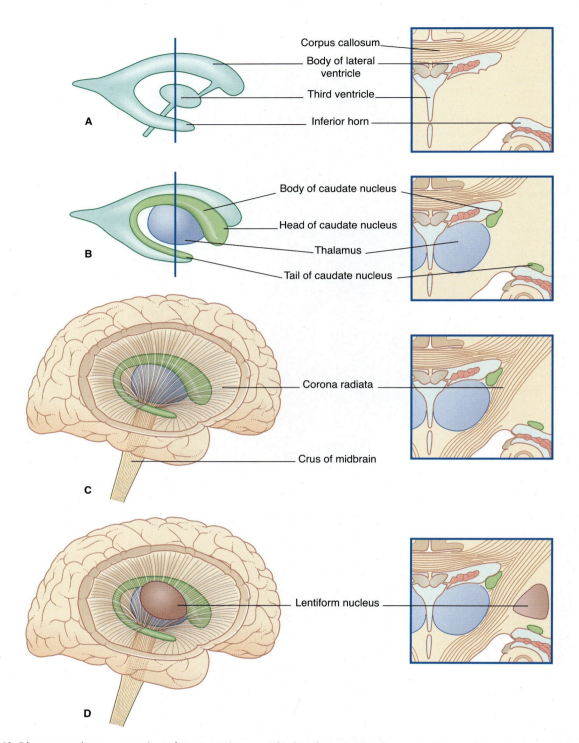

Figure 2.10 Diagrammatic reconstruction of corpus striatum and related structures. The vertical lines on the left in **A** and **B** indicate the level of the coronal sections on the right. **(A)** Ventricular system. **(B)** Thalamus and caudate nucleus in place. **(C)** Addition of projections to and from cerebral cortex. **(D)** Lentiform nucleus in place.

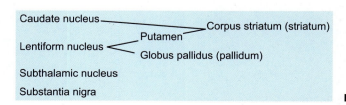

Figure 2.11 Nomenclature of basal ganglia.

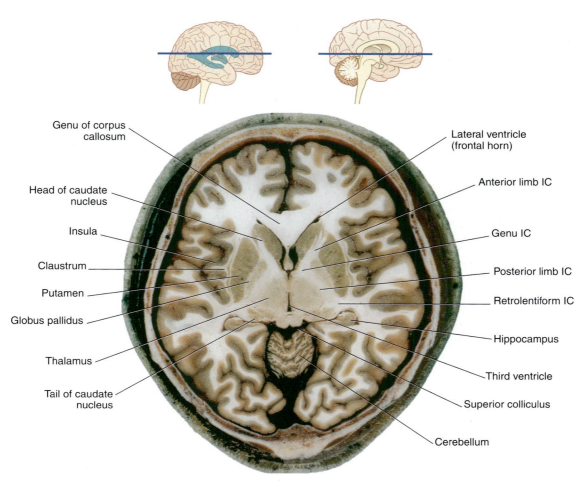

Genu of corpus callosum

Head of caudate nucleus

Insula

Claustrum

Putamen

Globus pallidus

Thalamus

Tail of caudate nucleus

Lateral ventricle (frontal horn)

Anterior limb IC

Genu IC

Posterior limb IC

Retrolentiform IC

Hippocampus

Third ventricle

Superior colliculus

Cerebellum

Figure 2.12 Horizontal section of fixed cadaver brain at the level indicated at top. IC, internal capsule. (From Liu et al. 2003, with permission of Shantung Press of Science and Technology.)

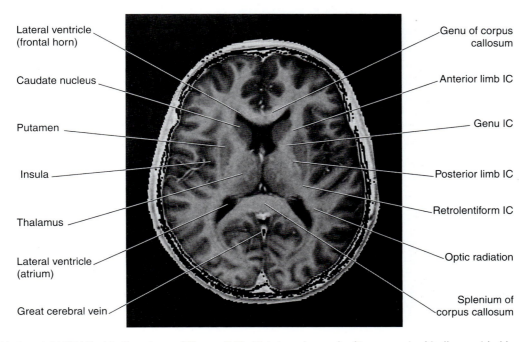

Lateral ventricle (frontal horn)

Caudate nucleus

Putamen

Insula

Thalamus

Lateral ventricle (atrium)

Great cerebral vein

Genu of corpus callosum

Anterior limb IC

Genu IC

Posterior limb IC

Retrolentiform IC

Optic radiation

Splenium of corpus callosum

Figure 2.13 Horizontal MRI 'slice' in the plane of Figure 2.12. IC, internal capsule. (From a series kindly provided by Professor J. Paul Finn, Director, Magnetic Resonance Research, Department of Radiology, David Geffen School of Medicine at UCLA, California, USA.)

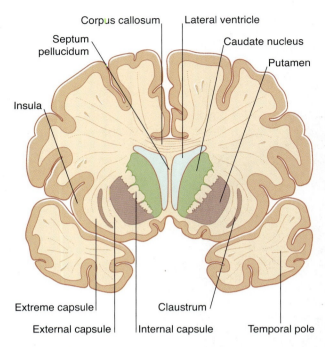

Figure 2.14 Drawing of a coronal section through the anterior limb of the internal capsule.

Cerebral commissures

Corpus callosum

The **corpus callosum** is much the largest of the commissures linking matching areas of the left and right cerebral cortex (Figure 2.21). From the **body**, some fibers pass laterally and upward, intersecting the corona radiata. Other fibers pass laterally and then bend downward as the **tapetum** to reach the lower parts of the temporal and occipital lobes. Fibers traveling to the medial wall of the occipital lobe emerge from the splenium on each side and form the **occipital (major) forceps**. The **frontal (minor) forceps** emerges from each side of the genu to reach the medial wall of the frontal lobe.

Minor commissures

The **anterior commissure** interconnects the anterior parts of the temporal lobes, as well as the two olfactory tracts.

The **posterior commissure** and the **habenular commissure** lie directly in front of the pineal gland.

The **commissure of the fornix** contains some fibers traveling from one hippocampus to the other by way of the two crura.

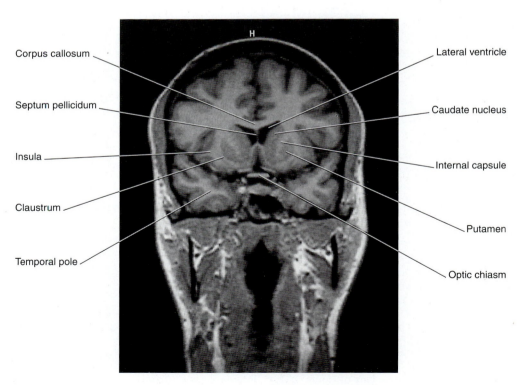

Figure 2.15 Coronal MRI 'slice' at the level indicated at top. (From a series kindly provided by Professor J. Paul Finn, Director, Magnetic Resonance Research, Department of Radiology, David Geffen School of Medicine at UCLA, California, USA.)

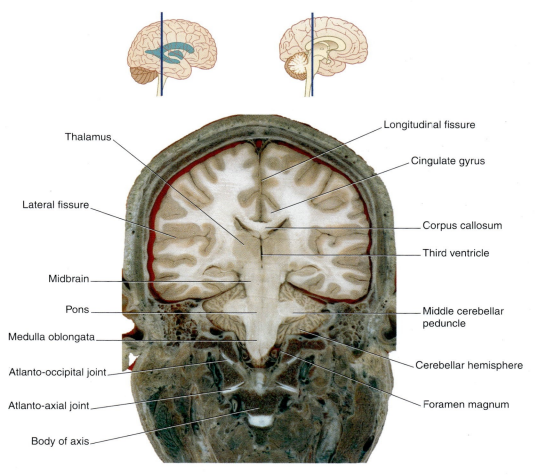

Figure 2.16 Coronal section of fixed cadaver brain at the level indicated at top. (From Liu et al. 2003, with permission of Shantung Press of Science and Technology.)

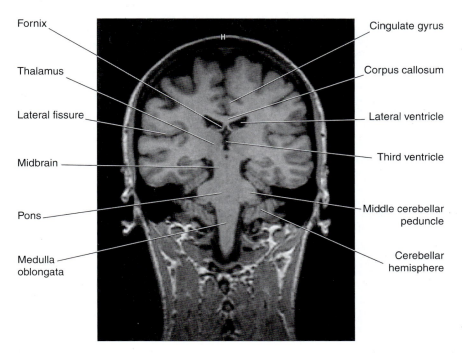

Figure 2.17 Coronal MRI 'slice' at the level indicated at top. (From a series kindly provided by Professor J. Paul Finn, Director, Magnetic Resonance Research, Department of Radiology, David Geffen School of Medicine at UCLA, California, USA.)

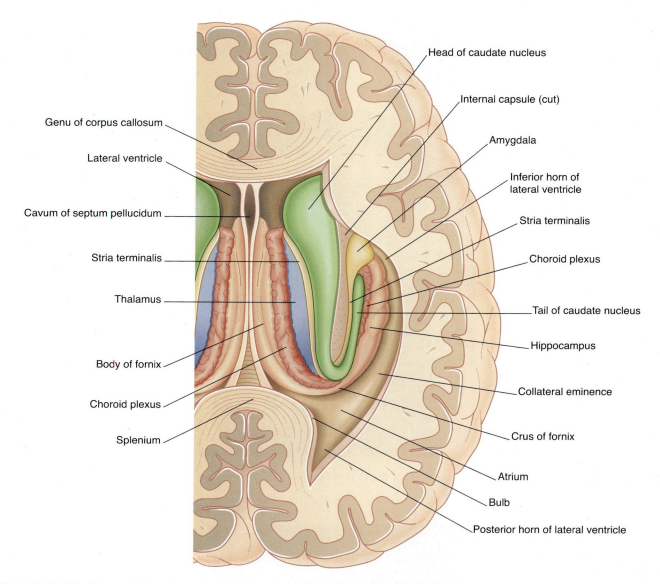

Head of caudate nucleus

Internal capsule (cut)

Amygdala

Inferior horn of lateral ventricle

Stria terminalis

Choroid plexus

Tail of caudate nucleus

Hippocampus

Collateral eminence

Crus of fornix

Atrium

Bulb

Posterior horn of lateral ventricle

Genu of corpus callosum

Lateral ventricle

Cavum of septum pellucidum

Stria terminalis

Thalamus

Body of fornix

Choroid plexus

Splenium

Figure 2.18 Tilted view of the ventricular system, showing the continuity of structures in the body and inferior horn of the lateral ventricle. *Note*: The amygdala, stria terminalis, and tail of caudate nucleus occupy the roof of the inferior horn; the hippocampus occupies the floor. (The choroid plexus is 'reduced' in order to show related structures.)

Lateral and third ventricles

The **lateral ventricle** consists of a **body** within the parietal lobe, and **anterior** (frontal), **posterior** (occipital), and **inferior** (temporal) **horns** (Figure 2.22). The anterior limit of the central part is the **interventricular foramen**, located between the thalamus and anterior pillar of the fornix, through which it communicates with the third ventricle. The central part joins the occipital and temporal horns at the **atrium** (Figures 2.23 and 2.24).

The relationships of the lateral ventricle are listed below.

- **Anterior horn**. Lies between head of caudate nucleus and septum pellucidum. Its other boundaries are formed by the corpus callosum: trunk above, genu in front, rostrum below.

- **Body**. Lies below the trunk of the corpus callosum and above the thalamus and anterior part of the body

of the fornix. Medially is the septum pellucidum, which tapers away posteriorly where the fornix rises to meet the corpus callosum. The **septum pellucidum** is formed of the thinned-out walls of the two cerebral hemispheres. Its bilateral origin may be indicated by a central cavity (**cavum**).

- **Posterior horn**. Lies posterolateral to the splenium and medial to the tapetum of the corpus callosum. On the medial side, the forceps major forms the **bulb** of the posterior horn.

- **Inferior horn**. Lies below the tail of the caudate nucleus and, at the anterior end, the **amygdala** (Figure 2.18), a nucleus belonging to the limbic system. The hippocampus and its associated structures occupy the full length of the floor.

- Outside these is the **collateral eminence**, created by the collateral sulcus.

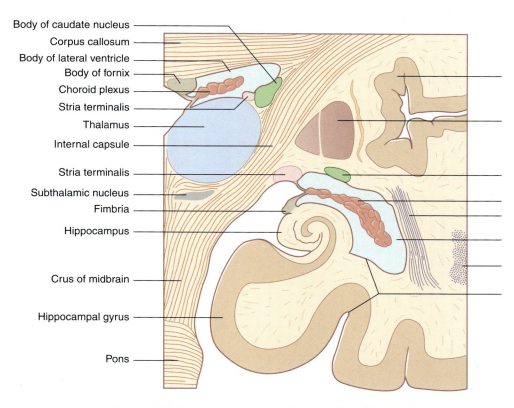

Body of caudate nucleus
Corpus callosum
Body of lateral ventricle
Body of fornix
Choroid plexus
Stria terminalis
Thalamus
Internal capsule
Stria terminalis
Subthalamic nucleus
Fimbria
Hippocampus
Crus of midbrain
Hippocampal gyrus
Pons

Figure 2.19 Coronal section through the body and inferior horn of the lateral ventricle.

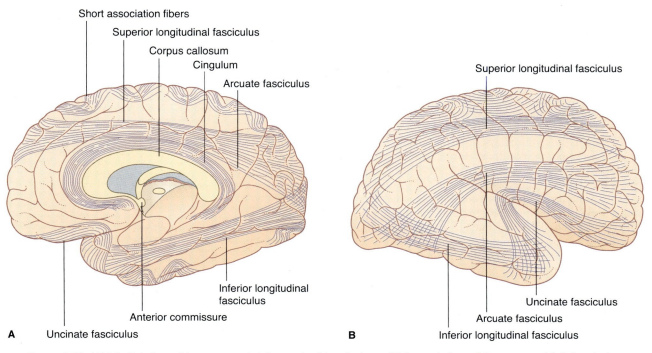

Short association fibers
Superior longitudinal fasciculus
Corpus callosum
Cingulum
Arcuate fasciculus

Superior longitudinal fasciculus

Inferior longitudinal fasciculus

Anterior commissure

A Uncinate fasciculus

Uncinate fasciculus

Arcuate fasciculus

B Inferior longitudinal fasciculus

Figure 2.20 (A) Medial view of 'transparent' right cerebral hemisphere. (B) Lateral view of 'transparent' left hemisphere.

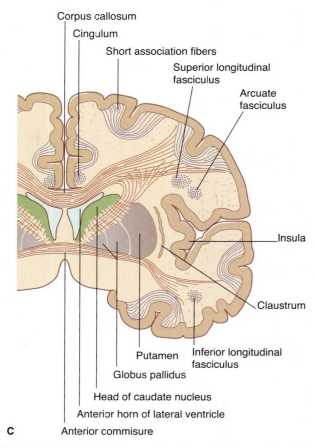

C

Figure 2.20 (*Cont'd*) (**C**) Coronal section, showing position of short and long association fiber bundles.

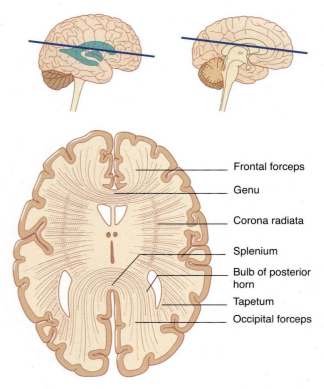

Figure 2.21 Horizontal section through genu and splenium of corpus callosum. Fibers passing laterally from the trunk intersect the corona radiata.

The **third ventricle** is the cavity of the diencephalon. Its boundaries are shown in Figure 2.6. A **choroid plexus** hangs from its roof, which is formed of a double layer of pia mater called the **tela choroidea**. Above this are the fornix and corpus callosum. In the side walls are the thalamus and hypothalamus. The anterior wall is formed by the anterior commissure, the **lamina terminalis**, and the **optic chiasm**. In the floor are the **infundibulum**, the **tuber cinereum**, the **mammillary bodies** (also spelt 'mamillary'), and the upper end of the midbrain. The **pineal gland** and adjacent commissures form the posterior wall. The pineal gland is often calcified, and the **habenular commissure** is sometimes calcified, as early as the second decade of life, thereby becoming detectable even on plain radiographs of the skull. The pineal gland is sometimes displaced to one side by a tumor, hematoma, or other mass (space-occupying lesion) within the cranial cavity.

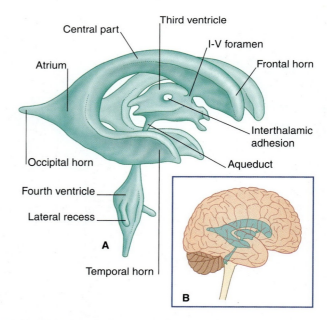

Figure 2.22 Ventricular system. (**A**) Isolated cast. (**B**) Ventricular system in situ.

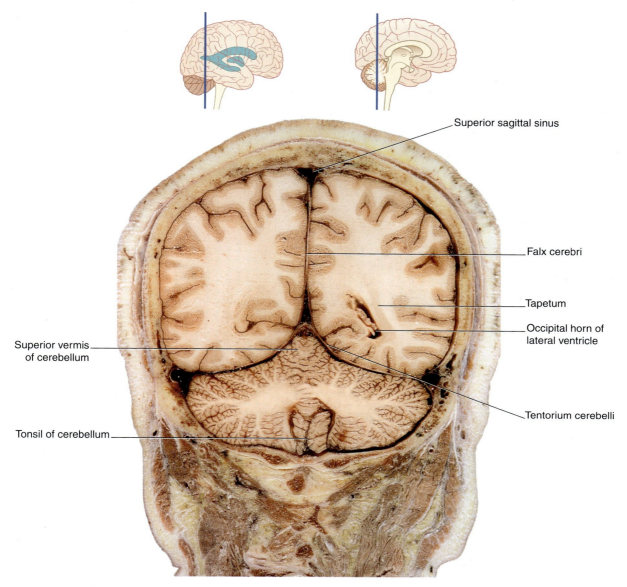

Superior sagittal sinus

Falx cerebri

Tapetum

Occipital horn of
lateral ventricle

Superior vermis
of cerebellum

Tentorium cerebelli

Tonsil of cerebellum

Figure 2.23 Coronal section of fixed cadaver brain at the level indicated at top. (From Liu et al. 2003, with permission of Shantung Press of Science and Technology.)

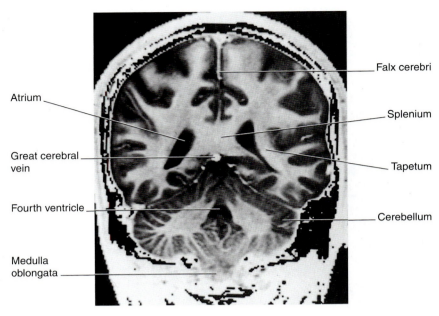

Atrium

Great cerebral vein

Fourth ventricle

Medulla oblongata

Falx cerebri

Splenium

Tapetum

Cerebellum

Figure 2.24 Coronal MRI 'slice' at the level indicated at top. (From a series kindly provided by Professor J. Paul Finn, Director, Magnetic Resonance Research, Department of Radiology, David Geffen School of Medicine at UCLA, California, USA.)

Box 2.1 Brain planes

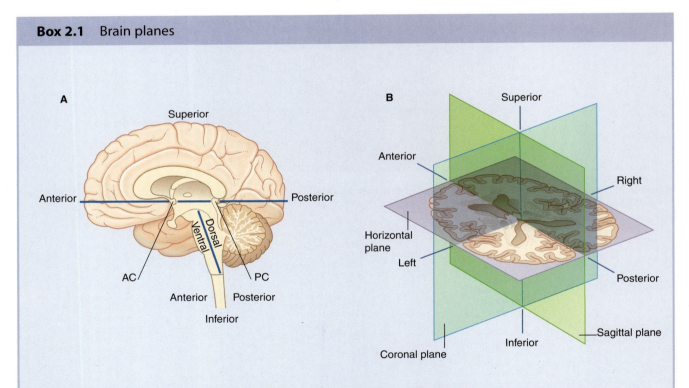

Figure Box 2.1.1 (A) Planes of reference for the central nervous system as a whole. In this presentation, only the brainstem (owing to its obliquity) differs from the standard for gross anatomy. However, some authors use the terms *ventral* and *dorsal* instead of *anterior* and *posterior* with respect to the spinal cord, and some use the terms *rostral* and *caudal* to signify *superior* and *inferior* with respect to spinal cord and/or brainstem. The horizontal line represents the bicommissural plane. AC, PC, anterior and posterior commissures. **(B)** The brain sectioned in the bicommissural plane. (Adapted from Kretschmann and Weinrich 1998, with permission of Thieme and the authors.)

Box 2.2   Magnetic resonance imaging

Magnetic resonance imaging of the CNS is immensely useful for detection of tumors and other space-occupying lesions (masses). When properly used, it is quite safe, even for young children and pregnant women. As will be shown later on, it can be adapted to the study of normal brain physiology in healthy volunteers.

The original name for the technique is *nuclear MRI*, because it is based on the behavior of atomic nuclei in applied magnetic fields. The simplest atomic nucleus is that of the element hydrogen, consisting of a single proton, and this is prevalent in many substances (e.g. water) throughout the body.

Nuclei possess a property know as *spin* (Figure Box 2.2.1), and it may be helpful to visualize this as akin to a spinning gyroscope. Normally, the direction of the spin (the axis of the gyroscope in our analogy) for

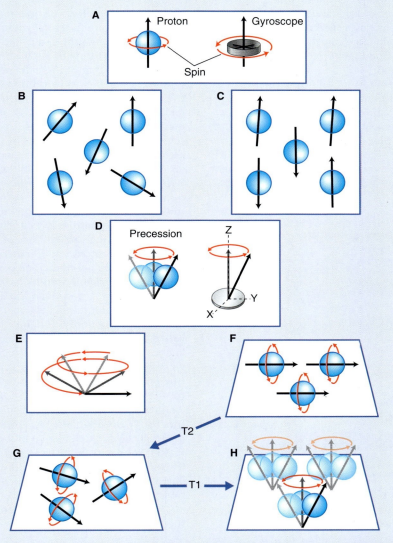

Figure Box 2.2.1   Basis of nuclear magnetic resonance and its manipulation by radio waves. **(A)** The proton of the hydrogen nucleus is in a constant state of spin analogous to that of a gyroscope. **(B)** At rest, the orientation of the axes of the spinning protons is random. **(C)** When the external magnet is switched on, all the axes become oriented along its longitudinal, z axis. The great majority are parallel, with a small minority antiparallel, as indicated. **(D)** At the same time, the magnetic moments immediately precess around the axis like a wobbling gyroscope, being oriented in an intermediate state between the z axis of the magnetic field and the x–y axis at right angles to it. **(E)** An excitatory radio-frequency pulse at right angles to the axis of the external magnetic field tips the net magnetic moment along a 'snail shell' spiral into the x–y plane. **(F)** While the radio-frequency transceiver pulse is 'on', the nuclei are precessing in phase. **(G)** Switching off the radio frequency allows the nuclei to dephase immediately, with a brief T_2 time constant. **(H)** Conical precession is resumed under the influence of the external magnet with a longer, T_1 time constant. (The assistance of Professor Hugh Garavan, Department of Psychology, Trinity College, Dublin, is gratefully appreciated.)

Box 2.2   *Continued*

any given nucleus is random. Spin produces a magnetic moment (vector) that makes it behave like a tiny dipole (north and south) magnet. In the absence of any external magnetic field, the dipoles are randomly arranged.

In the presence of a magnetic field, however, the dipoles will orient themselves along the direction of the magnetic field z (vertical) line.

The cylindric external magnet of an MRI machine (Figure Box 2.2.2) is immensely powerful, capable of lifting the weight of several cars at one time. When the magnet is switched on, individual nuclear magnetic moments undergo a process called *precession*—analogous to the wobbling of a gyroscope—whereby they adopt a cone-shaped spin around the z axis of the external magnetic field.

Excitatory pulses are transmitted from radio-frequency coils set at right angles to the z axis of the external magnetic field. The effect is to tilt the net nuclear magnetic moment into the x–y axis, with all the nuclei precessing 'in phase'. When the radio-frequency coils are switched off, the nuclei 'dephase' while still in the x–y axis, and then *relax* back to vertical alignment. The time constant involved is called T_2. The external

magnet then restores the conical precession around the z axis; the time constant here is much slower and is called T_1.

Because the spinning, precessing nuclei behave like little magnets, if they are surrounded by a coil of wire, they will induce a current in that coil that can then be measured. As it happens, the radiotransmitter coil is able to receive and measure this current, hence term *transceiver* in the diagram.

This is the basic principle of nuclear magnetic resonance. However, to be able to construct an actual image, we require to *spatially resolve* the detected signal. This can be achieved by introducing *gradient coils*. Superimposition of a second magnetic field, set at right angles to that of the main magnet, causes the resonant frequency to be disturbed along the axis of the new field, the proton spin being highest at one end and lowest at the other end. The magnetic resonance machine in fact contains three gradient coils, one being set in each of the three planes of space. The three coils are activated sequentially, allowing three-dimensional localization of tissue signals. In this way, it is possible to 'slice' through the patient, detecting the signal emitted from different components in each selected plane of the patient, and building up an image piece by piece.

The varying densities within the magnetic resonance images reflect the varying rates of dephasing and of relaxation of protons in different locations. The protons of the cerebrospinal fluid, for example, are free to resonate at maximum frequency, whereas in the white matter they are largely bound to lipid molecules. The gray matter has intermediate values, some protons being protein-bound. The radio-frequency pulses can be varied to exploit these differences. Almost all the images shown in textbooks (including this one) are T_1-weighted, favoring the very weak signal provided by free protons during the relaxation period. This accounts for the different densities of CSF, gray matter, and white matter, the last being strongest. The reverse is true for T_2-weighted images. T_2-weighted images are especially useful in detection of lesions in the white matter. For example, they can indicate an increase in free protons resulting from patchy loss of myelin sheath lipid in multiple sclerosis (see Ch. 6), or local edema of brain tissue resulting from a vascular stroke.

The standard orientation of coronal and axial slices is shown in Figure Box 2.2.3.

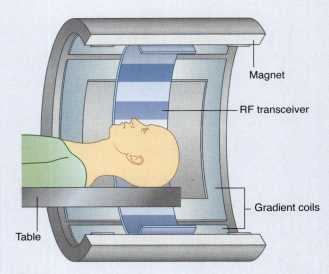

Magnet

RF transceiver

Gradient coils

Table

Figure Box 2.2.2 The MRI machine. Outermost is the magnet. Innermost is the radio-frequency transceiver. In between are the gradient coils.

Box 2.2 *Continued*

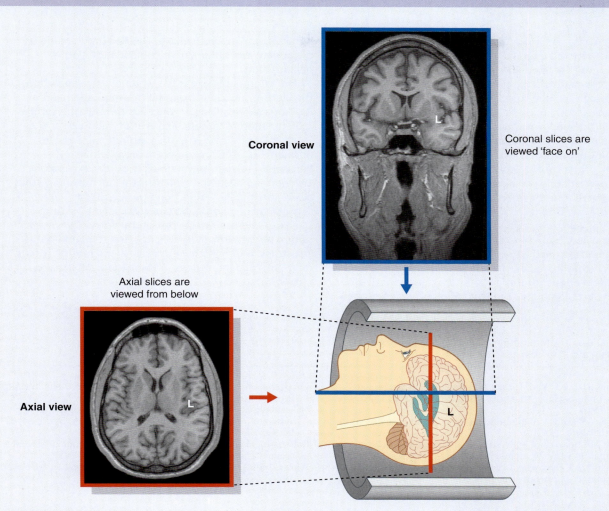

Coronal view

Coronal slices are viewed 'face on'

Axial slices are viewed from below

Axial view

Figure Box 2.2.3 Standard orientation of magnetic resonance images. Coronal sections are viewed from in front. Axial sections are viewed from below. (Sagittal sections not represented because they are not ambiguous.)

REFERENCES

Jones DK. Fundamentals of diffusion MRI imaging. In: Gillard JH, Waldman AD, Barker PB, eds. Fundamentals of neuroimaging. Cambridge: University Press; 2005:54–85.

Mitchell DG, Cohen MS. MRI principles. 2nd edn. Philadelphia: Saunders; 2004.

Saper CB, Iversen S, Frackoviak R. In: Kandel ER, Schwarz JH, Jessell TJ, eds. Principles of neural science. 4th edn. New York: McGraw-Hill; 2000:370–375.

Core Information

On the lateral surface of the cerebrum, four lobes are defined by the lateral and central sulci and an imaginary T-shaped line. The frontal lobe has six named gyri, the parietal lobe has seven, the occipital lobe five, the temporal lobe four. The insula is in the floor of the lateral sulcus.

On the medial surface, the corpus callosum comprises splenium, trunk, genu, and rostrum. The septum pellucidum stretches from the corpus callosum to the trunk of the fornix. Separating fornix from thalamus is the choroidal fissure through which the choroid plexus is inserted into the lateral ventricle. The third ventricle has the fornix in its roof, thalamus and hypothalamus in its side walls, infundibulum, tuber cinereum, and mammillary bodies in its floor. Behind it is the pineal gland, often calcified.

The basal ganglia comprise the corpus striatum (caudate and lentiform nuclei), subthalamic nucleus, and substantia nigra. The lentiform nucleus comprises putamen and globus pallidus. The striatum is made up of caudate and putamen, the pallidum of globus pallidus alone.

The internal capsule is the white matter separating the lentiform nucleus from the thalamus and head of caudate nucleus. The CST descends through the corona radiata, internal capsule, and crus of midbrain.

Association fibers (e.g. the longitudinal, arcuate, uncinate fasciculi) link different areas within a hemisphere. Commissural fibers (e.g. corpus callosum, anterior and posterior commissures) link matching areas across the midline. Projection fibers (e.g. corticothalamic, corticobulbar, corticospinal) pass to thalamus and brainstem. The lateral ventricles have a central part and frontal, occipital, and temporal horns. Structures determining ventricular shape include corpus callosum, caudate nucleus, thalamus, amygdala, and hippocampus.

REFERENCES

DeArmond SJ, Fusco MM, Dewey MM. Structure of the human brain: a photographic atlas. 2nd edn. Oxford: Oxford University Press; 1976.

England MA, Wakely J. A colour atlas of the brain and spinal cord. London: Wolfe; 1991.

Katada K. MR imaging of brain surface structures: surface anatomy scanning (SAS). Neuroradiology 1990; 32(5):439–448.

Kretschmann H-J, Weinrich W. Cranial neuroimaging and clinical neuroanatomy. Stuttgart: Thieme; 2004.

Liu S, et al, eds. Atlas of human sectional anatomy. Jinan: Shantung Press of Science and Technology; 2003.

Niewenhuys R, Voogd J, van Huijzen C. The human central nervous system: a synopsis and atlas. 3rd edn. New York: Springer-Verlag; 1988.

Roberts M, Hanaway J, Morest DK. Atlas of the human brain in section. 2nd edn. Philadelphia: Lea & Febiger; 1987.

Wicke L. Atlas of radiologic anatomy. Philadelphia: Lea & Febiger; 1994.

STUDY GUIDELINES
1 Become familiar with the locations of the ascending and descending pathways at the eight levels shown. Box 3.1 deserves special attention, because it indicates why certain pathways cross the midline and others do not. The brainstem crossings are formally addressed in Chapter 14.
2 Get to grips with the nomenclature used for sections of the midbrain.
3 Note that the cerebellum (part of the hindbrain) is considered *after* the spinal cord for the sake of continuity of motor and sensory pathways.
4 Become able to divide the 31 spinal nerves into five groups.
5 Relate the three cerebellar peduncles to the fourth ventricle as seen in cross-sections.

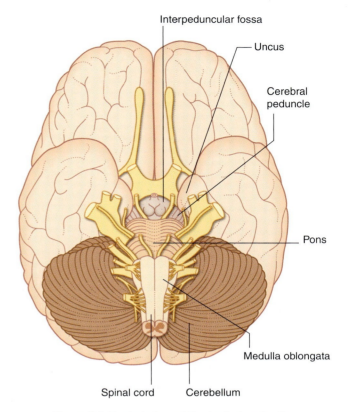

Figure 3.1 Ventral view of the brainstem in situ.

The midbrain connects the diencephalon to the hindbrain. As explained in Chapter 1, the hindbrain is made up of the pons, medulla oblongata, and cerebellum. The medulla oblongata joins the spinal cord within the foramen magnum of the skull.

In this chapter, the cerebellum (part of the hindbrain) is considered *after* the spinal cord, for the sake of continuity of motor and sensory pathway descriptions.

BRAINSTEM

Ventral view (Figures 3.1 and 3.2A)
Midbrain
The ventral surface of the midbrain shows two massive **cerebral peduncles** bordering the **interpeduncular fossa**. The **optic tracts** wind around the midbrain at its junction with the diencephalon. Lateral to the midbrain is the uncus of the temporal lobe. The **oculomotor nerve** (III) emerges from the medial surface of the peduncle. The **trochlear nerve** (IV) passes between the peduncle and the uncus.

Pons
The bulk of the pons is composed of **transverse fibers** that raise numerous surface ridges. On each side, the pons is marked off from the middle cerebellar peduncle

by the attachment of the **trigeminal nerve** (V). The **middle cerebellar peduncle** plunges into the hemisphere of the cerebellum.

At the lower border of the pons are the attachments of the **abducens** (VI), **facial** (VII), and **vestibulocochlear** (VIII) nerves (see Table 3.1).

Medulla oblongata
The **pyramids** are alongside the anterior median fissure. Just above the spinomedullary junction, the fissure is

Table 3.1 The cranial nerves

Number	Name
I	Olfactory, enters the olfactory bulb from the nose
II	Optic
III	Oculomotor
IV	Trochlear
V	Trigeminal
VI	Abducens
VII	Facial
VIII	Vestibulocochlear
IX	Glossopharyngeal
X	Vagus
XI	Accessory
XII	Hypoglossal

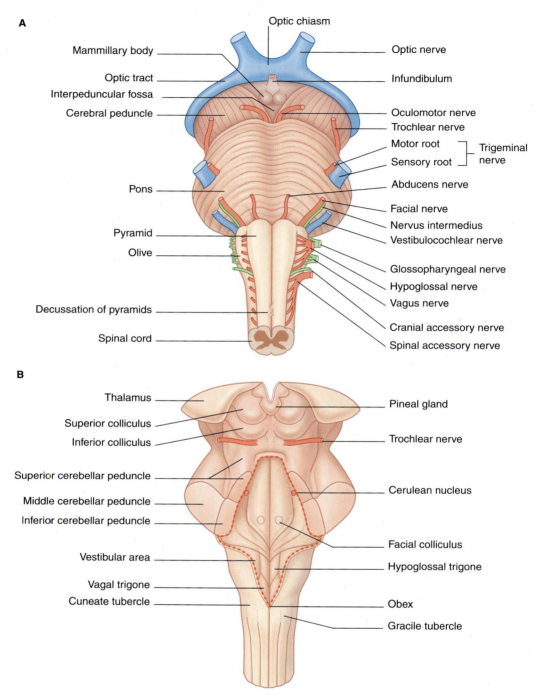

A

Optic chiasm

Mammillary body

Optic tract

Interpeduncular fossa

Cerebral peduncle

Pons

Pyramid

Olive

Decussation of pyramids

Spinal cord

Optic nerve

Infundibulum

Oculomotor nerve

Trochlear nerve

Motor root ⎤
⎦ Trigeminal
Sensory root ⎦ nerve

Abducens nerve

Facial nerve

Nervus intermedius

Vestibulocochlear nerve

Glossopharyngeal nerve

Hypoglossal nerve

Vagus nerve

Cranial accessory nerve

Spinal accessory nerve

B

Thalamus

Superior colliculus

Inferior colliculus

Superior cerebellar peduncle

Middle cerebellar peduncle

Inferior cerebellar peduncle

Vestibular area

Vagal trigone

Cuneate tubercle

Pineal gland

Trochlear nerve

Cerulean nucleus

Facial colliculus

Hypoglossal trigone

Obex

Gracile tubercle

Figure 3.2 (A) Anterior and **(B)** posterior view of the brainstem.

invaded by the **decussation of the pyramids**, where fibers of the two pyramids intersect while crossing the midline. Lateral to the pyramid is the **olive**, and behind the olive is the **inferior cerebellar peduncle**. Attached between pyramid and olive is the **hypoglossal nerve** (XII). Attached between olive and inferior cerebellar peduncle are the **glossopharyngeal** (IX), **vagus** (X), and **cranial accessory** (XIc) nerves. The **spinal accessory nerve** (XIs) arises from the spinal cord and runs up through the foramen magnum to join the cranial accessory.

Dorsal view (Figure 3.2B)

The roof or **tectum** of the midbrain is composed of four colliculi. The **superior colliculi** belong to the visual system, and the **inferior colliculi** belong to the auditory system. The **trochlear nerve** (IV) emerges below the inferior colliculus on each side.

The diamond-shaped **fourth ventricle** lies behind the pons and upper medulla oblongata, under cover of the cerebellum. The upper half of the diamond is bounded by the **superior cerebellar peduncles**, which are attached to the midbrain. The lower half is bounded by the **inferior cerebellar peduncles**, which are attached to the medulla oblongata. The middle cerebellar peduncles enter from the pons and overlap the other two.

Near the midline in the midregion of the ventricle is the **facial colliculus**, which is created by the facial nerve curving around the nucleus of the abducens nerve. The **vestibular area**, and the **vagal** and **hypoglossal trigones** overlie the corresponding cranial nerve nuclei. The **obex** is the posterior apex of the ventricle.

Below the fourth ventricle, the medulla oblongata shows a pair of **gracile tubercles** flanked by a pair of **cuneate tubercles**.

Sectional views

In the midbrain, the central canal of the embryonic neural tube is represented by the **aqueduct**. Behind the pons and upper medulla oblongata (Figure 3.3), it is represented by the fourth ventricle, which is tent-shaped in this view. The central canal resumes at midmedullary level; it is continuous with the central canal of the spinal cord, although movement of cerebrospinal fluid into the cord canal is negligible.

The intermediate region of the brainstem is called the **tegmentum**, which in the midbrain contains the paired **red nucleus**. Ventral to the tegmentum in the pons is the **basilar region**. Ventral to it in the medulla oblongata are the pyramids.

The tegmentum of the entire brainstem is permeated by an important network of neurons, the **reticular formation**. The tegmentum also contains *ascending sensory pathways* carrying general sensory information from the trunk and limbs. Illustrated in Figures 3.4–3.7 are the *posterior column–medial lemniscal* (PCML) *pathways*, which inform the brain about the position of the limbs in space. At

spinal cord level, the label **PCML** is used because these pathways occupy the **posterior columns** of white matter in the cord. In the brainstem, the label PCML is used because they continue upward as the **medial lemnisci**.

The most important *motor pathways* from a clinical standpoint are the **corticospinal tracts** (CSTs), the pathways for execution of voluntary movements. The CSTs are placed ventrally, occupying the crura of the midbrain, the basilar pons, and the pyramids of the medulla oblongata.

Note that, in the medulla oblongata, the PCML and CST *decussate*: one of each pair intersects with the other to gain the contralateral (opposite) side of the neuraxis (brainstem–spinal cord). The four most important decussations are illustrated in Box 3.1.

In the following account of seven horizontal sections of the brainstem, the positions of the cranial nerve nuclei are not included.

Midbrain   (Figure 3.4)

The main landmarks have already been identified. The medial lemniscal component of PCML occupies the lateral part of the tegmentum (*upper section*), on its way to a sensory nucleus of the thalamus immediately above this level. The CST has arisen in the cerebral cortex, and it is descending in the midregion of the cerebral crus on the same side.

The *decussation of the superior cerebellar peduncles* straddles the midline at the level of the inferior colliculi (*lower section*).

Pons   (Figure 3.5)

In the *upper section*, the cavity of the fourth ventricle is bordered laterally by the superior cerebellar peduncles, which are ascending (arrows) to decussate in the lower midbrain. In the floor of the ventricle is the central gray matter.

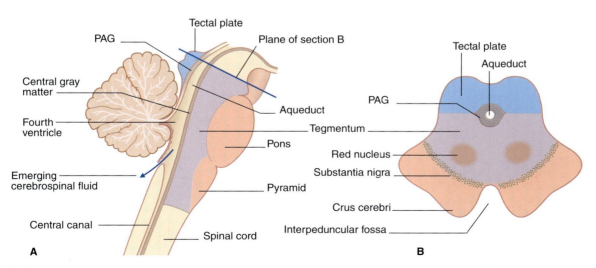

A   **B**

Figure 3.3 (A) Sagittal section of the brainstem. Cerebrospinal fluid descends along the aqueduct into the fourth ventricle and emerges into the subarachnoid space via three apertures including the median aperture containing the arrow. **(B)** Transverse section of midbrain at the level indicated in (A). The substantia nigra separates the tegmentum from the two crura cerebri. The interpeduncular fossa is so called because the entire midbrain is said to be made up of a pair of cerebral peduncles. PAG, periaqueductal gray matter.

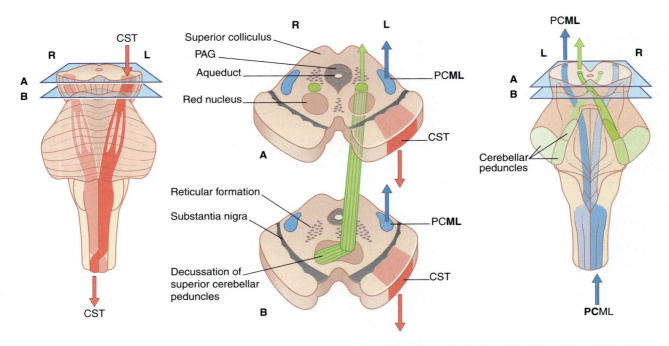

Figure 3.4 Transverse sections of midbrain. **(A)** At level of superior colliculi. **(B)** At level of inferior colliculi. In this and following diagrams, the corticospinal tract (CST) and posterior column–medial lemniscal (PCML) pathway connected to the *right* cerebral hemisphere are highlighted. PAG, periaqueductal gray matter.

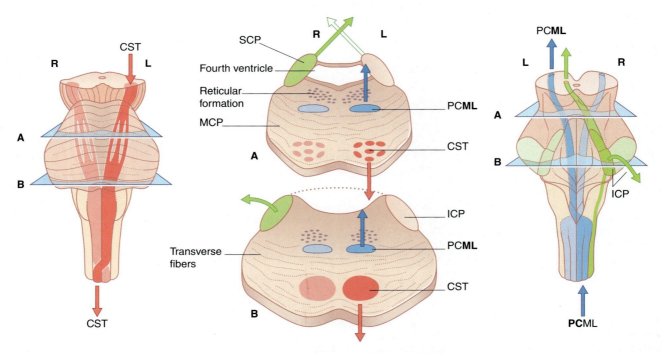

Figure 3.5 Transverse sections of pons. **(A)** Upper pons. **(B)** Lower pons. SCP, MCP, ICP, superior, middle, inferior cerebellar peduncles. CST, corticospinal tract; PCML, posterior column–medial lemniscal pathway.

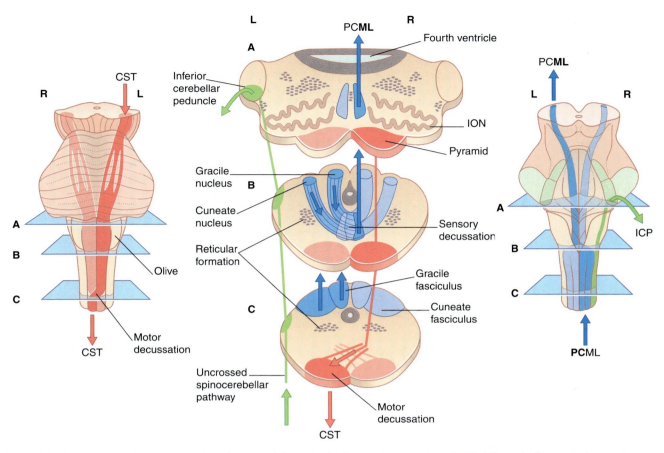

Figure 3.6 Transverse sections of medulla oblongata. **(A)** Level of inferior olivary nucleus (ION). **(B)** Level of sensory decussation. **(C)** Level of motor decussation. CST, Corticospinal tract; ICP, inferior cerebellar peduncle; PCML, posterior column–medial lemniscal pathways.

Box 3.1   Four decussations

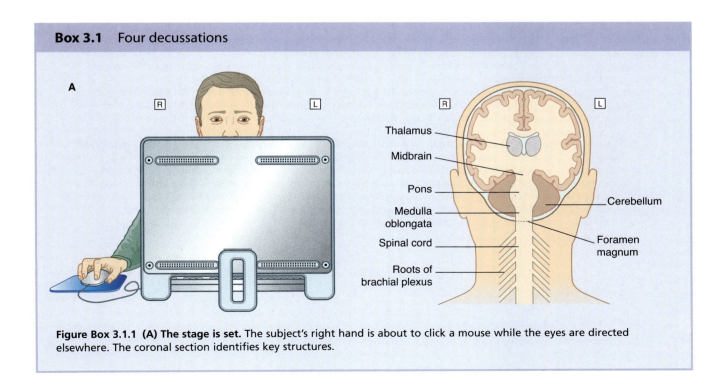

Figure Box 3.1.1 (A) The stage is set. The subject's right hand is about to click a mouse while the eyes are directed elsewhere. The coronal section identifies key structures.

Box 3.1 *Continued*

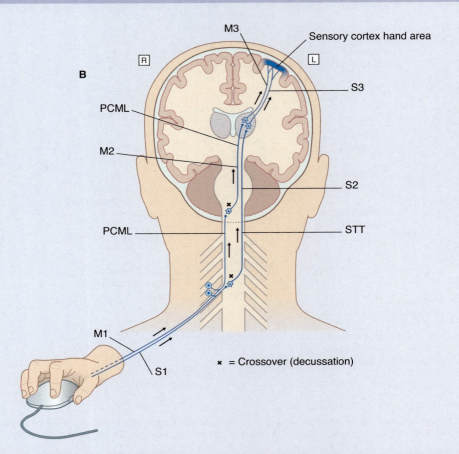

Figure Box 3.1.1 (B) Afferents. The left parietal lobe constructs a map of the right hand in relation to the mouse, based on information sent to the left somatic sensory cortex (postcentral gyrus) from the skin and deep tissues. The information is relayed by three successive sets of neurons from the skin and by another set of three from the deep tissues. The first set in each case is composed of *first-order* or *primary afferent* neurons. These neurons are called *unipolar*, because each axon emerges from a single point (or pole) of the cell body and divides in a T-shaped manner to provide continuity of impulse conduction from tissue to central nervous system. The primary afferent neurons terminate by forming contacts known as *synapses* on the *multipolar* (more or less star-shaped) cells of the *second-order* (secondary) set. The axons of the second-order neurons project across the midline before turning up to terminate on *third-order* (tertiary) multipolar neurons projecting to the postcentral gyrus.

Primary afferents activated by contacts with the skin of the hand (S1) terminate in the posterior horn of the gray matter of the spinal cord. Second-order cutaneous afferents (S2) cross the midline in the anterior white commissure and ascend to the thalamus within the *spinothalamic tract* (STT), to be relayed by third-order neurons to the hand area of the sensory cortex.

The most significant deep tissue sensory organs are *neuromuscular spindles* (muscle spindles) contained within skeletal muscles. The primary afferents supplying the muscle spindles of the intrinsic muscles of the hand belong to large unipolar neurons whose axons (labeled M1) ascend ipsilaterally (on the same side of the spinal cord) within the posterior funiculus, as already seen in *Figure 3.7*. They synapse in the nucleus cuneatus in the medulla oblongata. The multipolar second-order neurons send their axons across the midline in the sensory decussation (seen in *Figure 3.6*).The axons ascend (M2) through pons and midbrain before synapsing on third-order neurons (M3) projecting from thalamus to sensory cortex.

PCML, posterior column–medial lemniscal pathway.

Box 3.1   *Continued*

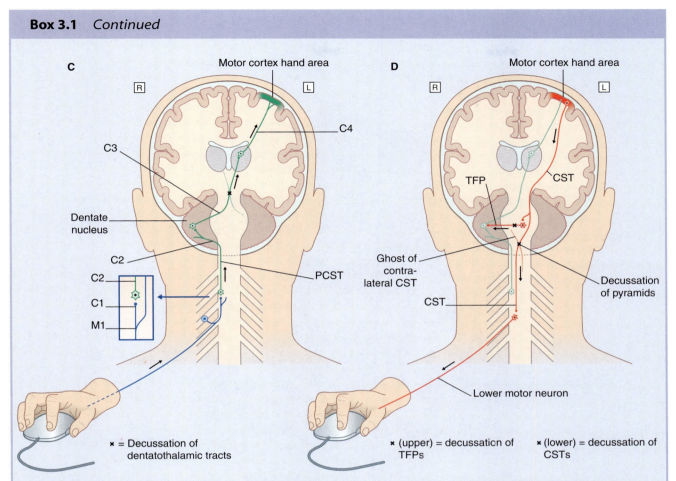

C

Motor cortex hand area

R   L

C4

C3

Dentate
nucleus

C2

C2

C1

M1

PCST

× = Decussation of
dentatothalamic tracts

D

Motor cortex hand area

R   L

TFP   CST

Ghost of
contra-
lateral CST

CST

Decussation
of pyramids

Lower motor neuron

× (upper) = decussation of
TFPs

× (lower) = decussation of
CSTs

Figure Box 3.1.1 (C) Cerebellar control. Before the brain sends an instruction to click the mouse, it requires information on the current state of contraction of the muscles. This information is constantly being sent from the muscles to the cerebellar hemisphere on the same side. As indicated in the diagram, M1 neurons are dual-purpose sensory neurons. At their point of entry to the posterior funiculus, they give off a branch, here labeled C1, to a *spinocerebellar* neuron that projects (C2) to the ipsilateral cerebellum. From here, a *cerebellothalamic* neuron (C3) is shown projecting across the midbrain to the contralateral thalamus, where a further neuron (C4) relays information to the hand area of the *motor* cortex in the precentral gyrus.

   (D) Motor output. Multipolar neurons in the left motor cortex now fire impulses along the upper motor neurons that constitute the corticospinal tract (CST), which crosses to the opposite side in the motor decussation, as already noted in *Figure 3.6*. The CST synapses on *lower motor neurons* projecting from the anterior horn of the spinal gray matter to activate flexor muscles of the index finger and local stabilizing muscles.

   Note that a copy of the outgoing message is sent to the right cerebellar hemisphere by way of *transverse fibers of the pons* (TFP) originating in multipolar neurons located on the left side of the pons.

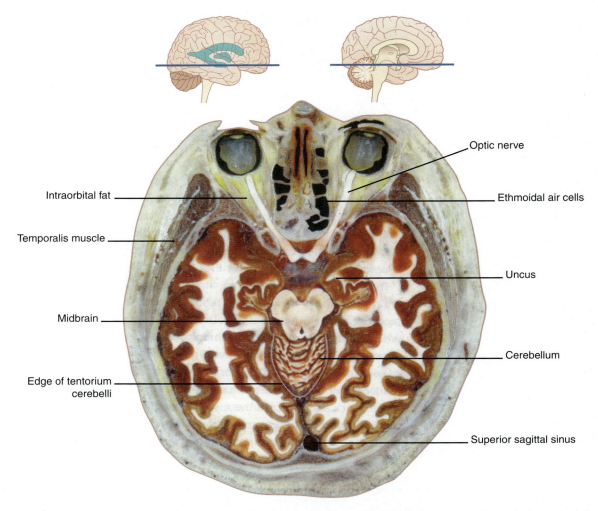

Figure 3.7 Horizontal section of fixed cadaver, taken at level of the midbrain. The cerebellum is seen through the tentorial notch. (From Liu et al. 2003, with permission of Shantung Press of Science and Technology.)

The medial lemniscus occupies the ventral part of the tegmentum on each side. The basilar region contains millions of **transverse fibers**, some of which separate the CST into individual fascicles. The transverse fibers enter the cerebellum via the middle cerebellar peduncles and *appear* to form a bridge (hence, *pons*) connecting the cerebellar hemispheres. But the *individual* transverse fibers arise on one side of the pons and cross to enter the contralateral cerebellar hemisphere. The transverse fibers belong to the giant *corticopontocerebellar pathway*, which travels from the cerebral cortex of one side to the contralateral cerebellar hemisphere, as depicted in Box 3.1.

The *lower section* contains the inferior cerebellar peduncle, about to plunge into the cerebellum. The CST bundles have reunited prior to entering the medulla oblongata.

Medulla oblongata (Figure 3.6)

Follow the CST from above down. It descends through sections A and B as the **pyramid**. In C, it intersects with its opposite number in the *motor decussation*, prior to entering the contralateral side of the spinal cord.

Follow the PCML pathway from below upward. In section C, it takes the form of the **gracile** and **cuneate fasciculi**, known in the spinal cord as the posterior columns of white matter. In section B, the posterior columns terminate in the **gracile** and **cuneate nuclei**. From these nuclei, fresh sets of fibers swing around the central gray matter and intersect with their opposite numbers in the *sensory decussation*. Having crossed the midline, the fibers turn upward. In section A, they form the medial lemniscal component of PCML.

On the left side of the medulla is shown the *uncrossed, posterior spinocerebellar tract*. Its function (non-conscious) is to inform the cerebellum of the state of activity of the ipsilateral (same side) skeletal muscles in the trunk and limbs.

The upper third of the medulla shows the wrinkled **inferior olivary nucleus**, which creates the olive of gross anatomy.

Sections of brainstem in situ are in Figures 3.7–3.11.

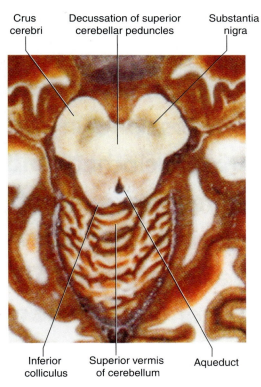

Crus cerebri   Decussation of superior cerebellar peduncles   Substantia nigra

Inferior colliculus   Superior vermis of cerebellum   Aqueduct

Figure 3.8 Enlargement from Figure 3.7.

SPINAL CORD

General features

The spinal cord occupies the upper half of the vertebral canal. Thirty-one pairs of spinal nerves are attached to it by means of **anterior** and **posterior nerve roots** (Figure 3.12A). The cord shows **cervical** and **lumbar enlargements** that accommodate nerve cells supplying the upper and lower limbs.

Internal anatomy

In transverse sections, the cord shows butterfly-shaped gray matter surrounded by three columns or **funiculi** of white matter (Figure 3.12B): an **anterior funiculus** in the interval between the **anterior median fissure** and the emerging **anterior nerve roots**; a **lateral funiculus** between the anterior and **posterior nerve roots**; and a **posterior funiculus** between the posterior roots and the **posterior median septum**.

The gray matter consists of **central gray matter** surrounding a minute central canal, and **anterior and posterior gray horns** on each side. At the levels of attachment of the 12 thoracic and upper two or three lumbar nerve roots, a **lateral gray horn** is present as well. Posterior nerve roots enter the posterior gray horn, and anterior nerve roots emerge from the anterior gray horn.

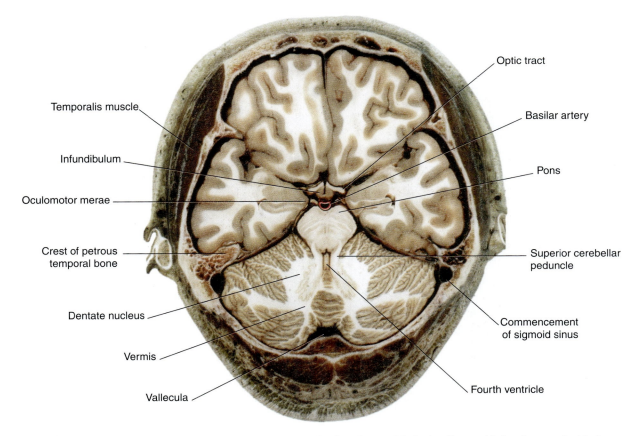

Temporalis muscle

Infundibulum

Oculomotor merae

Crest of petrous temporal bone

Dentate nucleus

Vermis

Vallecula

Optic tract

Basilar artery

Pons

Superior cerebellar peduncle

Commencement of sigmoid sinus

Fourth ventricle

Figure 3.9 Horizontal section taken at the level of upper pons. The fourth ventricle is slot-like at this level; on each side is a superior cerebellar peduncle traveling upward and medially from dentate nucleus toward the contralateral thalamus. (From Liu et al. 2003, with permission of Shantung Press of Science and Technology.)

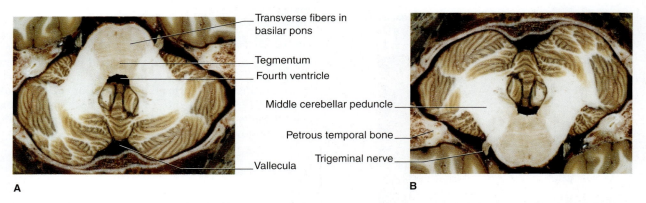

Transverse fibers in basilar pons

Tegmentum

Fourth ventricle

Middle cerebellar peduncle

Petrous temporal bone

Trigeminal nerve

Vallecula

A

B

Figure 3.10 Horizontal section taken through the middle of the pons. **(A)** In axial brain scans, the pons would be in the position shown, i.e. in the roof of the fourth ventricle. **(B)** In standard anatomic descriptions including histologic sections (cf. Ch. 17), the pons occupies the floor of the fourth ventricle, as shown here. Note the massive size of the middle cerebellar peduncles. (From Liu et al. 2003, with permission of Shantung Press of Science and Technology.)

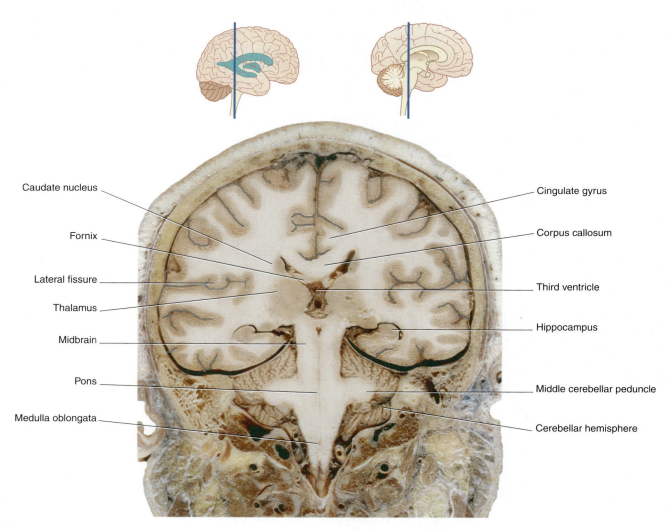

Caudate nucleus

Fornix

Lateral fissure

Thalamus

Midbrain

Pons

Medulla oblongata

Cingulate gyrus

Corpus callosum

Third ventricle

Hippocampus

Middle cerebellar peduncle

Cerebellar hemisphere

Figure 3.11 Coronal section of brainstem and cerebellum at the level shown at top. Note that the section passes through the tegmentum of the midbrain. The spinal and trigeminal lemnisci are entering the posterior–posterolateral nuclei of thalamus. The periaqueductal gray matter is sectioned longitudinally; the aqueduct itself is seen below the third ventricle.

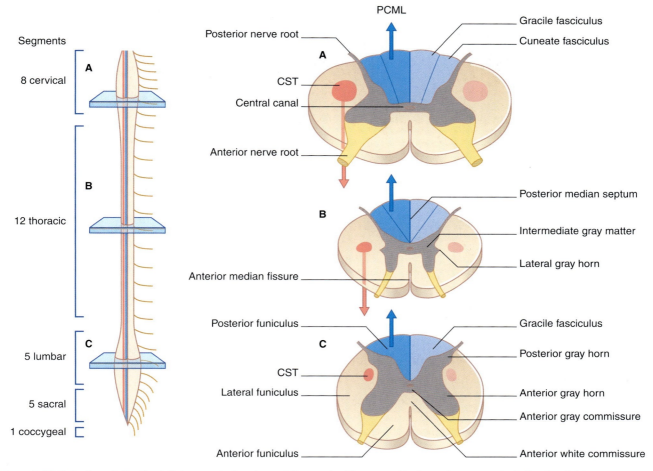

Figure 3.12 Spinal cord. On the left is an anterior view of the cord with nerve attachments enumerated. On the right are **(A)** cervical enlargement level, **(B)** thoracic level, and **(C)** lumbar enlargement level, all showing the arrangement of the largest motor and sensory pathways in the white matter, namely corticospinal tract (CST) and posterior column–medial lemniscal pathway (PCML) comprising gracile and cuneate fasciculi.

Axons pass from one side of the spinal cord to the other in the **anterior white** and **gray commissures** deep to the anterior median fissure.

The CST descends the cord within the lateral funiculus. Its principal targets are neurons in the anterior gray horn concerned with activation of skeletal muscles. *Special note*: In Chapter 16, it will be seen that a small, *anterior* CST separates from the main bundle and descends within the anterior funiculus. Accordingly, the proper name of the bundle depicted here is the *lateral* CST.

In the cord, the PCML pathway is represented by the **gracile** and **cuneate fasciculi**. The fasciculi are composed of the *central processes of peripheral sensory neurons* supplying muscles, joints, and skin. Processes entering from the lower part of the body form the gracile ('slender') fasciculus; those from the upper part form the cuneate ('wedge-shaped') fasciculus (Figures 3.13 and 3.14).

CEREBELLUM

The cerebellum is made up of two hemispheres connected by the **vermis** in the midline (Figure 3.15). The vermis is distinct only on the undersurface, where it occupies the floor of a deep groove, the **vallecula**. The hemispheres show numerous deep **fissures**, with **folia** between. About 80% of the cortex (surface gray matter) is hidden from view on the surfaces of the folia.

The oldest part of the cerebellum (present even in fishes) is the flocculonodular lobe consisting of the **nodule** of the vermis and the **flocculus** in the hemisphere on each side. More recent is the **anterior lobe**, which is bounded posteriorly by the **fissura prima** and contains the **pyramis** and the **uvula**. Most recent is the **posterior lobe**. A prominent feature of the posterior lobe is the **tonsil**. This tonsil lies directly above the foramen magnum of the skull; if the intracranial pressure is raised (e.g. by a brain tumor), one or both tonsils may descend into the foramen and pose a threat to life by compressing the medulla oblongata.

The white matter contains several deep nuclei. The largest of these is the **dentate nucleus** (Figure 3.16).

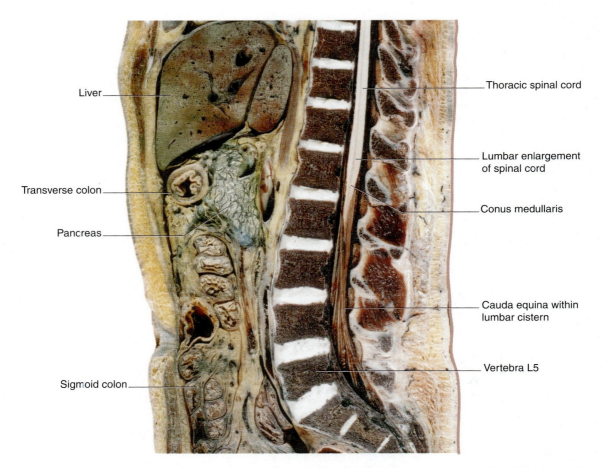

Liver

Transverse colon

Pancreas

Sigmoid colon

Thoracic spinal cord

Lumbar enlargement of spinal cord

Conus medullaris

Cauda equina within lumbar cistern

Vertebra L5

Figure 3.13 Midline sagittal section of fixed cadaver, displaying spinal cord and cauda equina in situ. It should be borne in mind that the cauda equina contains not only the motor and sensory nerve roots of the lumbosacral plexus supplying the lower limbs, but also the autonomic motor nerves supplying smooth muscle of hindgut (sigmoid colon and rectum), bladder, uterus, and erectile tissues. (From Liu et al. 2003, with permission of Shantung Press of Science and Technology.)

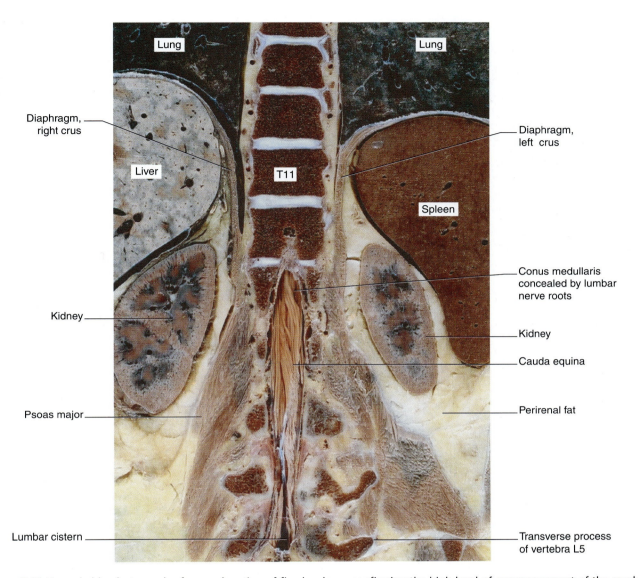

Figure 3.14 Remarkable photograph of coronal section of fixed cadaver, confirming the high level of commencement of the cauda equina as viewed from in front. In clinical context, this photograph is a reminder of the hazard to somatic (notably sciatic) and parasympathetic (notably to bladder and rectum) nerves incurred by crush fractures of lumbar vertebrae. (From Liu et al. 2003, with permission of Shantung Press of Science and Technology.)

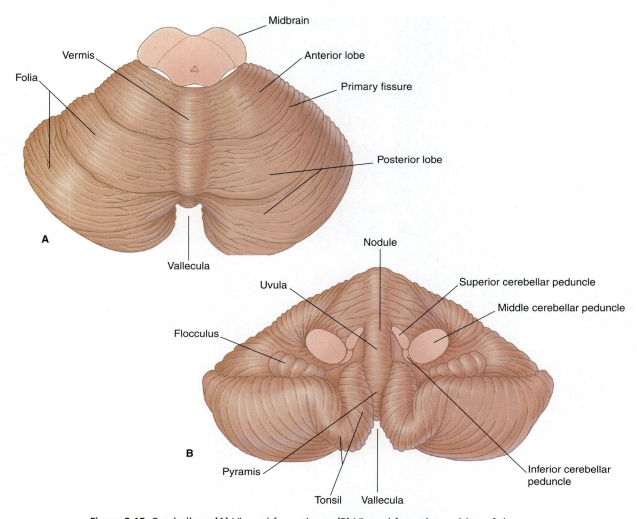

Figure 3.15 Cerebellum. **(A)** Viewed from above. **(B)** Viewed from the position of the pons.

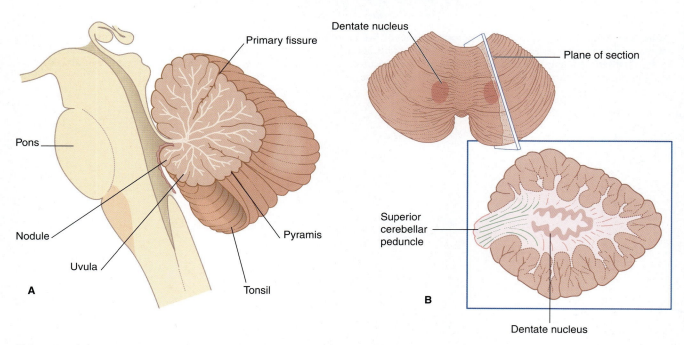

Figure 3.16 **(A)** Sagittal section of hindbrain. **(B)** Oblique section of cerebellum.

Core Information

Brainstem

The midbrain comprises tectum, tegmentum, and a crus cerebri on each side. The cerebral aqueduct is surrounded by periaqueductal gray matter. The tegmentum contains the red nucleus at the level of the upper part of the midbrain, and elements of the reticular formation at all levels of the brainstem. The largest component of the pons is the basilar region containing millions of transverse fibers belonging to the corticopontocerebellar pathways. The most prominent structure in the medulla oblongata is the inferior olivary nucleus.

The CST descends in the crus of midbrain, basilar pons, and medullary pyramid. Its principal component, the lateral CST, enters the pyramidal decussation and descends the spinal cord in the opposite lateral funiculus. Most of its fibers terminate in the anterior gray horn.

The posterior columns of the spinal cord comprise the gracile and cuneate fasciculi, which terminate in the lower medulla by synapsing upon neurons of the corresponding nuclei. A second set of fibers traverse the sensory decussation before ascending, as the medial lemniscus, to the contralateral sensory thalamus.

The posterior spinocerebellar tract carries information about ipsilateral muscular activity. It enters the inferior cerebellar peduncle. The cerebellum responds by sending signals through the superior cerebellar peduncle of that side to the contralateral motor thalamus via the decussation in the lower midbrain.

Spinal cord

The spinal cord occupies only the upper half of the vertebral canal, the sacral nerve roots being attached to it at the level of the first lumbar vertebra. In all, 31 pairs of roots are attached. The gray matter is most abundant at the levels of attachment of the brachial and lumbosacral plexuses. Anterior and posterior horns are present at all levels, and lateral horns at the level of thoracic and upper lumbar root attachments. The white matter comprises anterior, lateral, and posterior funiculi. Axons cross the midline in the gray commissures and in the white commisure. In general, propriospinal pathways are innermost, motor pathways are intermediate, and sensory pathways are outermost.

Cerebellum

The hemispheres are deeply fissured and are linked by the vermis. The oldest part is the flocculonodular lobe. More recent is the anterior lobe. Most recent is the posterior lobe, which includes the tonsils. The white matter contains several nuclei, including the dentate nucleus.

REFERENCES

DeArmond SJ, Fusco MM, Dewey MM. Structure of the human brain: a photographic atlas. 2nd edn. Oxford: Oxford University Press; 1976.

England MA, Wakely J. A colour atlas of the brain and spinal cord. London: Wolfe; 1991.

Katada K. MR imaging of brain surface structures: surface anatomy scanning (SAS). Neuroradiology 1990; 32(5):439–448.

Kretschmann H-J, Weinrich W. Cranial neuroimaging and clinical neuroanatomy. Stuttgart: Thieme; 2004.

Liu S, et al, eds. Atlas of human sectional anatomy. Jinan: Shantung Press of Science and Technology; 2003.

Niewenhuys R, Voogd J, van Huijzen C. The human central nervous system: a synopsis and atlas. 3rd edn. New York: Springer-Verlag; 1988.

Roberts M, Hanaway J, Morest DK. Atlas of the human brain in section. 2nd edn. Philadelphia: Lea & Febiger; 1987.

Wicke L. Atlas of radiologic anatomy. Philadelphia: Lea & Febiger; 1994.

Menchnges

STUDY GUIDELINES

1 Be able to compare the structure of the dura mater with that of the pia–arachnoid.

2 Be able to follow a drop of blood from the superior sagittal sinus to the internal jugular vein, and from an ophthalmic vein to the sigmoid sinus.

3 Name the nerves supplying (a) the supratentorial dura and (b) the infratentorial dura.

4 Identify the different vessels responsible for extradural, subdural, and subarachnoid bleeding.

5 Appreciate the mechanism of papilledema and why spinal tap (lumbar puncture) should not be undertaken in its presence.

6 Trace a drop of cerebrospinal fluid from a lateral ventricle to (a) its point of entry into the bloodstream, (b) to an in situ lumbar puncture needle.

7 Know about a major cause of hydrocephalus (a) in infancy, (b) in adults, and why both are examples of 'outlet obstruction'.

The meninges surround the central nervous system (CNS) and suspend it in the protective jacket provided by the cerebrospinal fluid (CSF). The meninges comprise the tough **dura mater** or **pachymeninx** (*Gr.* 'thick membrane'), and the **leptomeninges** (*Gr.* 'slender membranes') consisting of the **arachnoid mater** and **pia mater**. Between the arachnoid and the pia is the **subarachnoid space** filled with CSF.

CRANIAL MENINGES

Dura mater

The terminology used to describe the cranial dura mater varies among different authors. It seems best to regard it as a single, tough layer of fibrous tissue that is fused with the endosteum (inner periosteum) of the skull, except where it is reflected into the interior of the vault or is stretched across the skull base. Wherever it separates from the periosteum, the intervening space contains venous sinuses (Figure 4.1).

Two great dural folds extend into the cranial cavity and help to stabilize the brain. These are the **falx cerebri** and the **tentorium cerebelli**.

The falx cerebri occupies the longitudinal fissure between the cerebral hemispheres. Its attached border extends from the crista galli of the ethmoid bone to the upper surface of the tentorium cerebelli. Along the vault of the skull, it encloses the **superior sagittal sinus**. Its free border contains the **inferior sagittal sinus** that unites with the **great cerebral vein** to form the **straight sinus**. The straight sinus travels along the line of attachment of falx cerebri to tentorium cerebelli and meets the superior sagittal sinus at the **confluence of the sinuses**.

The crescentic **tentorium cerebelli** arches like a tent above the posterior cranial fossa, being lifted up by the falx cerebri in the midline. The attached margin of the tentorium encloses the **transverse sinuses** on the inner surface of the occipital bone and the **superior petrosal sinuses** along the upper border of the petrous temporal bone. The attached margin reaches to the posterior clinoid processes of the sphenoid bone. Most of the blood from the superior sagittal sinus usually empties into the right transverse sinus (Figure 4.2).

The free margin of the tentorium is U-shaped. The tips of the U are attached to the anterior clinoid processes. Just behind this, the two limbs of the U are linked by a sheet of dura, the **diaphragma sellae**, which is pierced by the pituitary stalk. Laterally, the dura falls away into the middle cranial fossae from the limbs of the U, creating the **cavernous sinus** on each side (Figure 4.3). Behind the sphenoid bone, the concavity of the U encloses the midbrain.

The cavernous sinus receives blood from the orbit via the ophthalmic veins. The superior petrosal sinus joins the transverse sinus at its junction with the **sigmoid sinus**. The sigmoid sinus descends along the occipital bone and discharges into the bulb of the internal jugular vein. The bulb receives the inferior petrosal sinus that descends along the edge of the occipital bone.

The tentorium cerebelli divides the cranial cavity into a **supratentorial compartment** containing the forebrain, and an **infratentorial compartment** containing the hindbrain. A small **falx cerebelli** is attached to the undersurface of the tentorium cerebelli and to the internal occipital crest of the occipital bone.

Innervation of the cranial dura mater

The dura mater lining the supratentorial compartment of the cranial cavity receives sensory innervation from the trigeminal nerve. That lining the anterior cranial fossa and anterior part of the skull vault is supplied by the ophthalmic nerve; that lining the middle cranial fossa and midregion of the vault is mainly supplied by the **nervus spinosus** (Figure 4.2). This nerve leaves the mandibular outside the foramen ovale, to return via the foramen spinosum and accompany the **middle meningeal artery** and its branches. Stretching or inflammation of the supratentorial dura gives rise to frontal or parietal headache.

The dura mater lining the infratentorial compartment is supplied by branches of upper cervical spinal nerves entering the foramen magnum (Figure 4.2). Occipital and

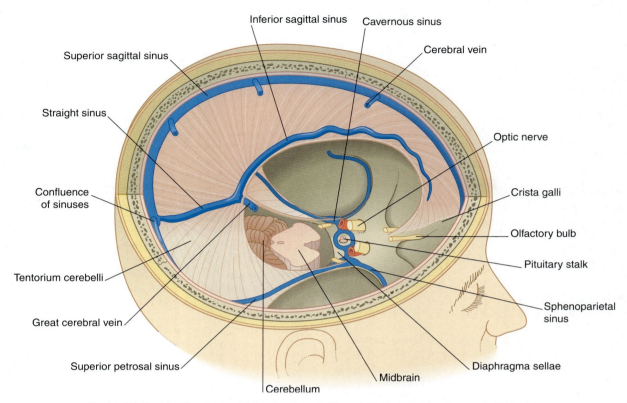

Inferior sagittal sinus

Cavernous sinus

Cerebral vein

Superior sagittal sinus

Straight sinus

Optic nerve

Confluence of sinuses

Crista galli

Olfactory bulb

Pituitary stalk

Tentorium cerebelli

Great cerebral vein

Sphenoparietal sinus

Superior petrosal sinus

Diaphragma sellae

Midbrain

Cerebellum

Figure 4.1 Dural reflections and venous sinuses. The midbrain occupies the tentorial notch.

posterior neck pains accompany disturbance of the infra-tentorial dura. Acute meningitis involving posterior cranial fossa meninges is associated with *neck rigidity* and often with *head retraction* brought about by reflex contraction of the posterior nuchal muscles, which are supplied by cervical nerves. Violent occipital headache follows *subarachnoid hemorrhage* (Ch. 35), where free blood swirls around the hindbrain.

Meningeal arteries

Embedded in the endosteum of the skull are several **meningeal arteries** whose main function is to supply the diploë (bone marrow). Much the largest is the middle meningeal artery, which ramifies over the inner surface of the temporal and parietal bones. Tearing of this artery, with its accompanying vein, is the usual source of an *extradural hematoma* (Clinical Panel 4.1).

Arachnoid mater

The arachnoid (*Gr.* 'spidery') is a thin, fibrocellular layer in direct contact with the dura mater (Figure 4.4). The outermost cells of the arachnoid are bonded to one another by tight junctions that seal the **subarachnoid space**. Innumerable **arachnoid trabeculae** cross the space to reach the pia mater.

Pia mater

The pia mater invests the brain closely, following its contours and lining the various sulci (Figure 4.4). Like the arachnoid, it is fibrocellular. The cellular component of the pia is external and is permeable to CSF. The fibrous component occupies a narrow **subpial space** that is continuous with **perivascular spaces** around cerebral blood vessels penetrating the brain surface.

Note: Although the subarachnoid and subpial spaces are proven, there is no sign of any 'subdural space' in properly fixed material. Such a space can be created, however, by leakage of blood into the cellular layer of the dura mater following a tear of a cerebral vein at its point of anchorage to the fibrous layer. (See *Subdural hematoma* in Clinical Panel 4.1.)

Subarachnoid cisterns

Along the base of the brain and the sides of the brainstem, pools of CSF occupy subarachnoid *cisterns* (Figures 4.5 and 4.6). The largest of these is the **cisterna magna**, in the interval between the cerebellum and the medulla oblongata. More rostrally are the **cisterna pontis** ventral to the pons, the **interpeduncular cistern** between the cerebral peduncles, and the **cisterna ambiens** at the side of the midbrain. The complete list of cisterns is in Table 4.1.

Sheath of the optic nerve

The optic nerve is composed of CNS white matter, and it has a complete meningeal investment. The dura mater fuses with the scleral shell of the eyeball; the subarachnoid space is a tubular *cul de sac* (dead end). The central vessels of the retina pierce the meninges to enter it (Figure 4.7). Any sustained elevation of intracranial

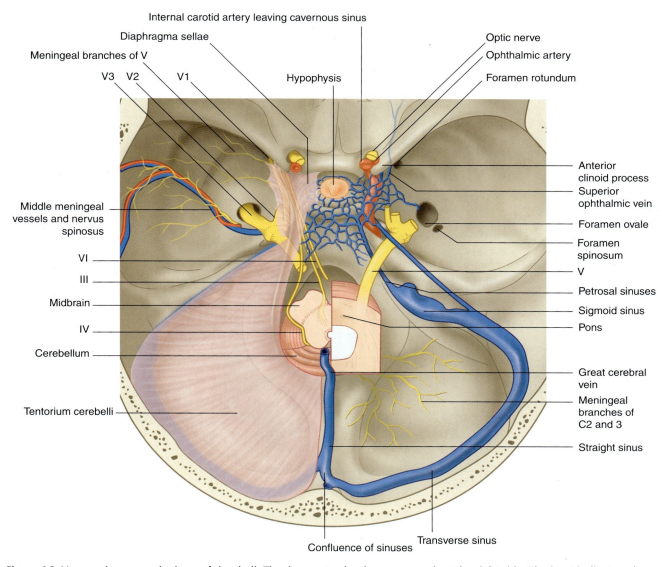

Figure 4.2 Venous sinuses on the base of the skull. The dura mater has been removed on the right side. The inset indicates where grooves for sinuses are seen on the dry skull. *Note:* On the left, the midbrain is seen at the level of the tentorial notch. On the right, a lower level section shows the trigeminal nerve attached to the pons.

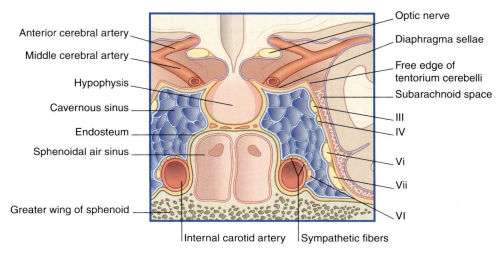

Figure 4.3 Coronal section of the cavernous sinus.

Clinical Panel 4.1 Extradural and subdural hematomas

An *extradural (epidural) hematoma* is typically caused by a blow to the side of the head severe enough to cause a fracture with associated tearing of the anterior or posterior branch of the middle meningeal artery. Following the initial *concussion* of the brain, with loss of consciousness, there may be a *lucid interval* of several hours. Onset of increasing headache and drowsiness signals *cerebral compression* produced by expansion of the hematoma. Coma and death will supervene unless the hematoma is drained though a burr hole. The favored site of access is the H-shaped suture complex known as the *pterion*, which overlies the anterior branch of the middle meningeal artery (Figure CP 4.1.1).

Subdural hematomas are caused by rupture of superficial cerebral veins in transit from the brain to an intracranial venous sinus. An *acute subdural hematoma* most often follows severe head injury in children. It must always be suspected where a child remains unconscious after a head injury. Child-battering is a possible explanation if this situation arises in the home. A *subacute subdural hematoma* may follow head injury at any age. Symptoms and signs of raised intracranial pressure (described in Ch. 6) develop up to 3 weeks after the injury.

Chronic subdural hematomas occur in older people, where the transit veins have become brittle and made taut by shrinkage of the aging brain. Head

Figure CP 4.1.1 Side view of skull. The circle encloses the pterion.

injury may be mild or even absent. A significant number of these patients are alcoholics with reduced blood clotting. Presenting symptoms are variable and include personality changes, headaches, and epileptic seizures.

Table 4.1 Subarachnoid cisterns

Cistern	Location
Posterior cerebellomedullary (cisterna magna)	Between cerebellum and dorsal surface of medulla oblongata
Lateral cerebellomedullary	Along each side of the medulla
Chiasmatic	Behind and above the optic chiasm
Cistern of lateral cerebral fossa	Along the lateral sulcus (Sylvian fissure)
Interpeduncular	Interpeduncular fossa
Ambient (cisterna ambiens)	On each side of the midbrain
Quadrigeminal	Surrounding the great cerebral vein dorsal to the midbrain colliculi (quadrigeminal bodies)

pressure will be transmitted to the subarachnoid sleeve surrounding the nerve. The **central vein** will be compressed, resulting in swelling of the retinal tributaries of the vein and edema of the optic papilla, where the optic nerve begins. The condition is known as *papilledema* (Figure 4.8). It can be recognized on inspection of the retina with an ophthalmoscope.

SPINAL MENINGES (Figure 4.9)

The spinal dural sac is like a test tube, attached to the rim of the foramen magnum and reaching down to the level of the second sacral vertebra. The outer surface of the tube is adherent to the posterior longitudinal ligament of the vertebrae in the midline; elsewhere, it is surrounded by fat containing the epidural, **internal vertebral venous plexus** (Ch. 14).

The internal surface of the dura is lined with arachnoid mater. The pia mater lines the surface of the spinal cord and is attached to the dura mater at regular intervals by the serrated **denticulate** (toothed) **ligament**.

Because the spinal cord reaches only to first or second lumbar vertebral level, a large **lumbar cistern** is created, containing the free-floating motor and sensory roots of the sacral and lower lumbar spinal nerves (Ch. 14). The lumbar cistern may be tapped to procure samples of CSF

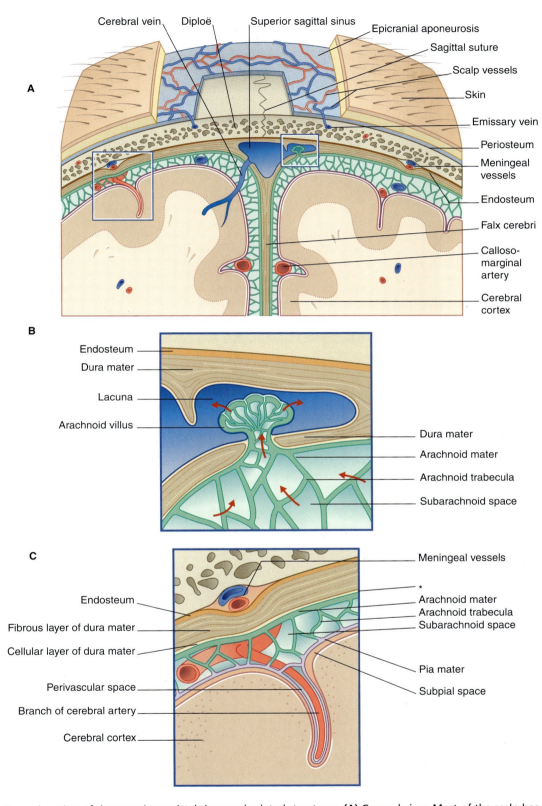

Figure 4.4 Coronal section of the superior sagittal sinus and related structures. **(A)** General view. Most of the scalp has been removed to show two emissary veins transferring blood from the diploë into scalp veins on the surface of the epicranial aponeurosis. On the right, the diploë is being fed and drained by meningeal vessels. Also seen is a cerebral vein draining into the superior sagittal sinus. **(B)** Enlargement from (A), showing an arachnoid granulation transferring cerebrospinal fluid from the subarachnoid space to a lacuna connected to the superior sagittal sinus. **(C)** Enlargement from (A), showing an artery sequentially surrounded by subarachnoid, subpial, and perivascular space extracellular fluid. The asterisk marks the potential space between dura and arachnoid for spread of subdural blood from a torn cerebral vein. Note the extradural position of the meningeal vessels.

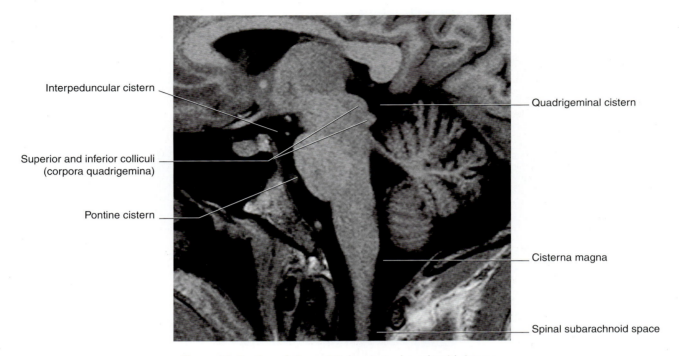

Interpeduncular cistern

Superior and inferior colliculi
(corpora quadrigemina)

Pontine cistern

Quadrigeminal cistern

Cisterna magna

Spinal subarachnoid space

Figure 4.5 Portion of *Figure 2.8* showing subarachnoid cisterns.

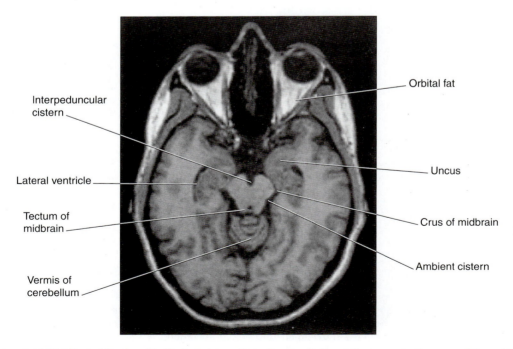

Interpeduncular
cistern

Lateral ventricle

Tectum of
midbrain

Vermis of
cerebellum

Orbital fat

Uncus

Crus of midbrain

Ambient cistern

Figure 4.6 Horizontal MRI 'slice' at the level indicated at top. Note the proximity of the uncus to the crus of the midbrain (cf. uncal herniation, *Clinical Panel 6.2*). (From a series kindly provided by Professor J. Paul Finn, Director, Magnetic Resonance Research, Department of Radiology, David Geffen School of Medicine at UCLA, California, USA.)

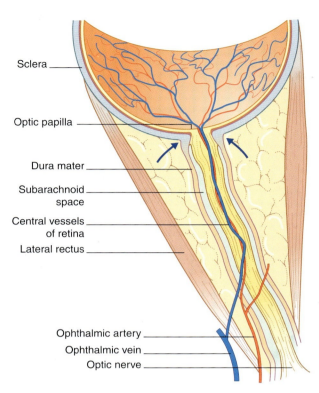

Figure 4.7 Horizontal section of the left orbit. The subarachnoid space extends forward to the level of fusion of dura mater with the scleral coat of the eyeball (*arrows*).

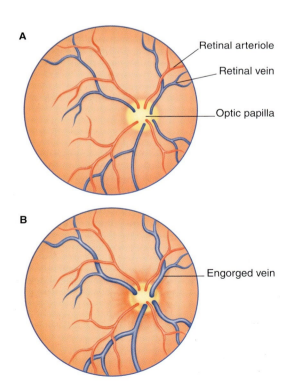

Figure 4.8 Fundus oculi as seen with an ophthalmoscope. **(A)** Normal. **(B)** Papilledema resulting from raised intracranial pressure.

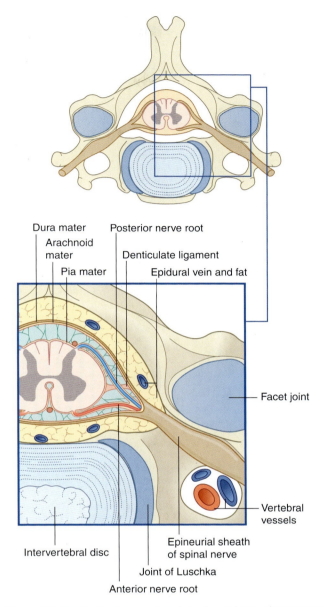

Figure 4.9 Contents of cervical vertebral canal. The dura mater blends with the epineurium of the spinal nerve trunk.

for analysis (Clinical Panel 4.2) or to deliver a spinal anesthetic (Ch. 14).

CIRCULATION OF CEREBROSPINAL FLUID (Figure 4.10)

The principal source of the CSF is the secretion of the choroid plexuses into the ventricles of the brain. From the lateral ventricles, the CSF enters the third through the interventricular foramen. It descends to the fourth ventricle through the aqueduct and squirts into the subarachnoid space through the median and lateral apertures. (Flow within the central canal of the spinal cord is negligible.)

Clinical Panel 4.2 Lumbar puncture (Figure CP 4.2.1)

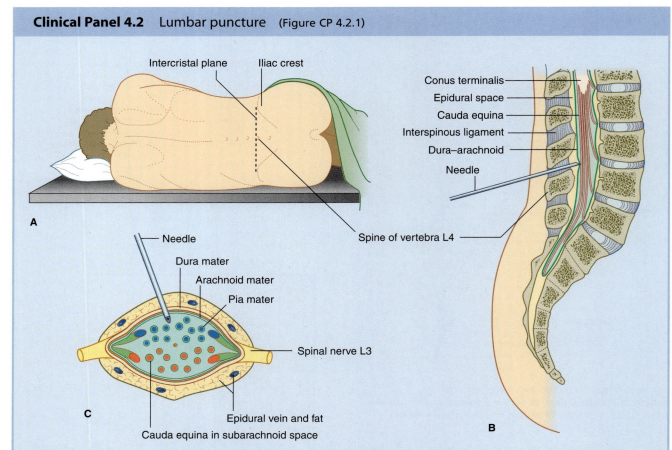

Figure CP 4.2.1 Lumbar puncture (spinal tap). **(A)** The patient lies on one side, curled forward to open the interspinous spaces of the lumbar region. The spine of vertebra L4 is identified in the intercristal (supracristal) plane at the level of the tops of the iliac crests. **(B)** Under aseptic conditions, a lumbar puncture needle is introduced obliquely above the spine of vertebra L4, parallel to the plane of the spine. The needle is passed through the interspinous ligament. A slight 'give' is perceived when the needle pierces the dura–arachnoid mater and enters the subarachnoid space. **(C)** Transverse section showing the cauda equina floating in the subarachnoid space. The anterior and posterior roots of spinal nerve L3 are coming together as they leave the lumbar cistern.

Within the subarachnoid space, some of the CSF descends through the foramen magnum, reaching the lumbar cistern in about 12 h. From the subarachnoid space at the base of the brain, the CSF ascends through the tentorial notch and bathes the surface of the cerebral hemispheres before being returned to the blood through the **arachnoid granulations** (Figure 4.4). The arachnoid granulations are pinhead pouches of arachnoid mater projecting through the dural wall of the major venous sinuses—especially the superior sagittal sinus and the small venous **lacunae** that open into it. CSF is transported across the arachnoid epithelium in giant vacuoles.

As much as a quarter of the circulating CSF may not reach the superior sagittal sinus. Some enters small arach-noid villi projecting into spinal veins exiting intervertebral foramina, and some drains into lymphatics in the adventitia of arteries at the base of the brain and in the epineurium of cranial nerves. These lymphatics drain into cervical lymph nodes. Bimanual downward massage of the sides of the neck is sometimes used to enhance this drainage in patients suffering from cerebral edema.

About 300 mL of CSF is secreted by the choroid plexuses every 24 h. Another 200 mL is produced from other sources, as described in Chapter 5. Blockage of flow through the ventricular system or cranial subarachnoid space will cause back-up within the ventricular system: a state of *hydrocephalus* (Clinical Panel 4.3).

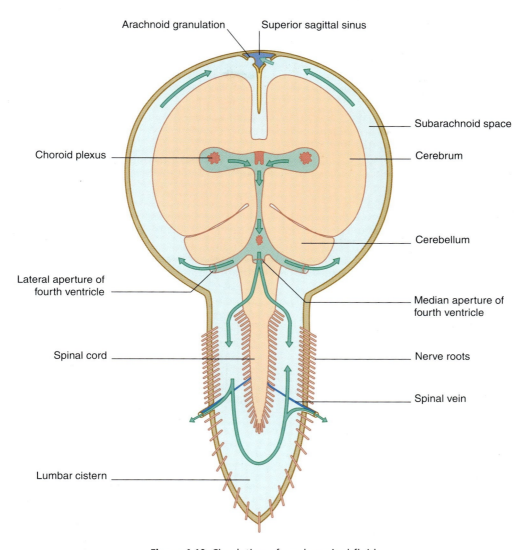

Arachnoid granulation   Superior sagittal sinus

Choroid plexus

Lateral aperture of
fourth ventricle

Spinal cord

Lumbar cistern

Subarachnoid space

Cerebrum

Cerebellum

Median aperture of
fourth ventricle

Nerve roots

Spinal vein

Figure 4.10 Circulation of cerebrospinal fluid.

Clinical Panel 4.3   Hydrocephalus

Hydrocephalus (*Gr.* 'water in the brain') denotes accumulation of CSF in the ventricular system. With the exception of overproduction of CSF by a rare papilloma of the choroid plexus, hydrocephalus results from obstruction of the normal CSF circulation, with consequent dilatation of the ventricles. (The term is not used to describe the accumulation of fluid in the ventricles and subarachnoid space in association with senile atrophy of the brain.)

In the great majority of cases, hydrocephalus is caused by obstruction of the foramina opening the fourth ventricle to the subarachnoid space. A major cause of outlet obstruction in *infancy* is the *Arnold–Chiari malformation*, in which the cerebellum is partly extruded into the vertebral canal during fetal life because the posterior cranial fossa is underdeveloped. In untreated cases, the child's head may become as

large as a football and the cerebral hemispheres paper-thin. The condition is nearly always associated with spina bifida (Ch. 14). Early treatment is essential to prevent severe brain damage. The obstruction can be bypassed by means of a catheter having one end inserted into a lateral ventricle and the other inserted into the internal jugular vein.

A major cause of outlet obstruction in *adults* is displacement of the cerebellum into the foramen magnum by a space-occupying lesion (mass) such as a tumor or hematoma (see Ch. 6).

Meningitis can cause hydrocephalus at any age. The development of leptomeningeal adhesions may compromise CSF circulation at the level of the ventricular outlets, the tentorial notch, and/or the arachnoid granulations.

Core Information

Meninges

The meninges comprise dura, arachnoid, and pia mater. The subarachnoid space contains CSF.

The cranial dura mater shows two large folds: the falx cerebri and tentorium cerebelli. The attached edge of the falx encloses the superior sagittal sinus, which usually enters the right transverse sinus. The free edge of the falx encloses the inferior sagittal sinus, which joins the great cerebral vein, forming the straight sinus that enters the confluence of the sagittal and transverse sinuses. The attached edge of the tentorium encloses the transverse sinus, which descend as the sigmoid sinus to continue as the internal jugular vein. The midbrain is partly enclosed by the free edge of the tentorium, which is attached to the anterior clinoid processes of the sphenoid bone and provides a U-shaped gap for passage of the midbrain. Dura drapes from each side of the U into the middle cranial fossa, creating the cavernous sinus. This sinus receives blood from the ophthalmic veins and drains via petrosal sinuses into each end of the sigmoid sinus.

The supratentorial dura mater is innervated by the trigeminal nerve, the infratentorial dura by upper cervical nerves. The meningeal vessels run extradurally to supply the diploë; if torn by skull fracture, they may form an extradural hematoma compressing the brain. A subdural hematoma may be caused by leakage from a cerebral vein in transit to the superior sagittal sinus.

Cerebrospinal fluid

Pools of CSF at the base of the brain include the cisterna magna, the cisterna pontis, the interpeduncular cistern, and the cisterna ambiens. CSF also extends along the meningeal sheath of the optic nerve, and raised intracranial pressure may compress the central vein of the retina, causing papilledema. The spinal dural sac extends down to S2 vertebral level. The lumbar cistern contains spinal nerve roots and is accessible for lumbar puncture (spinal tap). CSF secreted by the choroid plexuses escapes into the subarachnoid space through the three apertures of the fourth ventricle. Some descends to the lumbar cistern. The CSF ascends through the tentorial notch and the cerebral subarachnoid space to reach the superior sagittal sinus and its lacunae via the arachnoid granulations. Blockage of CSF flow anywhere along its course leads to hydrocephalus.

REFERENCES

Brinker T, Ludemann W, von Rautenfeld BD, et al. Dynamic properties of lymphatic pathways for the absorption of cerebrospinal fluid. Acta Neuropathol (Berlin) 1997; 94:493–498.

Hutchings M, Weller RO. Anatomical relationships of the pia mater to cerebral blood vessels in man. J Neurosurg 1986; 65:316–325.

Nicholas DS, Weller RO. The fine anatomy of the human spinal meninges. J Neurosurg 1988; 69:276–282.

Vandenabeele L, Creemers J, Lambrichts I. Ultrastructure of the human spinal arachnoid mater and dura mater. J Anat 1996; 189:417–430.

Blood supply of the brain

5

STUDY GUIDELINES
Because interpretation of the symptoms caused by cerebrovascular accidents requires prior understanding of brain function, Clinical Panels on this subject are placed in the final chapter.

A Clinical Panel on blood–brain barrier pathology is placed in the present chapter because the symptoms are of a general nature.

1 On simple outline drawings of the lateral, medial, and inferior surfaces of a cerebral hemisphere, learn to shade in the territories of the three cerebral arteries.

2 Identify the main sources of arterial supply to the internal capsule.

3 Become familiar with carotid and vertebral angiograms.

4 Be able to list the territories supplied by the vertebral and basilar arteries.

5 Identify the two blood–brain barriers. Be able to understand why, for example, shallow breathing following abdominal surgery may tip a patient into coma.

INTRODUCTORY NOTE

The brain is absolutely dependent on a continuous supply of oxygenated blood. It controls the delivery of blood by sensing the momentary pressure changes in its main arteries of supply, the internal carotids. It controls the arterial oxygen tension by monitoring respiratory gas levels in the internal carotid artery and in the cerebrospinal fluid (CSF) beside the medulla oblongata. The control systems used by the brain are exquisitely sophisticated, but they can be brought to nothing if a distributing artery ruptures spontaneously or is rammed shut by an embolus.

ARTERIAL SUPPLY OF THE FOREBRAIN

The blood supply to the forebrain is derived from the two **internal carotid arteries** and from the **basilar artery** (Figure 5.1).

Each internal carotid artery enters the subarachnoid space by piercing the roof of the cavernous sinus. In the subarachnoid space, it gives off **ophthalmic**, **posterior communicating**, and **anterior choroidal arteries** before dividing into the **anterior** and **middle cerebral arteries**.

The basilar artery divides at the upper border of the pons into the two **posterior cerebral arteries**. The **cerebral arterial circle** (*circle of Willis*) is completed by a linkage of the posterior communicating artery with the posterior cerebral on each side, and by linkage of the two anterior cerebrals by the **anterior communicating artery**.

The choroid plexus of the lateral ventricle is supplied from the **anterior choroidal** branch of the internal carotid artery and by the **posterior choroidal** branch from the posterior cerebral artery.

Dozens of fine **central (perforating) branches** are given off by the constituent arteries of the circle of Willis. They enter the brain through the **anterior perforated substance** beside the optic chiasm and through the **posterior perforated substance** behind the mammillary bodies. They have been classified in various ways but can be conveniently grouped into short and long branches. **Short central branches** arise from all the constituent arteries and from the two choroidal arteries. They supply the optic nerve, chiasm, and tract, and the hypothalamus. **Long central branches** arise from the three cerebral arteries. They supply the thalamus, corpus striatum, and internal capsule. They include the **striate branches** of the anterior and middle cerebral arteries.

Anterior cerebral artery   (Figure 5.2)

The anterior cerebral artery passes above the optic chiasm to gain the medial surface of the cerebral hemisphere. It forms an arch around the genu of the corpus callosum, making it easy to identify in a carotid angiogram (see later). Close to the anterior communicating artery, it gives off the **medial striate artery**, also known as the *recurrent artery of Heubner* (*pron.* 'Hoibner'), which contributes to the blood supply of the internal capsule. Cortical branches of the anterior cerebral artery supply the medial surface of the hemisphere as far back as the parietooccipital sulcus (Table 5.1). The branches overlap on to the orbital and lateral surfaces of the hemisphere.

Middle cerebral artery   (Figure 5.3)

The middle cerebral artery is the main continuation of the internal carotid, receiving 60–80% of the carotid blood flow. It immediately gives off important central branches, then passes along the depth of the lateral fissure to reach the surface of the insula. There it usually breaks into upper and lower divisions. The upper division supplies the frontal lobe, the lower division supplies the parietal and

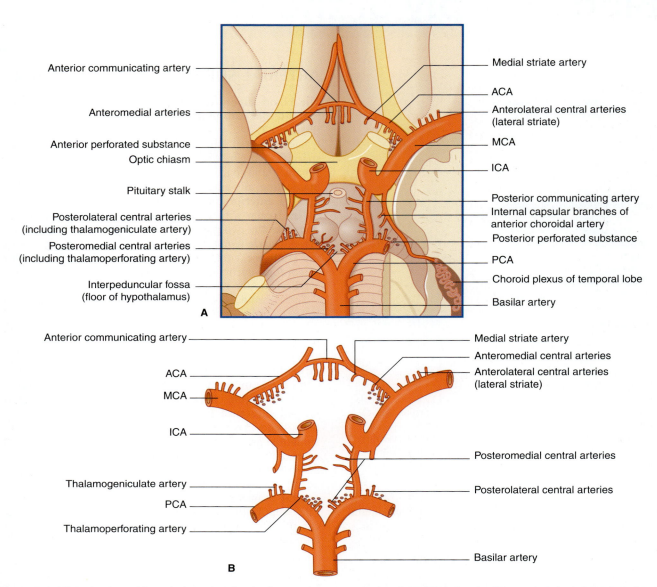

Figure 5.1 (A) Brain viewed from below, showing background structures related to the circle of Willis. Part of the left temporal lobe (to right of picture) has been removed to show the choroid plexus in the inferior horn of the lateral ventricle. (B) The arteries comprising the circle of Willis. The four groups of central branches are shown; the thalamoperforating artery belongs to the posteromedial group, and the thalamogeniculate artery belongs to the posterolateral group. ACA, MCA, PCA, anterior, middle, posterior cerebral arteries; ICA, internal carotid artery.

Table 5.1 Named cortical[a] branches of the anterior cerebral artery

Branch	Territory
Orbitofrontal	Orbital surface of frontal lobe
Polar frontal	Frontal pole
Callosomarginal	Cingulate and superior frontal gyri; paracentral lobule
Pericallosal	Corpus callosum

[a]The term *cortical* is conventional. *Terminal* is better, because these arteries also supply the underlying white matter.

temporal lobes and the midregion of the optic radiation. Named branches and their territories are listed in Table 5.2. Overall, the middle cerebral supplies two-thirds of the lateral surface of the brain.

The **central branches** of the middle cerebral include the **lateral striate** arteries (Figure 5.4). These arteries supply the corpus striatum, internal capsule, and thalamus. Occlusion of one of the lateral striate arteries is the chief cause of classic *stroke*, where damage to the pyramidal tract in the posterior limb of the internal capsule causes *hemiplegia*, a term denoting paralysis of the contralateral arm, leg, and lower part of face.

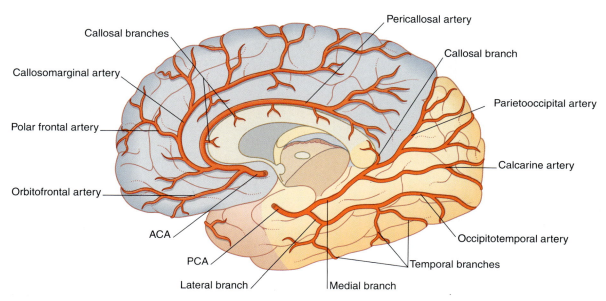

Figure 5.2 Medial view of the right hemisphere, showing the cortical branches and territories of the three cerebral arteries. ACA, PCA, anterior, posterior cerebral arteries.

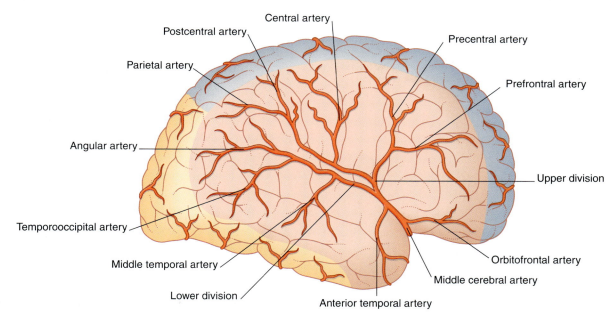

Figure 5.3 Lateral view of right cerebral hemisphere, showing the cortical branches and territories of the three cerebral arteries.

Note: Additional information on the blood supply of the internal capsule is provided in Chapter 35.

Posterior cerebral artery (Figures 5.2 and 5.5)

The two posterior cerebral arteries are the terminal branches of the basilar. However, in embryonic life they arise from the internal carotid, and in about 25% of individuals the internal carotid persists as the primary source of blood on one or both sides, by way of a large posterior communicating artery.

Close to its origin, each posterior cerebral artery gives branches to the midbrain and a **posterior choroidal artery** to the choroid plexus of the lateral ventricle. Additional, **central branches** are sent into the posterior perforated substance (Figure 5.1). The main artery winds around the midbrain in company with the optic tract. It supplies the splenium of the corpus callosum and the cortex of the occipital and temporal lobes. Named cortical branches and their territories are given in Table 5.3.

The central branches, called **thalamoperforating** and **thalamogeniculate**, supply the thalamus, subthalamic nucleus, and optic radiation.

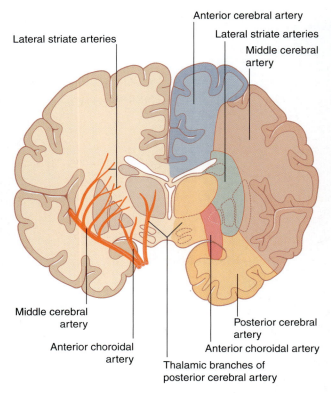

Figure 5.4 Distribution of perforating branches of the middle cerebral, anterior choroidal, and posterior cerebral arteries (schematic). The anterior choroidal artery arises from the internal carotid.

Table 5.2 Cortical branches of the middle cerebral artery

Origin	Branch(es)	Territory
Stem	Frontobasal	Orbital surface of frontal lobe
	Anterior temporal	Anterior temporal cortex
Upper division	Prefrontal	Prefrontal cortex
	Precentral	Premotor areas
	Central	Pre- and postcentral gyri
	Postcentral	Postcentral and anterior parietal cortex
	Parietal	Posterior parietal cortex
Lower division	Middle temporal	Midtemporal cortex
	Temporooccipital	Temporal and occipital cortex
	Angular	Angular and neighboring gyri

Table 5.3 Named cortical branches of the posterior cerebral artery

Branch	Artery	Territory
Lateral	Anterior temporal	Anterior temporal cortex
	Posterior temporal	Posterior temporal cortex
	Occipitotemporal	Posterior temporal and occipital cortex
Medial	Calcarine	Calcarine cortex
	Parietooccipital	Cuneus and precuneus
	Callosal	Splenium of corpus callosum

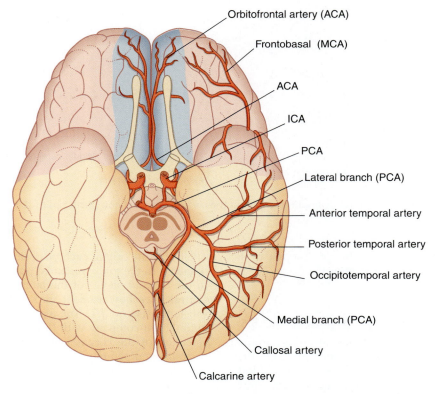

Figure 5.5 View from below the cerebral hemispheres, showing the cortical branches and territories of the three cerebral arteries. ACA, MCA, PCA, anterior, middle, posterior cerebral arteries; ICA, internal carotid artery.

Note: Additional information on the central branches is provided in Chapter 35.

Neuroangiography

The cerebral arteries and veins can be displayed under general anesthesia by rapid injection of a radiopaque dye into the internal carotid or vertebral artery, followed by serial radiography every 2 s. The dye completes its journey through the arteries, brain capillaries, and veins in about 10 s. The *arterial phase* of the journey yields either a *carotid angiogram* or a *vertebrobasilar angiogram*. Improved vascular definition in radiographs of the arterial phase or of the *venous phase* can be procured by a process of *subtraction*, whereby positive and negative images of the overlying skull are superimposed on one another, thereby virtually deleting the skull image.

A relatively recent technique, *three-dimensional angiography*, is based on simultaneous angiography from two slightly separate perspectives.

Arterial phases of carotid angiograms are shown in Figures 5.6–5.8.

Figure 5.9 was taken at the *parenchymal* phase, when the dye is filling a web of minute terminal branches of the anterior and middle cerebral arteries, some of these anastomosing on the brain surface but most occupying the parenchyma, i.e cortex and subjacent white matter.

ARTERIAL SUPPLY TO HINDBRAIN

The brainstem and cerebellum are supplied by the vertebral and basilar arteries and their branches (Figure 5.10).

The two **vertebral arteries** arise from the subclavian arteries and ascend the neck in the foramina transversaria of the upper six cervical vertebrae. They enter the skull through the foramen magnum and unite at the lower border of the pons to form the **basilar artery**. The basilar artery ascends to the upper border of the pons and divides into two posterior cerebral arteries (Figures 5.11 and 5.12).

All the primary branches of the vertebral and basilar arteries give branches to the brainstem.

Vertebral branches

The **posterior inferior cerebellar artery** supplies the side of the medulla before giving branches to the cerebellum. **Anterior** and **posterior spinal arteries** supply the ventral and dorsal medulla, respectively, before descending through the foramen magnum.

Basilar branches

The **anterior inferior cerebellar** and **superior cerebellar arteries** supply the side of the pons before giving branches

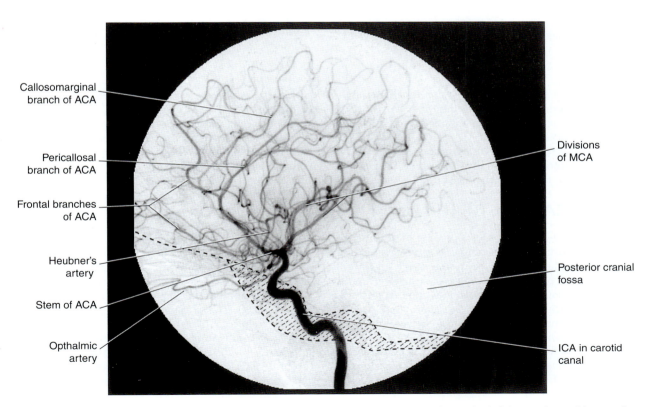

Figure 5.6 Arterial phase of a carotid angiogram, lateral view. Contrast medium injected into the left internal carotid artery is passing through the anterior and middle cerebral arteries (ACA, MCA). The base of the skull is shown in hatched outline. ICA, internal carotid artery. (From an original series kindly provided by Dr. Michael Modic, Department of Radiology, The Cleveland Clinic Foundation.)

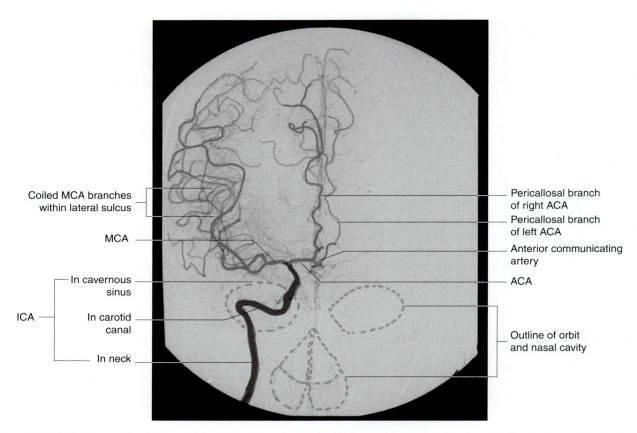

Figure 5.7 Arterial phase of a right carotid angiogram, anteroposterior view. Note some perfusion of left anterior cerebral artery (ACA) territory (via the anterior communicating artery). ICA, internal carotid artery; MCA, middle cerebral artery. (Angiogram kindly provided by Dr. Pearse Morris, Director, Interventional Neuroradiology, Wake Forest University School of Medicine, Winston-Salem, North Carolina, USA.)

to the cerebellum. The anterior inferior cerebellar usually gives off the **labyrinthine artery** to the inner ear.

About a dozen **pontine arteries** supply the full thickness of the medial part of the pons.

The midbrain is supplied by the **posterior cerebral artery**, and by the **posterior communicating artery** linking the posterior cerebral to the internal carotid.

VENOUS DRAINAGE OF THE BRAIN

The venous drainage of the brain is of great importance in relation to neurosurgical procedures. It is also important to the professional neurologist, because a variety of clinical syndromes can be produced by venous obstruction, venous thrombosis, and congenital arteriovenous communications. In general medical practice, however, problems (other than subdural hematomas, Ch. 4) caused by cerebral veins are rare in comparison with arterial disease.

The cerebral hemispheres are drained by superficial and deep cerebral veins. Like the intracranial venous sinuses, they are devoid of valves.

Superficial veins

The **superficial cerebral veins** lie in the subarachnoid space overlying the hemispheres. They drain the cerebral cortex and underlying white matter, and empty into intracranial venous sinuses (Figures 5.13A, 5.14, and 5.15).

The upper part of each hemisphere drains into the superior sagittal sinus. The middle part drains into the cavernous sinus (as a rule) by way of the **superficial middle cerebral vein**. The lower part drains into the transverse sinus.

Deep veins (Figure 5.13B)

The deep cerebral veins drain the corpus striatum, thalamus, and choroid plexuses.

A **thalamostriate vein** drains the thalamus and caudate nucleus. Together with a **choroidal vein**, it forms the **internal cerebral vein**. The two internal cerebral veins unite beneath the corpus callosum to form the **great cerebral vein** (of Galen).

A **basal vein** is formed beneath the anterior perforated substance by the union of **anterior** and **deep middle cerebral veins**. The basal vein runs around the crus cerebri and empties into the great cerebral vein.

Finally, the great cerebral vein enters the midpoint of the tentorium cerebelli. As it does so, it unites with the **inferior sagittal sinus** to form the **straight sinus**. The straight sinus empties in turn into the **left transverse sinus**.

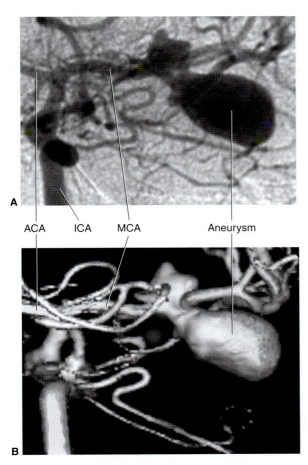

A

ACA   ICA   MCA   Aneurysm

B

Figure 5.8 **(A)** Excerpt from a conventional carotid angiogram, anteroposterior view, showing an aneurysm attached to the middle cerebral artery. **(B)** Excerpt from a three-dimensional image of the same area. ACA, MCA, anterior and middle cerebral arteries; ICA, internal carotid artery. (Originals kindly provided by Dr. Pearse Morris, Director, Interventional Neuroradiology, Wake Forest University School of Medicine, Winston-Salem, North Carolina, USA.)

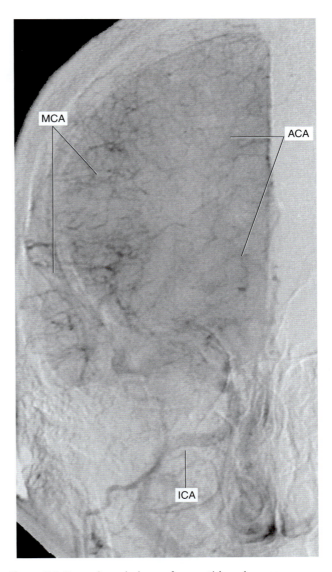

MCA

ACA

ICA

Figure 5.9 Parenchymal phase of a carotid angiogram, anteroposterior view. ACA, MCA, anterior, middle, cerebral arteries; ICA, internal carotid artery. (Angiogram kindly provided by Dr. Pearse Morris, Director, Interventional Neuroradiology, Wake Forest University School of Medicine, Winston-Salem, North Carolina, USA.)

REGULATION OF BLOOD FLOW

Blood flow in the cerebral vessels is primarily controlled by *autoregulation*, which is defined as the capacity of a tissue to regulate its own blood supply.

The most powerful source of autoregulation in the central nervous system (CNS) is the *H^+ ion concentration* in the extracellular fluid (ECF) surrounding the arterioles within the brain parenchyma. Generalized relaxation of arteriolar smooth muscle tone is produced by hypercapnia (excess plasma $P\mathrm{CO_2}$). On the other hand, hypocapnia causes arteriolar constriction.

A second powerful source of autoregulation is the *intraluminal pressure* within the arterioles. Any increase in pressure elicits a direct, myogenic response. When other factors are controlled (in animal experiments), the myogenic response is sufficient to maintain steady-state perfusion of the brain within a systemic blood pressure range of 80–180 mmHg (11–24 kPa).

Local blood flow increases within cortical areas and deep nuclei involved in particular motor, sensory, or cognitive tasks. The local arteriolar relaxation can be accounted for by a rise in K^+ levels caused by propagation of action potentials, and by a rise in H^+ caused by increased cell metabolism.

A large number of vasoactive substances have been identified in neural networks surrounding the cerebral conducting arteries and the arterioles. A specific role is difficult to assign to any of them within the physiologic range of blood flow.

THE BLOOD–BRAIN BARRIER

The nervous system is isolated from the blood by a barrier system that provides a stable and chemically optimal environment for neuronal function. The neurons and

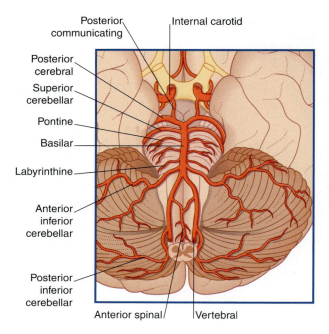

Figure 5.10 Arterial supply of hindbrain.

neuroglia are bathed in *brain extracellular fluid*, which accounts for 15% of total brain volume.

The extracellular compartments of the CNS are shown diagrammatically in Figure 5.16. As previously described (Ch. 4), CSF secreted by the choroid plexuses circulates through the ventricular system and the subarachnoid space before passing through the arachnoid villi into the dural venous sinuses. In addition, CSF diffuses passively through the ependyma–glial membrane lining the ventricles and enters the brain extracellular spaces. It adds to the ECF produced by the capillary bed and by cell metabolism, and it diffuses through the pia–glial membrane into the subarachnoid space. This 'sink' movement of fluid compensates for the absence of lymphatics in the CNS.

Metabolic water is the only component of the CSF that does not pass through the blood–brain barrier. It carries with it any neurotransmitter substances that have not been recaptured following liberation by neurons, and it accounts for the presence in the subarachnoid space of transmitters and transmitter metabolites that could not penetrate the blood–brain barrier.

Relative contributions to the CSF obtained from a spinal tap are approximately as follow:

- choroid plexuses, 60%
- capillary bed, 30%
- metabolic water, 10%.

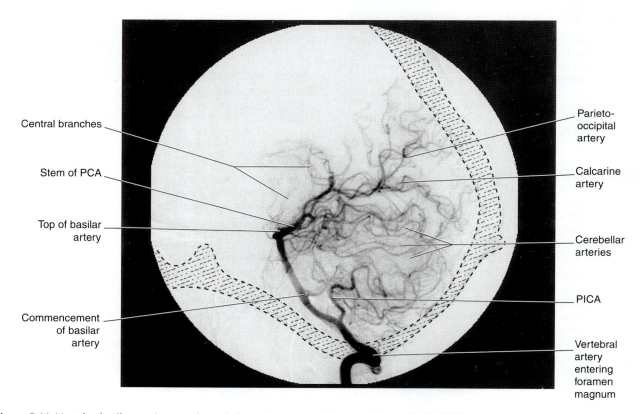

Figure 5.11 Vertebrobasilar angiogram, lateral view. Contrast medium was injected into the left vertebral artery. Basilar supply to the upper half of the cerebellum is somewhat obscured by overlying posterior parietal branches of the posterior cerebral artery. PCA, posterior cerebral artery; PICA, posterior inferior cerebellar artery. (From an original series kindly provided by Dr. Michael Modic, Department of Radiology, The Cleveland Clinic Foundation.)

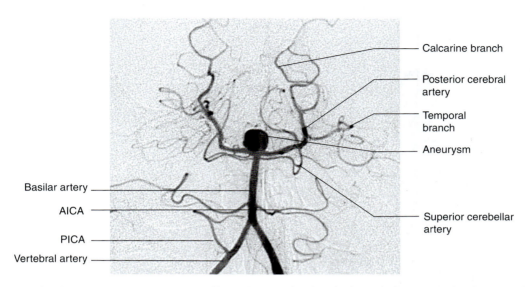

Figure 5.12 Vertebrobasilar angiogram, Townes's view (from above and in front), showing the vertebrobasilar arterial system. Note the large aneurysm arising from the bifurcation point of the basilar artery and accounting for the patient's persistent headache. AICA, anterior inferior cerebellar artery; PICA, posterior inferior cerebellar artery. (Angiogram kindly provided by Dr. Pearse Morris, Director, Interventional Neuroradiology, Wake Forest University School of Medicine, Winston-Salem, North Carolina, USA.)

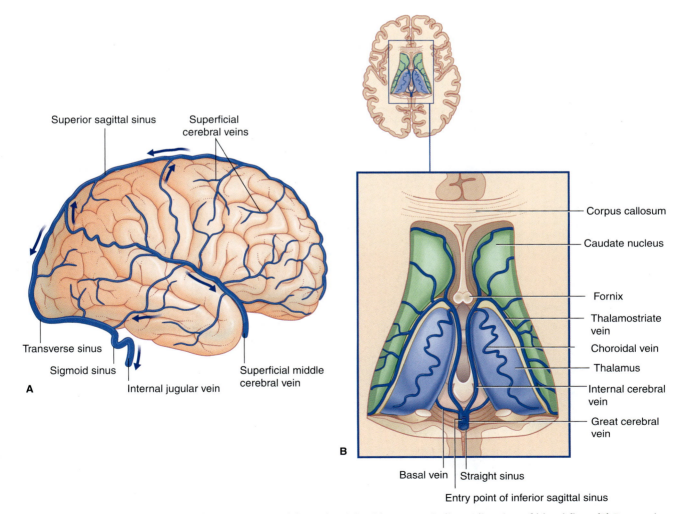

Figure 5.13 Cerebral veins. **(A)** Superficial veins viewed from the right side; arrows indicate direction of blood flow. **(B)** Deep veins viewed from above.

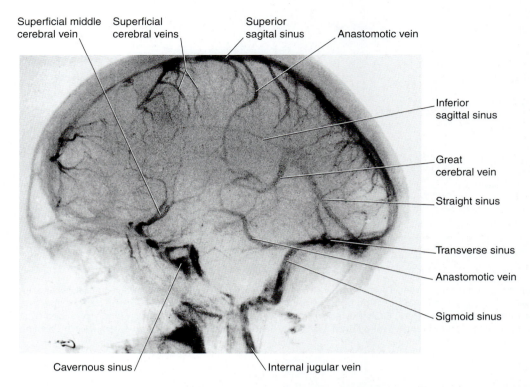

Superficial middle cerebral vein
Superficial cerebral veins
Superior sagital sinus
Anastomotic vein
Inferior sagittal sinus
Great cerebral vein
Straight sinus
Transverse sinus
Anastomotic vein
Sigmoid sinus
Cavernous sinus
Internal jugular vein

Figure 5.14 Internal carotid angiogram, venous phase, lateral view. The dye is draining into the dural venous sinuses. (Photograph kindly provided by Dr. James Toland, Department of Radiology, Beaumont Hospital, Dublin, Ireland.)

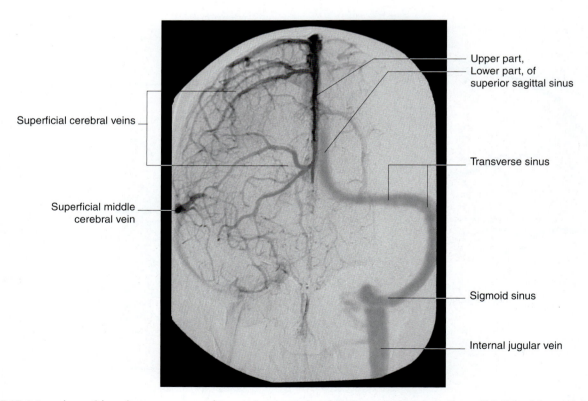

Upper part,
Lower part, of superior sagittal sinus
Superficial cerebral veins
Transverse sinus
Superficial middle cerebral vein
Sigmoid sinus
Internal jugular vein

Figure 5.15 Internal carotid angiogram, venous phase, anteroposterior view. Same patient as in Figure 5.6; this picture taken circa 8 s later. The vascular pattern is unusual, in that the left rather than the right transverse sinus is dominant. (Angiogram kindly provided by Dr. Pearse Morris, Director, Interventional Neuroradiology, Wake Forest University School of Medicine, Winston-Salem, North Carolina, USA.)

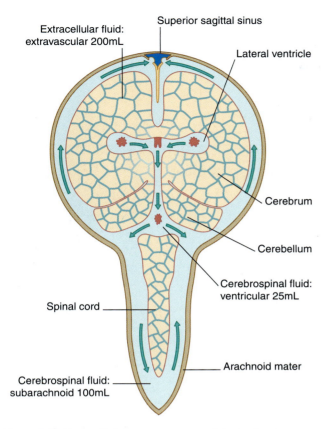

Figure 5.16 Extracellular compartments of the brain. *Arrows* indicate circulation of cerebrospinal fluid.

The blood–brain barrier has two components. One is at the level of the choroid plexus, the other resides in the CNS capillary bed.

Blood–CSF barrier   (Figure 5.17)

The blood–CSF barrier resides in the specialized ependymal lining of the choroid plexuses. This choroidal epithelium differs from the general ependymal epithelium in three ways.

1 Cilia are almost completely replaced by microvilli.
2 The cells are bonded by tight junctions. These pericellular belts of membrane fusion are the actual site of the blood–CSF barrier.
3 The epithelium contains numerous enzymes specifically involved in transport of ions and metabolites.

Blood–ECF barrier   (Figure 5.18)

The blood–ECF barrier resides in the CNS capillary bed, which differs from that of other capillary beds in three ways.

1 The endothelial cells are bonded by tight junctions.
2 Pinocytotic vesicles are rare, and fenestrations are absent.
3 The cells contain the same transport systems as those of the choroidal epithelium.

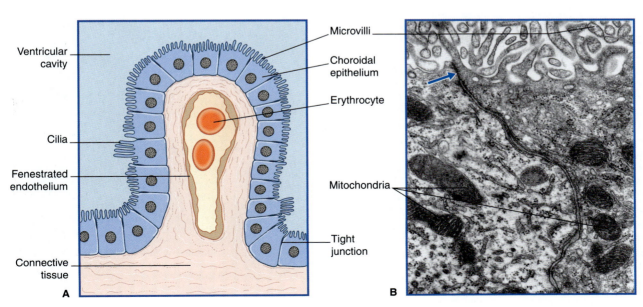

Figure 5.17 (A) Diagram of blood–cerebrospinal fluid barrier. **(B)** Ultrastructure of choroidal epithelium. The epithelial cells are rich in mitochondria and granular endoplasmic reticulum. Apical regions of adjacent cells are bonded by a tight junction (arrow). (From Pannese 1994, with permission of Thieme.)

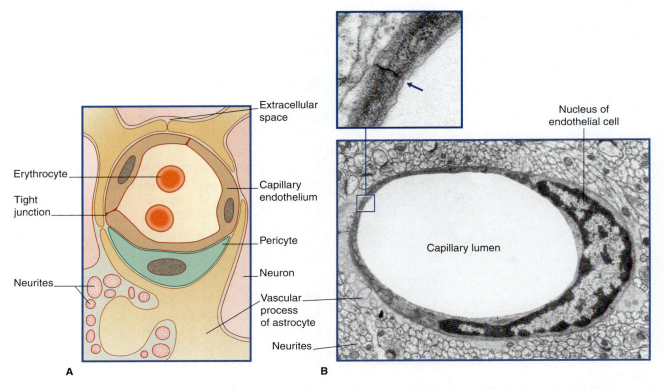

Figure 5.18 (A) Diagram of blood–extracellular fluid barrier. (Astrocytes are described in Ch. 6.) **(B)** Central nervous system capillary. In this transverse section, a single capillary completely surrounds the lumen, its edges being sealed by a tight junction (inset). Outside its basement membrane, the capillary is invested with an astrocytic sheath. (From Pannese 1994, with permission of Thieme.)

Roles of microvascular pericytes

Pericytes are in cytoplasmic continuity with the endothelial cells, by way of gap junctions. Tissue culture studies have provided strong evidence for their primary roles in capillary angiogenesis during development and in the production and maintenance of the tight junctions.

Pericytes express receptors for vasoactive mediators, including norepinephrine (noradrenaline), vasopressin, and angiotensin II, all indicative of a role in cerebrovascular autoregulation. In the presence of chronic hypertension, they strengthen the capillary bed by undergoing hypertrophy, hyperplasia, and internal production of cytoplasmic contractile protein filaments.

Pericytes are equipped for a hemostatic function, having an appropriate membrane surface for assembly of the prothrombin complex.

Pericytes are also phagocytic, and possess immunoregulatory cytokines.

The surface area of the brain capillary bed is about the size of a tennis court! This huge area accounts for the brain's consumption of 20% of basal oxygen intake by the lungs. The density of the cortical capillary bed is demonstrated in the latex cast shown in Figure 5.19.

Functions of the blood–brain barrier

• Modulation of the entry of metabolic substrates. Glucose, in particular, is a fundamental source of energy for neurons. The level of glucose in the brain

Figure 5.19 Latex injection cast of the blood vessels in human postmortem brain. The convoluted whitish threads represent cortical capillaries. (From Duvernoy et al. 1981, with permission.)

ECF is more stable than that of the blood, because the specific carrier becomes saturated when blood glucose rises and becomes hyperactive when it falls.

- Control of ion movements. Na^+–K^+ ATPase in the barrier cells pumps sodium into the CSF and pumps potassium out of the CSF into the blood.

- Prevention of access to the CNS by toxins and by peripheral neurotransmitters escaping into the bloodstream from autonomic nerve endings.

For some clinical notes concerning the blood–brain barrier, see Clinical Panel 5.1.

Clinical Panel 5.1   Blood–brain barrier pathology

The following five conditions are associated with breakdown of the blood–brain barrier.

1 Patients suffering from hypertension are liable to attacks of *hypertensive encephalopathy* should the blood pressure exceed the power of the arterioles to control it. The pressure may then open the tight junctions of the brain capillary endothelium. Rapid exudation of plasma causes *cerebral edema* with severe headache and vomiting, sometimes progressing to convulsions and coma.

2 In patients with severe *hypercapnia* brought about by reduced ventilation of the lungs (as in pulmonary or heart disease, or after surgery), relaxation of arteriolar muscle may be sufficient to induce cerebral edema even if the blood pressure is normal. In this case, the edema may be expressed by mental confusion and drowsiness progressing to coma.

3 *Brain injury*, whether from trauma or spontaneous hemorrhage, leads to edema owing to the osmotic effects of tissue damage (and other factors).

4 *Infections* of the brain or meninges are accompanied by breakdown of the blood–brain barrier, perhaps because of the large-scale emigration of leukocytes through the brain capillary bed. The breakdown can be exploited because the porous capillary walls will permit the passage of non–lipid-soluble antibiotics.

5 The capillary bed of *brain tumors* is fenestrated. As a result, radioactive tracers too large to penetrate healthy brain capillaries can be detected within tumors.

Core Information

Arteries

The circle of Willis comprises the anterior communicating artery and two anterior cerebral arteries, the internal carotids, two posterior communicating arteries, and the two posterior cerebral arteries.

The anterior cerebral artery gives off Heubner's artery to the anteroinferior internal capsule, then arches around the corpus callosum and supplies the medial surface of the hemisphere as far back as the parietooccipital sulcus, with overlap on to the lateral surface.

The middle cerebral artery enters the lateral sulcus and supplies two-thirds of the lateral surface of the hemisphere. Its central branches include the leak-prone lateral striate supplying the upper part of the internal capsule.

The posterior cerebral artery arises from the basilar artery; it supplies the splenium of corpus callosum and the occipital and temporal cortex.

The vertebral arteries enter the foramen magnum. They supply spinal cord, posterior-inferior cerebellum and medulla oblongata before uniting to form the basilar artery. The basilar artery supplies the anterior-inferior and superior cerebellum, the pons and inner ear, before dividing into posterior cerebral arteries.

Veins

Superficial cerebral veins drain the cerebral cortex and empty into dural venous sinuses. The internal cerebral veins drain the thalami and unite as the great cerebral vein. The great veins drain the corpus striatum via the basal vein before entering the straight sinus.

Autoregulation

Hypercapnia causes arteriolar dilatation, hypocapnia causes constriction. A rise of intraluminal pressure produces a direct, myogenic response by arteriolar walls.

Blood–brain barrier

A blood–CSF barrier resides in the choroidal epithelium (modified ependyma) of the ventricles. A blood–ECF barrier resides in the endothelium of the brain capillary bed.

REFERENCES

Balabanonov B, Dore-Duffy P. Role of the microvascular pericyte in the blood–brain barrier. J Neurosci Res 1998; 53:637–644.

Duvernoy HM, Delon S, Vannson JL. Cortical blood vessels of the human brain. Brain Res Bull 1981; 7:519–530.

Duvernoy HM. Human brainstem vessels. New York: Springer-Verlag; 1978.

Gloger S, Gloger A, Vogt H, et al. Computer-assisted 3D reconstruction of the terminal branches of the cerebral arteries. Neuroradiology 1994; 36:173–180; 181–187; 251–257.

Kapp JP. The cerebral venous system and its disorders. Orlando: Grune & Stratton; 1984.

Pannese E. Neurocytology. Fine structure of neurons, nerve processes and neuroglial cells. New York: Thieme; 1994

Sage MR, Wilson AJ. The blood–brain barrier: an important concept in neuroimaging. Am J Neuroradiol 1994; 94:601–622.

Scremin IU. Cerebral vascular system. In: Paxinos G, Mai JK, eds. The human nervous system. 2nd edn. Amsterdam: Elsevier; 2004.

Wahl M, Schilling L. Regulation of cerebral blood flow—a brief review. Acta Neurochir 1993; 59:3–10.

Neurons and neuroglia: overview

STUDY GUIDELINES

1 Appreciate the challenge faced by many neurons in having to deliver and retrieve materials over enormous distances, and the economy of transmitter recycling at nerve endings.

2 Appreciate how a healthy transport system can spread disease in the nervous system.

3 Appreciate the lock and key analogy used in pharmacology.

4 Draw an axodendritic synapse, then add another axon dividing to exert both pre- and postsynaptic inhibition.

5 Understand why demyelinating disorders compromise conduction.

6 Draw up a structure–function list for neuroglial cells.

7 Gliomas will obviously interfere with brain function in the region they grow. Try to understand how they may exert 'distance effects'.

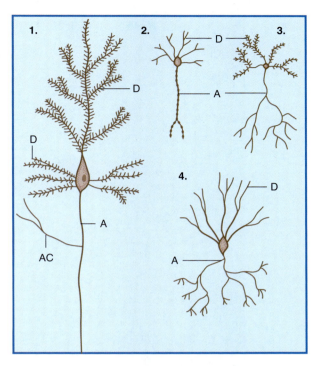

Figure 6.1 Profiles of neurons from the brain. **(1)** Pyramidal cell, cerebral cortex. **(2)** Neuroendocrine cell, hypothalamus. **(3)** Spiny neuron, corpus striatum. **(4)** Basket cell, cerebellum. Neurons 1 and 3 show dendritic spines. A, axon; AC, axon collateral; D, dendrites.

Nerve cells, or **neurons**, are the structural and functional units of the nervous system. They generate and conduct electrical changes in the form of nerve impulses. They communicate chemically with other neurons at points of contact called **synapses**. **Neuroglia** (literally, 'nerve glue') is the connective tissue of the nervous system.

Neuroglial cells outnumber neurons by about five to one. They have important nutritive and supportive functions.

NEURONS

Billions of neurons form a shell, or **cortex**, on the surface of the cerebral and cerebellar hemispheres. In this general context, **nuclei** are aggregates of neurons buried within the white matter.

In the central nervous system (CNS), almost all neurons are multipolar, their cell bodies or **somas** having multiple poles or angular points. At every pole but one, a **dendrite** emerges and divides repeatedly (Figure 6.1). On some neurons, the shafts of the dendrites are smooth. On others, the shafts show numerous short **spines** (Figure 6.2). The dendrites receive synaptic contacts from other neurons, from some on the spines and from others on the shafts.

The remaining pole of the soma gives rise to the **axon**, which conducts nerve impulses. Most axons give off *collateral* branches (Figure 6.3). *Terminal* branches synapse on target neurons.

Most synaptic contacts between neurons are either *axodendritic* or *axosomatic*. Axodendritic synapses are usually excitatory in their effect on target neurons, whereas most axosomatic synapses have an inhibitory effect.

Internal structure of neurons

All parts of neurons are permeated by **microtubules** and **neurofilaments** (Figure 6.4). The soma contains the nucleus and the cytoplasm or **perikaryon** (*Gr.* 'around the nucleus'). The perikaryon contains clumps of granular endoplasmic reticulum known as *Nissl bodies* (Figure 6.5), also Golgi complexes, free ribosomes, mitochondria, and smooth endoplasmic reticulum (SER).

Intracellular transport

Turnover of membranous and skeletal materials takes place in all cells. In neurons, fresh components are continuously synthesized in the soma and moved into the axon and

Polyribosomes

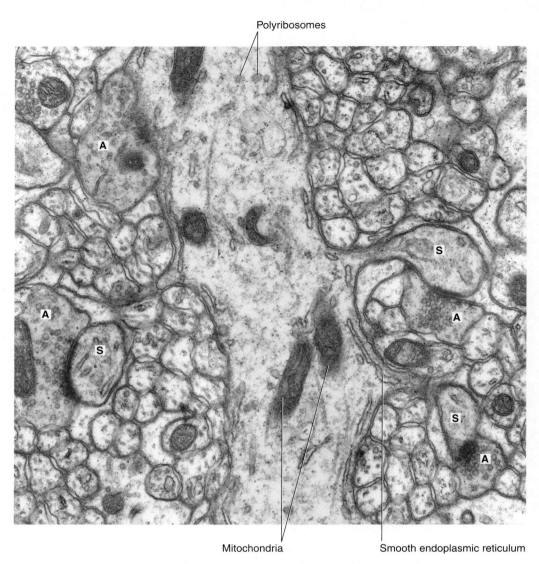

Mitochondria   Smooth endoplasmic reticulum

Figure 6.2 Dendritic spines. This section is taken from the cerebellum, where the dendrites of the giant cells of Purkinje are studded with spines. In this field, three spines (S) are in receipt of synaptic contacts by axonic boutons (A). A fourth axon (top left) is synapsing on the shaft of the dendrite. (From Pannese 1994, with permission of Thieme.)

dendrites by a process of *anterograde transport*. At the same time, worn-out materials are returned to the soma by *retrograde transport* for degradation in lysosomes (see also *target recognition*, later).

Anterograde transport is of two kinds: rapid and slow. Included in *rapid* transport (at a speed of 300–400 mm/day) are free elements such as synaptic vesicles, transmitter substances (or their precursor molecules), and mitochondria. Also included are lipid and protein molecules (including receptor proteins) for insertion into the plasma membrane. Included in *slow* transport (at 5–10 mm/day) are the skeletal elements, and soluble proteins including some of those involved in transmitter release at nerve endings. Microtubules seem to be largely constructed within the axon. They are exported from the soma in preassembled short sheaves that propel one another along the initial segment of the axon; further progress is mainly

by a process of elongation (up to 1 mm apiece) performed by the addition of tubulin polymers at their distal ends, with some disassembly at their proximal ends. The bulk movement of neurofilaments slows down to almost zero distally; there, the filaments are refreshed by the insertion of filament polymers moving from the soma by slow transport.

Retrograde transport of worn-out mitochondria, SER, and plasma membrane (including receptors therein) is fairly rapid (150–200 mm/day). In addition to its function in waste disposal, retrograde transport is involved in *target cell recognition*. At synaptic contacts, axons constantly 'nibble' the plasma membrane of target neurons by means of endocytotic uptake of protein-containing *signaling endosomes*. These proteins are known as *neurotrophins* ('neuron foods'). They are brought to the soma and incorporated into Golgi complexes there. In addition, uptake

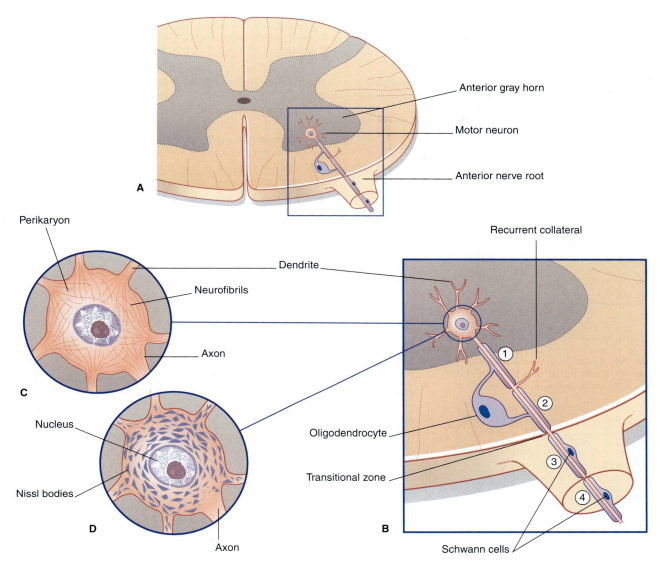

Figure 6.3 **(A)** Motor neuron in the anterior gray horn of the spinal cord. **(B)** Enlargement from (A). Myelin segments 1 and 2 occupy central nervous system white matter and have been laid down by an oligodendrocyte; a recurrent collateral branch of the axon originated from the node. Myelin segments 3 and 4 occupy peripheral nervous system and have been laid down by Schwann cells; the node at the transitional zone is bounded by an oligodendrocyte and a Schwann cell. **(C)** Neurofibrils (matted neurofilaments) are seen after staining with silver salts. **(D)** Nissl bodies (clumps of granular endoplasmic reticulum) are seen after staining with a cationic dye such as thionin.

of target cell 'marker' molecules is important for cell recognition during development. It may also be necessary for viability later on, because adult neurons shrink and may even die if their axons are severed proximal to their first branches.

The longest-known neurotrophin is nerve growth factor, on which the developing peripheral sensory and autonomic systems are especially dependent. Adult brain neurons synthesize brain-derived neurotrophic factor (BDNF) in the soma and send it to their nerve endings by anterograde transport. Animal studies have shown that BDNF maintains the general health of neurons in terms of metabolic activity, impulse propagation, and synaptic transmission.

Transport mechanisms

Microtubules are the supporting structures for neuronal transport. Microtubule-binding proteins, in the form of ATPases, propel organelles and molecules along the outer surface of the microtubules. Distinct ATPases are used for anterograde and retrograde work. Retrograde transport of signaling endosomes is performed by the *dynein* ATPase. Failure of dynein performance has been linked to motor neuron disease, described in Chapter 16.

Neurofilaments do not seem to be involved in the transport mechanism. They are rather evenly spaced, having side arms that keep them apart and provide skeletal stability by attachment to proteins beneath the axolemmal membrane. Neurofilament numbers are in direct proportion

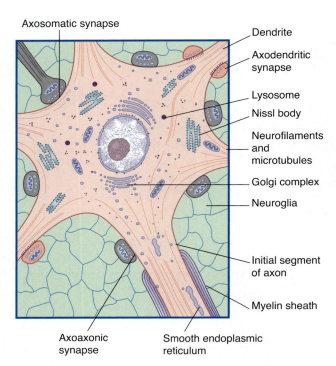

Axosomatic synapse

Dendrite

Axodendritic synapse

Lysosome

Nissl body

Neurofilaments and microtubules

Golgi complex

Neuroglia

Initial segment of axon

Myelin sheath

Axoaxonic synapse

Smooth endoplasmic reticulum

Figure 6.4 Ultrastructure of a motor neuron. Stems of five dendrites are included, also three excitatory synapses (*red*) and five inhibitory synapses.

to axonal diameter, and the filaments may in truth *determine* axonal diameter.

Some points of clinical relevance are highlighted in Clinical Panel 6.1.

SYNAPSES

Synapses are the points of contact between neurons.

Electrical synapses

Electrical synapses are scarce in the mammalian nervous system. They consist of gap junctions (nexuses) between dendrites or somas of contiguous neurons, where there is cytoplasmic continuity through 1.5-nm channels. No transmitter is involved, and there is no synaptic delay. They permit electrotonic changes to pass from one neuron to another. Being tightly coupled, modulation is not possible. Their function is to ensure synchronous activity of neurons having a common action. An example is the inspiratory center in the medulla oblongata, where all the cells exhibit synchronous discharge during inspiration. A second example is among neuronal circuits controlling *saccades*, where the gaze darts from one object of interest to another.

Chemical synapses

Conventional synapses are *chemical*, depending for their effect on the release of a *transmitter substance*. The typical chemical synapse comprises a **presynaptic membrane**, a **synaptic cleft**, and a **postsynaptic membrane** (Figure 6.6). The presynaptic membrane belongs to the terminal bouton, the **postsynaptic** membrane to the target neuron. Transmitter substance is released from the bouton by exocytosis, traverses the narrow synaptic cleft, and activates receptors in the postsynaptic membrane. Underlying the postsynaptic membrane is a **subsynaptic web**, in which numerous biochemical changes are initiated by receptor activation.

The bouton contains **synaptic vesicles** loaded with transmitter substance, together with numerous mitochondria and sacs of SER (Figure 6.7). Following conventional methods of fixation, *presynaptic dense projections* are visible, and microtubules seem to guide the synaptic vesicles to active zones in the intervals between the projections.

Clinical Panel 6.1 Clinical relevance of neuronal transport

Tetanus
Wounds contaminated by soil or street dust may contain *Clostridium tetani*. The toxin produced by this organism binds to the plasma membrane of nerve endings, is taken up by endocytosis, and is carried to the spinal cord by retrograde transport. Other neurons upstream take in the toxin by endocytosis—notably Renshaw cells (Ch. 15), which normally exert a braking action on motor neurons through the release of an inhibitory transmitter substance, glycine. Tetanus toxin prevents the release of glycine. As a result, motor neurons go out of control, particularly those supplying the muscles of the face, jaws, and spine. These muscles exhibit prolonged, agonizing spasms. About half of the patients who show these classic signs of tetanus die of exhaustion within a few

days. Tetanus is entirely preventable by appropriate and timely immunization.

Viruses and toxic metals
Retrograde axonal transport has been blamed for the passage of viruses from the nasopharynx to the central nervous system, also for the uptake of toxic metals such as lead and aluminum. Viruses, in particular, may be spread widely through the brain by means of retrograde transneuronal uptake.

Peripheral neuropathies
Defective anterograde transport seems to be involved in certain 'dying back' neuropathies in which the distal parts of the longer peripheral nerves undergo progressive atrophy.

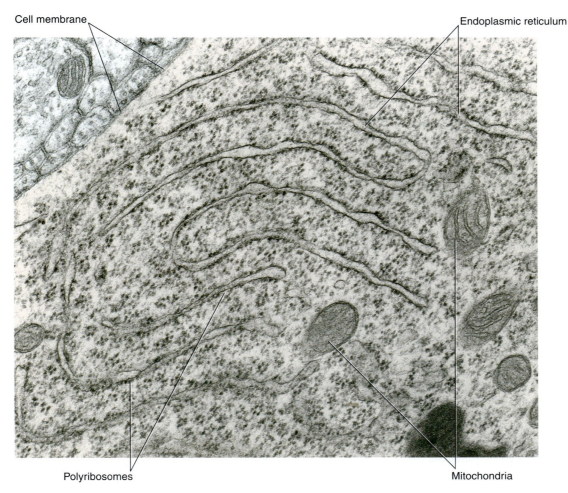

Cell membrane

Endoplasmic reticulum

Polyribosomes

Mitochondria

Figure 6.5 Nissl substance in the soma of a motor neuron. The endoplasmic reticulum has a characteristic stacked arrangement. Polyribosomes are studded along the outer surface of the cisternae; many others lie free in the cytoplasm. (*Note*: Faint color tones have been added here and later for ease of identification.) (From Pannese 1994, with permission of Thieme.)

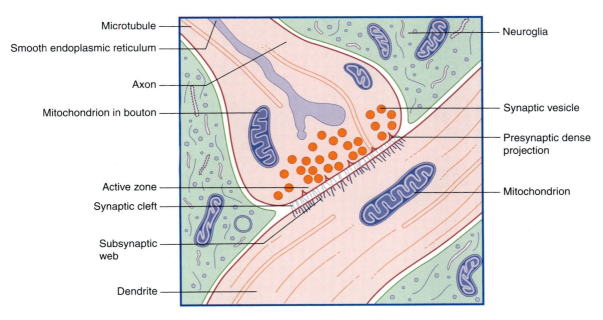

Microtubule

Smooth endoplasmic reticulum

Axon

Mitochondrion in bouton

Active zone

Synaptic cleft

Subsynaptic web

Dendrite

Neuroglia

Synaptic vesicle

Presynaptic dense projection

Mitochondrion

Figure 6.6 Ultrastructure of an axodendritic synapse.

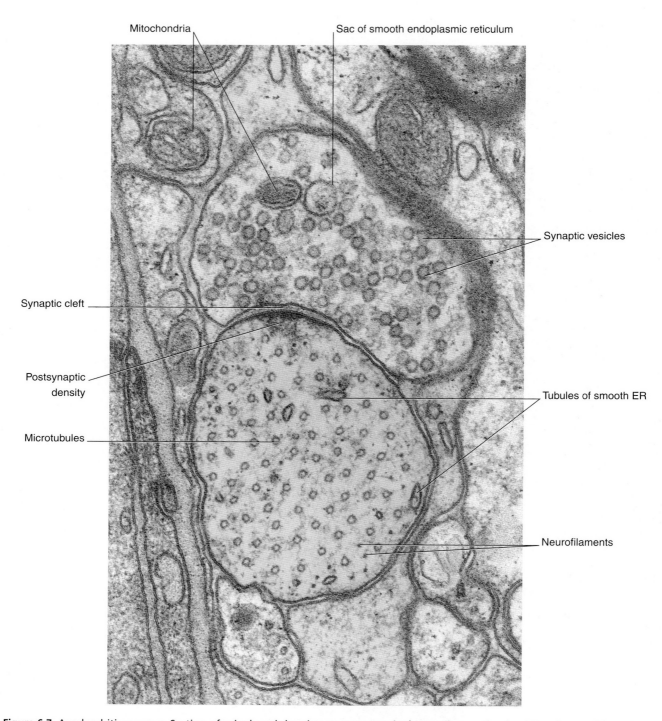

Mitochondria

Sac of smooth endoplasmic reticulum

Synaptic vesicles

Synaptic cleft

Postsynaptic density

Tubules of smooth ER

Microtubules

Neurofilaments

Figure 6.7 Axodendritic synapse. Section of spinal cord showing an axon terminal synapsing on the dendrite of a possible motor neuron. The spherical synaptic vesicles together with the asymmetric morphology (strong postsynaptic density) indicate an excitatory synapse. The dendrite is cut transversely, as are the numerous microtubules; some of the neurofilaments can also be seen. The synapse is invested by a protoplasmic astrocyte. (From Pannese 1994, with permission of Thieme.)

Receptor activation

Transmitter molecules cross the synaptic cleft and activate receptor proteins that straddle the postsynaptic membrane (Figure 6.8). The activated receptors initiate ionic events that either depolarize the postsynaptic membrane (excitatory postsynaptic effect) or hyperpolarize it (inhibitory postsynaptic effect). The voltage change passes over the soma in a decremental wave called *electrotonus*, and alters the resting potential of the first part or **initial segment** of the axon. (See Ch. 7 for details of the ionic

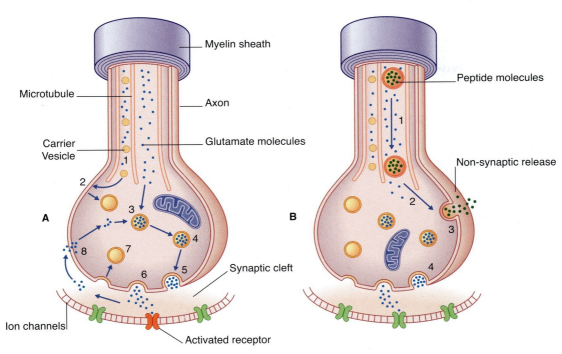

Figure 6.8 Dynamic events at two types of nerve terminal. **(A)** Small molecule transmitter, exemplified by a glutamatergic nerve ending. **(1)** Carrier vesicles containing synaptic vesicle membrane proteins are rapidly transported along microtubules and stored in the plasma membrane of the terminal bouton. At the same time, enzymes and glutamate molecules are conveyed by slow transport. **(2)** Vesicle membrane proteins are retrieved from the plasma membrane and form synaptic vesicles. **(3)** Glutamate is taken into the vesicles, where it is stored and concentrated. **(4)** Loaded vesicles approach the presynaptic membrane. **(5)** Following depolarization, the 'docked' vesicles undergo exocytosis. **(6)** Released transmitter diffuses across the synaptic cleft and activates specific receptors in the postsynaptic membrane. **(7)** Vesicular membranes are retrieved by means of endocytosis. **(8)** Some glutamate is actively transported back into the bouton for recycling. **(B)** Neuropeptide cotransmission. The example here is peptide *substance P* cotransmission with glutamate, a combination found at the central end of unipolar neurons serving pain sensation. **(1)** The vesicles and peptide precursors (propeptides) are synthesized in Golgi complexes in the perikaryon and taken to the terminal bouton by rapid transport. **(2)** As they enter the bouton, peptide formation is being completed, whereupon the vesicle approaches the plasma membrane. **(3)** Following membrane depolarization, the vesicular contents are sent into the intercellular space by means of exocytosis. **(4)** Glutamate is simultaneously released into the synaptic cleft.

events.) If excitatory postsynaptic potentials are dominant, the initial segment will be depolarized to threshold and generate action potentials.

In the CNS, the commonest excitatory transmitter is glutamate; the commonest inhibitory one is γ-aminobutyric acid (GABA). In the peripheral nervous system, the transmitter for motor neurons supplying striated muscle is *acetylcholine*; the main transmitter for sensory neurons is *glutamate*.

The sequence of events involved in *glutamatergic* synaptic transmission is shown in Figure 6.8A. In the case of peptide *cotransmission* with glutamate, release of (one or more) peptides is *non-synaptic*, as shown in Figure 6.8B.

Many sensory neurons liberate one or more peptides in addition to glutamate; the peptides may be liberated from any part of the neuron, but their usual role is to modulate (raise or lower) the effectiveness of the transmitter.

A further kind of transmission is known as *volume* transmission. This kind is typical of *monoamine (biogenic amine)* neurons, which fall into two categories. One category synthesizes a *catecholamine*, namely *norepinephrine (noradrenaline)* or *dopamine*, both synthesized from the amino acid tyrosine. The other synthesizes *serotonin*,

derived from tryptophan. As illustrated in Figure 6.9, for dopamine, the transmitter is liberated from *varicosities* (where they are also synthesized) as well as from synaptic contacts. The transmitter enters the extracellular fluid of the CNS and activates specific receptors up to 100 μm away before being degraded. The monoamine neurons have enormous territorial distribution, and deviation from normal function is implicated in a variety of ailments including Parkinson disease, schizophrenia, and major depression.

Nitric oxide (a gaseous molecule) within glutamatergic neurons is also associated with volume transmission. Excess nitric oxide liberation is *cytotoxic*, notably in areas rendered avascular by cerebral arterial thrombosis. Glutamate itself is also potentially cytotoxic.

In the context of volume transmission, the conventional kind is called 'wiring' to indicate its relatively fixed nature.

Lock and key analogy for drug therapy

The receptor may be likened to a lock, the transmitter being the key that operates it. The transmitter output of certain neurons may falter as a consequence of age or

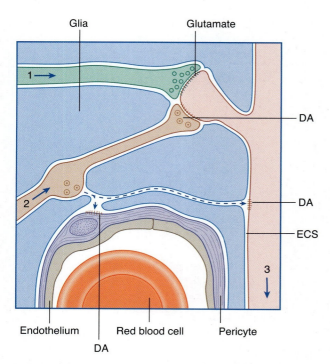

Figure 6.9 Volume transmission in the brain. The axons of a glutamatergic neuron **(1)** and of a dopaminergic neuron **(2)** are making conventional synaptic contacts on the spine of a spiny stellate cell **(3)** in the striatum. Dopamine (DA) is also escaping from a varicosity and diffusing through the extracellular space (ECS) to activate dopamine receptors on the dendritic shaft and on the wall of a capillary pericyte (see Ch. 5).

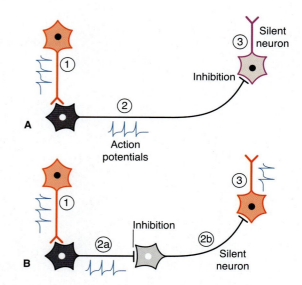

Figure 6.10 Disinhibition. **(A)** Excitatory neuron 1 is activating inhibitory neuron 2 with consequent silencing of neuron 3 by neuron 2. **(B)** Interpolation of a second inhibitory neuron (2b) has the opposite effect on neuron 3, because 2b is silenced. Neuron 3 (spontaneously active unless inhibited) is released.

disease, and a duplicate key can often be provided in the form of a drug that mimics the action of the transmitter. Such a drug is called an *agonist*. On the other hand, excessive production of a transmitter may be countered by a *receptor blocker*—the equivalent of a dummy key that will occupy the lock without activating it.

Inhibition versus disinhibition

Spontaneously active neurons are often held in check by inhibitory neurons (usually GABAergic), as shown in Figure 6.10A. The inhibitory neurons may be silenced by others of the same kind, leading to *disinhibition* of the target cell (Figure 6.10B). Disinhibition is a major feature of neuronal activity in the basal ganglia (Ch. 33).

Less common chemical synapses

Two varieties of **axoaxonic** synapses are recognized. In both cases, the boutons belong to inhibitory neurons. One variety occurs on the initial segment of the axon, where it exercises a powerful veto on impulse generation (Figure 6.11). In the second kind, the boutons are applied to excitatory boutons of other neurons, and they inhibit transmitter release. The effect is called *presynaptic inhibition*, any conventional contact being *postsynaptic* in this context (Figure 6.12).

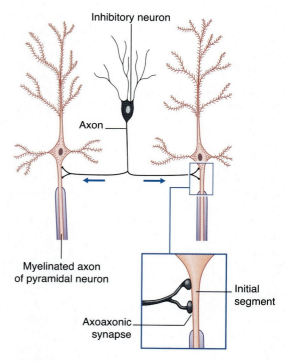

Figure 6.11 Axoaxonic synapses in the cerebral cortex. Arrows indicate direction of impulse conduction.

Dendrodendritic (D-D) synapses occur between dendritic spines of contiguous spiny neurons, and alter the electrotonus of the target neuron rather than generating nerve impulses. In *one-way* D-D synapses, one of the two spines contains synaptic vesicles. In *reciprocal* synapses,

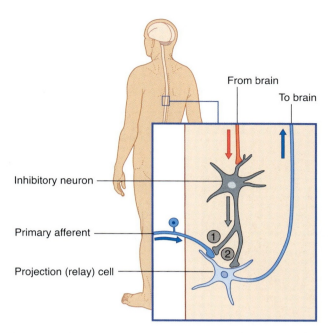

Figure 6.12 (1) Presynaptic and **(2)** postsynaptic inhibition of a spinal neuron projecting to the brain. *Arrows* indicate directions of impulse conduction (relay cell may be silenced by inhibitory cell activity).

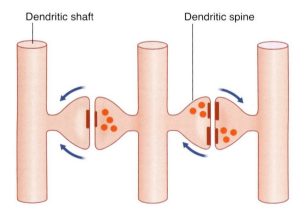

Figure 6.13 Dendrodendritic excitation. The dendrites belong to three separate neurons. On the right is a reciprocal synapse. *Arrows* indicate direction of electrotonic waves.

both do. Excitatory D-D synapses are shown in Figure 6.13. Inhibitory D-D synapses are numerous in relay nuclei of the thalamus (Ch. 25).

Somatodendritic and **somatosomatic** synapses have also been identified, but they are scarce.

NEUROGLIAL CELLS OF THE CENTRAL NERVOUS SYSTEM

Four different types of neuroglial cell are found in the CNS: astrocytes, oligodendrocytes, microglia, and ependymal cells.

Astrocytes

Astrocytes are bushy cells with dozens of fine radiating processes. The cytoplasm contains abundant intermediate filaments. This confers a degree of rigidity on these cells, which helps to support the brain as a whole. Glycogen granules, which are also abundant, provide an immediate source of glucose for the neurons.

Some astrocyte processes form **glial-limiting membranes** on the inner (ventricular) and outer (pial) surfaces of the brain. Other processes invest synaptic contacts between neurons (Figure 6.14). In addition, *vascular processes* invest brain capillaries.

Astrocytes use specific channels (Ch. 8) to mop up K$^+$ ion accumulation in the extracellular space during periods of intense neuronal activity. They participate in recycling certain neurotransmitter substances following release, notably the chief excitatory CNS transmitter, *glutamate*, and the chief inhibitory transmitter, *GABA*.

Astrocytes can multiply at any time. As part of the healing process following CNS injury, proliferation of astrocytes and their processes results in dense glial scar tissue (*gliosis*). More importantly, spontaneous local proliferation of astrocytes may give rise to a brain tumor (Clinical Panel 6.2).

Oligodendrocytes

Oligodendrocytes are responsible for wrapping myelin sheaths around axons in the white matter. In the gray matter, they form **satellite cells** that seem to participate in ion exchange with neurons.

Myelination

Myelination commences during the middle period of gestation, and continues well into the second decade. A single oligodendrocyte lays myelin on upward of three dozen axons by means of a spiraling process whereby the inner and outer faces of the plasma membrane form the alternating **major** and **minor dense lines** seen in transverse sections of the myelin sheath (Figure 6.15). Some cytoplasm remains in **paranodal pockets** at the ends of each myelin segment. In the intervals between the glial wrappings, the axon is relatively exposed, at **nodes**.

Myelination greatly increases the speed of impulse conduction, because the depolarization process jumps from node to node (see Ch. 9). During myelination, K$^+$ ion channels are deleted from the underlying axolemma. For this reason, demyelinating diseases such as multiple sclerosis (Clinical Panel 6.3) are accompanied by progressive failure of impulse conduction.

Unmyelinated axons abound in the gray matter. They are fine (0.2 μm or less in diameter) and not individually ensheathed.

Microglia (Figure 6.14)

The weight of evidence indicates that microglial cells are of mesodermal original and are derived from bone marrow monocytes. *Resting* microglial cells are minute (hence the name), but when *activated* by inflammation or by myelin sheath breakdown, they enlarge and become motile phagocytes.

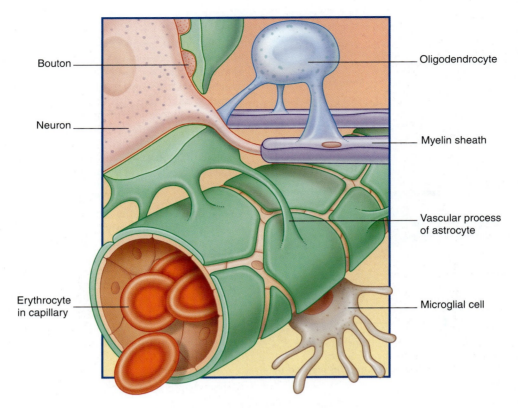

Bouton

Neuron

Erythrocyte
in capillary

Oligodendrocyte

Myelin sheath

Vascular process
of astrocyte

Microglial cell

Figure 6.14 Three neuroglial cell types.

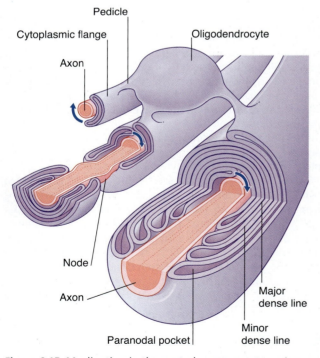

Pedicle

Cytoplasmic flange

Oligodendrocyte

Axon

Node

Axon

Paranodal pocket

Major
dense line

Minor
dense line

Figure 6.15 Myelination in the central nervous system. *Arrows* indicate movement of the growing edge of the cytoplasmic flange of the oligodendrocyte.

Ependyma

Ependymal cells line the ventricular system of the brain. Cilia on their free surface help the propulsion of cerebro-spinal fluid through the ventricles.

Clinical Panel 6.2   Gliomas

Brain tumors most commonly originate from neuroglial cells, especially astrocytes.

General symptoms produced by expanding brain tumors are indicative of *raised intracranial pressure*. They include headache, drowsiness, and vomiting. Radiologic investigation may reveal displacement of midline structures to the opposite side. Tumors below the tentorium (usually cerebellar) are likely to block the exit of cerebrospinal fluid from the fourth ventricle, in which case ballooning of the ventricular system will add to the intracranial pressure.

Local symptoms depend on the position of the tumor. For example, clumsiness of an arm or leg may be caused by a cerebellar tumor on the same side, and motor weakness of an arm or leg may be caused by a cerebral tumor on the opposite side.

Progression

Expansion of a tumor may cause one or more brain hernias to develop, as shown in Figure CP 6.2.1.

1. *Subfalcal herniation* (in the interval between falx cerebri and corpus callosum) seldom causes specific symptoms.

2. *Uncal herniation* is the term used to denote displacement of the uncus of the temporal lobe into the tentorial notch. Compression of the ipsilateral crus cerebri by the uncus (Figure CP 6.2.2) may give rise to contralateral motor weakness. Alternatively, compression of the contralateral crus against the sharp edge of the tentorium cerebelli may cause *ipsilateral* motor weakness.

3. *Pressure coning:* a cone of cerebellar tissue (the tonsil) may descend into the foramen magnum, squeezing the medulla oblongata and causing death from respiratory or cardiovascular failure by inactivation of *vital centers* in the reticular formation (Ch. 24).

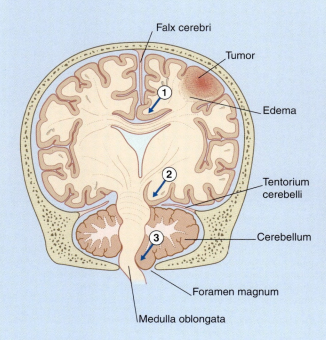

Figure CP 6.2.1 Brain herniations. For numbers, see text.

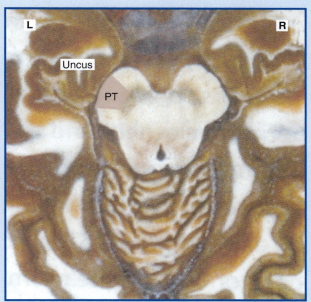

Figure CP 6.2.2 Enlargement from *Figure 3.7* emphasizing the proximity of the uncus to the pyramidal tract (PT).

Clinical Panel 6.3   Multiple sclerosis

Multiple sclerosis (MS) is the commonest neurologic disorder of young adults in the temperate latitudes north and south of the equator. It is more prevalent in women, with a female:male ratio of 3:2. The peak age of onset is around 30 years, the range being 15–45.

Multiple sclerosis is a *primary demyelinating disease*: the initial feature is the development of plaques (patches) of demyelination in the white matter. The denuded axons also undergo large-scale degeneration, probably initiated by failure of the sodium pump, as described in Chapter 7. Impulse conduction in neighboring myelinated fibers is also compromised by edema (inflammatory exudate). Over time, the plaques are progressively replaced by glial scar tissue. Old plaques feel firm (sclerotic) in postmortem slices of the brain.

Common locations of early plaques are the cervical spinal cord, upper brainstem, optic nerve, and periventricular white matter (Figure CP 6.3.1) including that of the cerebellum. MS is not a systems disease: it is not anatomically selective, and a plaque may involve parts of adjacent motor and sensory pathways.

Presenting symptoms can be correlated with lesion sites as follows.

- *Motor weakness*, usually in one or both legs, signifies a lesion involving the corticospinal tract.
- *Clumsiness* in reaching and grasping usually accompanies a lesion in the cerebellar white matter.
- *Numbness or tingling*, often spreading up from the legs to the trunk, may be caused by a lesion in the posterior white matter of the spinal cord. Tingling ('pins and needles') is attributed to spontaneous firing of partially demyelinated sensory fibers.
- *Diplopia* (double vision) may be produced by a plaque within the pons or midbrain affecting the function of one of the ocular motor nerves.
- A *scotoma* (patch of blindness in the visual field of one eye) is produced by a plaque within the optic nerve.

- *Urinary retention* (failure of the bladder to empty) can be caused by interruption of the central autonomic pathway descending from the brainstem to the lower part of the cord.

The usual course of the disease is one of remissions and relapses, with an overall slow progression and development of multiple disabilities.

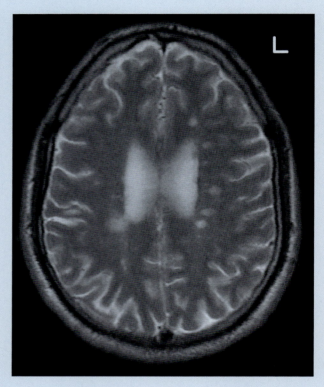

Figure CP 6.3.1 An MRI scan of a 28-year-old man with multifocal demyelination secondary to multiple sclerosis. Axial T_2-weighted image showing multiple lesions of high signal intensity in the white matter. On the left side of the brain, at least five of these plaques are periventricular. (Kindly provided by Dr. Joe Walsh, Department of Radiology, University College Hospital, Galway, Ireland.)

Core Information

Neurons

The multipolar neuron of the central nervous system (CNS) comprises soma, dendrites, and axon; the axon gives off collateral and terminal branches. The soma contains rough and smooth endoplasmic reticulum, Golgi complexes, neurofilaments, and microtubules. Microtubules pervade the entire neuron; they are involved in anterograde transport of synaptic vesicles, mitochondria, and membranous replacement material, and in retrograde transport of marker molecules and degraded organelles.

The three kinds of chemical neuronal interaction are synaptic (e.g. glutamatergic), non-synaptic (e.g. peptidergic), and volume (e.g. monoaminergic, serotonergic).

Anatomic varieties of chemical synapse include axodendritic, axosomatic, axoaxonic, and dendrodendritic. Structure includes pre- and postsynaptic membranes, synaptic cleft, and subsynaptic web.

Electrical synapses via gap junctions render some neuronal groups electrically coupled, for synchronous activation.

Neuroglia

Astrocytes have supportive, nutritive, and retrieval functions. They are the main source of brain tumors. Oligodendrocytes form CNS myelin sheaths, which are subject to destruction in demyelinating diseases. Microglia are potential phagocytes.

REFERENCES

Alter CA, Cal N, Bliven T, et al. Anterograde transport of brain-derived neurotrophic factor and its role in the brain. Nature 1997; 389:856–859.

Calakos N, Scheller RH. Synaptic vesicle biogenesis, docking, and fusion: a molecular description. Physiol Rev 1996; 76:1–29.

Federoff S. Development of microglia. In: Kettenmann H, Ransom BR, eds. Neuroglia. New York: Oxford University Press; 1995:163–184.

Gehrmann J, Matsumoto Y, Kreutzberg DW. Microglia: intrinsic immunoeffector cells of the brain. Brain Res Rev 1995; 20:269–287.

Golding DW. Synaptic, non-synaptic and parasynaptic exocytosis. BioEssays 1994; 16:503–508.

Grafstein B. Axonal transport: function and mechanisms. In: Waxman S, Kocsis JD, Stys PK, eds. The axon. New York: Oxford University Press; 1995:185–199.

Grafstein B. Intracellular traffic in nerve cells. Brain Res Bull 1999; 50:311–322.

Hirokawa N, Reiko T. Molecular motors in neuronal development, intracellular transport and diseases. Curr Opin Neurobiol 2004; 14:564–573.

Howe CH, Mobley WC. Long-distance retrograde neurotrophic signaling. Curr Opin Neurobiol 2005; 15:40–48.

Jessen KR. Glial cells. Int J Biochem Cell Biol 2004; 36:1861–1867.

Lee RMKW. Morphology of cerebral arteries. Pharmacol Ther 1995; 66:149–173.

Mercer JA, Albanesci JP, Brady ST. Molecular motors and cell motility in the brain. Brain Pathol 1994; 4:167–179.

Norenberg MD. Astrocyte responses to injury. J Neuropathol Exp Neurol 1994; 53:213–220.

Pannese E. Neurocytology. Fine structure of neurons, nerve processes and neuroglial cells. New York: Thieme; 1994.

Privat A, Giminez-Robotta M, Ridet J-L. Morphology of astrocytes. In: Kettenmann H, Ransom BR, eds. Neuroglia. New York: Oxford University Press; 1995:3–22.

Stys PK. General mechanisms of axonal damage and its repair. J Neurol Sci 2005; 233:3–13.

Torrealba F, Carrasco MA. A review on electron microscopy and neurotransmitter systems. Brain Res Rev 2004; 47:5–17.

Volknandt W. The synaptic vesicle and its targets. Neuroscience 1995; 64:277–300.

Zoli M, Torri C, Ferrari R, et al. The emergence of the volume transmission concept. Brain Res Rev 1998; 26:136–147.

STUDY GUIDELINES

1 The information in this chapter underpins the science of clinical electrophysiology.

2 Appreciate that neuronal membranes carry an electrical charge based on passive ion diffusion along specific ion channels and regulated by a sodium–potassium pump, and that action potentials are abrupt changes in membrane voltage caused by activation of voltage-gated ion channels.

3 Note that impulse propagation is an all-or-nothing event and of the same magnitude all the way along the nerve fiber and its branches.

4 Myelination dramatically increases the speed of impulse conduction.

5 Examples of clinical neurophysiology related to the peripheral nervous system await your attention in Chapter 12.

STRUCTURE OF THE PLASMA MEMBRANE

In common with cells elsewhere, the plasma membrane of neurons is a double layer (bilayer) of phospholipids made up of phosphate heads facing the aqueous media of the extracellular and intracellular spaces, and paired lipid tails forming a fatty membrane in between (Figure 7.1). The phosphate layer is water-soluble (*hydrophilic*, or *polar*) and the double lipid layer is water-insoluble (*hydrophobic*, or *non-polar*).

Both the extracellular and the intracellular fluids are aqueous salt solutions in which many soluble molecules dissociate into positively or negatively charged atoms or groups of atoms called *ions*. Ions and molecules in aqueous solutions are in a constant state of agitation, being subject to *diffusion*, whereby they tend to move from areas of higher concentration to areas of lower concentration. In addition to passing down their concentration gradients by diffusion, ions are influenced by electrical gradients. Positively charged ions including Na^+ and K^+ are called *cations* because, in an electrical field, they migrate to the cathode. Negatively charged ions including Cl^- are called *anions* because these migrate to the anode. Like charges (e.g. Na^+ and K^+) repel one another, unlike charges (e.g. Na^+ and Cl^-) attract one another.

The cell membrane can be regarded as an electric *capacitor*, because it comprises outer and inner layers carrying ionic charges of opposite kind, with a (fatty) insulator in between. Away from the membrane, the voltage in the tissue fluid is brought to zero (0 mV) by the neutralizing effect of chloride anions on sodium and other cations, and the voltage in the cytosol away from the membrane is brought to zero by the neutralizing effect of anionic proteins on K^+ cations.

Ion channels

Ion channels are membrane-spanning proteins having a central pore that permits passage of ions across the cell membrane. Most channels are selective for a particular ion, for example Na^+ *or* K^+ *or* Cl^-.

Several channel categories are recognized, of which the first three are of immediate relevance.

- *Passive* (non-gated) channels are open at all times, permitting ions to move across the membrane.
- *Voltage-gated* channels contain a voltage-sensitive string of amino acids that cause the channel pore to open or close in response to changes in membrane voltage.
- *Channel pumps* are energy-driven ion exporters and/or importers designed to maintain steady-state ion concentrations. The Na^+–K^+ exchange pump (usually referred to as the *sodium pump*) is vital to maintenance of the resting membrane potential.
- *Transmitter-gated* channels abound in postsynaptic membranes. Some are activated directly by transmitter molecules, others indirectly (see Ch. 8).
- *Transduction* channels are activated by peripheral sensory stimulation. Sensory nerve endings exhibit different stimulus specificities in different locations, for example mechanical in muscle; tactile, thermal, or chemical in skin; acoustic in the cochlea; vestibular in the labyrinth; electromagnetic in the retina; gustatory in the tongue; olfactory in the upper part of the nasal mucous membrane.

Figure 7.2 depicts the three passive channels concerned with generating the resting potential.

The existence of distinct channels for Na^+, K^+, and Cl^- ions would result in zero voltage difference across the membrane if passive diffusion of the three ions were equally free. However, the number of sodium channels is relatively small, and movement of the Na^+ ion is relatively slow because of its relatively large 'hydration shell' of H_2O molecules. In effect, the membrane is many times more permeable to K^+ and Cl^- than to Na^+.

Resting membrane potential

The membrane potential of the resting (inactive) neuron is generated primarily by differences in concentration of the sodium (Na^+) and potassium (K^+) ions dissolved in the aqueous environments of extracellular fluid (ECF) and cytosol. In Table 7.1, it can be seen that potassium is 20 times more concentrated in the cytosol; sodium is 10 and chloride 11.5 times more concentrated in the ECF.

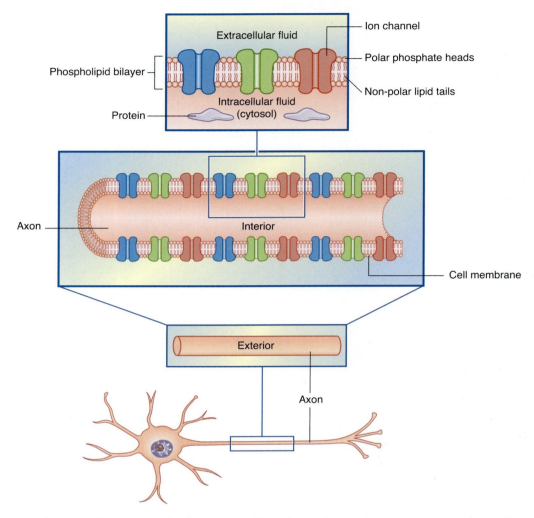

Figure 7.1 Structure of the neuronal cell membrane. The only membrane proteins shown here are ion channels.

In Figure 7.3, a voltmeter is connected to electrodes inserted into the ECF surrounding an axon. One of the electrodes has been inserted into a glass pipette having a minute tip. On the left side of the figure, both electrode tips are in the ECF, and there is no voltage difference; a zero value is recorded. On the right side, the pipette has been lowered, puncturing the plasma membrane of the axon and admitting the intracellular fluid of the cytosol. The electrical charge now reveals a *potential* (voltage) difference of −70 mV. In practice, the membrane potential ranges from −60 mV to −80 mV in different neurons. These values represent the *resting* membrane potential, i.e. when impulses are not being conducted.

Resting membrane permeability
Potassium ions

From what has been mentioned, it is clear that K^+ concentrations on either side of the cell membrane would be the same were there no constraint. In fact, there are two electrical constraints at the level of the ion pore, namely the attraction exerted by the protein anions on the inside, and the repulsion exerted by the Na^+ cations on the outside. The equilibrium state exists when the concentration gradient is exactly balanced by the voltage gradient; the potential difference at this point is expressed as E_k, the potassium equilibrium potential. This can be expressed by means of the *Nernst equation*, which uses principles of thermodynamics to convert the concentration gradient of an ion to an equivalent voltage gradient.

Table 7.1 Ionic concentrations in cytosol and extracellular fluid

Ion	Concentration (mmol/L)		Equilibrium potential (mV)
	Cytosol	Extracellular fluid	
K^+	100	5	−90
Na^+	15	150	+60
Cl^-	13	50	−70

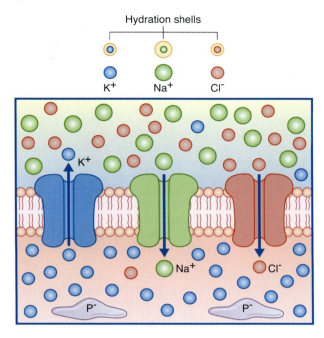

Figure 7.2 In the resting state, Na⁺ and Cl⁻ ions are concentrated external to the membrane, because of the slow inward passage of the hydrated Na⁺ ions through their channels, combined with their attraction to Cl⁻ ions. K⁺ ions are concentrated on the inside, because of their attraction to protein anions (P⁻). The arrows are directed down the concentration gradients of the respective ions.

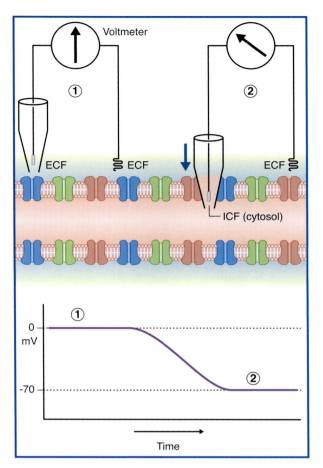

Figure 7.3 The resting membrane potential. **(1)** The two electrodes of a voltmeter are inserted into the extracellular fluid (ECF) surrounding an axon. The left electrode tip occupies a micropipette. There is no voltage difference registering, hence the zero value in the record below. **(2)** On the right, the pipette has been lowered (arrow), puncturing the plasma membrane to sample the intracellular fluid (ICF) immediately beneath. A voltage difference of −70 mV is recorded.

$$E_K = \frac{RT}{FZ_K} \log_e \frac{[K^+]_o}{[K^+]_i}$$

E_k = equilibrium potential for potassium expressed as volts
R = universal gas constant (8.31 J/mol/°absolute)
T = temperature in degrees Kelvin (310° at 37°C)
F = Faraday (96 500 C/per mole of charge)
Z_k = valence of potassium (+1)
E = natural logarithms
$[K^+]_o$ = potassium concentration outside the cell membrane
$[K^+]_i$ = potassium concentration inside the cell membrane

Converting the natural log to $\log_{10}$ and resolving the numeric fractions yields

$$Ek = 61.5 \times \log_{10} \times 5/100 = -90 \text{ mV}.$$

Repeating the exercise for sodium and chloride yields

$$ENa = +60 \text{ mV and } ECl = -70 \text{ mV}.$$

The value of the resting potential can be calculated using the *Goldman equation*, from the relative distributions of the three principal ions involved (Table 7.1), and their membrane permeabilities.

$$RP = 62 \log \frac{P_{K^+}[K^+]_o + P_{Na^+}[K^+]_o + P_{Cl^-}[Cl^-]_o}{P_{K^+}[K^+]_i + P_{Na^+}[K^+]_i + P_{Cl^-}[Cl^-]_i}$$

RP = resting potential
P = the three membrane permeabilities
O and I refer to outside and inside the cell. The concentrations of the negative chloride ions are shown inverted because $-\log (X/Y) = \log (Y/X)$
Brackets signify concentration

The Goldman equation is nothing more than the Nernst equation for each of the three ions, with each concentration multiplied by the conductance of that ion. The effect of the chloride ion on resting potential is insignificant, because its equilibrium potential is the same as the resting potential. The sum of the fractions for K⁺ and Na⁺ yields an outcome of −70 mV, as required by the known value.

Sodium pump

The resting potential needs to be stabilized, because of the tendency of Na⁺ ions to leak inward and K⁺ to leak outward along their concentration gradients. Stability is assured by the Na⁺–K⁺ pump making appropriate corrections for their passive flows. This channel is capable of simultaneously extruding Na⁺ and importing K⁺. Three sodium ions are exported for every two potassium ions imported (Figure 7.4). In both cases, the movement is *against* the existing concentration gradient. The required energy for this activity is provided by the ATPase enzyme that converts ATP to ADP. The greater the amount of Na⁺ in the cytosol, the greater is the activity of the enzyme.

As mentioned in Chapter 6, the axonal degeneration occurring in multiple sclerosis is attributable to failure of the sodium pump along the denuded axolemma. This leads to Na⁺ overload, which in turn promotes excess Ca²⁺ release from intracellular stores.

RESPONSE TO STIMULATION: ACTION POTENTIALS

Neurons typically interact at chemical synapses, where liberation of a transmitter substance is produced by the arrival of *action potentials*, or *spikes*, at the synaptic boutons. The transmitter crosses the synaptic cleft and activates *receptors* embedded in the membranes of target neurons. The receptors activate *transmitter-gated* ion channels to alter the level of polarization of the target neuron. Receptors activated by an inhibitory transmitter cause the membrane potential to increase beyond the resting value of −70 mV, perhaps to −80 mV or more, a process known as *hyperpolarization*. Excitatory transmitters cause the membrane potential to diminish, a process of *depolarization*.

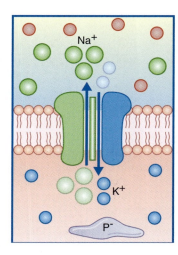

Figure 7.4 The Na⁺–K⁺ pump. The diagram indicates simultaneous expulsion of three sodium ions for every two potassium ions imported.

Electrotonic potentials

The initial target cell response to excitatory stimulation takes the form of local, *graded* or *electrotonic potentials* (*EPTs*). Positive ETPs on multipolar neurons are usually on one or more dendrites in receipt of excitatory synapses. At a low frequency of stimulation, small, decremental waves of depolarization extend for 50–100 μm along the affected dendrites, dying away after 2 or 3 ms (Figure 7.5). With increasing frequency, the waves undergo stepwise *temporal summation* to form progressively larger waves continuing on over the surface of the soma. *Spatial summation* occurs when waves traveling along two or more dendrites coalesce on the soma (Figure 7.6). About 15 mV of depolarization, to a value of −55 mV, brings the neuron to *threshold* (firing level) at its most sensitive region, or *trigger point*, in the initial segment of the axon (Figure 7.7). The initial segment is the first region to 'give way' at threshold voltage, because it is exceptionally rich in voltage-gated sodium channels. When the level of depolarization (the *generator potential*) reaches threshold, nerve impulses in the form of *action potentials* are suddenly fired off.

In sensory neurons of cranial and spinal nerves, the trigger zone generates what is known as the *receptor potential*. The trigger zone of sensory neurons is exceptionally rich in the sensation-specific transduction channels defined earlier.

In the case of myelinated nerve fibers, the trigger point is easily identified: in multipolar neurons, it is immediately proximal to the first myelin segment, and in peripheral sensory neurons it is immediately distal to the final segment.

Negative excitatory postsynaptic potentials are elicited by inhibitory transmitters. They, too, are decremental.

The shape of action potentials

A single action potential is depicted in Figure 7.8. The *spike* segment of the potential commences when the local response reaches threshold value at −55 mV. The rising phase of depolarization passes beyond zero to include an *overshoot* phase reaching about +35 mV. Overshoot phase includes the rising and falling phases above zero potential. The falling phase planes out in a brief *after-depolarization*, prior to an *undershoot* phase of *after-hyperpolarization* where the membrane potential reaches about −75 mV before returning to baseline.

It should be pointed out that standard representations such as this figure show the voltage changes plotted against a time base. When direction is substituted for time, it becomes obvious that the time-based picture matches the sequence in a peripheral sensory neuron. For all multipolar neurons, the representation should be the reverse (Figure 7.7).

When the local response to stimulation has depolarized the membrane to threshold, the sudden increase in depolarization is brought about by the opening of voltage-gated sodium channels (Figure 7.9). Sodium entry produces further depolarization, and the positive feedback causes the remaining Na⁺ channels of the trigger zone to open, driving the membrane charge momentarily into a charge reversal (overshoot) of +35 mV, approaching the Nernst potential for sodium. At this point, the sodium channels

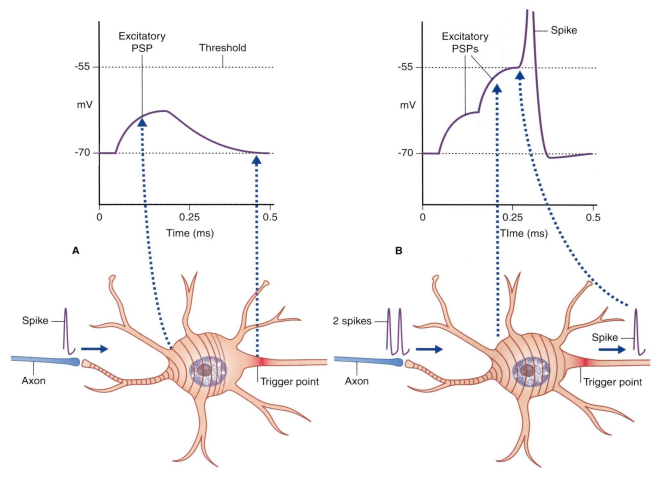

Figure 7.5 Temporal summation. **(A)** A sensory axon (blue) delivers a single spike to a motor neuron, sufficient to elicit an excitatory postsynaptic potential (PSP) that dies away. **(B)** The axon delivers two spikes that undergo temporal summation to reach firing threshold at the initial segment of the axon, which responds by generating a spike that will pass along the motor axon.

commence a progressive inactivation, and the voltage-gated potassium channels are simultaneously triggered to open. Current flow switches from Na$^+$ inflow to K$^+$ outflow. The hyperpolarization phase is explained by the voltage-gated sodium channels being completely inactivated prior to closure of the potassium channels. Any remaining discrepancy is adjusted by activity of the Na$^+$–K$^+$ pump.

Close analyses of the sodium channels involved have revealed a dual mechanism of operation, as indicated in Figure 7.10. In the resting state, an *activation* gate in the midregion of both Na$^+$ and K$^+$ channel pores is closed. The sodium channel is the first to respond at threshold, by opening its activation gate and allowing a torrential inflow of Na$^+$ ions down the concentration gradient. One millisecond later, a second, *inactivation* gate, in the form of a flap of globular protein, seals the exit into the cytosol while the K$^+$ channel pore is opening. When the membrane potential approaches normality, the sodium gating reverts to its resting inactive state.

The action potential response to depolarization is *all or none*, a term signifying that if it occurs at all, it is total. In this respect, it is quite unlike the graded potentials that summate to initiate action potentials. Action potentials are also distinguished from graded potentials in being *non-decremental*; they are propagated at full strength along the nerve fiber all the way to the nerve endings, which in the case of lower limb neurons may be more than a meter away from the parent somas.

During the rising and early falling phases of the action potential, the neuron passes through an *absolute refractory period* where it is incapable of initiating a second impulse because too many voltage-gated channels are already open (Figure 7.11). This is followed by a *relatively refractory period*, where stimuli in excess of the standard 15-mV requirement can elicit a response. It is quite common for the generator potential to reach up to 35 mV, triggering impulses at 50–100 impulses per second, expressed as 50–100 Hz (Hertz = times per second).

Propagation

Reversal of potential at the trigger zone is propagated (conducted) along the axon in accordance with the electrotonic circuit shown in Figure 7.12. The positive internal

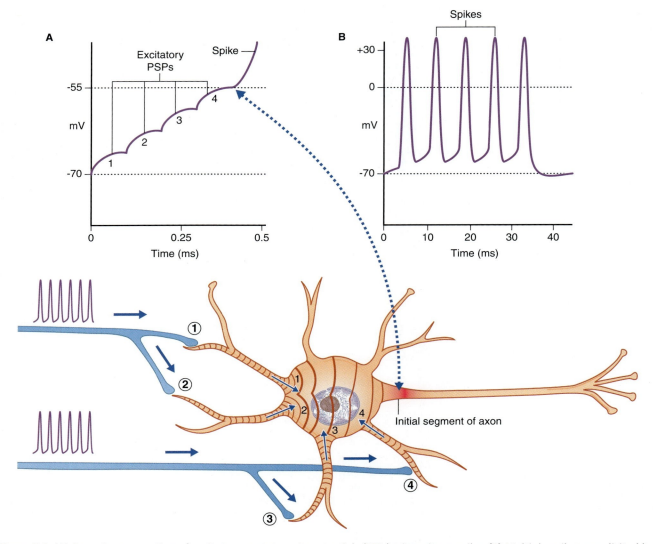

Figure 7.6 (A) Stepwise summation of excitatory postsynaptic potentials (PSPs) triggering a spike. **(B)** Multiple spikes are elicited by generator potentials of sufficient strength. *Arrow* indicates the region enlarged in (A).

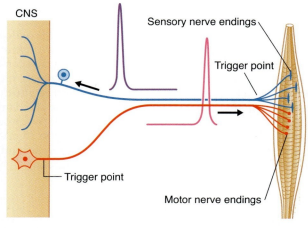

Figure 7.7 Shape of action potentials for motor and sensory nerves supplying skeletal muscle. CNS, central nervous system.

membrane charge passes in both directions within the axoplasm, while positive outside charge passes in both directions within the ECF to neutralize the negative external potential. The membrane immediately proximal is sufficiently refractory to resist depolarization, whereas that immediately distal undergoes a local response progressing to firing level. This process continues distally along the stem axon and its branches, thereby conducting the action potential all the way to the nerve terminals.

Whereas conduction along unmyelinated nerve fibers is *continuous*, along myelinated fibers it is *saltatory* ('jumping'). Myelin sheaths are effective insulators over-lying the internodal segments, whereas Na^+ channels are very abundant at the nodes. Accordingly, spike potentials are generated at each successive node, the positive current traveling along the axoplasm of the internode before exiting at the next node in line. As the current travels back through the ECF to recharge the depolarized patch of

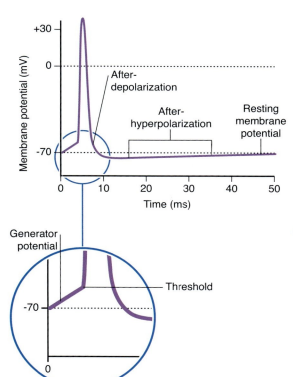

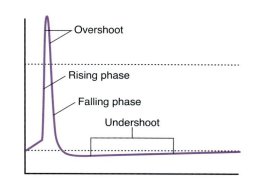

Figure 7.8 Principal features of the action potential.

membrane, withdrawal of positive charge causes the next node to depolarize.

Conduction velocities

In the case of unmyelinated nerve fibers, conduction velocity is proportionate to axonal diameter, because (a) the greater the volume of axoplasm, the more rapid is the longitudinal current flow; and (b) wider axons have a greater surface membrane area, with proportionate increase in ion channel numbers permitting faster membrane depolarization and voltage recovery. Diameters range from 0.5 to 2 μm, and velocities from 0.5 to 2 m/s.

Myelinated nerve fibers range in external diameter (i.e. including the myelin sheath) from 2 to 25 μm. In addition to the two axonal size benefits mentioned, wider myelinated fibers possess longer internodal myelin segments. The spikes are accordingly further apart, with increased conduction velocity reminiscent of a runner with a longer stride. Conduction velocities for various kinds of peripheral nerve fibers are given in Chapter 9.

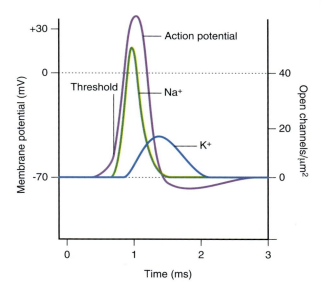

Figure 7.9 Changes in sodium and potassium conductances responsible for the action potential.

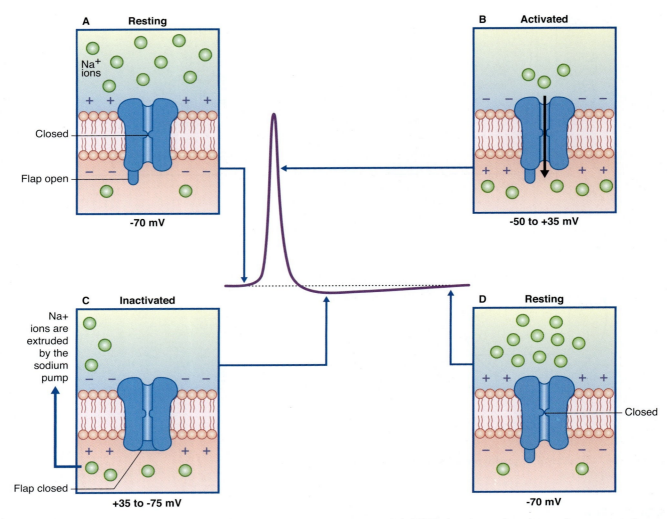

A   Resting

Na$^+$ ions

Closed

Flap open

-70 mV

B   Activated

-50 to +35 mV

C   Inactivated

Na+ ions are extruded by the sodium pump

Flap closed

+35 to -75 mV

D   Resting

Closed

-70 mV

Figure 7.10 Voltage-gated sodium channel behavior during an action potential. **(A)** During the resting phase prior to onset, the midregion of the channel pore is closed and the inactivation flap is open. **(B)** When the threshold level is crossed, activation of the channel opens the pore completely, with a time limit of 1 ms. **(C)** The pore is closed by the inactivation flap. **(D)** Restoration of the resting potential causes the midregion to close and the flap to open.

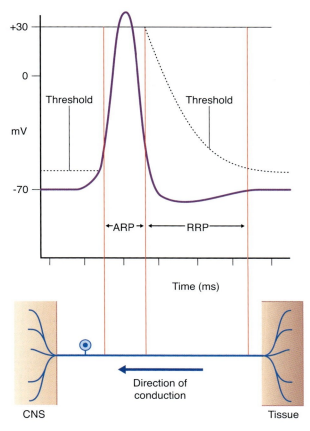

Figure 7.11 Refractory periods. ARP, absolute refractory period; CNS, central nervous system; RRP, refractory period.

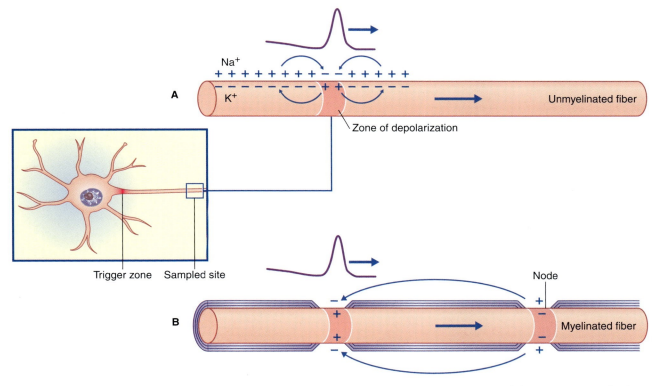

Figure 7.12 Current flow during impulse propagation, represented as movement of positive charges. **(A)** Continuous conduction along an unmyelinated fiber. **(B)** Saltatory conduction along a myelinated fiber.

Core Information

Ions are electrically charged atoms or groups of atoms. Na^+ and K^+ are cations, Cl^- and P^- (proteins) are anions. Cell membranes are charged capacitors carrying a resting potential (voltage) of −70 mV.

Passive ion channels for Na^+, K^+, and Cl^- are open at all times, and the ions diffuse down their concentration gradient through their respective channels. However, Na^+ channels are relatively scarce, whereas K^+ and Cl^- are numerous. K^+ ions are abundant in the cytosol, being attracted by protein anions in the cytoskeleton and repelled by the Na^+ ions outside. A Na^+–K^+ channel pump maintains the resting steady state of the membrane potential.

The initial response of a multipolar neuron to an excitatory stimulus takes the form of decremental waves of positive electrotonus. Temporal and/or spatial summation of such waves produces a generator potential at the initial segment of the axon. At a threshold value of 50 mV, action potentials are fired off along the nerve fiber. On the other hand, inhibitory stimulation takes the form of negative electrotonic waves that summate to produce overall hyperpolarization.

The action potential (spike) is characterized by a rising phase from baseline up to + 35 mV, a falling phase down to baseline followed by a hyperpolarization phase down to −75 mV with return to baseline. Triggering the rapid depolarization is activation of voltage-gated ion channels, whereby the ion channel is briefly (1 ms) opened completely, allowing massive Na^+ inflow tipping the potential to +35 mV, whereupon the channels are shut by inactivation gates and voltage-gated K^+ channels are opened, with a current switch from Na^+ inflow to K^+ outflow.

For about 1 ms following impulse initiation, the trigger zone initial segment is absolutely refractory to further stimulation; for the following 3 ms, it is relatively refractory.

Action potentials are propagated in an all-or-none manner and at full strength along the fiber and its branches. Propagation is continuous along unmyelinated axons, saltatory (from node to node) along myelinated axons. Saltatory conduction is much faster. The widest fibers have the longest internodal segments and the fastest conduction rates.

REFERENCES

Koester J, Siegelbaum SA. Peripheral signaling: the action potential. In: Kandel ER, Schwarz JH, Jessell TJ. Principles of neural science. 4th edn. New York: McGraw-Hill; 2000:150–170.

Transmitters and receptors

STUDY GUIDELINES

1 This chapter dwells on chemical neurotransmission. The information is intended to set the stage for learning about therapeutic applications, notably in relation to the autonomic nervous system and to emotional or behavioral disorders including the psychoses.

2 The numbered flow diagrams are meant to convey a sense of motion. It may be a good idea to 'go with the flow' by following the possible paths taken by a given transmitter molecule (e.g. five for catecholamines).

3 Learn to distinguish between the nature of fast (ionotropic) and slow (metabotropic) receptors.

4 The transmitter or receptor agonists and antagonists mentioned here will come up in later chapters.

ELECTRICAL SYNAPSES

Electrical synapses are scarce in the mammalian nervous system. As seen in Figure 8.1, they consist of gap junctions (nexuses) between dendrites or somas of contiguous neurons, where there is cytoplasmic continuity through 1.5-nm channels. No transmitter is involved, and there is no synaptic delay.

The gap junctions are bridged by tightly packed ion channels, each comprising mirror image pairs of connexons, which are transmembrane protein groups (*connexins*) disposed in hexagonal format around an ion pore. Wedge-shaped connexin subunits bordering each ion pore are closely apposed when the neurons are inactive. Action potentials passing along the cell membrane cause the subunits to rotate individually, creating a pore large enough to permit free diffusion of ions and small molecules down their concentration gradients.

The overall function of these gap junctions is to ensure synchronous activity of neurons having a common action. An example is the inspiratory center in the medulla oblongata, where all the cells exhibit synchronous discharge during inspiration. A second example is among neuronal circuits controlling *saccades*, where the gaze darts from one object of interest to another.

CHEMICAL SYNAPSES

Transmitter liberation (Table 8.1 and Figure 8.2)

At resting nerve terminals, synaptic vesicles accumulate in the active zones, where they are tethered to the presynaptic densities by strands of *docking proteins* including actin. With the arrival of action potentials, myriads of voltage-gated calcium channels located in the presynaptic membrane are opened, leading to instant flooding of the active zone with Ca^{2+} ions. These ions cause contraction of actin filaments to bring the nearest vesicles into direct contact with the presynaptic membrane.

When the vesicle protein *synaptophysin* contacts the vesicle membrane, it becomes embedded in it and hollows out to create a *fusion pore* through which the vesicle discharges its contents into the synaptic cleft. The act of expulsion appears to be a function of another vesicle protein, *calmodulin*, activated by Ca^{2+} ions. These and other local protein responses to calcium entry are extremely rapid, the time elapsing between calcium entry and transmitter expulsion being less than 1 ms. In the case of small synaptic vesicles such as those containing glutamate or γ-aminobutyric acid (GABA), single spikes are sufficient to yield some transmitter release. In the case of peptidergic neurons, impulse frequencies of 10 Hz or more are required to induce typically slow (delay of 50 ms or more) transmitter release from the large, dense-cored vesicles.

Target cell receptor binding

Transmitter molecules bind with receptor protein molecules in the postsynaptic membrane. The two great categories of receptor are known as *ionotropic* and *metabotropic*. Each category contains some receptors whose activation leads to opening of ion pores and others whose activation leads to closure.

Ionotropic receptors

Ionotropic receptors are characterized by the presence of an ion channel within each receptor macromolecule

Table 8.1 Some named proteins involved in transmitter transport and vesicle recycling

Named protein	Function
Actin	Brings vesicle into contact with presynaptic membrane
Synaptophysin	Creates the membrane fusion pore
Calmodulin	Expels vesicle content into synaptic cleft
Clathrin	Withdraws vesicle membrane from synaptic cleft
Dynamin	Pinches the neck of the developing vesicle to complete its separation
Ligand	Receptor protein that binds with the transmitter molecule

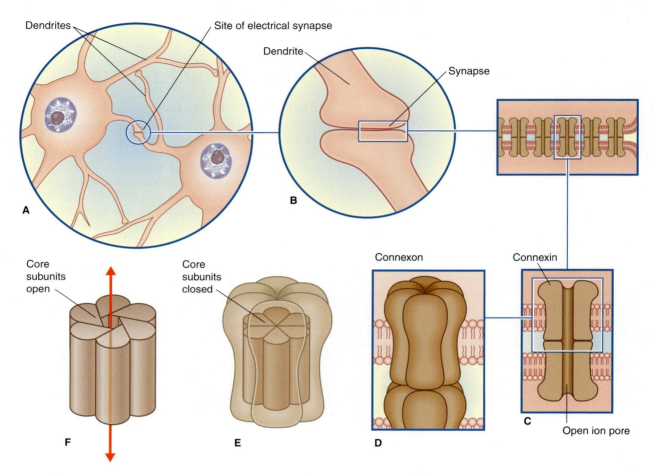

Figure 8.1 Structure of an electrical synapse. **(A)** Synaptic contact between two dendrites. **(B)** Enlargement from (A). **(C)** The gap junction between the cell membranes is bridged by close-packed ion channels. **(D)** Each ion channel comprises a mirror image pair of connexons. **(E)** Each connexon comprises six connexins, each having a wedge-shaped subunit bordering the ion channel. **(F)** The subunits open the ion channel by synchronous rotation.

(Figure 8.6). The transmitter binds with its specific receptor facing the synaptic cleft, causing it to change its conformation so as to open or close its ion pore. Ionotropic receptor channels are said to be transmitter-gated, or *ligand*-gated (from the Latin *ligandum*, 'binding'), signifying their capacity to bind a transmitter molecule or a drug substitute.

In Figure 8.3A, the excitatory channel has been opened by the transmitter, causing a major influx of sodium and a minor efflux of potassium; the excitatory postsynaptic potential (EPSP) has depolarized the cell membrane almost to firing point. In Figure 8.3B, the membrane can be hyperpolarized only to −70 mV, the chloride equilibrium potential. A greater level of depolarization requires a second messenger system (see below).

Ionotropic receptors are called *fast* receptors on account of their immediate and brief effects on ion channels.

Metabotropic receptors

Metabotropic receptors are so called because many are capable of generating multiple metabolic effects within the cytoplasm of the neuron. The receptor macromolecule is an intrinsic membrane protein devoid of an ion channel.

Its function is initiated by one or other of two subunits (α subunit, β subunit) that detaches in response to transmitter activation and moves along the inner surface of the membrane. The subunits are called *G proteins*, because most bind with guanine nucleotides such as guanine triphosphate (GTP) or guanine diphosphate. Their action on ion channels is usually indirect, via a *second messenger system*. However, some G proteins do not bind with nucleotides; instead, they activate ion channels *directly* (see later).

A G protein having a stimulatory effect is known as a G_s protein; a G_i protein is one with an inhibitory effect. Three second messenger systems are well recognized.

1 The cyclic AMP system, responsible for phosphorylation of proteins.

2 The phosphoinositol system, responsible for liberating calcium from cytoplasmic stores.

3 The arachidonic acid system, responsible for production of arachidonic acid metabolites.

Cyclic AMP system
(The abbreviation cyclic AMP represents cyclic adenosine monophosphate.) In the examples shown in Figure 8.4,

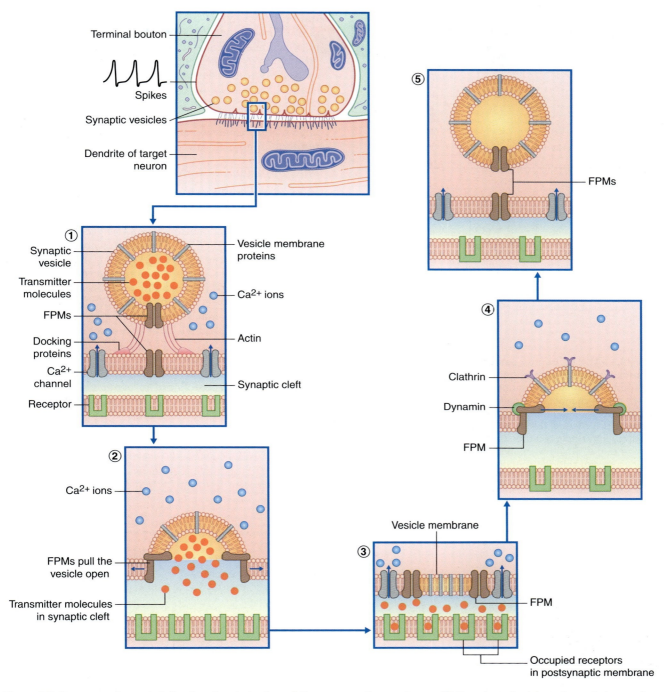

Figure 8.2 Sequence of events following depolarization of the presynaptic membrane. **(1)** Opening of calcium channels (*arrows*) causes synaptic vesicles to be pulled into contact with the presynaptic membrane by actin filaments. Matching pairs of fusion protein macromolecules (FPMs) in the vesicle and presynaptic membrane are aligned. **(2)** The FPMs separate (*outward arrows*), permitting transmitter molecules to enter the synaptic cleft. **(3)** Vesicle membrane is incorporated into the presynaptic membrane while transmitter is activating the specific receptors. **(4)** Clathrin molecules assist inward movement of vesicle membrane. Dynamin molecules (*green*) assist approximation of FPM pairs (*inward arrows*) and pinch the neck of the emerging vesicle. **(5)** The vesicle is now free for recycling.

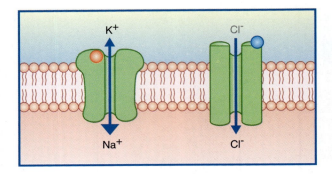

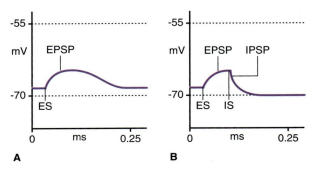

A   **B**

Figure 8.3 **(A)** A transmitter-gated excitatory ionotropic receptor. Binding of the transmitter (*red*, in this example representing glutamate) has opened the pore of a 'mixed' Na^+–K^+ cation channel. A large influx of sodium ions has depolarized the membrane, as shown by the excitatory postsynaptic potential (EPSP). **(B)** A transmitter-gated inhibitory ionotropic receptor. Binding of the transmitter (*blue*, representing $GABA_A$) has opened the pore of a chloride channel. Inward chloride conductance has been increased, and the inhibitory postsynaptic potential (IPSP) restores the membrane potential to its resting state.

transmitter–receptor binding releases the alpha subunit of a G_s protein, leaving it free to link with GTP, which in turn facilitates *adenosine cyclase* to convert ATP to cyclic AMP. Protein kinase A in the membrane is stimulated by cyclic AMP to transfer phosphate ions from ATP to an ion channel, causing its pore to admit sodium ions, initiating depolarization of the target neuron. When the G_s protein is switched off, the membrane-attached enzyme *protein phosphatase* catalyzes extraction of the phosphate ions, resulting in pore closure.

Phosphoinositol system
In the example shown in Figure 8.5, activation of another kind of G_s protein alpha subunit causes the effector enzyme *phospholipase C* to split a membrane phospholipid (PIP_2) into a pair of second messengers: *diacylglycerol* (*DAG*) and *inositol phosphate* (IP_3). DAG activates protein kinase C, which initiates protein phosphorylation. IP_3 diffuses into the cytosol, where it opens calcium-gated channels, mainly in nearby membranes of smooth endoplasmic reticulum. The Ca^{2+} ions activate certain calcium-dependent enzymes downstream, and may open calcium-gated K^+ (outward) and Cl^- (inward) channels with excitatory effect.

Arachidonic acid system
This is described later, in connection with histamine.

Metabotropic receptors are in general *slow* receptors. Membrane channel effects may continue for hundreds of milliseconds after a single stimulus.

Gene transcription effects
It is also well established that the reflex responses to repetitive stimuli may be either progressively increased (a state of *sensitization*, usually induced by noxious stimuli)

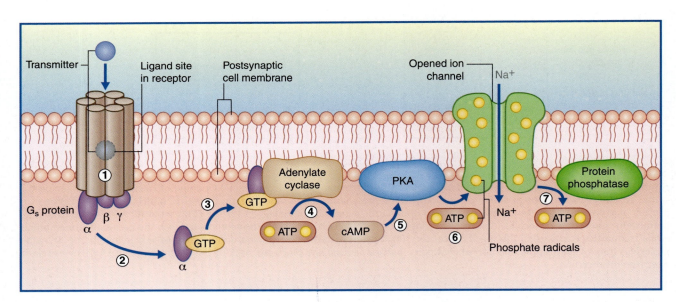

Figure 8.4 The cyclic AMP (cAMP) system. The diagram shows the basic steps along the path from a G_s protein-linked receptor via cyclic AMP to an ion channel. **(1)** Transmitter is activating the receptor macromolecule. **(2)** G_s protein α subunit is freed to bind with guanosine triphosphate (GTP). **(3)** GTP links the unit to adenylate cyclase. **(4)** Adenylate cyclase catalyzes synthesis of cyclic AMP from ATP. **(5)** Cyclic AMP activates protein kinase A (PKA). **(6)** PKA transfers phosphate groups from ATP to a sodium ion channel, causing its pore to open and Na^+ ions to rush into the cytosol, with a depolarizing effect. **(7)** Following inactivation of the G_s protein, dephosphorylation of the channel by a phosphatase enzyme allows the channel pore to close.

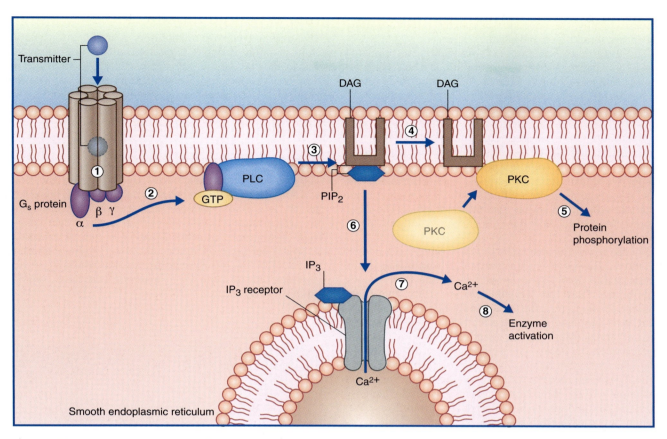

Figure 8.5 The phosphoinositol system. The steps indicate the dual function of this system. **(1)** The transmitter activates the receptor macromolecule. **(2)** The G_s protein α subunit is freed to bind with guanosine triphosphate (GTP), which links it to phospholipase C (PLC). **(3)** PLC moves along the membrane and splits the enzyme phospholipase C (PLC) into diacylglycerol (DAG) and inositol triphosphate (IP$_3$). **(4)** DAG attracts enzyme protein kinase C (PKC) into the membrane, where **(5)** DAG is triggered to phosphorylate several proteins, potentially including ion channels. **(6)** IP$_3$ activates calcium channels in sacs of smooth endoplasmic reticulum. **(7)** Stored calcium spills out of the channel pore into the cytosol. **(8)** Ca^{2+}-dependent enzymes are activated.

or diminished (a state of habituation induced by harmless stimuli). Animal experiments involving reflex arcs involving sets of sensory, internuncial, and motor neurons have shown that a characteristic of sensitization is the development of additional synaptic contacts between the internuncials and the motor neurons, together with additional transmitter synthesis and release. Characteristic of habituation is a reduction of transmitter synthesis and release. All these changes result from *alterations of gene transcription*. Repetitive noxious stimuli cause cyclic AMP to increase its normal rate of activation of protein kinases involved in phosphorylation of proteins that regulate gene transcription. The outcome is increased production of proteins (including enzymes) required for transmitter synthesis, and of other proteins for construction of additional channels and synaptic cytoskeletons. Repetitive harmless stimuli merely reduce the rate of transmitter synthesis and release.

Gene transcription effects are especially important in the context of forming long-term memories (Ch. 34).

TRANSMITTERS AND MODULATORS

Several criteria should be fulfilled for a substance to be accepted as a neurotransmitter.

- The substance must be present within neurons, together with the molecules, including enzymes, required to synthesize it.
- The substance must be released following depolarization of the nerve endings that contain it, and this release must be induced by entry of calcium.
- The postsynaptic membrane must contain specific receptors that will modify the membrane potential of the target neuron.
- The isolated substance must exert the same effect when applied to a target neuron via a micropipette (microiontophoresis).
- Specific antagonist molecules, whether delivered through the circulation or by iontophoresis, must block the effect of the putative ('thought to be') transmitter.
- The physiologic mode of termination of the transmitter effect must be identified, whether it be by enzymatic degradation or by active transport into the parent neuron or adjacent neuroglial cells.

Many transmitters limit their own rate of release by negative feedback activation of autoreceptors in the presynaptic membrane, having the effect of inhibiting further release. Ideally, the existence of specific inhibitory auto-receptors should be established.

The term *neuromodulator* (L. *modulare*, 'to regulate') has been subject to several interpretations. The most satisfactory appears to derive from the terms *amplitude modulation* and *frequency modulation* in electrical engineering, signifying superimposition of one wave or signal onto another. Figure 8.6 represents a sympathetic and a parasympathetic nerve ending, close to a pacemaker cell (modified cardiac myocyte). This neighborly arrangement of nerve endings is common in the heart, and allows the respective transmitters to modulate each other's activity. The sympathetic nerve ending liberates norepinephrine (noradrenaline), which has a stimulatory effect. The three modulators shown exert their effects via second messenger systems.

The figure caption also refers to *autoreceptors* and *heteroreceptors*. Receptors for a particular transmitter often occur in the presynaptic as well as in the postsynaptic membrane. They are called *autoreceptors*. These are activated by high transmitter concentration in the synaptic cleft, and they have a negative feedback effect, inhibiting further transmitter release from the synaptic bouton. *Heteroreceptors* occupy the plasma membrane of neurons that do not liberate the specific transmitter. In the example shown, activity at sympathetic nerve endings is accompanied by inhibition of parasympathetic activity through the medium of heteroreceptors located on parasympathetic nerve endings.

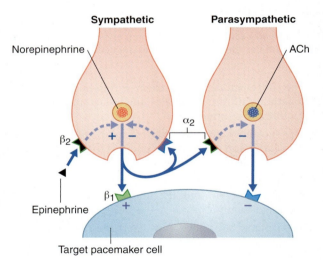

Figure 8.6 Neuromodulation occurs at nerve endings in the sinuatrial node of the heart, where sympathetic and parasympathetic nerve endings often occur in pairs. In this representation, the sympathetic system is the more active, releasing the transmitter norepinephrine, which depolarizes cardiac pacemaker cells via β_1 pacemaker membrane receptors. Circulating epinephrine exerts positive modulation on the nerve ending by increasing transmitter release via β_2 presynaptic membrane heteroreceptors. Inhibitory modulation of excess norepinephrine release is expressed via α_2 presynaptic membrane autoreceptors. At the same time, release of the inhibitory transmitter acetylcholine (ACh) from the parasympathetic bouton is inhibited via α_2 heteroreceptors.

Fate of neurotransmitters

The ultimate fate of transmitters released into synaptic clefts is highly variable. Some transmitters are inactivated within the cleft, some diffuse away into the cerebrospinal fluid via the extracellular fluid, and some are recycled either by direct uptake or indirectly via glial cells.

The principal transmitters and modulators are shown in Table 8.2. Respective receptor types are in Table 8.3.

Amino acid transmitters

The most prevalent excitatory transmitter in the brain and spinal cord is the amino acid *L-glutamate* (Figure 8.7). As an important example, *all* neurons projecting into the white matter from the cerebral cortex, regardless of their destinations in other areas of cortex or brainstem or spinal cord, are excitatory and use glutamate as transmitter. Glutamate is derived from α-ketoglutarate; it also provides the substrate for formation of the most common inhibitory transmitter, GABA.

Table 8.2 Main types of transmitters and modulators, with examples of each[a]

Type	Example(s)
Amino acids	Glutamate
	γ-Aminobutyric acid (GABA)
	Glycine
Biogenic amines	Acetylcholine
	Monoamines
	Catecholamines (dopamine, norepinephrine [noradrenaline], epinephrine [adrenaline])
	Serotonin
	Histamine
Neuropeptides	Vasoactive intestinal polypeptide
	Substance P
	Enkephalin
	Endorphins
	Many others
Adenosine	—
Gaseous	Nitric oxide

[a]The five monoamines contain a single amine group. Catecholamines also contain a catechol nucleus.

Table 8.3 Receptor types activated by different neurotransmitters

Ionotropic receptors	Metabotropic receptors
Glutamate (AMPA/K)	Glutamate (mGluR)
GABA$_A$	GABA$_B$
Acetylcholine (nicotinic)	Acetylcholine (muscarinic)
Glycine	Dopamine (D$_1$, D$_2$)
Serotonin (5-HT$_3$)	Serotonin (5-HT$_1$, 5-HT$_2$)
Purines	Norepinephrine (noradrenaline) (α_1, α_2), epinephrine (adrenaline)
	Histamine (H$_1$, H$_2$, H$_3$)
	All neuropeptides
	Adenosine

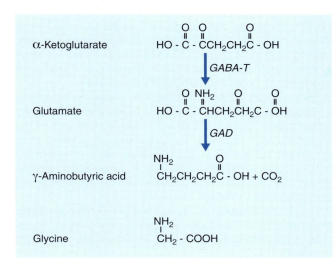

Figure 8.7 The three amino acid transmitters. Glutamate is derived from α-ketoglutarate by the enzyme GABA transaminase (GABA-T); γ-aminobutyric acid (GABA) is derived from glutamate by glutamic acid decarboxylase (GAD). Glycine is the simplest amino acid.

GABA is widely distributed in the brain and spinal cord, being the transmitter in approximately one-third of all synapses. Millions of GABAergic neurons form the bulk of the caudate and lentiform nuclei, and they are also concentrated in the hypothalamus, periaqueductal gray matter, and hippocampus. Moreover, GABA is the transmitter for the large Purkinje cells, which are the *only* output cells of the cerebellar cortex, projecting to the dentate and other cerebellar nuclei.

GABA is synthesized from glutamate by the enzyme *glutamic acid decarboxylase*.

A third amino acid transmitter, *glycine*, is the same molecule that is used in the synthesis of proteins in all tissues. It is the simplest of the amino acids, being synthesized from glucose via serine. It is an inhibitory transmitter largely confined to internuncial neurons of the brainstem and spinal cord.

Glutamate

Glutamate acts on specific receptors of both ionotropic and metabotropic kinds. The three ionotropic ones, named after synthetic agonists that activate them, are known as AMPA, kainate, and NMDA (referring to amino-methylisoxazole propionic acid, kainate, and *N*-methyl-D-aspartate, respectively). Kainate receptors are scarce; they occur in company with AMPA, hence the common use of either AMPA–K or non-NMDA to include both.

Ionotropic glutamate receptors

Activation of AMPA–K receptor channels in the post-synaptic membrane allows an immediate inrush of Na^+ together with a small outward movement of K^+ ions (Figure 8.8), generating the early component of the EPSP in the target neuron. Should this component depolarize the target cell membrane from $-65\,mV$ to $-50\,mV$, this will suffice to generate electrostatic repulsion of Mg^{2+} cations that plug the NMDA receptor ion pore at rest. Na^+ ions enter via the pore and generate action potentials. Significantly, Ca^{2+} ions also enter, and the extended period of depolarization (up to 500 ms from a single action potential) allows activation of calcium-dependent enzymes with knock-on capacity to modify the structure

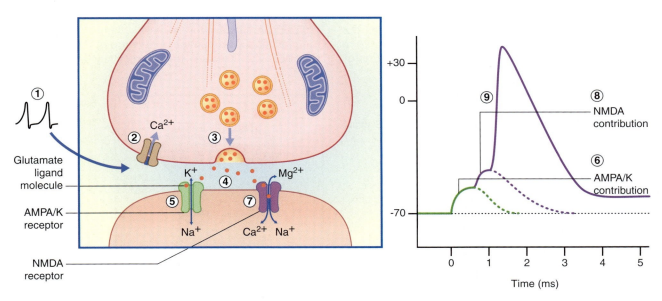

Figure 8.8 Ionotropic glutamate receptors. **(1)** On arrival of action potentials at the nerve terminal, **(2)** calcium channels open, and **(3)** Ca^{2+} ions cause synaptic vesicles to be pulled against the cell membrane. **(4)** Glutamate molecules undergo exocytosis into the synaptic cleft. **(5)** Transmitter binding to the AMPA–K receptor opens the ion channel, permitting large influx of sodium and a small efflux of potassium. **(6)** The resulting excitatory postsynaptic potential (EPSP) produces some 20 mV of depolarization, which **(7)** permits the glutamate ligand to activate the NMDA receptor by expulsion of its Mg^{2+} 'plug'. Both sodium and calcium ions are admitted. **(8)** The NMDA-induced EPSP is sufficient to **(9)** trigger action potentials having an extended repolarization period.

and even the number of synaptic contacts in the target cell. The phenomenon of *activity-dependent synaptic modification* is especially detectable in experimental studies of cultured slices of rat hippocampus, and is likely to be important in the generation of short-term memory traces. (For example, the anesthetic drug *ketamine*, which blocks the NMDA channel, also blocks memory formation.) A characteristic effect of repetitive activation of the NMDA receptor is *long-term potentiation*, represented by above-normal EPSP responses even some days after 'training'. (See *long-term depression*, later.)

The role of NMDA receptors in the phenomenon called *glutamate excitotoxicity* has been demonstrated in vascular strokes produced in experimental animals. The mass death of neurons in this kind of experiment is thought to be the result of degradation caused by excess calcium influx in accordance with the following sequence: ischemia > excess calcium influx > activation of calcium-dependent proteases and lipases > degradation of proteins and lipids > cell death. Ischemic damage may be less severe if an NMDA antagonist drug is administered soon after the initial insult.

Metabotropic glutamate receptors

More than 100 different metabotropic glutamate receptors have been identified, all of them being intrinsic membrane proteins, most occupying postsynaptic membranes and having an excitatory function, others occupying presynaptic membranes where they act as inhibitory autoreceptors.

GABA

Two major classes of GABA receptor are recognized, one being ionotropic and the other metabotropic.

Ionotropic GABA receptors

Termed *GABA$_A$*, these are especially abundant in the limbic lobe of the brain. Each is directly linked to a chloride ion channel (Figure 8.9). Following activation of

GABA$_A$ receptors, channel pores are opened and Cl$^-$ ions diffuse down their concentration gradient from synaptic cleft to cytosol. Hyperpolarization up to −80 mV or more is brought about by summation of successive inhibitory postsynaptic potentials (IPSPs) (Figure 8.10).

The *sedative, hypnotic* barbiturates and benzodiazepines, for example diazepam (Valium), exert their effects by activating the natural receptor. So, too, does ethanol. (Loss of social control under the influence of ethanol may follow the release of target excitatory neurons normally held in check by tonic GABAergic activity.) Some volatile anesthetics also bind with the natural receptor, prolonging the open state of the ion channel.

The chief antagonist at the receptor site is the convulsant drug bicuculline. Another convulsant is picrotoxin, which binds with protein subunits that choke the ion pore when activated.

Metabotropic GABA receptors

Termed *GABA$_B$*, these are relatively uniformly distributed throughout the brain. They are also found within peripheral autonomic nerve plexuses. Although most of their G proteins operate via second messengers, a significant number act *directly* on a special class of postsynaptic K$^+$ channels known as *GIRK* channels (G protein inwardly rectifying K$^+$ channels). As shown in Figure 8.11, transmitter-binding releases the βγ subunit, which *expels* K$^+$ ions through the GIRK channel, thereby producing an IPSP.

The response properties of the target neuronal receptors are slower and weaker than those of GABA$_A$ ionophores, requiring higher-frequency stimulation to be activated. This has led to the belief that they may be extrasynaptic in position rather than facing the synaptic cleft. This is indicated in Figure 8.12, where the belief is supported by the existence of another type of G-direct channel in extrasynaptic locations. This is a calcium channel that is also voltage-gated, and therefore participates in provision of the calcium ions needed to draw synaptic vesicles against the presynaptic membrane. Activation of a G–Ca^{2+} ligand site closes Ca^{2+} channels, thereby reducing the effectiveness

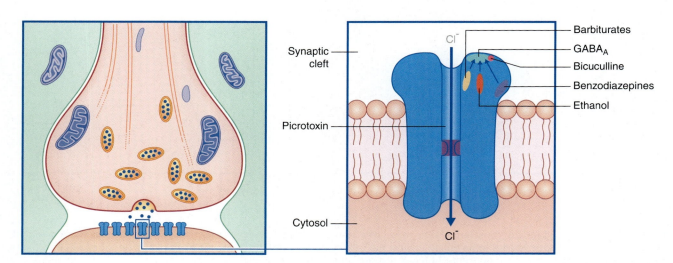

Figure 8.9 Drugs and the ionotropic GABA$_A$ receptor. *Green* signifies agonist effect, *red* signifies antagonist effect. Barbiturates, benzodiazepines, and ethanol exert their hyperpolarizing effect via the natural receptor. Bicuculline antagonizes the receptor, whereas picrotoxin closes the ion pore by direct action locally.

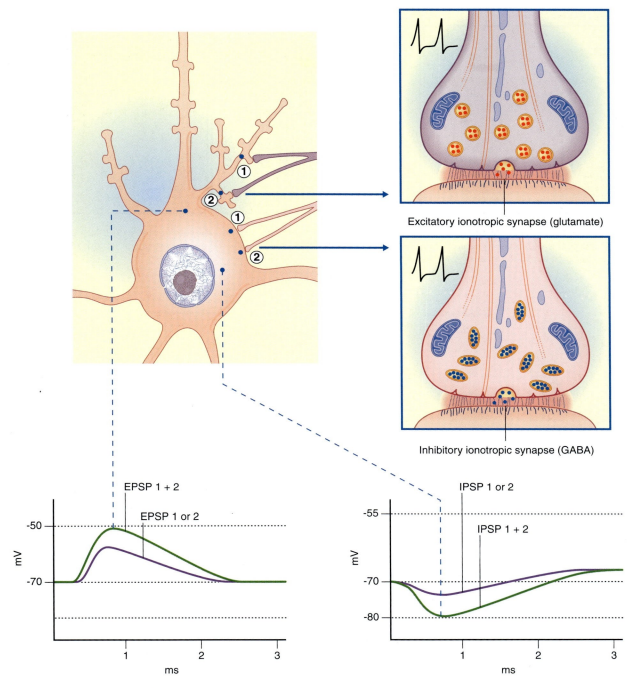

Excitatory ionotropic synapse (glutamate)

Inhibitory ionotropic synapse (GABA)

EPSP 1 + 2

EPSP 1 or 2

IPSP 1 or 2

IPSP 1 + 2

Figure 8.10 Glutamatergic and GABAergic synapses on a multipolar neuron with spiny dendrites. Spatial summation of postsynaptic potentials is illustrated for each pair of synapses.

of action potentials, with inhibitory effect on the parent neuron and on any nearby glutamatergic neurons.

In clinical disorders that involve excessive tonic muscle reflex responses (the state of *spasticity*, Ch. 16), the muscle relaxant *baclofen* is sometimes injected into the subarachnoid space surrounding the spinal cord. The drug seeps into the cord and inhibits release of glutamate from the terminals of muscle afferents, mainly by diminishing the massive calcium entry associated with excessively frequent action potentials.

Recycling of glutamate and GABA

The two routes for recycling are indicated for glutamate in Figure 8.13 and for GABA in Figure 8.14. On the left of each diagram, some transmitter molecules are retrieved from the synaptic cleft by a membrane transporter protein and reincorporated into a synaptic vesicle. On the right, transmitter molecules are being recycled through an adjacent astrocyte. Glutamate is converted to glutamine by glutamine synthetase during transit through astrocytes. Following intercellular transport into the bouton, glutamate

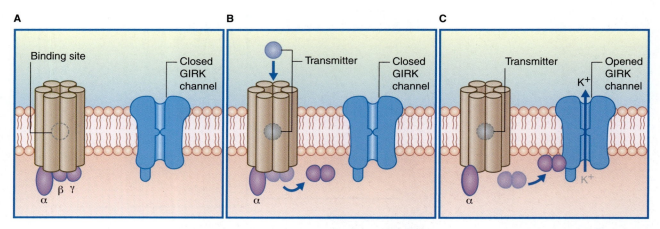

Figure 8.11 G-protein direct opening of a GIRK channel in a postsynaptic membrane. **(A)** Inactive state. **(B)** GABA activation of the receptor causes the G protein βγ subunit to be transferred to the GIRK channel. **(C)** The βγ subunit causes K⁺ ions to be expelled, thereby leading to hyperpolarization of the cell membrane.

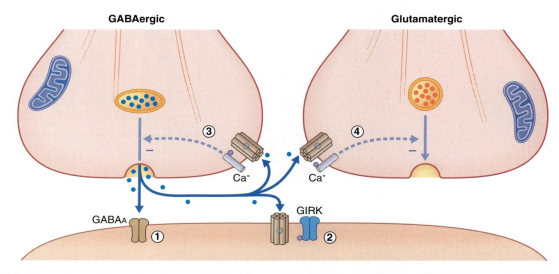

Figure 8.12 Following transmitter release from a GABAergic neuron. **(1)** Transmitter binding with GABA$_A$ receptors has a hyperpolarizing effect on the target membrane by opening chloride channels. **(2)** Binding with GIRK GABA$_A$ receptors works in the same direction by opening inwardly rectifying potassium channels (GIRKs). **(3)** Binding with GABA$_A$ autoreceptors dampens transmitter release from the parent neuron by closing ligand–G protein-operated calcium channels (Ca⁻). **(4)** Binding with GABA$_A$ heteroreceptors on neighboring glutamatergic boutons has the same Ca⁻ effect.

is reassembled by glutaminase and then repacked into a synaptic vesicle. GABA is converted to glutamate by GABA transaminase during transit. Following return to the bouton, glutamate is coverted (by glutamate decarboxylase) to GABA prior to storage in vesicles.

The remarkable autoimmune disorder known as *stiff person syndrome*, caused by blockade of glutamate decarboxylase, is described in Chapter 29.

Glycine

Glycine is synthesized from glucose via serine. Its main function as a transmitter is to provide tonic negative feedback on to motor neurons in the brainstem and spinal cord. Inactivation of glycine, for example by strychnine

poisoning, results in agonizing convulsions (Clinical Panel 8.1).

Recycling
Glycine is rapidly taken up by an axonal transporter and re-stored in synaptic vesicles.

Biogenic amine transmitters
Acetylcholine

Acetylcholine (ACh) is a highly significant transmitter. In the central nervous system (CNS), activity of cholinergic neurons projecting from the basal region of the forebrain to the hippocampus is essential for learning and memory; degeneration of these neurons is consistently associated

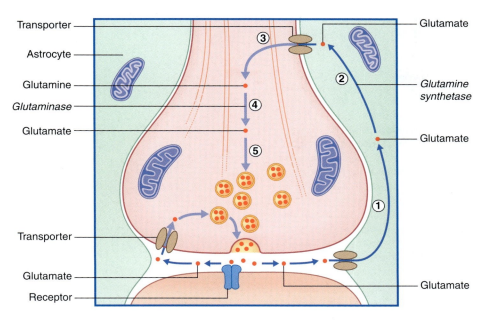

Figure 8.13 Glutamate reuptake and resynthesis. **(1)** Glutamate is taken up by an astrocyte, and **(2)** is converted to glutamine by glutamine synthetase. **(3)** The glutamine is transported back into the nerve terminal, **(4)** where it is converted to glutamate by glutaminase and **(5)** returned to a synaptic vesicle.

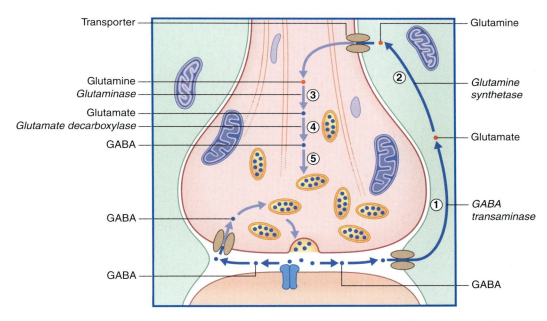

Figure 8.14 GABA reuptake and resynthesis. On the left, GABA molecules are being recycled intact. On the right, GABA is taken up by an astrocyte, then **(1)** GABA is converted to glutamate by GABA transaminase. **(2)** Glutamate is converted to glutamine by glutamine synthetase. **(3)** Glutamine is taken up by the axon and converted to glutamate by glutaminase. **(4)** Glutamate is converted to GABA by glutamate decarboxylase and **(5)** returned to a synaptic vesicle.

with the onset of Alzheimer disease. In the peripheral nervous system (PNS), all motor neurons to skeletal muscle are cholinergic; all preganglionic neurons supplying the ganglia of the sympathetic and parasympathetic systems are cholinergic, as is the postganglionic nerve supply of the parasympathetic system to cardiac muscle, the smooth muscle of the intestine and bladder, and the smooth muscles of the eye involved in accommodation for close-up vision.

Acetylcholine is formed when an acetyl group is transferred to choline from acetyl coenzyme A (CoA) by the enzyme *choline acetyltransferase* (Figure 8.15), which is unique to cholinergic neurons. The choline is actively transported into the neuron from the extracellular space.

Clinical Panel 8.1 Strychnine poisoning

Strychnine is a glycine receptor blocker. The victim of strychnine poisoning suffers agonizing convulsions because of liberation of α motor neurons from the tonic inhibitory control of Renshaw cells (Figure CP 8.1.1). The convulsions resemble those induced by the tetanus toxin, described in Chapter 6. This is no surprise, because tetanus toxin prevents the release of glycine from Renshaw cells. Postmortem studies of normal human brain, using radiolabeled strychnine, have shown glycine receptors to be especially abundant on internuncial neurons in the nucleus of the trigeminal nerve supplying the jaw muscles, and in the nucleus of the facial nerve supplying the muscles of facial expression. These two muscle groups are especially affected in both types of convulsive attack.

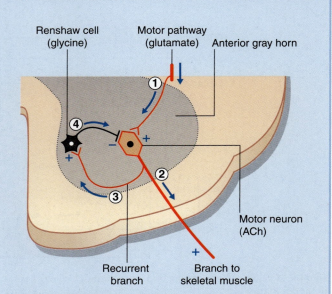

Figure CP 8.1.1 The negative feedback loop whereby Renshaw cells inhibit excessive firing by motor neurons. ACh, acetylcholine. **(1)** Fiber from a descending motor pathway is exciting a spinal motor neuron. **(2)** The motor neuron is stimulating muscle contraction. **(3)** A recurrent branch stimulates a Renshaw cell. **(4)** The Renshaw cell provides sufficient inhibition to prevent overactivity of the motor neuron.

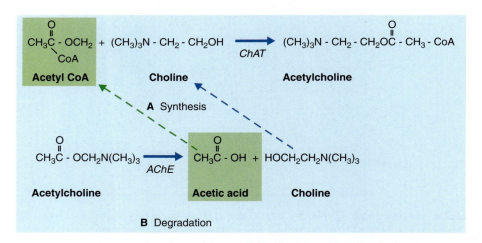

Figure 8.15 (A) Synthesis of acetylcholine (ACh) from acetyl coenzyme A (CoA) and choline catalyzed by choline acetyltransferase (ChAT). **(B)** Degradation of ACh catalyzed by acetylcholinesterase (AChE). Dashed arrows indicate recycling of acetic acid and choline.

CoA is synthesized in mitochondria that are concentrated in the nerve terminal and also provide the enzyme. Following release, ACh is degraded in the synaptic cleft by *acetylcholinesterase (AChE)*, yielding choline and acetic acid. These molecules are largely recaptured and recycled to form fresh transmitter.

Some steps in synthesis, degradation, and recycling of ACh are also shown in Figure 8.16.

Both ligand-gated and G protein-coupled ACh receptors are recognized. The former are called *nicotinic*, because they were first discovered to be activated by nicotine extracted from the tobacco plant. The latter are called

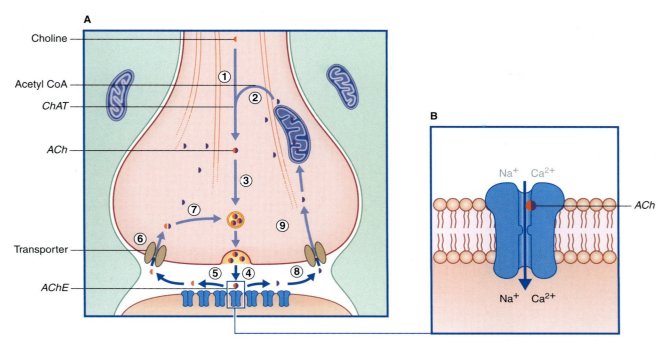

Figure 8.16 (A) History of acetylcholine molecules in the central nervous system. The postsynaptic membrane shown contains nicotinic receptors (nAChR). **(1)** Choline taken up from the extracellular fluid is sent to the nerve ending. **(2)** The choline is acetylated by acetyl coenzyme A released by mitochondria, the reaction being catalyzed by choline acetyltransferase (ChAT). **(3)** Completed ACh molecules are taken up by synaptic vesicles. **(4)** Liberated ACh bonds briefly with its receptor. **(5)** Acetylcholinesterase (AChE) hydrolyzes the molecule. **(6)** The choline moiety is transported back into the cytosol. **(7)** Formation of a fresh molecule of ACh is mediated by the transferase, en route to a synaptic vesicle. **(8)** The acetate moiety is transported into the cytosol. **(9)** Mitochondria use the acetic acid to produce fresh acetyl CoA. **(B)** The ligand-gated nicotinic receptor, indicating the inrush of sodium and potassium ions associated with ACh–receptor binding.

muscarinic because activated by *muscarine* extracted from the poisonous mushroom *Amanita muscaria*.

Nicotinic receptors

Nicotinic receptors are found in the neuromuscular junctions of skeletal muscle, in all autonomic ganglia, and in the CNS. Activation by ACh causes the ion pore to open, with immediate inrush of calcium and sodium ions and prompt depolarization of the target neuron.

The nicotinic receptor is considered further in relation to the innervation of skeletal muscle (Ch. 9).

Muscarinic receptors

The G protein-gated muscarinic receptors are especially numerous (a) in the temporal lobe of the brain, where they are involved in the formation of memories; (b) in autonomic ganglia; (c) on cardiac muscle fibers, including the modified muscle of the conducting tissue; (d) on the smooth muscle of intestine and bladder; and (e) on the secretory cells of sweat glands. Five subtypes of receptor have been identified, numbered M_1 to M_5.

Broadly, M_1, M_3, and M_5 receptors are excitatory, the enzyme cascades allowing calcium influx with consequent depolarization; M_2 and M_4 are inhibitory autoreceptors that produce potassium efflux with hyperpolarization.

Cholinergic effects in the heart and other viscera are described in Chapter 13.

Recycling of acetylcholine

Following hydrolysis in the synaptic cleft, the choline and acetate moieties are recaptured by specific transporters (Figure 8.16).

Monoamines

Catecholamines

As indicated in Table 8.1, the catecholamines comprise dopamine, norepinephrine, and epinephrine (adrenaline). As shown in Figure 8.17, all three are derived from the amino acid tyrosine.

The transmitters are synthesized in the nerve terminals, the requisite tyrosine and enzymes having been sent there by rapid transport. Newly synthesized transmitter must be packaged immediately into a synaptic vesicle by a monoamine transporter protein lodged in the vesicular membrane, because the catabolic enzyme *monoamine oxidase (MAO)* permeates the cytosol. On release, most of the transmitter binds with one or more specific receptors in the postsynaptic membrane and (where present) with an autoreceptor in the presynaptic membrane. Of the remainder, some is inactivated by *catechol-O-methyl transferase (COMT)*, an enzyme liberated from the postsynaptic membrane into the synaptic cleft (Figure 8.18). The rest is taken up by a specific uptake transporter, and is either collected by a vesicular protein transporter or is inactivated by MAO.

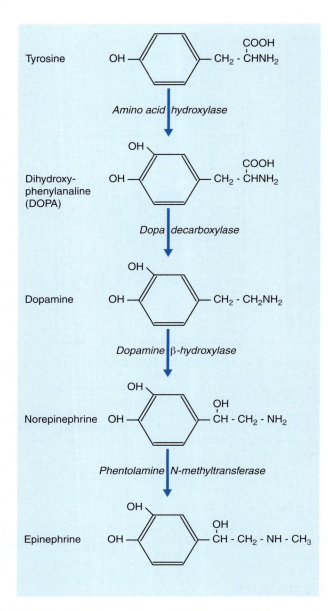

Figure 8.17 Synthetic pathway of the catecholamine transmitters. The catechol ring is in *blue*.

Dopamine Dopamine is of particular interest in the clinical contexts of Parkinson disease, drug addiction, and schizophrenia. It is synthesized from tyrosine in two steps (Figure 8.18), being converted to DOPA (dihydroxy-phenylalanine) by *amino acid hydroxylase*, and from DOPA to dopamine by *dopa decarboxylase*, an enzyme restricted to catecholaminergic neurons. The two main sets of dopaminergic neurons are located in the midbrain. They are the substantia nigra and the ventral part of the tegmentum called the ventral tegmental area (VTA).

The substantia nigra belongs functionally to the basal ganglia (Ch. 28). A dopaminergic *nigrostrial pathway* projects from substantia nigra to striatum (caudate nucleus and putamen). This pathway controls a motor loop of neurons feeding forward to the motor cortex. Degeneration of neurons in the substantia nigra is a classic feature of

Parkinson disease, in which normal movements are disrupted by rigidity of the musculature and/or tremor.

The VTA projects groups of dopaminergic neurons into the forebrain. One group, called *mesocortical*, projects to the prefrontal cortex; overactivity of this system has been invoked to explain some clinical features of schizophrenia (Ch. 29). The other, called *mesolimbic*, projects to several limbic nuclei including the nucleus accumbens (bedded in the ventral striatum); dopamine liberation within the nucleus accumbens appears to be the basis of the *dopamine rush*, or *dopamine high*, associated with several kinds of drug addiction (Ch. 34).

Receptors Dopamine receptors are all G protein-coupled. D_1 and D_2 receptors are recognized, each having more than one subtype. D_1 receptors activate G_s proteins and are excitatory, activating adenylate cyclase with consequent receptor phosphorylation. D_2 receptors activate G_i proteins and are inhibitory; they may inactivate adenylate cyclase, and may also promote hyperpolarization by opening GIRK ion channels and/or inhibiting voltage-gated Ca^{2+} channels. Both kinds are numerous in the striatum, where they are required for the proper execution of learned motor programs including locomotion (Ch. 28).

Norepinephrine In the CNS, noradrenergic neurons are concentrated in the cerulean nucleus (locus ceruleus) in the floor of the fourth ventricle. From here, they project to all parts of the gray matter of the brain and spinal cord (see Figure 24.4). These neurons are important for regulation of the sleep–wake cycle and the control of mood. In the PNS, norepinephrine is liberated by sympathetic nerve endings, notably throughout the cardiovascular system, where it maintains the blood pressure. It is an integral component of the 'fight or flight' response to danger. Formation of norepinephrine takes place in neurons containing *dopamine β–hydroxylase*. This enzyme is remarkable in being restricted to the inner surface of the membrane of synaptic vesicles.

Receptors All are G protein-gated. They are broadly grouped into alpha and beta sets, with two principal subtypes within each. Some details are provided in Table 8.4.

Catecholamine recycling Recycling of dopamine and norepinephrine occurs via specific reuptake transporters, as indicated in Figure 8.18. The figure indicates that not all the molecules are recycled into synaptic vesicles within the parent neuron. Any of three other fates are possible: some are metabolized in or near the synaptic cleft by the enzyme *COMT*; others are carried for up to 100 μm by the extracellular fluid, perhaps bonding to isolated specific membrane heteroreceptors on other neurons as depicted in Figure 6.6 ('volume transmission'); and others achieve reuptake only to be metabolized by the enzyme *MAO* liberated by nearby mitochondria.

Epinephrine Neuronal production of epinephrine in the CNS appears to be confined to a group of cells in the upper lateral part of the medulla oblongata. Only these contain the enzyme (*phentolamine N-methyltransferase*) that provides the final link in the catecholamine chain (Figure 8.17). Some of these neurons project upward to the hypothalamus, others to the lateral gray horn of the spinal cord. Their functions are not yet clear.

In the PNS, the *chromaffin cells* of the adrenal medulla liberate epinephrine as a hormone into the capillary bed.

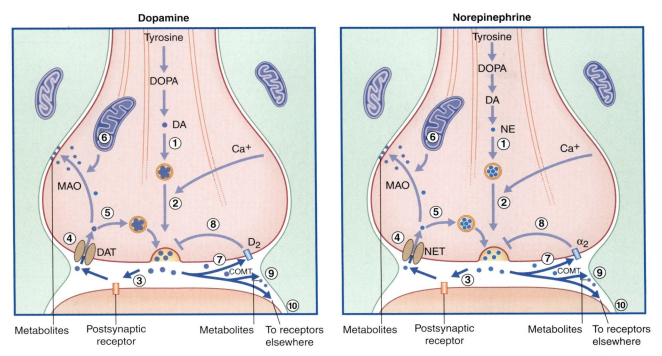

Figure 8.18 History of the main catecholamine transmitter molecules. **(1)** Transmitter dopamine (DA) or norepinephrine (NE) molecules are transported into synaptic vesicles. **(2)** Following depolarization-induced calcium entry to the cytosol, synaptic vesicles are pulled into contact with the presynaptic membrane of the terminal bouton. **(3)** Liberated transmitter molecules have three possible fates. Most bind with G protein-coupled receptors in the postsynaptic membrane, initiating second messenger events. **(4)** Specific transmitter reuptake transporters (DAT, NET) return transmitter molecules to the cytosol. **(5)** Some of these molecules are repackaged for further use. **(6)** Surplus molecules in the cytosol are degraded by mitochondria-derived monoamine oxidase (MAO) enzyme. The metabolites float away in extracellular fluid destined to weep through ventricular walls into the cerebrospinal fluid, where some metabolites may be detected. **(7)** A second group of transmitter molecules within the synaptic cleft are broken down by catechol-*O*-methyl transferase (COMT). **(8)** A third group function as D_2 or α_2 autoreceptors, inhibiting further transmitter release. **(9)** A second set of metabolites drifting away in the extracellular fluid. **(10)** Some intact transmitter molecules drift away to activate receptors elsewhere.

Table 8.4 Main features of noradrenergic receptors

Receptor type	Location on neuron	Regions found	Second messenger	Effects
α_1	Postsynaptic	Smooth muscle, brain	Phosphoinositol	Excitatory. Opens calcium channels.
α_2	Mainly presynaptic	—	G protein direct	Inhibitory. Opens GIRK channels.
β_1	Postsynaptic	Heart, brain	Adenylate cyclase +	Excitatory. Opens calcium channels.
β_2	Postsynaptic	Smooth muscle, liver, brain	Adenylate cyclase ±	Inhibitory to smooth muscle. Glycogen breakdown + in liver. Excitatory in brain.

The epinephrine augments sympathetic effects on the circulatory and other systems during the alarm response to danger. As shown in Figure 13.6, the chromaffin cells are modified sympathetic ganglion cells receiving synaptic contacts from preganglionic cholinergic neurons. One function of circulating epinephrine, illustrated in Figure 13.5, is to boost norepinephrine output at sympathetic nerve terminals by activating β_2 heteroreceptors there.

Serotonin
In the medical literature, more has been written about serotonin than about any other transmitter. Depletion of serotonin has a well-established connection with depression. Abnormalities of serotonin metabolism have

been implicated in other behavioral disorders, including anxiety states, obsessive-compulsive disorders, and bulimia.

As indicated in Figures 21.2 and 21.5, serotonergic cell bodies occupy the midregion or *raphe* (seam) of the brainstem. Their axonal ramifications are quite prodigious, penetrating to every region of the gray matter of brain and spinal cord.

Serotonin, commonly referred to as 5-HT because it is 5-hydroxytryptamine, is derived from the dietary amino acid tryptophan, present in the circulation. It is actively transported across the blood–brain barrier into the brain extracellular fluid, then transported into serotonergic neurons. Formation of serotonin from tryptophan is a

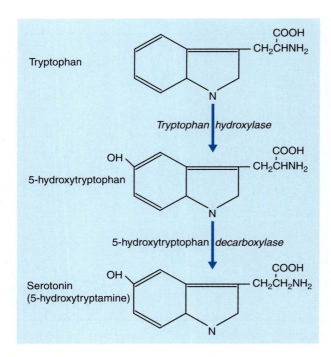

Figure 8.19 Synthesis of serotonin from tryptophan.

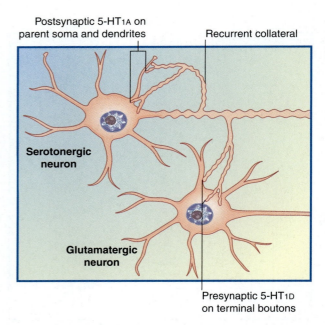

Figure 8.20 Serotonergic autoreceptors. 5-HT$_{1A}$ reduce both excitability and serotonin synthesis. 5-HT$_{1D}$ reduce serotonin release. (After Nestler et al. 2001, with permission of McGraw-Hill.)

two-step process (Figure 8.19). Tryptophan is converted to 5-hydroxytryptophan by the enzyme *tryptophan hydroxylase*, and this is converted to serotonin by *5-hydroxytrytophan decarboxylase*.

Receptors Seven groups of receptors have been identified. Ongoing research seeks to refine drug therapy so as to target individual receptors thought to be responsible, by either over- or underactivity, for specific disorders, notably in the psychiatric domain.

Table 8.5 provides some details for a selected shortlist of serotonin receptors. The table includes a reference to receptors on the somas and dendrites of the parent cell. These are targets of recurrent axon collaterals, as indicated in Figure 8.20.

Recycling Recycling follows the same general pattern as for the catecholamines. Here again, the final step of transmitter synthesis takes place within the terminal bouton, and the released molecules may activate either presynaptic autoreceptors on the parent neuron or isolated heteroreceptors on other neurons nearby. There appears to be no degradatory enzyme in or near the synaptic cleft comparable with COMT, but MAO is present within the parent bouton (as shown in Figure 8.21).

Monoamines and abnormal emotional or behavioral states

Abnormal monoamine function has been implicated in a great variety of abnormal emotional or behavioral states, including depression, insomnia, anxiety disorders, panic attacks, and specific phobias. Brain areas involved are discussed in Chapters 23 and 29.

Histamine

Histamine is synthesized from histidine by *histidine decaboxylase*, as shown in Figure 8.22. The somas of histaminergic neurons appear to be confined to the posterior part of the hypothalamus, where they occupy the

Table 8.5 Some serotonin receptors of clinical interest				
Receptor type	**Neuronal location**	**Second messenger**	**Effect**	**Activity contributes to**
5-HT$_{1A}$	Parent cell somas and dendrites	Inhibits cyclic AMP	Inhibitory	Anxiety, depression, pain.
5-HT$_{1D}$	Presynaptic autoreceptors	Inhibits cyclic AMP, opens GIRK channels	Inhibitory	Vasoconstriction. Activation relieves migraine.
5-HT$_{2A}$	Target cell somas and dendrites	Stimulates phosphoinositol	Excitatory	Overactivity causes hallucinations.
5-HT$_{2C}$	Target cell somas and dendrites	Stimulates phosphoinositol	Excitatory	Overactivity causes increased appetite.
5-HT$_3$	Target cell somas and dendrites in area postrema	No: it is ionotropic	Excitatory	Stimulation causes emesis (vomiting).

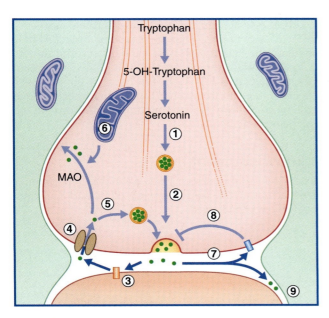

Figure 8.21 History of serotonin transmitter molecules in the central nervous system. **(1)** Serotonin is transported into synaptic vesicles. **(2)** Transmitter molecules undergo exocytosis into the synaptic cleft. **(3)** A postsynaptic serotonin receptor is being targeted. **(4)** The (therapeutically important) serotonin reuptake transporter returns transmitter to the terminal cytosol. **(5)** Some transmitter is repackaged into synaptic vesicles. **(6)** Some is degraded instead by (therapeutically important) monoamine oxidase (MAO). **(7)** Some activates presynaptic autoreceptors. **(8)** A 5-HT$_{1D}$ presynaptic autoreceptor is retarding further transmitter release. **(9)** Some diffuses through the extracellular space, using 'volume transmission' to activate receptors on other neurons.

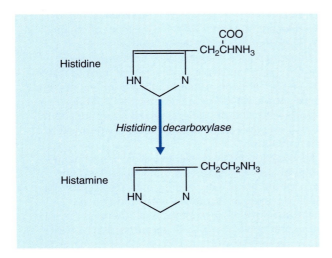

Figure 8.22 Synthesis of histamine from histidine.

small *tuberomammillary nucleus* shown in Figure 26.1. However, their axons extend widely, mainly to all parts of the cerebral cortex. The main function of histaminergic neurons is to participate with cholinergic and serotonergic neurons in maintaining the awake state. These neurons are active in the awake state and silent during sleep (see

Ch. 24). Activation is a function of the peptide *orexin* locally produced by lateral hypothalamic neurons. The compulsive daytime sleep disorder known as *narcolepsy* (Ch. 30) appears to result from failure of orexin production.

Receptors H$_1$, H$_2$, and H$_3$ receptors on target neurons are known. All three activate G proteins. H$_1$ and H$_2$ receptors activate G$_s$ proteins, acting via arachidonic acid production. The acid is processed further to yield prostaglandins and other endoperoxides, some of which modify cyclic AMP activity and others that bind directly to ion channels. The H$_3$ group are inhibitory autoreceptors.

In clinical practice, *antihistamines* are widely prescribed for the purpose of blocking H$_2$ receptors involved in gastric acid secretion and various allergic responses. Drowsiness has been a common side effect of such drugs prior to production of cimetidine and others that are not able to cross the blood–brain barrier.

Neuropeptides

More than 50 neuropeptides have been isolated. All of them are linear chains of amino acids linked by peptide bonds. Peptide precursor chains (called propeptides) are passed through the Golgi complex and budded off in large, dense-cored vesicles that are rapidly transported to the nerve endings, where peptide formation is completed. As previously illustrated in Figure 6.5, peptides undergo non-synaptic release and may travel some distance to reach their receptors.

Receptors

These are all G protein-coupled. In general, they are cotransmitters, and their function is to modulate the effect of principal, small-molecule transmitters such as glutamate or ACh. Calcium channels are relatively scarce outside the synaptic cleft, and peptide liberation characteristically requires relatively high-frequency action potentials. An example is mentioned in Chapter 13: sweat glands are supplied by cholinergic neurons having vasoactive intestinal polypeptide (VIP) as cotransmitter. At low-frequency stimulation, ACh alone is sufficient to provide routine 'insensible perspiration', which is also invisible. *Sweating* for any length of time requires local vasodilatation in addition to an abundance of ACh, and this is provided by VIP, a potent dilator of arterioles.

Within the CNS, naturally occurring *opioid* (opium-like) peptides, called *endorphins*, are highly significant in relation to the control of pain perception, as discussed in Chapter 24.

Adenosine

Adenosine, derived from ATP, is a well-established excitatory cotransmitter with ACh in parasympathetic neurons innervating smooth and cardiac muscle. In the brain, adenosine is an inhibitory cotransmitter with glutamate. Adenosine receptors are G protein-coupled. Those on presynaptic terminals reduce glutamate release, and on postsynaptic dendrites tend to hyperpolarize by opening K$^+$ and Cl$^-$ membrane channels. Adenosine-containing compounds are sedative. Adenosine receptor antagonists have the opposite effect, increasing alertness and providing temporary improvement in cognitive function. The antagonists are *methylxanthenes*, comprising caffeine found in coffee, theophylline in tea, and theobromine in cocoa.

Nitric oxide

Nitric oxide is not a 'classical' transmitter, but is a lipid- and water-soluble gaseous radical that diffuses briefly and rapidly across cell membranes, including those of neurons. It is synthesized from arginine by the enzyme *nitric oxide synthase* in response to Ca^{2+} entry following depolar-

ization; it activates guanylate cyclase and increases cyclic AMP in target cells, thereby enabling cyclic AMP to modulate the activity of conventional neurotransmitters. In the autonomic nervous system, it is a powerful smooth muscle relaxant (Ch. 13). In the brain, it appears to be especially relevant to memory formation by eliciting long-term potentiation in glutamatergic neurons in the hippocampus (Ch. 34).

Acknowledgment

The assistance of Professor Brian Leonard, Department of Pharmacology, National University of Ireland, Galway, is gratefully acknowledged.

Core Information

Electrical synapses are gap junctions designed to ensure synchronous activity of groups of neurons. The gaps are bridged by tightly packed ion channels. Protein subunits surrounding the individual ion channels are apposed when the neurons are silent. In response to passage of action potentials along the cell membrane, they separate instant diffusion of ions from one cytosol to another.

At chemical synapses, transmitter molecules are expelled into the synaptic cleft and united with their specific target receptors in the manner already summarized in Table 8.1.

Ionotropic receptors are ligand-gated. Each is either excitatory through admission of Na^+ ions or inhibitory through admission of Cl^- ions. Metabotropic receptors have no ion pore. Their receptor macromolecule responds to transmitter activation by detaching a G-protein subunit that usually binds to guanine triphosphate or guanine diphosphate, which in turn activates the cyclic AMP, phosphoinositol, or arachidonic acid system.

Amino acid transmitters include glutamate, GABA, and glycine. Biogenic amine transmitters and modulators include acetylcholine (ACh) and the monoamines, i.e. catecholamines (dopamine, norepinephrine, epinephrine), serotonin, and histamine. Neuropeptides include vasoactive intestinal polypeptide (VIP), substance P, enkephalin, and endorphins. Also prevalent are adenosine and nitric oxide.

Glutamate activation of target AMPA–K receptors produces the early component of the excitatory postsynaptic potential, which in turn opens NMDA receptors, producing action potentials through entry of Na^+, and long-term potentiation through entry of Ca^{2+}. Excitotoxicity caused by excessive Ca^{2+} influx may cause target cell necrosis.

GABA activation of target $GABA_A$ (ionotropic) receptors generates inhibitory postsynaptic potentials by causing Cl^- influx. These receptors are also activated by barbiturates, benzodiazepines, alcohol, and some volatile anesthetics. Activation of $GABA_B$ (metabotropic)

receptors leads to hyperpolarization indirectly by depressing cyclic AMP formation, and directly by expelling K^+ ions through GIRK channels.

Glycine released by Renshaw cells provides tonic negative feedback on to motor neurons. Strychnine and tetanus convulsions are caused by inactivation of glycine.

Acetylcholine target receptors are either nicotinic (causing entry of Na^+ and Ca^{2+}) or muscarinic. The latter include excitatory M_1, M_3, and M_5 receptors, and inhibitory M_2 and M_4 autoreceptors.

Dopamine is relevant to Parkinson disease by the nigrostriatal pathway, to drug addiction and to schizophrenia by mesocortical and mesolimbic pathways. Target receptors are all G protein-coupled. D_1 receptors are excitatory via cyclic AMP activation. D_2 receptors are inhibitory via cyclic AMP or Ca^{2+} channel inactivation and/or activation of GIRK channels.

Norepinephrine is liberated by noradrenergic neurons. Main source within the central nervous system (CNS) is the cerulean nucleus; in the peripheral nervous system, postganglionic sympathetic fibers. Target receptors are all G protein-gated and are grouped into α and β subtypes, some of each being excitatory and others inhibitory.

Serotonin is highly relevant to clinical psychology and psychiatry. It is synthesized mainly in the raphe nuclei of the brainstem. Seven groups of receptors have been identified. $5\text{-}HT_{1A}$ serves autoinhibition via somatodendritic autoreceptors, $5\text{-}HT_{1D}$ serves autoinhibition via presynaptic receptors, $5\text{-}HT_{2A}$ excites target neurons via phosphoinositol stimulation, $5\text{-}HT_{2C}$ stimulates excitatory ionotropic channels in the area postrema (vomiting center).

Histaminergic neurons project from the hypothalamic tuberomammillary nucleus to all parts of the cerebral cortex. They help to maintain the state of arousal.

Neuropeptides include VIP, substance P, enkephalin, and endorphins. In general, they are cotransmitters having a modulatory effect. Their target receptors are all G protein-coupled.

Core Information *(Continued)*

Adenosine is derived from ATP. In the autonomic nervous system, it is an excitatory cotransmitter with ACh. In the CNS, it is inhibitory, and adenosine-containing compounds are sedative.

Nitric oxide is a lipid- and water-soluble gaseous radical synthesized from arginine in response to Ca^{2+} entry following depolarization. It activates guanylate cyclase and increases cyclic AMP in target neurons, thereby modulating the activity of conventional transmitters. It is a peripheral vasodilator, and in the hippocampus it participates in memory formation by eliciting long-term potentiation.

REFERENCES

An S, Seong D. Regulation of exocytosis in neurons and neuroglial cells. Curr Opin Neurobiol 2004; 14:522–530.

Bear MF, Connors BW, Paradiso MA. Neuroscience: exploring the brain. Baltimore: Williams & Wilkins; 1996.

Leonard BE. Fundamentals of psychopharmacology. 3rd edn. Chichester: Wiley; 2003.

Mayer ML. Glutamate receptor ion channels. Curr Opin Neurobiol 2005; 15:282–288.

Nestler EJ, Hyman SE, Malenka RC. Molecular pharmacology: a foundation for clinical neuroscience. New York: McGraw-Hill; 2001.

Schwartz JH. Elementary interaction between neurons: synaptic transmission. In: Kandel ER, Schwarz JH, Jessell TM, eds. Principles of neural science. 4th edn. New York: McGraw-Hill; 2000:175–316.

Stevens CF. Presynaptic function. Curr Rev Neurobiol 2004; 14:341–345.

Peripheral nerves

STUDY GUIDELINES

1 Understand where the limb plexuses fit into the general scheme of the spinal nerves.

2 Be able to name the connective tissue sheath surrounding a nerve trunk, a nerve fascicle, a nerve fiber.

3 Understand how myelin sheaths are composed of cell membranes.

4 Appreciate the functional significance of the nodes of Ranvier, and why thicker nerve fibers conduct faster.

5 Understand why nerves regenerate better following a crush injury than a cut.

6 Recognize the two kinds of transneuronal atrophy.

7 To appreciate the clinical importance of peripheral nerves, we suggest you pay a preliminary visit to the Clinical Panels in Chapter 12.

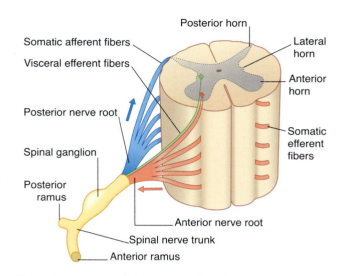

Figure 9.1 Segment of thoracic spinal cord with attached nerve roots. Arrows indicate directions of impulse conduction. Green indicates sympathetic outflow.

GENERAL FEATURES

The peripheral nerves comprise the cranial and spinal nerves linking the brain and spinal cord to the peripheral tissues. The spinal nerves are formed by the union of **anterior** and **posterior nerve roots** at their points of exit from the vertebral canal (Figure 9.1). The swelling on each posterior root is a **spinal** or **posterior root ganglion**. The **spinal nerve** is only 1 cm long and occupies an intervertebral foramen. On emerging from the foramen, it divides into **anterior** and **posterior rami**.

The posterior rami supply the erector spinae muscles and the overlying skin of the trunk. The anterior rami supply the muscles and skin of the side and front of the trunk, including the muscles and skin of the limbs; they also supply sensory fibers to the parietal pleura and parietal peritoneum.

The cervical, brachial, and lumbosacral plexuses are derived from anterior rami, which form the *roots* of the plexuses. The term *root* therefore has two different meanings depending on the context. (Details of the plexuses are in standard anatomy texts.)

The neurons contributing to peripheral nerves are partly contained within the central nervous system (CNS) (Figure 9.2). The cells giving rise to the motor (*efferent*) nerves to skeletal muscles are **multipolar** *alpha* and *gamma* **neurons** of similar configuration to the one depicted in

Figure 6.4; in the spinal cord, they occupy the anterior horn of gray matter. Further details are in Chapter 10. Those giving rise to posterior nerve roots are **unipolar neurons** whose cell bodies lie in posterior root ganglia and whose sensory (afferent) central processes enter the posterior horn of gray matter.

The spinal nerves supply *somatic efferent fibers* to the skeletal muscles of the trunk and limbs, and *somatic afferent fibers* to the skin, muscles, and joints. They all carry *visceral efferent*, autonomic fibers, and some spinal nerves contain visceral afferent fibers as well.

MICROSCOPIC STRUCTURE OF PERIPHERAL NERVES

Figure 9.3 illustrates the structure of a typical peripheral nerve. It is not possible to designate individual nerve fibers as motor or sensory on the basis of structural features alone.

Peripheral nerves are invested with **epineurium**, a loose, vascular connective tissue sheath surrounding the fascicles (bundles of fibers) that make up the nerve. Nerve fibers are exchanged between fascicles along the course of the nerve.

Each fascicle is covered by **perineurium**, composed of several layers of pavement epithelium bonded by tight junctions. Surrounding the individual Schwann cells is a network of reticular collagenous fibers, the **endoneurium**.

Less than half of the nerve fibers are enclosed in myelin sheaths. The remaining, unmyelinated fibers travel in deep gutters along the surface of Schwann cells.

The term *nerve fiber* is usually used in the context of nerve impulse conduction, where it is equivalent to *axon*. An anatomic definition is possible for a myelinated fiber:

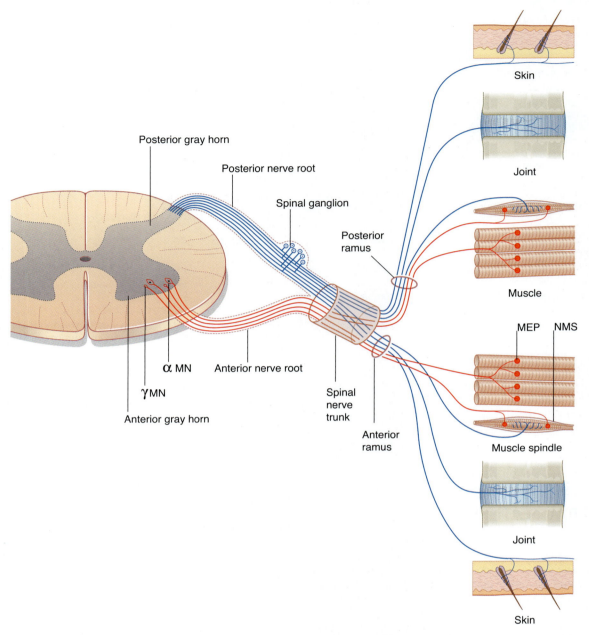

Figure 9.2 Composition and distribution of a cervical spinal nerve. *Note*: The sympathetic component is not shown. MEP, motor end plate; MN, multipolar neurons; NMS, neuromuscular spindle.

it comprises axon, myelin, and neurolemmal sheaths, and endoneurium. It is not possible for unmyelinated axons because they share *neurolemmal* (Schwann cell) and endoneurial sheaths.

Myelin formation

The **Schwann cell** is the representative neuroglial cell of the peripheral nervous system (PNS). It forms chains of neurolemmal cells along the nerves. Modified Schwann cells form **satellite cells** in posterior root ganglia and in autonomic ganglia, and **teloglia** at encapsulated sensory nerve endings (Ch. 11).

If an axon is to be myelinated, it receives the simultaneous attention of a sequence of Schwann cells along its length. Each one encloses the axon completely, creating a 'mesentery' of plasma membrane, the **mesaxon** (Figure 9.4). The mesaxon is displaced progressively, being rotated around the axon. Successive layers of plasma membrane come into apposition to form the **major** and **minor dense lines** (Figure 9.4).

Paranodal pockets of cytoplasm persist at the ends of the myelin segments, on each side of the nodes of Ranvier. (Louis Ranvier identified them 80 years before CNS nodes were demonstrated with the electron microscope in the early 1960s.) The paranodal pockets may be responsible

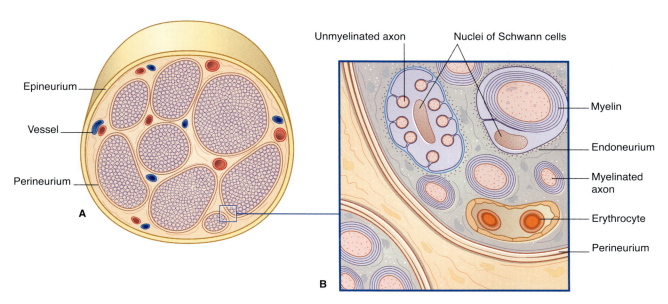

Figure 9.3 Transverse section of a nerve trunk. **(A)** Light microscopy. **(B)** Electron microscopy.

for maintaining the dense population (about 10^5) of Na$^+$ channels in the nodal plasma membrane.

Myelin expedites conduction

Along unmyelinated fibers, impulse conduction is *continuous* (uninterrupted). Its maximum speed is 15 m/s. Along myelinated fibers, excitable membrane is confined to the nodes of Ranvier, because myelin is an electrical insulator. Impulse conduction is called *saltatory* ('jumping'), because it leaps from node to node. Speed of conduction is much greater along myelinated fibers, with a maximum of 120 m/s. The number of impulses that can be conducted by myelinated fibers is also much greater than by unmyelinated ones.

The larger the myelinated fiber, the more rapid the conduction, because larger fibers have longer internodal segments and the nerve impulses take longer 'strides' between nodes. A 'rule of six' can be used to express the ratio between size and speed: a fiber of 10 μm external diameter (i.e. including myelin) will conduct at 60 m/s, one of 15 μm at 90 m/s, and so on.

In physiologic recordings, peripheral nerve fibers are classified in accordance with conduction velocities and other criteria. Motor fibers are classified into groups A, B, and C in descending order. Sensory fibers are classified into types I–IV. In practice, there is some interchange of usage: for example, unmyelinated sensory fibers are usually called C fibers rather than type IV.

Details of diameters and sources are given in Tables 9.1 and 9.2. Examples are in Figure 9.5.

The electron micrograph in Figure 9.6 illustrates a myelinated peripheral nerve fiber with attendant Schwann cell, that in Figure 9.7 a group of unmyelinated fibers bedded in the cytoplasm of a Schwann cell, and Figure 9.8 a nodal region within the CNS.

Central nervous system–peripheral nervous system transitional region

Close to the brainstem and spinal cord, peripheral nerves enter the *CNS–PNS transitional region* (Figure 9.9). Astrocyte processes reach out of the CNS into the endoneurial compartments of peripheral nerve rootlets and interdigitate with the Schwann cells. In unmyelinated fibers, the astrocytes burrow into the space between axons and Schwann cells. In myelinated fibers, nodes are bounded by Schwann cell myelin (showing some transitional features) on the peripheral side, and by oligodendrocytic myelin centrally.

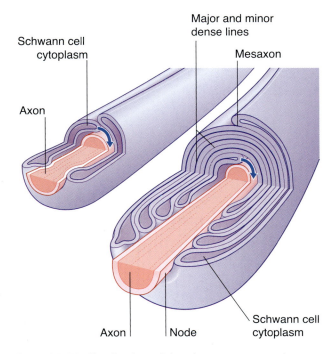

Figure 9.4 Myelination in peripheral nervous system. Arrows indicate movement of flange of Schwann cell cytoplasm.

Table 9.1 Classification of peripheral nerve fibers

Nerve type	Number	Letter	Diameter (mm)	Conduction velocity (m/s)
Myelinated				
Large	I	Aα	12–20	70–120
Medium	II	Aβ	6–12	35–70
Small	III	Aγ	3–6	10–40
Small	—	Aδ	2–5	5–35
Unmyelinated	IV	C	0.2–1.5	0.5–2

Table 9.2 Sources of peripheral nerve fiber types

Fiber type	Origin
Sensory	
Ia	Muscle spindle annulospiral endings
Ib	Golgi tendon organs
II (Aβ)	Muscle spindle flower spray endings; touch or pressure receptors in skin and elsewhere
III (Aδ)	Follicular endings; fast pain and thermal receptors
IV (C)	Slow pain, itch, touch receptors
Motor	
Aα	Alpha motor neurons
Aγ	Gamma motor neurons

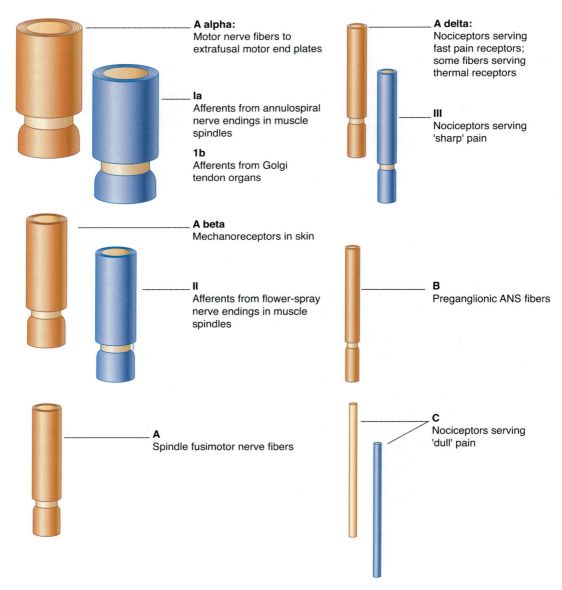

A alpha:
Motor nerve fibers to extrafusal motor end plates

Ia
Afferents from annulospiral nerve endings in muscle spindles

1b
Afferents from Golgi tendon organs

A beta
Mechanoreceptors in skin

II
Afferents from flower-spray nerve endings in muscle spindles

A
Spindle fusimotor nerve fibers

A delta:
Nociceptors serving fast pain receptors; some fibers serving thermal receptors

III
Nociceptors serving 'sharp' pain

B
Preganglionic ANS fibers

C
Nociceptors serving 'dull' pain

Figure 9.5 Size–function relationships in peripheral nerves. ANS, autonomic nervous system.

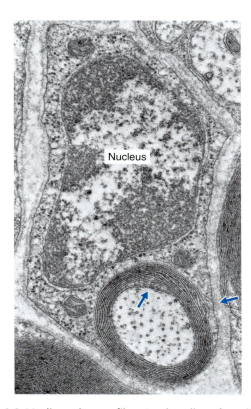

Figure 9.6 Myelinated nerve fiber. Ten lamellae of myelin form a continuous spiral from the outer to the inner mesaxon of the Schwann cell (arrows). A basal lamina surrounds the Schwann cell. (From Pannese 1994, with permission of Thieme.)

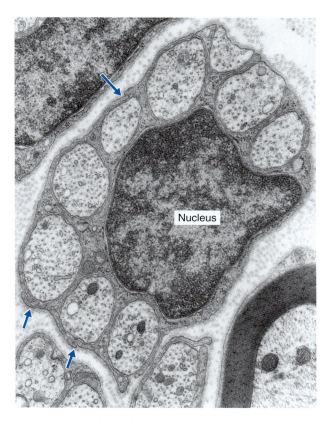

Figure 9.7 Unmyelinated nerve fibers. Nine unmyelinated fibers are lodged in the cytoplasm of this Schwann cell. Mesaxons (arrows indicate examples) are detectable where the axons are fully embedded. Two incompletely embedded axons at top right are covered by the basal lamina of the Schwann cell. (From Pannese 1994, with permission of Thieme.)

DEGENERATION AND REGENERATION

When nerves are cut or crushed, their axons degenerate distal to the lesion, because axons are pseudopodial outgrowths and depend on their parent cells for survival. In the PNS, regeneration is vigorous and it is often complete. In the CNS, on the other hand, it is neither vigorous nor complete.

Wallerian degeneration of peripheral nerves

The principal events in peripheral nerve degeneration are represented in Figure 9.10 and described in the caption. Following a crush or cut injury to a nerve, the axons and myelin sheaths distal to the cut break up to form 'ellipsoids' during the first 48 h—mainly because of Ca^{2+}-activated release of proteases by Schwann cells. The debris is cleared by monocytes that enter the damaged endoneurial sheaths from the blood and become macrophages. In addition to their phagocytic function, the macrophages are mitogenic to Schwann cells and participate with Schwann cells in provision of trophic (feeding) and tropic (guidance) factors for regenerating axons.

The end result of degeneration is a shrunken nerve skeleton with intact connective tissue and perineurial sheaths, and a core of intact, multiplying Schwann cells.

Regeneration of peripheral nerves

The principal events in regeneration of a peripheral nerve are summarized in Figure 9.10B. Following a clean cut, axons begin to sprout from the face of the proximal stump within a few hours, but in the more common crush or tear injuries seen clinically, the axons die back for 1 cm or more and sprouting may be delayed for a week. Successful regeneration requires that the axons make contact with Schwann cells of the distal stump. Failure to make contact leads to production of a *pseudoneuroma* consisting of whorls of regenerating axons trapped in scar tissue at the site of the initial injury. Following amputation of a limb, an *amputation pseudoneuroma* can be a source of severe pain.

Two reparative events are in simultaneous progress within hours of the injury. In the proximal stump, multiple branchlets begin to extend distally, their tips exhibiting swellings called **growth cones**; in the distal stump, Schwann cells send processes in the direction of the growth cones. The cones are surmounted by antenna-like **filopodia**, and these develop surface receptors that become anchored temporarily to complementary *cell surface adhesion molecules* in Schwann cell basement membranes. Filaments of actin within the filopodia become attached to

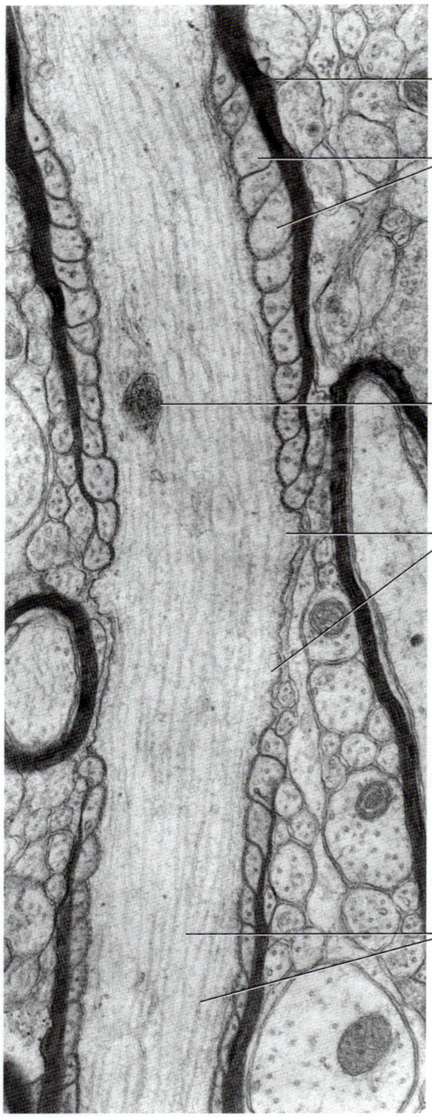
Myelin sheath
Paranodal pockets
Mitochondrion
Nodal region
Microtubules, neurofilaments, SER channels

Figure 9.8 Central nervous system nodal region. The myelin sheaths taper as they approach the nodal region, successive wrappings terminating in paranodal pockets of oligodendrocyte cytoplasm. The nodal region is about 10 mm long and lacks any basal lamina. The longitudinal streaks are created by microtubules, neurofilaments and elongated sacs of smooth endoplasmic reticulum (SER). (From Pannese 1994, with permission of Thieme.)

the surface receptors; from these points of anchorage, they are able to exert onward traction on the growth cones.

Growth cones are mitogenic to Schwann cells, which divide further before wrapping the larger axons with myelin lamellae.

Regeneration proceeds at about 5 mm/day in the larger nerve trunks, slowing down to 2 mm/day in the finer branches. Not surprisingly, the functional outlook is better after a crush injury (endoneurium preserved) than after complete severance. At the same time, filopodia of motor and sensory axons 'recognize' Schwann cell basement membranes previously occupied by axons of similar kind.

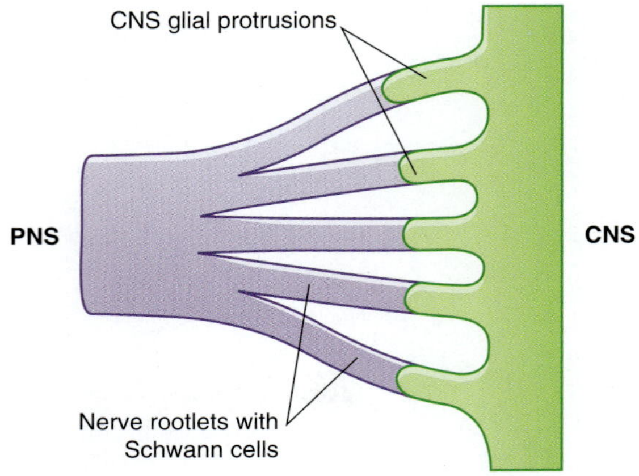
CNS glial protrusions
PNS
CNS
Nerve rootlets with Schwann cells

Figure 9.9 Central nervous system (CNS)–peripheral nervous system (PNS) transitional zone.

When nerve trunks have been completely severed, it is common practice to wait about 3 weeks before attempting repair. By that time, the connective tissue sheaths will have thickened a little and will be better able to hold suture material than would freshly injured, edematous sheaths. Moreover, the trimming of the nerves required before insertion of sutures creates a second axotomy, on the axons emerging from the proximal stump. In animal experiments, a second axotomy induces a more vigorous and sustained regenerative response.

Upstream effects of nerve section are as follow.

- Within a few days of axotomy, Nissl bodies can no longer be identified by cationic dyes in parent cells in the dorsal root ganglia and spinal gray matter (Figure 9.11). The phenomenon is known as *chromatolysis* ('loss of color'). Electron microscopy reveals that the granular endoplasmic reticulum is in fact increased in amount. But it is now dispersed throughout the perikaryon, with accumulations located deep to the plasma membrane.
- The nucleus becomes eccentric because of osmotic changes in the perikaryon.
- Parent motor neurons become isolated from synaptic contacts in the gray matter by the intrusion of neuroglial cells into all the synaptic clefts.
- In monkeys, it has been demonstrated that, following transection of sensory nerves, 30–40% of their posterior nerve root terminals undergo Wallerian degeneration. Because their terminals are in central gray matter, they do not regenerate; however, some of their synaptic sites are taken over by *collateral sprouts* given off by healthy neighbors. Overall, this observation may account for the usually incomplete recovery of sensory function in patients.

Degeneration in the central nervous system

Following injury to the white matter, distal degeneration occurs after the manner of peripheral nerves. However,

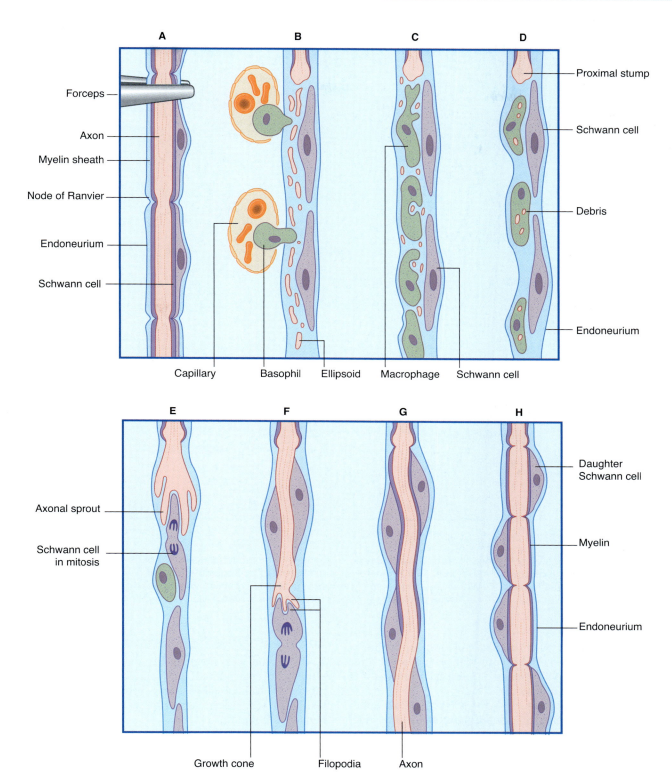

Figure 9.10 Events in degeneration of a single myelinated nerve fiber. **(A)** Intact fiber, showing its four components. The fiber is being pinched at its upper end. **(B)** The myelin and axon have broken up into 'ellipsoids' and droplets. Monocytes are entering the endoneurial tube from the blood. **(C)** The droplets are being engulfed by monocytes. **(D)** Clearance of debris is almost complete. Schwann cells and endoneurium remain intact. Events in regeneration. **(E)** An axonal sprout has entered the distal stump. The sprout is mitogenetic to each Schwann cell it encounters. **(F)** The growth cone is extending distally along the surface of Schwann cells. **(G)** Myelination is commencing along the proximal part of the regenerating axon. **(H)** When regeneration is complete, the fiber has a normal appearance, but the myelin segments are shorter than the originals.

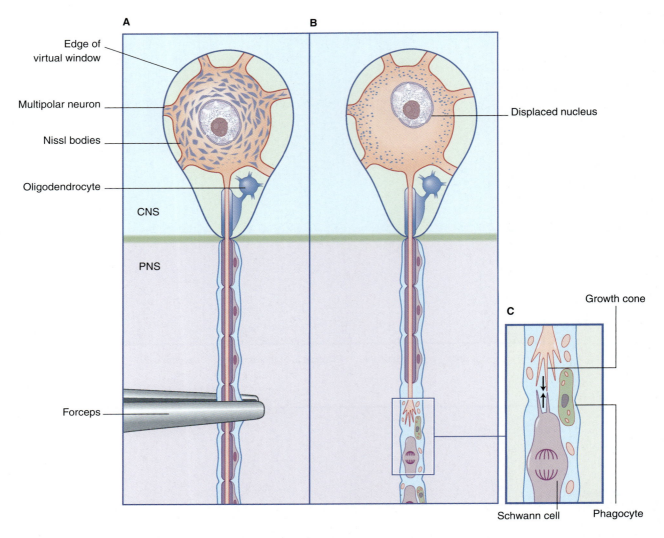

Figure 9.11 Schematic representation of some events following crush injury to a peripheral nerve. **(A)** Motor neuron seen through a virtual window in the central nervous system. **(B)** Chromatolysis is characterized by fragmentation and dispersal of Nissl bodies and displacement of the nucleus. **(C)** Within the crushed area, clearance of debris permits growth cone filopodia to establish contact with proximal extensions of a Schwann cell (arrows). CNS, central nervous system; PNS, peripheral nervous system.

clearance of debris by microglial cells and fresh monocytes is much slower. Debris can still be identified after 6 months in the CNS, whereas in peripheral nerves it is virtually cleared in 6 days.

Chromatolysis is unusual in the CNS. Instead, large-scale necrosis (death) of injured neurons is the rule. Neurons that survive may appear wasted, with permanent isolation from synaptic contacts.

Transneuronal atrophy

Central nervous system neurons have a trophic (sustaining) effect on one another. If the main input to a group of neurons is destroyed, the group is likely to waste away and die. This is known as *orthograde transneuronal atrophy*. It is comparable with the atrophy that occurs in skeletal muscle when its motor nerve is cut. In some situations, *retrograde transneuronal degeneration* takes place in neurons upstream to those destroyed by a lesion.

End result of central nervous system injury

If the lesion has been small, the neuronal debris is ultimately replaced by a glial scar composed of astrocyte processes. A large lesion may result in cystic cavities walled by scar tissue, containing cerebrospinal fluid and hemolyzed blood.

Regeneration in the central nervous system

Remarkable levels of functional recovery are often observed after CNS lesions. However, injured motor and sensory pathways do not reestablish their original connections. They regenerate for a few millimeters at most, and such synapses as develop are on other neurons close to the site of injury. Adult CNS neurons (in laboratory animals at least) do have regenerative capacity, as witnessed by their liberal sprouting and invasion of the endoneurial tubes of implanted peripheral nerves. The principal deter-

Innervation of muscles and joints | 10

STUDY GUIDELINES

1 Try to appreciate the vast significance of motor units with respect to movements of all kinds, and to appreciate the functional significance of their sizes and muscle chemistries.

2 Learn to sketch a motor end plate, indicating locations of transmitter, receptor, and hydrolytic enzyme.

3 Sketch an intrafusal muscle fiber, indicating locations of two motor and three sensory nerve endings.

4 Appreciate the functional significance of coactivation of alpha and gamma motor neurons during voluntary movements.

5 With the essentials of this chapter still in mind, consider a first read through the Electromyography section of Chapter 12.

In gross anatomy, the nerves to skeletal muscles are branches of mixed peripheral nerves. The branches enter the muscles about one-third of the way along their length, at *motor points* (Figure 10.1). Motor points have been identified for all major muscle groups for the purpose of

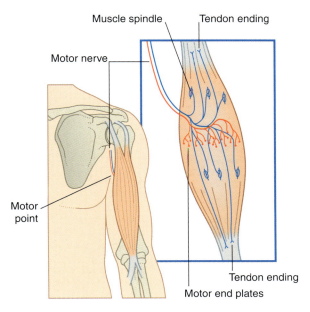

Figure 10.1 Pattern of innervation of skeletal muscle.

functional electrical stimulation by physical therapists, in order to increase muscle power.

Only 60% of the axons in the nerve to a given muscle are motor to the muscle fibers that make up the bulk of the muscle. The rest are sensory in nature, although the largest sensory receptors—the neuromuscular spindles—have a motor supply of their own.

MOTOR INNERVATION OF SKELETAL MUSCLE

The nerve of supply branches within the muscle belly, forming a plexus from which groups of axons emerge to supply the muscle fibers (Box 10.1 and Figure 10.1). The axons supply single motor end plates placed about half-way along the muscle fibers (Figure 10.2A).

A *motor unit* comprises a motor neuron in the spinal cord or brainstem together with the *squad* of muscle fibers it innervates. In large muscles (e.g. the flexors of the hip or knee), each motor unit contains 1200 muscle fibers or more. In small muscles (e.g. the intrinsic muscles of the hand), each unit contains 12 muscle fibers or less. Small units contribute to the finely graded contractions used for delicate manipulations.

There are three different types of skeletal muscle fiber.

1 *Slow-twitch, oxidative fibers* are small, rich in mitochondria and blood capillaries (hence, red). They exert small forces and are fatigue-resistant. They are deeply placed and suited to sustained postural activities, including standing.

2 *Fast, glycolytic (FG) fibers* are large, mitochondria-poor, and capillary-poor (hence, white). They produce brief, powerful contractions. They predominate in superficial muscles.

3 *Intermediate (fast, oxidative–glycolytic, FOG) fibers* have properties intermediate between the other two. Every muscle contains all three kinds of fiber, but a given motor unit contains only one kind. The fibers of each unit interdigitate with those of other units.

Motor end plates

At the **myoneural junction**, the axon divides into a handful of branchlets that groove the surface of the muscle fiber (Figure 10.2B). The underlying sarcolemma is thrown into **junctional folds**. The basement membrane of the muscle fiber traverses the synaptic cleft and lines the folds. The underlying sarcoplasm shows an accumulation of nuclei, mitochondria, and ribosomes known as a **sole plate**.

Each axonal branchlet forms an elongated terminal bouton containing thousands of synaptic vesicles loaded with acetylcholine (ACh). Synaptic transmission takes place at **active zones** facing the crests of the junctional folds (Figure 10.2C). Vesicular ACh is extruded at great speed by exocytosis into the synaptic cleft. The ACh diffuses through the basement membrane to bind with ACh receptors in the sarcolemma.

Box 10.1   Muscle fiber: internal details

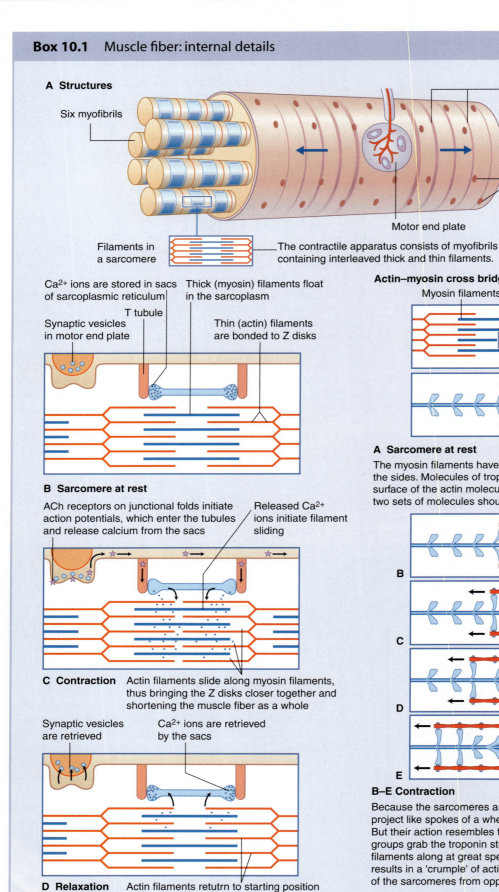

A Structures

Six myofibrils

Waves of depolarization

T tubule openings in plasma membrane

Motor end plate

Filaments in a sarcomere

The contractile apparatus consists of myofibrils containing interleaved thick and thin filaments.

Ca²⁺ ions are stored in sacs of sarcoplasmic reticulum

Synaptic vesicles in motor end plate

T tubule

Thick (myosin) filaments float in the sarcoplasm

Thin (actin) filaments are bonded to Z disks

B Sarcomere at rest

ACh receptors on junctional folds initiate action potentials, which enter the tubules and release calcium from the sacs

Released Ca²⁺ ions initiate filament sliding

C Contraction   Actin filaments slide along myosin filaments, thus bringing the Z disks closer together and shortening the muscle fiber as a whole

Synaptic vesicles are retrieved

Ca²⁺ ions are retrieved by the sacs

D Relaxation   Actin filaments retutrn to starting position

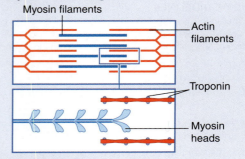

Actin–myosin cross bridges

Myosin filaments

Actin filaments

Troponin

Myosin heads

A Sarcomere at rest

The myosin filaments have globular heads projecting from the sides. Molecules of troponin are studded along the surface of the actin molecules. In the resting state, these two sets of molecules should not be in contact.

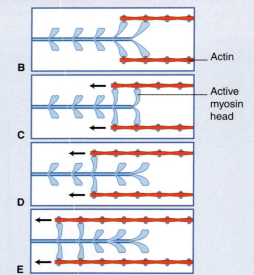

B

Actin

C

Active myosin head

D

E

B–E Contraction

Because the sarcomeres are cylindric, the myosin heads project like spokes of a wheel rather than oars of a boat. But their action resembles that of pairs of oars: successive groups grab the troponin studs and flick the actin filaments along at great speed. Maximal muscle contraction results in a 'crumple' of actin filaments arriving in the middle of the sarcomeres from opposite halves.

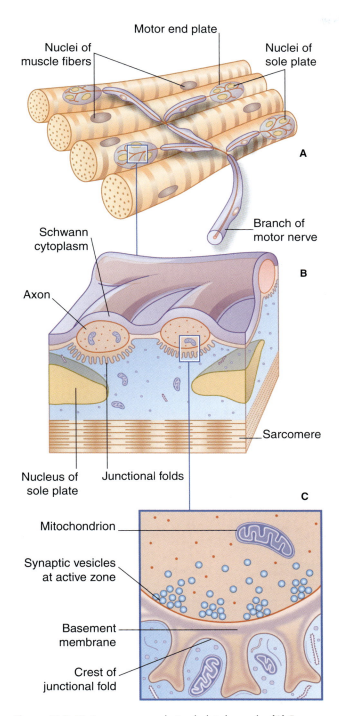

Figure 10.2 Motor nerve supply to skeletal muscle. **(A)** Four motor end plates supplied from a single axon. **(B)** Enlargement from (A). **(C)** Enlargement from (B), showing active zones.

Activation of the receptors leads to depolarization of the sarcolemma. The depolarization is led into the interior of the muscle fiber by **T tubules**. The sarcoplasmic reticulum liberates Ca^{2+} ions that initiate contraction of the sarcomeres.

Acetylcholinesterase enzyme is concentrated in the basement membrane, and about 30% of released ACh is hydrolyzed without reaching the postsynaptic membrane. Following hydrolysis, the choline moiety is returned to the axoplasm.

Also in terminal boutons are some dense-cored vesicles containing one or more peptides (Figure 12.2C). Best known is *calcitonin gene-related peptide*, a potent vasodilator.

Details of the muscle fiber contraction process are in Box 10.2.

Motor units in the elderly

Progressive wasting of muscles seen in the elderly is mainly due to loss of motor neurons from the spinal cord and brainstem, often due in part to low-grade peripheral neuropathy arising from vascular disease and/or nutritional deficiency. Electromyographic records taken from contracting muscles show *giant motor unit potentials* during the seventh and eighth decades. The extra-large potentials result from takeover of vacated motor end plates of lost motor neurons by collateral sprouts from the axons of adjacent healthy motor units. Details of electromyography, and of clinical neuromuscular disorders, are in Chapter 12.

SENSORY INNERVATION OF SKELETAL MUSCLE

Neuromuscular spindles

Muscle spindles are up to 1 cm in length and vary in number from a dozen to several hundred in different muscles. They are abundant (a) in the antigravity muscles along the vertebral column, femur, and tibia; (b) in the muscles of the neck; and (c) in the intrinsic muscles of the hand. All these muscles are rich in slow, oxidative muscle fibers. Spindles are scarce where FG or FOG fibers predominate.

Muscle spindles contain up to a dozen **intrafusal muscle fibers** (Figure 10.3). (Ordinary muscle fibers are extrafusal in this context.) The larger intrafusal fibers emerge from the **poles** (ends) of the spindles and are anchored to connective tissue (perimysium). Smaller ones are anchored to the collagenous spindle capsule. At the spindle **equator** (middle), the sarcomeres are replaced almost entirely by nuclei, in the form of 'bags' (in wide fibers) or 'chains' (in slender fibers).

Innervation

Muscle spindles have both a motor and a sensory nerve supply. The motor fibers, called *fusimotor*, are in the Aγ size range, in contrast to the Aα fibers supplying extrafusal muscle. The fusimotor axons divide to supply the striated segments at both ends of the intrafusal muscles (Figure 10.3). A single *primary* sensory fiber of type Ia caliber applies *annulospiral* wrappings around the bag or chain segments of the intrafusal muscle fibers. *Secondary*, 'flower spray' sensory endings on one or both sides of the primary are supplied by type II fibers.

Activation

Muscle spindles are *stretch receptors*. Ion channels in the surface membrane of the sensory terminals are opened by

Box 10.2   Muscle fiber contraction

This flow diagram shows the sequential events taking place during contraction of a single striated muscle fiber.

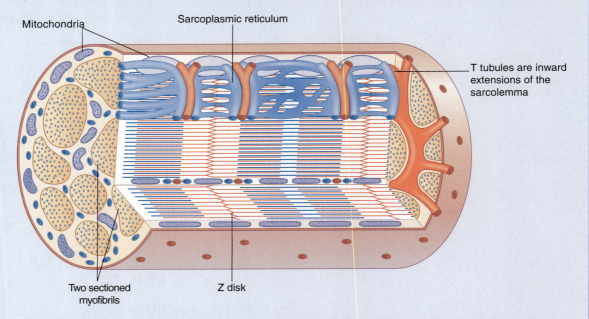

Mitochondria

Sarcoplasmic reticulum

T tubules are inward extensions of the sarcolemma

Two sectioned myofibrils

Z disk

Muscle fiber diagram revealing internal details. T tubules join up to form rings, completing 'triads' of rings around the sarcomeres along with neighboring pairs of sarcoplasmic sacs, having the function of releasing calcium ions from the sacs when the tubules are depolarized.

Muscle fiber actin comprises a twisted pair of polymerized actin monomers, a double strand of tropomyosin, and a troponin molecular complex at regular intervals.

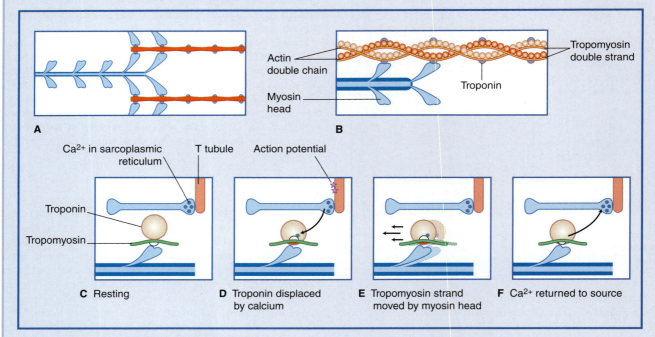

Actin double chain

Tropomyosin double strand

Myosin head

Troponin

A

B

Ca^{2+} in sarcoplasmic reticulum

T tubule

Action potential

Troponin

Tropomyosin

C Resting

D Troponin displaced by calcium

E Tropomyosin strand moved by myosin head

F Ca^{2+} returned to source

Excitation–contraction coupling:
Calcium liberated by tubule depolarization causes displacement of troponin, with exposure of actin binding sites to which myosin heads attach and pull the thin fiber toward the center of the sarcomere. The required energy is provided by ATPase enzyme contained in the myosin heads. Calcium is returned to source and the muscle actively relaxes by using ATPase to detach the myosin heads from the binding sites.

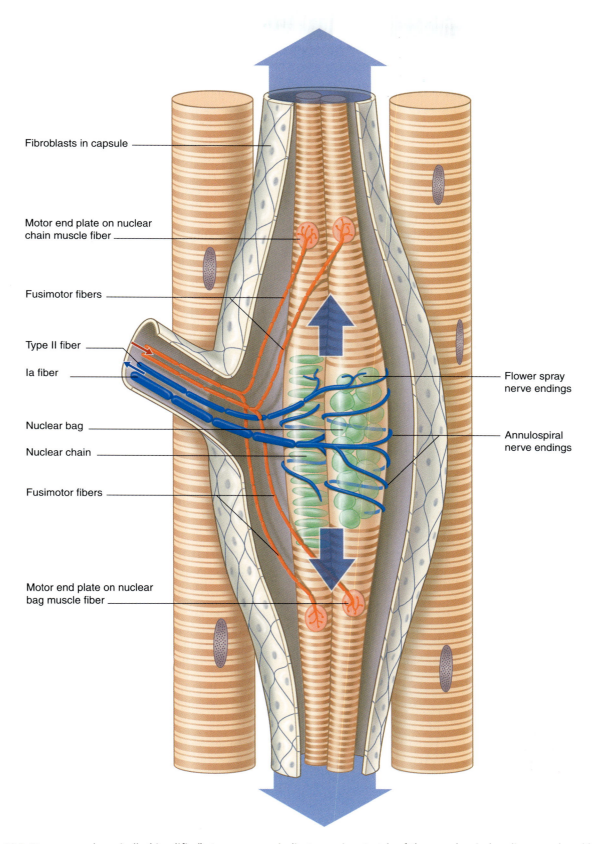

Fibroblasts in capsule

Motor end plate on nuclear chain muscle fiber

Fusimotor fibers

Type II fiber

Ia fiber

Nuclear bag

Nuclear chain

Fusimotor fibers

Motor end plate on nuclear bag muscle fiber

Flower spray nerve endings

Annulospiral nerve endings

Figure 10.3 Neuromuscular spindle (simplified). *Large arrows* indicate passive stretch of the annulospiral endings produced by lengthening of the relaxed muscle as a whole. *Medium arrows* indicate active stretch of annulospiral endings produced by activity of fusimotor nerve fibers. Active stretch more than compensates for the unloading effect of simultaneous extrafusal muscle fiber contraction. *Small arrows* indicate directions of impulse conduction to and from the spindle when the parent muscle is in use.

stretch, creating positive electronic waves that summate close to the final heminode of the parent sensory fiber. Summation produces a *receptor potential* that will fire off nerve impulses when it reaches threshold.

Muscle spindles may be stretched either *passively* or *actively*.

Passive stretch

Passive stretch of muscle spindles occurs when an entire muscle belly is passively lengthened. For example, in eliciting a tendon reflex such as the *knee jerk*, the spindles in the belly of the quadriceps muscle are passively stretched when the tendon is struck. The type Ia and type II fibers discharge to the spinal cord, where they synapse on the dendrites of α motor neurons (Figure 10.4). (α Motor neurons are so called because they give rise to axons of Aα diameter.) The response to the *positive feedback* from spindles is a twitch of contraction in the extrafusal muscle fibers of quadriceps. The spindles, because they lie in parallel with the extrafusal muscle, are passively shortened; they are described as being *unloaded*.

Tendon reflexes are monosynaptic reflexes. They have a *latency* (stimulus–response interval) of about 15–25 ms.

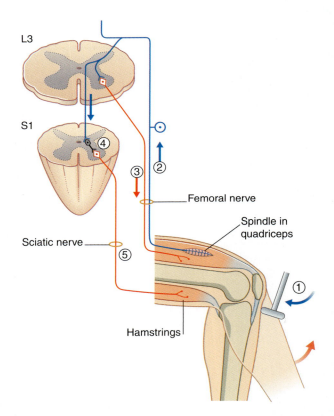

Figure 10.4 Patellar reflex, including reciprocal inhibition. Arrows indicate nerve impulses. **(1)** A tap to the patellar ligament stretches the spindles in quadriceps femoris. **(2)** Spindles discharge excitatory impulses to the spinal cord. **(3)** Alpha motor neurons respond by eliciting a twitch in quadriceps, with extension of the knee. **(4 and 5)** Ia inhibitory internuncials respond by suppressing any activity in the hamstrings.

In addition to exciting homonymous motor neurons (i.e. motor neurons supplying the same muscles), the spindle afferents *inhibit* the α motor neurons supplying the antagonist muscles, through the medium of inhibitory internuncial (interposed) neurons (Figure 10.4). This effect is called *reciprocal inhibition*. The inhibitory neurons involved are called *Ia internuncials*.

Information coding

Spindle primary afferents are most active *during* the stretching process. The more rapid the stretch, the more impulses they fire off. They therefore encode the *rate* of muscle stretch.

Spindle secondary afferents are more active than the primaries when a given position is held. The greater the degree of *maintained* stretch, the more impulses they fire off. They therefore encode the *degree* of muscle stretch.

Active stretch

Active stretch is produced by the fusimotor neurons, which elicit contraction of the striated segments of the intrafusal muscle fibers. Because the connective tissue attachments are relatively fixed, the intrafusal fibers stretch the spindle equators by pulling them in the direction of the spindle poles (Figure 10.5). (This could be called a 'Christmas cracker' effect.)

During all voluntary movements, α and α motor neurons are coactivated by the corticospinal (pyramidal) tract. As a result, the spindles are *not* unloaded by extrafusal muscle contraction. Through ascending connections, the spindle afferents on both sides of the relevant joints are able to keep the brain informed about contractions and relaxations during any given movement.

Tendon endings

Golgi tendon organs are found at muscle–tendon junctions (Figure 10.6). A single Ib caliber nerve fiber forms elaborate sprays that intertwine with tendon fiber bundles enclosed within a connective tissue capsule.

A dozen or more muscle fibers insert into the intra-capsular tendon fibers, which are *in series* with the muscle fibers. The bulbous nerve endings are activated by the tension that develops during muscle contraction. Because the rate of impulse discharge along the parent fiber is related to the applied tension, tendon endings signal the *force* of muscle contraction.

The Ib afferents exert *negative feedback* on to the homonymous motor neurons, in contrast to the positive feedback exerted by muscle spindle afferents. The effect is called *autogenetic inhibition*, and the reflex arc is disynaptic because of the interpolation of an inhibitory neuron (Figure 10.7). If need be, there follows *reciprocal excitation* of motor neurons supplying antagonist muscles.

An important function of tendon organ afferents is to dampen (restrict) the inherent tendency of moving limb segments to oscillate (sway to and fro). Dampening intro-duces an element known to physiologists as *joint stiffness*. Paradoxically, when 1b afferents are allowed too much freedom, as in *Parkinson's disease* (Ch. 28), they reinforce

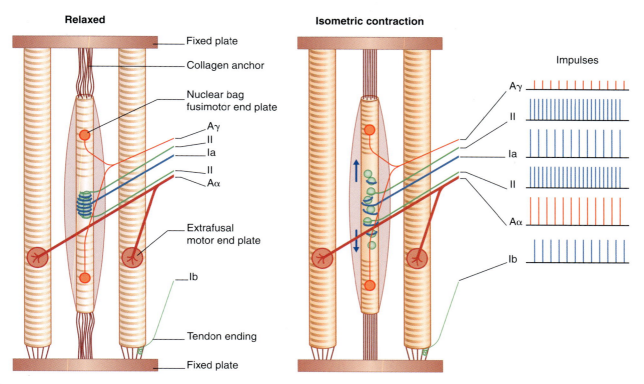

Figure 10.5 Active stretch of a muscle spindle under isometric conditions. The term *isometric* means 'same length'. Certainly, the extrafusal muscle fibers would remain isometric when pulling with both ends rigidly fixed. The muscle spindle also remains isometric, being anchored to the plate indirectly via connective tissue. But the striated elements of each intrafusal fiber *do not* remain isometric: they shorten because the spindle equator is elastic and yields to stress. The primary and secondary spindle afferents, applied to the equator, respond to the 'active' stretch produced by fusimotor activity by discharging afferent impulses to the CNS, with consequent reinforcement of extrafusal contraction via the gamma loop.

the inherent tendency to oscillate, and contribute to the characteristic *resting tremor* that is most obvious in the forearm (pronation–supination) and in the fingers ('pill-rolling' of the thumb pad by adjacent fingers).

Free nerve endings

Muscles are rich in freely ending nerve fibers, distributed to the intramuscular connective tissue and investing fascial envelopes. They are responsible for pain sensation caused by direct injury or by accumulation of metabolites including lactic acid.

INNERVATION OF JOINTS

Freely ending unmyelinated nerve fibers are abundant in joint ligaments and capsules, and in the outer parts of intraarticular menisci. They mediate pain when a joint is strained, and they operate an excitatory reflex to protect the capsule. For example, the anterior wrist capsule is supplied by the median and ulnar nerves; if it is suddenly stretched by forced extension, motor fibers in these nerves are reflexly activated and cause wrist flexion.

Animal experiments have shown that, when a joint is inflamed, more freely ending nerve fibers are excited than is the case when a healthy joint capsule is stretched. It seems that some nerve endings are *only* stimulated by inflammation.

Encapsulated nerve endings in and around joint capsules include Ruffini endings that signal tension, lamellated endings responsive to pressure, and Pacinian corpuscles responsive to vibration (see Ch. 11).

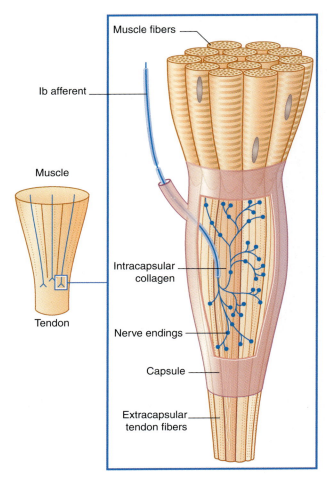

Figure 10.6 Golgi tendon organ.

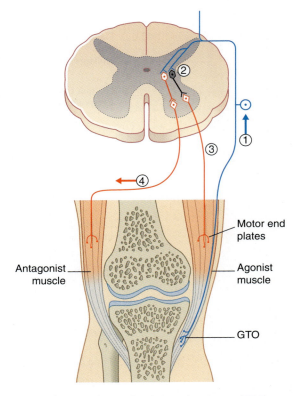

Figure 10.7 Reflex effects of Golgi tendon organ (GTO) stimulation. **(1)** Agonist contraction excites GTO afferent, which **(2)** excites inhibitory internuncial synapsing on **(4)** homonymous motor neuron, and **(3)** excites excitatory internuncial synapsing on **(4)** motor neuron supplying antagonist.

Core Information

Muscle
A motor unit comprises a motor neuron and the group of muscle fibers it supplies. Each unit contains only one histochemical type of muscle fiber. At the myoneural synapse, the terminal bouton (containing vesicular acetylcholine [ACh] quanta) is separated from sarcolemmal junctional folds (containing ACh receptors) by a basement membrane containing acetylcholinesterase.

Muscle spindles contain intrafusal muscle fibers activated simultaneously at both ends by γ fusimotor neurons. Sensory fibers of type Ia diameter provide primary annulospiral endings at the equator, and fibers of type II diameter provide secondary endings nearby; both kinds are stretch receptors. Stretch may be passive, for example by a tendon reflex, or active during fusimotor activity. Homonymous motor neurons are monosynaptically excited; antagonists are reciprocally inhibited via Ia internuncials. Spindle primaries signal the rate of muscle stretch; secondaries signal the degree. During voluntary movements, α and γ motor neurons are coactivated.

Golgi tendon organs signal the force of muscle contraction. They comprise encapsulated tendon tissue innervated by type 1b diameter afferents exerting disynaptic inhibition on homonymous motor neurons and reciprocal excitation of antagonists.

Free intramuscular nerve endings subserve pain sensation.

Joints
Free nerve endings abound in ligaments and capsules, and in the outer part of menisci. They mediate pain and operate an articular protective reflex. Encapsulated endings signal joint movement.

REFERENCES

Arvidsson U, Piehl F, Johnson H, et al. The peptidergic motoneurone. NeuroReport 1993; 4:1249–1256.

Burke D, Gandevia SC. Peripheral motor system. In: Paxinos G, ed. The human nervous system. San Diego: Academic Press; 1990:125–1412.

McCloskey DI, Gandevia SC. Aspects of proprioception. In: Gandevia SC, Burke D, Anthony M, eds. Science and practice in clinical neurology. Cambridge: Cambridge University Press; 1993:3–19.

McCloskey DI. Human proprioceptive sensation. J Clin Neurosci 1994; 1:173–177.

Salpeter MM. Vertebrate neuromuscular junctions: general morphology, molecular organization, and functional consequences. In: Salpeter MM, ed. The vertebrate neuromuscular junction. New York: Alan R. Liss; 1997:55–116.

STUDY GUIDELINES
1 Learn to define sensory unit, sensory overlap, receptive field, receptor adaptation.
2 Be able to state locations and properties of three kinds of encapsulated receptor.
3 Be able to sketch a hair follicle with its nerve palisade and rings.
4 Try to appreciate the huge numbers of mechanoreceptors that are deployed in the hands. Name two kinds of receptors used to discriminate textures, for example to read Braille.
5 Consider a quick preview of the Clinical Panel on peripheral neuropathies in Chapter 12.

From the cutaneous branches of the spinal nerves, innumerable fine twigs enter a **dermal nerve plexus** located in the base of the dermis. Within the plexus, individual nerve fibers divide and overlap extensively with others before terminating at higher levels of the skin. Because of overlap, the area of anesthesia resulting from injury to a cutaneous nerve (e.g. superficial radial, saphenous) is smaller than its anatomic territory.

SENSORY UNITS

A given stem fiber forms the same kind of nerve ending at all of its terminals. In physiologic recordings, the stem fiber and its family of endings constitute a *sensory unit*. Together with its parent unipolar nerve cell, the sensory unit is analogous to the motor unit described in Chapter 10.

The territory from which a sensory unit can be excited is its *receptive field*. There is an inverse relationship between the size of receptive fields and sensory acuity; for example, fields measure about 2 cm^2 on the arm, 1 cm^2 at the wrist, and 5 mm^2 on the finger pads.

Sensory units interdigitate so that different *modalities* of sensation can be perceived from a given patch of skin.

NERVE ENDINGS

Free nerve endings   (Figure 11.1A, B)

As they run toward the skin surface, many sensory fibers shed their perineural sheaths and then their myelin sheaths (if any) before branching further in a subepidermal network. The Schwann cell sheaths open to permit naked axons to terminate between collagen bundles (**dermal nerve endings**) or within the epidermis (**epidermal nerve endings**).

Functions

Some sensory units with free nerve endings are *thermo-receptors*. They supply either 'warm spots' or 'cold spots' scattered over the skin. Two kinds of *nociceptors* (pain-transducing) units with free endings are also found. One kind responds to severe mechanical deformation of the skin, for example pinching with a forceps. The parent fibers are finely myelinated (Aδ). The other kind comprises *polymodal nociceptors*; these are C-fiber units able to transduce mechanical deformation, intense heat (some also intense cold), and irritant chemicals.

C-fiber units are responsible for the axon reflex (Clinical Panel 11.1).

Follicular nerve endings   (Figure 11.1A, D)

Just below the level of the sebaceous glands, myelinated fibers apply a *palisade* of naked terminals along the outer root sheath epithelium of the hair follicles. Outside this is a *circumferential* set of terminals.

Each follicular unit supplies many follicles, and there is much territorial overlap. Follicular units are *rapidly adapting*: they fire when the hairs are being bent, but not when the bent position is held. Rapid adaptation accounts for our being largely unaware of our clothing except when donning or offing.

Merkel cell–neurite complexes
(Figure 11.1A, C)

Expanded nerve terminals are applied to **Merkel cells** (*tactile menisci*) in the basal epithelium of epidermal pegs and ridges. These *Merkel cell–neurite complexes* are *slowly adapting*. They discharge continuously in response to sustained pressure, for example while we hold a pen or wear spectacles, and they are markedly sensitive to the *edges* of objects held in the hand.

Encapsulated nerve endings

The *capsules* of the three nerve endings to be described comprise an outer coat of connective tissue, a middle coat of perineural epithelium, and an inner coat of modified Schwann cells (**teloglia**). All three are *mechanoreceptors*, transducing mechanical stimuli.

- **Meissner's corpuscles** are most numerous in the finger pads, where they lie beside the intermediate ridges of the epidermis (Figure 11.2A–C). In these ovoid receptors, several axons zigzag among stacks of teloglial lamellae. Meissner's corpuscles are rapidly adapting. Together with the slowly adapting Merkel

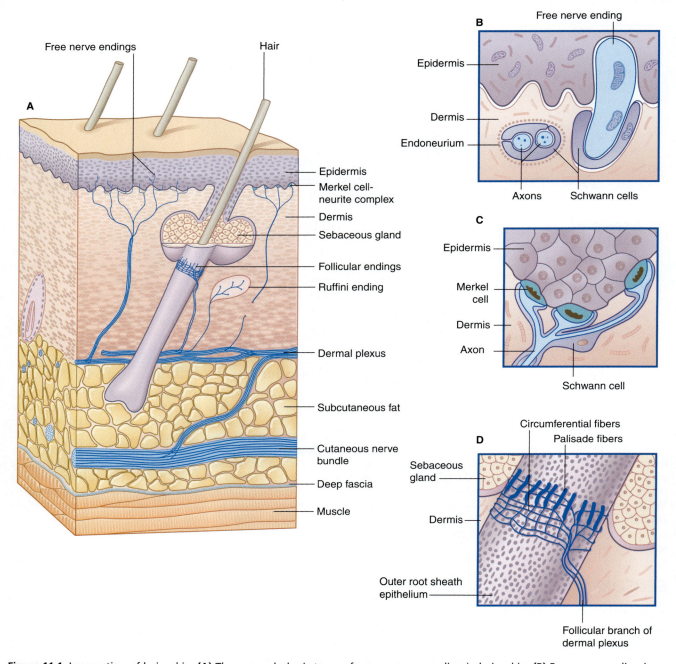

Figure 11.1 Innervation of hairy skin. **(A)** Three morphologic types of sensory nerve ending in hairy skin. **(B)** Free nerve ending in the basal layer of the epidermis. **(C)** Merkel cell–neurite complex. **(D)** Palisade and circumferential nerve endings on the surface of the outer root sheath of a hair follicle.

cell–neurite complexes, they provide the tools for delicate detective work on textured surfaces such as cloth or wood, or on embossed surfaces such as Braille text. Elevations as little as 5 μm in height can be detected!

- **Ruffini endings** are found in both hairy and glabrous skin (Figures 11.1A and 11.2D). They respond to *drag* (shearing stress) and are slowly adapting. Their structure resembles that of Golgi tendon organs,

having a collagenous core in which several axons branch liberally.

- **Pacinian corpuscles** (Figure 11.2B, E) are the size of rice grains. They number about 300 in the hand. They are subcutaneous, close to the underlying periosteum, and numerous along the sides of the fingers and in the palm. Inside a thin connective sheath are onion-like layers of perineural epithelium containing some blood capillaries. Innermost are several teloglial lamellae

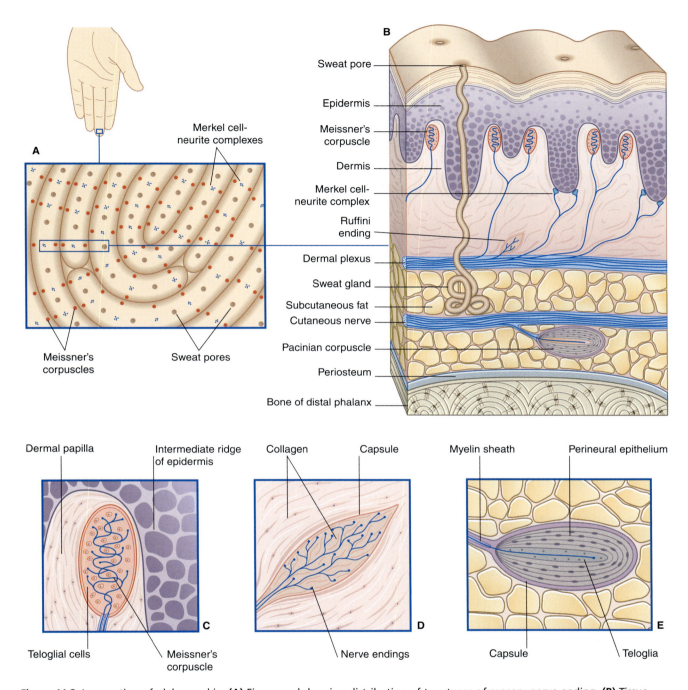

Figure 11.2 Innervation of glabrous skin. **(A)** Finger pad showing distribution of two types of sensory nerve ending. **(B)** Tissue block from (A) showing positions of four types of sensory nerve endings. **(C)** Meissner's corpuscle. **(D)** Nerve ending of Ruffini. **(E)** Pacinian corpuscle.

surrounding a single central axon that has shed its myelin sheath at point of entry. Pacinian corpuscles are rapidly adapting and are especially responsive to *vibration*—particularly to bone vibration. In the limbs, many corpuscles are embedded in the periosteum of the long bones.

Pacinian corpuscles discharge one or two impulses when compressed, and again when released. In the hands, they seem to function in group mode: when an object such as an orange is picked up, as many as 120 or more corpuscles are activated momentarily, with a momentary repetition when the object is released. For this reason, they have been called 'event detectors' during object manipulation.

The digital receptors are classified as follows by sensory physiologists.

- Merkel cell–neurite complexes = SA I
- Meissner's corpuscles = RA I
- Ruffini endings = SA II
- Pacinian corpuscles = RA II

Clinical Panel 11.1   Neurogenic inflammation: the axon reflex

When sensitive skin is stroked with a sharp object, a red line appears in seconds, owing to capillary dilatation in direct response to the injury. A few minutes later, a red *flare* spreads into the surrounding skin, owing to arteriolar dilatation, followed by a white *wheal* owing to exudation of plasma from the capillaries. These phenomena constitute the *triple response*. The flare and wheal responses are produced by *axon reflexes* in the local sensory cutaneous nerves. The sequence of events follows the numbers in Figure CP 11.1.1.

1   The noxious stimulus is transduced (converted to nerve impulses) by polymodal nociceptors.

2   As well as transmitting impulses to the central nervous system in the normal, orthodromic direction, the axons send impulses in an *antidromic* direction from points of bifurcation into the neighboring skin. The nociceptive endings respond to antidromic stimulation by releasing one or more peptide substances, notably substance P.

3   Substance P binds with receptors on the walls of arterioles, leading to arteriolar dilatation—the flare response.

4   Substance P also binds with receptors on the surface of mast cells, stimulating them to release *histamine*. The histamine increases capillary permeability and leads to local accumulation of tissue fluid—the wheal response.

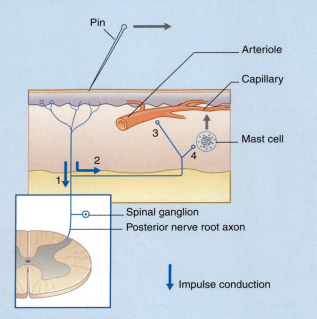

Figure CP 11.1.1 The axon reflex.

Clinical Panel 11.2   Leprosy

The leprosy bacillus enters the skin through minor abrasions. It travels proximally within the perineurium of the cutaneous nerves and kills off the Schwann cells. Loss of myelin segments ('segmental demyelination') blocks impulse conduction in the larger nerve fibers. Later, the inflammatory response to the bacillus compresses all the axons, leading to Wallerian degeneration of entire nerves and gross thickening of the connective tissue sheaths. Patches of anesthetic skin develop on the fingers, toes, nose, and ears. The protective function of skin sensation is lost, and the affected parts suffer injury and loss of tissue. Motor paralyses occur later on as a consequence of invasion of mixed nerve trunks proximal to the points of origin of their cutaneous branches.

When three-dimensional objects are being manipulated out of sight, significant contributions to perceptual evaluation are made by muscle afferents (especially from muscle spindles) and articular afferents from joint capsules. The cutaneous, muscular, and articular afferents relay information independently to the contralateral somatic sensory cortex. The three kinds of information serve the function of *tactile discrimination*. They are integrated (brought together at cellular level) in the posterior part of the contralateral parietal lobe, which is specialized for *spatial sense*, both tactile and visual. Spatial tactile sense is called *stereognosis*. In the clinic, stereognosis is tested by asking the patient to identify an object such as a key without looking at it.

Cutaneous sensory effects of peripheral neuropathies are described in Chapter 12. Clinical Panel 11.2 gives a short account of *leprosy*.

Core Information

Cutaneous nerves branch to form a dermal nerve plexus where individual afferent fibers branch and overlap. Each stem fiber and its terminal receptors constitute a sensory unit. The territory of a stem fiber is its receptive field.

Sensory units with free nerve endings include thermoreceptors and both mechanical and thermal nociceptors. Follicular units are rapidly adapting touch receptors, active only when hairs are in motion. Merkel cell–neurite complexes are slowly adapting edge detectors.

The encapsulated endings are mechanoreceptors. Meissner's corpuscles lie beside intermediate ridges in glabrous skin and are rapidly adapting. Ruffini endings lie near hair follicles and fingernails; they are slowly adapting drag receptors. Pacinian corpuscles are subcutaneous, rapidly adapting event detectors and vibration receptors.

Coded information from skin, muscles, and joints is integrated at the level of the posterior parietal lobe of the brain, yielding the faculties of tactile discrimination and stereognosis.

REFERENCES

Cunningham FO, FitzGerald MJT. Encapsulated nerve endings in hairy skin. J Anat 1972; 112:113–117.

Foreman JC. Peptides and neurogenic inflammation. Br Med Bull 1987; 43:386–400.

Iggo A. Sensory receptors in the skin of mammals and their sensory functions. Rev Neurol 1985; 141:611–615.

James LA. Peripheral mechanisms for touch and proprioception. Can J Physiol Pharmacol 1994; 72:484–487.

Johansson RS. How is grasping modified by somatosensory input? In: Humphrey DR, Freund H-J, eds. Motor control: concepts and issues. Chichester: John Wiley; 1991:331–355.

Mitsumoto H, Wilburn AJ. Causes and diagnosis of sensory neuropathies: a review. J Clin Neurophysiol 1994; 11:553–567.

Stark B, Carlstedt T, Hallin RG. Distribution of Pacinian corpuscles in the hand. J Hand Surg (Br) 1998; 23:370–372.

Torebjork AB, Vallbo AB, Ochoa JL. Intraneural microstimulation in man: its relation to specificity of tactile sensations. Brain 1987; 112:15011–15211.

STUDY GUIDELINES

1 A review of the electrical events described in Chapter 7 may be worthwhile. This chapter applies some of the basic principles described there.

2 Remember that when nerves are stimulated electrically through the skin, whether for the study of motor or sensory conduction, the waves of depolarization travel in both directions.

3 Nerve conduction studies (NCS) help to elucidate the nature and extent of neuropathies involving motor and/or sensory nerves. This is a quite complex area of investigation given the very large number of possible causes of peripheral neuropathies.

4 Electromyography (EMG), whereby a recording electrode is passed into the interior of selected muscles, is essential for detection of spontaneously generated abnormal waveforms. It is used to help define the etiology of muscle weakness and to monitor progress under therapy.

5 The second of the four Clinical Panels may require consultation of the peripheral nerve section of a gross anatomy textbook.

The primary concerns of clinical neurophysiology laboratories are twofold: assessment of the functional state of the peripheral nervous system (PNS), and assessment of cerebral cortical function. PNS assessment entails the use of *nerve conduction studies* by stimulation of selected peripheral nerves while recording the waveforms of their response, and of *electromyography* by recording the waveforms generated by selected muscles during voluntary contraction. The combination of NCS and EMG is referred to as *electrodiagnostic examination*.

NERVE CONDUCTION STUDIES

Nerve conduction studies are routinely employed as part of the clinical examination of suspected disorders of the PNS. Through stimulation of nerves allied to recording of muscle fiber depolarizations, it is possible to determine whether the disorder involves the nerve, neuromuscular junction, or muscle, also whether it is a focal or diffuse process involving sensory and/or motor axons, and whether it is primarily affecting myelin or axons.

Nerve conduction in the upper limb

The role model for detection of distributed (as distinct from focal) disorders within the peripheral neuromuscular system in general is the median nerve. The median, a mixed motor and sensory nerve, has three key advantages for electrophysiologic studies of a general nature.

1 It is readily accessible for stimulation and/or recording at elbow and wrist.

2 For motor NCS, the abductor pollicis brevis, supplied by the median, is readily available for either surface or needle electromyography.

3 For sensory NCS, the skin of the index finger is ideal for recording action potentials traveling antidromically following median stimulation at the elbow or wrist. (As noted in Ch. 11, *antidromic* means 'running against' the normal (orthodromic) direction of impulse conduction.)

Motor nerve conduction

Stimulation

A typical stimulating electrode is one with an anode and a cathode in the form of two blunt prongs that are applied to the skin surface overlying the nerve. In Figure 12.1, it has been placed over the median nerve at the wrist (just lateral to the cord-like palmaris longus tendon). The cathode is placed nearer to the recording site than

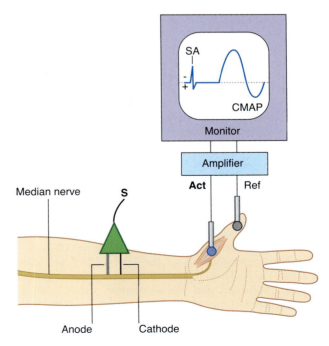

Figure 12.1 Basic setup for recording compound motor action potentials (CMAPs). The stimulating electrode (S) has been placed over the median nerve. The active electrode (Act) has been placed over the abductor pollicis brevis muscle and the reference (Ref) electrode has been placed distally. SA, stimulus artifact.

the anode in order to prevent any conduction block by the anode. When sufficient current is passed from cathode to anode, transmembrane ionic movements initiate impulse propagation in both directions along the nerve. Large myelinated nerve fibers lying nearest the cathode are the first to become depolarized; these include the Aα diameter axons of anterior horn motor neurons. A pulse of 20–40 mA with a duration of 0.1 ms is usually sufficient to activate all motor units in abductor pollicis brevis.

Recording

An *active* surface recorder, in the form of a disk in this situation, is placed over the midregion of the muscle where the motor end plates are concentrated. A second, *reference* electrode is placed over a neutral site a short distance away. The amplifier used to magnify evoked motor responses is designed to record the potential differences between the two sites. The setup is arranged so that, if the active electrode records a more negative response, this will take the form of an upward deflection on the monitor.

At low level of stimulation, the only on-screen change in the tracing will be a small *stimulus artifact* on an otherwise flat tracing. As the current increases, small compound motor action potentials (CMAPs) appear. These are produced by activation of large myelinated axons close to the stimulator; the depolarization wave traveling along each will in turn depolarize all the muscle fibers in the territory of that axon. In the case of the intrinsic muscles of the hand, including abductor pollicis brevis, each motor unit has an innervation ratio of two or three hundred muscle fibers per motor neuron. In large muscles not specialized for fine movements (e.g. deltoid, gastrocnemius), the minimum deflection on the monitor will be several times larger for two reasons: their motor innervation ratio is 1:1000 or more, and their larger muscle fibers generate action potentials of greater amplitude.

It should be emphasized that the on-screen waveform is *not* produced by the contraction process itself, but by the extracellular potentials generated by depolarization of the muscle membranes and filtered through the tissues and skin. However, while this distinction needs to be remembered, most disorders of muscle will also affect the surface membrane depolarization and hence lead to abnormalities of the waveform morphology.

Increasing the applied voltage activates additional motor units until all are activated by each pulse. The required stimulus is called *maximal*. For good measure, the final stimulus is often *supramaximal* at 5–10% above maximal. The final waveform observed constitutes the *compound motor action potential*. It is produced by *summation* of the individual muscle fiber potentials (Figure 12.2).

Routine measurements of the final CMAP are shown in Figure 12.3. They include the *latency* (time interval) between stimulus and depolarization onset, and the *amplitude* and *duration* of the negative phase of the waveform. (The final, positive phase is produced by inward ion movement during collective repolarization of the muscle fibers.)

Motor nerve conduction velocity

The setup required to determine MNCV for the median nerve is straightforward, as shown in Figure 12.4. Here,

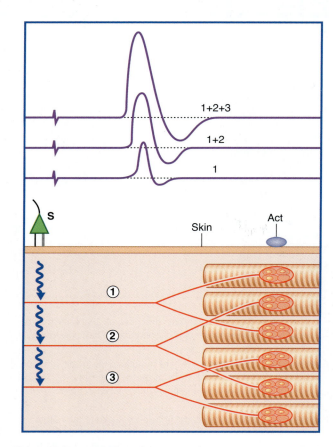

Figure 12.2 Summation of compound motor action potentials. Motor units are simply represented by interdigitating pairs of muscle fibers. Moderate **(1)**, medium **(2)**, and maximal **(3)** stimulation yield progressively larger waveforms on screen, despite being physiologically separate phenomena.

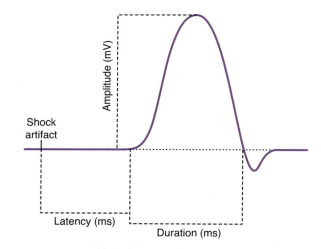

Figure 12.3 Routine compound motor action potential measurements.

the nerve has first been activated at the wrist (S1) to generate and store a 'wrist to muscle' velocity record. The stimulator has then been placed over the median nerve at the elbow (S2) to provide an 'elbow to muscle' record. Speed being the product of distance over time, the elbow-

Radicular nerve conduction

Nerve root pathology is known as *radiculopathy* (*L. radix*, 'root'). Radiculopathies are encountered:

- in the neck, where roots of spinal nerves C6 and C7 are especially prone to being pinched by osteophytes generated by cervical spondylosis, as explained in Clinical Panel 12.1;
- in the lower back, where the nerve roots of S1 are especially prone to compression by a prolapsed L5–S1 intervertebral disk, as explained in Clinical Panel 12.2; and
- as part of a generalized peripheral neuropathy.

The H response

Owing to their deep location, nerve conduction in spinal nerve roots can be assessed only indirectly, by activating sensorimotor reflex arcs at appropriate levels. The standard test, named after Hoffmann, who first described it, is known as the *H response* or *H wave* test. This is frequently used to assess overall conduction velocity in the S1 reflex arc—the same neurons that are evaluated clinically by the Achilles reflex (Figure 12.8). The tibial nerve is stimulated using the minimum current sufficient to elicit a muscle twitch. The objective here is to excite the largest myelinated afferent fibers, namely those serving annulospiral nerve endings in neuromuscular spindles, thereby eliciting a

---

Clinical Panel 12.1   Peripheral neuropathies

Peripheral neuropathies are amenable to several classifications, each with its own relevance.

- *Histologic* classification is touched on in the figure, where neuropathy originates in myelin sheaths of the top two nerve fibers and in the axon of the other three.
- *Anatomic* classification identifies nerve numbers and locations. Numbers include *mononeuropathy*, referring to a single spinal or cranial nerve (e.g. sciatica, trigeminal neuralgia, isolated peripheral nerve trauma); *mononeuropathy multiplex*, referring to more than one affected nerve trunk (e.g. right median and radial, left sural, right peroneal). Locational labels include *plexopathy*, referring to involvement of the cervical or the brachial or the lumbosacral plexus in an individual case), and *radiculopathy*, referring to nerve root pathology, most frequently caused by compression of a nerve root in the region of an intervertebral foramen. The terms *primary* and *secondary* are also anatomic. A *primary* neuropathy originates within nerve tissue (disease of myelin sheaths or of axons). A *secondary* neuropathy typically is caused by anterior horn cell disease leading to axonal degeneration followed by collapse of myelin sheaths.
- *Etiologic* classification identifies causative agencies. Headings include *toxins*, including lead and arsenic; *immune disorders*, including effects of viruses; *metabolic disorders*, including diabetes; *vitamin deficiencies* such as B_{12} in pernicious anemia or thiamine deficiency associated with alcoholism mainly of the B series; *genetic disorders* such as Charcot–Marie–Tooth disease.
- *Time course* classification may be condensed into *acute* and *subacute*, where the patient seeks help within days or weeks of onset, and *chronic*, where the patient may persevere for more than a year before seeking help.

One kind of acute polyneuropathy and two kinds of chronic polyneuropathy will now be described.

Guillain–Barré syndrome (GBS) is an acute, autoimmune, inflammatory neuropathy occurring in all countries and affecting individuals aged between 8 and 80. Possibly relevant antecedent events may include a mild viral infection or immunization, or a surgical procedure. GBS is sometimes referred to as *Landry's acute ascending paralysis*. Typical presentation is a predominantly motor peripheral failure involving both somatic and autonomic nerves, commencing in the feet and hands, and ascending to involve the muscles of the trunk, neck, and face.

Rarely, progress may be so rapid as to cause death within a few days from respiratory and/or circulatory collapse. Usually, the acute phase lasts 1–2 weeks, with motor weakness associated with diminution or loss of tendon reflexes. Electrodiagnostic examination may reveal reduced conduction velocity in motor nerves. Aching pain and tenderness occur in affected muscles, along with some cutaneous sensory loss. Reduced autonomic function may be demonstrated by fluctuating heart rate and blood pressure, and/or retention of urine requiring catherization for a few days.

Rapid recovery may be spontaneous in relatively mild cases, but many patients require either immune globulin injections or plasma exchange where this is available and safe. Where axons have degenerated in the acute phase, recovery may take more than a year and may be incomplete.

Chronic polyneuropathy originating in myelin sheaths
This type of neuropathy is associated with chronic vitamin deficiency, longstanding diabetes, or chronic hypothyroidism. The myelin sheaths of the peripheral nerves degenerate while the axons remain relatively intact. Because potassium channels are largely obliterated when the sheaths are initially laid down, and because saltatory conduction is lost along with the sheaths, sensory conduction is progressively impaired. As might be anticipated, the longest nerves are the most affected, yielding 'glove and stocking' paresthesia (Figure CP 12.1.1). (The term *paresthesia*

Clinical Panel 12.1 *Continued*

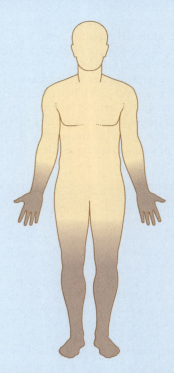

Figure CP 12.1.1 'Glove and stocking' paresthesia.

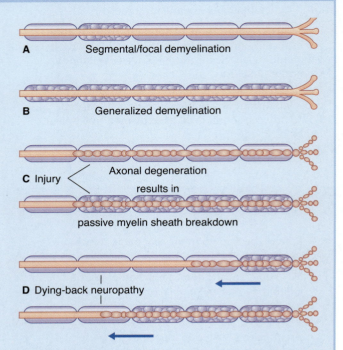

Figure CP 12.1.2 Histologic changes associated with some types of peripheral neuropathy. The figure does not distinguish between sensory and motor nerves. **(A)** In segmental or focal demyelination, impulse conduction is slowed or blocked, but in a nerve conduction study this will not be apparent if stimulator and recorder are both placed distal to the lesion site. **(B)** Generalized demyelination results in conduction slowing or block, although the axons may be preserved. **(C)** Damage to the axon causes the entire distal length of axon to break up into droplets, and the release of tension causes the myelin sheaths to follow suit (Wallerian degeneration). **(D)** In dying-back neuropathy, myelin breakdown is again secondary to axonal degeneration.

refers to sensations of numbness, pins and needles, and/or tingling.)

Initially, demyelination may be *segmental* or *focal*, as shown in Figure CP 12.1.2.

After several months, motor weakness and wasting may become evident in the muscles of the hands and feet. With further progression, *joint sense* and *vibration sense* may be lost there. (Joint sense refers to the ability to detect passive movement at a joint performed by a clinician; vibration sense refers to the ability to feel the buzz of a tuning fork applied to bone (see Ch. 16).

Chronic polyneuropathy originating in axons
The most prevalent form of chronic polyneuropathy is one that occurs in late middle age and is of unknown causation, hence the terms *idiopathic* ('private pathology') and *cryptogenic* ('hidden origin') applied to this condition. Biopsy studies have demonstrated that the initial pathology is *axonal*, in the form of a distal-

to-proximal *dying-back* process. In contrast to demyelinating neuropathies, the leading symptom here is *chronic pain* in the feet, sometimes later in the hands also. The pain is attributed to abnormal firing patterns in axons during the breakdown period, and is associated with reduced conduction rates and smaller spike amplitudes. At the same time, there is reduced sensitivity to pinprick. Peripheral motor weakness sets in gradually but is seldom debilitating.

Clinical Panel 12.2 Entrapment neuropathies

Peripheral nerves may be 'trapped' beneath ligamentous bridges or stretched at bony angulations, with consequent symptoms depending on the distribution of the affected nerves. Sensory disturbances caused by compression tend to be early and prominent, motor weakness later and at times severe.

Nerve conduction studies can be helpful in defining the nerve or nerves involved, extent of damage already done, 'monitoring' for progression, and, perhaps most importantly, confirming the clinical diagnosis. Because entrapment syndromes are more frequent in the presence of generalized polyneuropathies, this may be the most frequent predisposing condition to their development.

Upper limb

The most common *entrapment neuropathy* results from compression of the median nerve in the space between the overlying flexor retinaculum and the underlying flexor tendons within their common synovial sheath or *carpal tunnel syndrome*. Characteristic sensory symptoms are paresthesias in the affected hand and fingers, and bouts of pain that may extend from the hand up along the arm. These symptoms commonly occur during the night; by day, they are especially brought on by grasping or pinching actions in the workplace. Wringing (flicking) the affected hand may afford some relief. On examination, the skin overlying the distal phalanges that is supplied by the median nerve, namely those of thumb, index and middle finger, show reduced sensory acuity. The thenar eminence may appear flattened due to wasting of the abductor pollicis brevis, and that muscle may be weak, in response to *forward* (not outward) movement of the thumb against resistance. Tapping over the nerve at the wrist may elicit pain in the hand (*Tinel's sign*), but this is significant only if light tapping elicits severe pain.

Cervical spondylosis, described in Chapter 14, is a potential source of confusion. This disorder is another example of 'nerve entrapment', being caused by compression of one or more cervical spinal nerves by bony outgrowths next to apophyseal facet joints in the neck. Most commonly affected nerves are C6 (sensory to skin of lateral forearm, lateral hand, and entire thumb) and C7 (sensory to skin of outer three fingers front and back). Arm and forearm tendon reflexes may be diminished, and some motor weakness may be apparent in the distribution of the affected ventral roots. In addition to the more extensive cutaneous symptoms and signs, cervical spondylosis is unrelated to manual activities, seldom causes nocturnal symptoms, and has a relatively advanced age profile.

Confirmatory of carpal tunnel syndrome is a prolonged distal latency in the motor nerve conduction test (Figure 12.4) and/or in the sensory nerve conduction test (Figure 12.5).

Ulnar nerve entrapment may occur at the elbow or at the wrist. At the elbow, it may be compressed against the ulna by the fibrous arch linking the humeral and ulnar origins of the flexor carpi ulnaris muscle. The patient may be aware of having a sensitive 'funny bone' in the affected area, and/or paresthesia affecting the medial two fingers and the hypothenar skin area. In chronic cases, there may be weakness of flexor carpi ulnaris and of the flexor digitorum profundus contribution to the medial two fingers. Usually, the motor weakness is confined to the hand, and this may create diagnostic confusion because compression at the wrist can have the same effect. Wrist level compression occurs in the interval between the pisiform bone and the hook of the hamate. The sensory effects are confined to the inner two fingers, because the palmar branch of the nerve arises in the forearm and is spared. If only the motor branch to the hypothenar muscles is involved, weakness will be evident during abduction of the little finger against resistance. Involvement of the deep branch leads to weakness of abduction and adduction of index, middle, and ring fingers.

Lower limb

Meralgia paresthetica ('thigh pain with pins and needles') is an annoying condition affecting the lateral cutaneous nerve of the thigh where it pierces the inguinal ligament close to the anterior superior spine. The nerve may be pinched by tension of the ligament during extended periods of exercise, for example playing football. It is also associated with pregnancy, where increased tissue fluid may generate a carpal tunnel syndrome at the same time. Intermittent 'flicks' of pins and needles are experienced on the outside of the thigh, and skin sensitivity may be progressively reduced by degeneration of the nerve. Nerve conduction from the skin of the lower lateral thigh is retarded on the affected side, as revealed by sensory evoked potentials, i.e. stimulating the skin while recording electrical activity over the contralateral somatosensory cortex. However, this procedure is rarely attempted in currently asymptomatic individuals.

Sensory evoked potential techniques are described in Chapter 31.

Common peroneal nerve entrapment is a term used when the common peroneal nerve exhibits signs of compression at the level of the neck of the fibula. Here, the nerve passes through a tendinous arch formed by the peroneus longus. Rarely is it a true entrapment, however. Usually, the problem is one of frequent compression either during sleep or from habitual seating with the legs crossed, whereby the nerve is pressed against the lateral condyle of the femur of the other knee. Reduced nerve conduction affecting the superficial peroneal branch leads to weakness of eversion of the foot and sensory loss in the skin of lower leg and dorsum of foot. Affecting the deep

Clinical Panel 12.2 *Continued*

peroneal branch, it leads to weakness of dorsiflexion of the foot and toes, resulting in foot drop with characteristic slapping gait. Either branch may escape more or less completely; identification of individual affected muscles requires needle electromyography.

Iatrogenic entrapment is a well-known danger associated with application of a plaster cast following fracture of the tibia. The normal procedure is to insert protective padding before the plaster has hardened.

Tarsal tunnel syndrome results from compression of the tibial nerve and/or its plantar branches within the tarsal tunnel roofed by the flexor retinaculum of the ankle. The compression is often not from the retinaculum itself, but from an outside agency such as ill-fitting footwear or a tight plaster cast following fracture of the tibia. The result is pain in the ankle region and paresthesias in the sole of the foot.

Finally, paresthesias confined to the forefoot and two or three adjacent toes are likely to be caused by squeezing of planter digital nerves between adjacent metatarsal heads.

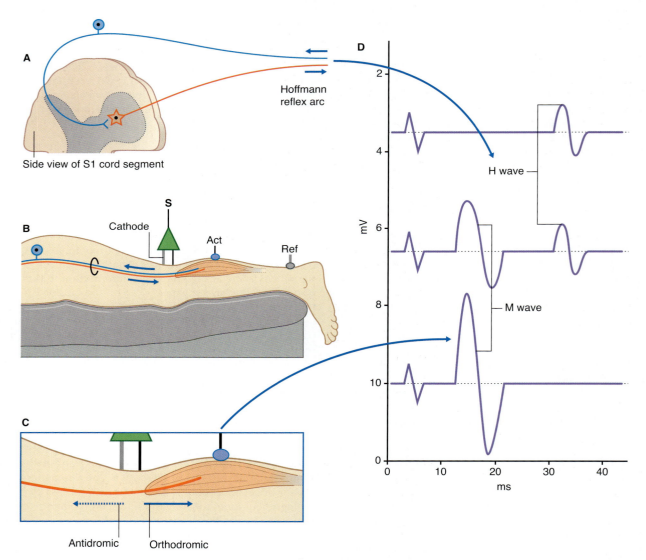

Figure 12.8 Anatomic background of the H and M waves. **(A)** The H wave is mediated by a monosynaptic reflex arc as shown. **(B)** Recording the S1 segment Achilles reflex arc compound motor action potential (CMAP). The stimulator overlies the tibial nerve, the recorder overlies the triceps surae muscle. Both limbs of the reflex arc are within the tibial component of the sciatic nerve. Note that the CMAP is maximal at the minimal (threshold) level of stimulation. **(C)** Increasing the current directly activates axons supplying the muscle, yielding the short-latency M wave. **(D)** Note that the H wave progressively disappears as stimulus intensity increases, because of cancellation of the orthodromic motor impulses in (A) by antidromic impulses newly generated by the cathode, represented by the dashed line in (C).

monosynaptic, minimal latency twitch in the triceps surae (gastrocnemius–soleus). The minimal latency is in fact quite long—up to 35 ms depending on the patient's overall height—because S1 segment of the spinal cord lies behind the body of vertebra L1, creating a 130- to 150-cm upandown (sic!) trip. Increasing the current reaches the point where the *M wave* appears (Figure 12.8D). The M wave is produced by direct orthodromic activation of the motor end plates. Antidromic conduction accounts for progressive canceling out of the action potentials descending in the efferent limb of the H reflex arc.

In the upper limb, the nerve roots of spinal nerve C6 may be tested by stimulating the median nerve and recording from flexor carpi radialis. C7 roots may be tested by stimulating the posterior cutaneous nerve of the forearm and recording from the triceps brachii.

ELECTROMYOGRAPHY

Electromyography is a technique in which an electrode incorporated into a fine needle is inserted into a muscle in order to sample the depolarization waveforms produced by voluntary contraction. There are several components to the test; when combined with the results of NCS, they provide valuable diagnostic information.

The test begins, like the NCS, with a clinical question, and the individual muscles chosen for EMG are based on the most probable clinical diagnosis provided by the history and physical examination. For example, if there is clinical evidence suggestive of a specific nerve injury, muscles are chosen that are supplied by that nerve. Recordings from adjacent muscles (or from the same muscle on the opposite side) during contraction would also be sampled to provide control waveforms for comparison. The results are combined with NCS to make a case for or against the provisional diagnosis.

Needle electrode

The recording electrode occupies the lumen of a fine needle (Figure 12.9). An insulation sleeve isolates the recording electrode from the barrel of the needle, which functions as a reference electrode. As in the case of NCS, the EMG record is based on the potential difference between the recording and reference electrodes. During muscle contractions, an extracellular record is taken of the low-voltage potentials that originate from the muscle membrane depolarizations.

The needle is passed through the skin and into the muscle in question. It is then pushed, in small increments, into various portions of the muscle; after each needle movement, the effect is observed. The moving needle will generate spiky, *insertional activity*, also known as *injury potentials*, caused by muscle membrane depolarizations, which cease when the electrode movement stops.

The normal electromyogram

The sensitivity settings on the machine are adjusted in order to record voluntary muscle contraction. Patients are asked to slightly contract the muscle; as they do, irregular waveforms appear, representing *motor unit action*

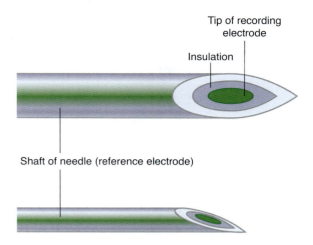

Figure 12.9 Structure of a conventional concentric electrode. The upper picture is a face-on view depicting the insulator separating the active (recording) electrode from the reference electrode.

potentials (MUAPs). Each of these individual waveforms represents activation of the muscle fibers that belong to an individual motor unit. While the electrode is stationary, all MUAPs that are of similar shape originate from the same individual anterior horn cell and reflect depolarization of that cell. Their shape, in normal situations, is similar to the familiar QRS of an electrocardiogram recording, and measurements are made of their amplitude, duration, and morphology. Each individual MUAP is a sample of the summated depolarizations of the fibers of a single motor unit. We must bear in mind that the electrode can record only from the muscle fibers that are the closest—not all fibers of a unit contribute to the observed response. As indicated in Figure 12.10, the overlap of the territories of

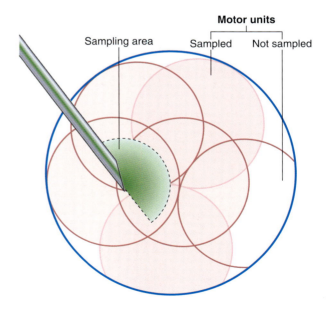

Figure 12.10 Sampling motor unit action potentials. This diagram represents the overlap of six motor units. The area (*green*) being sampled includes parts of five units; sampling is greatest close to the recording electrode.

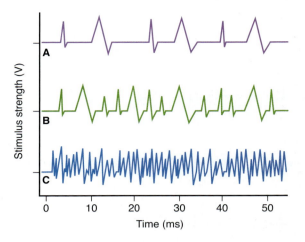

Figure 12.11 Motor unit activation at increasing strengths of voluntary contraction. **(A)** In this example, during weak contraction the recorder has picked up activity in two motor units, each with its characteristic shape. **(B)** Stronger contraction has recruited a third motor unit within reach of the electrode. **(C)** Substantially greater contraction has produced so much overlap of action potentials that the individual shape tracings are distorted ('interference pattern').

individual anterior horn cells permits several units to be sampled simultaneously.

The recorded waveforms tell us about the form and function of the motor units, and about changes under various pathologic conditions. Each depolarization of an anterior horn cell results in a virtually synchronous depolarization of all its target muscle fibers. The needle electrode records a summation of the individual action potentials closest to its exposed tip, to provide a MUAP. As long as the recording electrode remains stationary, the MUAP waveform will remain the same. On the monitor, they 'march across' at a frequency that is the same as the firing rate of the neurons being sampled. The stronger the voluntary muscle contraction, the greater the number of motor neurons recruited by the corticospinal tract and the more frequent the firing rate (Figure 12.11).

The recorded waveforms tell us about the form and function of the motor units, and about changes under various pathologic conditions. Each depolarization of an anterior horn cell results in a virtually synchronous depolarization of all its target muscle fibers. The needle electrode records a summation of the individual action potentials closest to its exposed tip, to provide a MUAP. As long as the recording electrode remains stationary, the MUAP waveform will remain the same. On the monitor, they 'march across' at a frequency that is the same as the firing rate of the neurons being sampled. The stronger the voluntary muscle contraction, the greater the number of motor neurons recruited by the corticospinal tract and the more frequent the firing rate (Figure 12.11).

Some clinical applications

Denervation of muscle

Skeletal muscle may become denervated as a result of:

- acute physical injury to its nerve of supply, for example laceration or acute crush injury;
- chronic pressure on its nerve of supply, known as *entrapment*, for example progressive squeezing of the median nerve in carpal tunnel syndrome, or of the ulnar nerve in its tunnel behind the medial epicondyle;
- death of α motor neurons in the anterior gray horn or cranial motor nuclei, in the course of *motor neuron disease* (Ch. 16);
- involvement of motor nerves in the course of an acute or chronic polyneuropathy.

Some abnormal MUAPs are illustrated and explained in Clinical Panel 12.3. Clinical Panel 12.4 is an account of the autoimmune disorder known as *myasthenia gravis* ('grave muscle weakness').

Reinnervation of muscle

The sequence of events is described in Figure 12.12.

Clinical Panel 12.3   Abnormal motor unit action potentials

(Myasthenia gravis is described in Clinical Panel 12.4.)

Fibrillation potentials   (Figure CP 12.3.1)
Fibrillation potentials are a characteristic feature of relaxed muscles in the early stages of atrophy. They take the form of abnormally small potentials, either triphasic or positive, occurring with great regularity at up to 15 Hz. They are not clinically visible, and the patient is not aware. The atrophy may be caused either by a *neuropathy* of any kind that results in denervation of the motor end plates in the muscle under examination, or by a *primary myopathy*, a degenerative change originating in the muscle fibers themselves, for example in various muscular dystrophies. *Denervation supersensitivity* has been invoked to account for fibrillations in both conditions. Loss of end-plate innervation is known to be associated with insertion of numerous new acetylcholine (ACh) receptors into the plasma membrane of denervated muscle fibers at some distance from their end plates, sufficient to evoke small localized action potentials by the minute amount of circulating ACh. In primary myopathies, the likely explanation is that of a deterioration of the muscle membrane leading to failure

of propagation of action potentials originating at the end plate; this appears sufficient to signal a requirement for additional receptors.

Fasciculation potentials   (Figure CP 12.3.1)
Fasciculation potentials are not infrequent among healthy individuals, being perceived as a localized 'wriggly' sensation in a relaxed individual muscle, usually after vigorous exercise. However, in association with motor neuron degeneration from any cause, they are indicative of spontaneous development of action potentials anywhere along the lower motor neuron pathway from anterior horn cell to muscle. They are often visible as a twitching and dimpling of the overlying skin. On the electromyographic record, they appear as somewhat misshapen motor unit action potentials (MUAPs) that appear infrequently and are not under voluntary control.

Prolonged polyphasic and giant compound MUAPs
The term *polyphasic* signifies an abnormally large number of positive and negative phases. Polyphasic MUAPs signify *reinnervation* of motor end plates

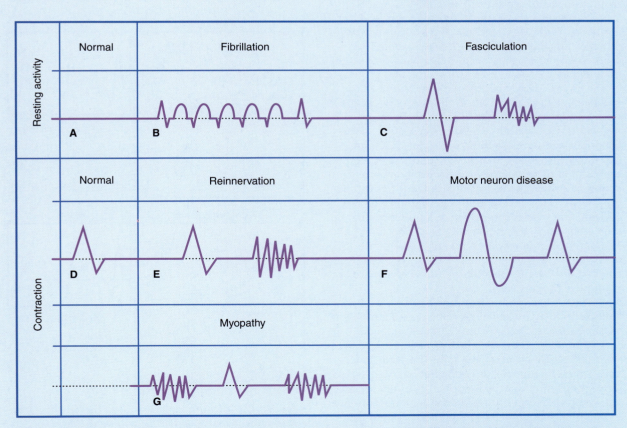

Figure CP 12.3.1 Characteristic waveforms under different conditions. Resting activity: **(A)** Normal resting muscle is silent. **(B)** Fibrillation potentials: low amplitude, high frequency. Contraction: **(C)** High-amplitude spikes accompanied by low-amplitude potentials. **(D)** Normal motor unit action potential (MUAP). **(E)** Reinnervation: normal plus polyphasic MUAPs. **(F)** Normal plus giant MUAPs. **(G)** Low-amplitude, mostly polyphasic MUAPs.

vacated by earlier degeneration of their nerve supply, followed by takeover by neighboring healthy axons. Figure 12.12 provides a basic explanation. In this figure, two separate motor neurons are each represented by a single parent axon supplying three muscle fibers. Following interruption of one parent axon, its vacated nerve sheaths exert a chemotropic effect, inducing the surviving stem and/or branch axons to issue collateral sprouts that eventually reinnervate the vacated end plates. The outcome is the production of *giant motor unit potentials* by the enlarged motor unit.

Giant MUAPs are called 'neuropathic' because they signify motor neuron pathology. As mentioned in Chapter 10, they occur in the elderly as a result of 'fall out' of spinal motor neurons. *Motor neuron disease* (Ch. 16) is associated with progressive loss of spinal and cranial nerve motor neurons on a much greater scale; even the neurons that provide reinnervation are eventually lost. *Radiculopathy* (Ch. 14) resulting from compression of nerve roots, and axonal polyneuropathy are other causes.

Clinical Panel 12.4   Myasthenia gravis

The acetylcholine (ACh) receptors of skeletal muscle normally undergo turnover with a half-life (50% loss rate) of 12 days. New receptors are constantly synthesized in Golgi complexes located around the nuclei of the sole plate and inserted into the sarcolemma of the junctional folds. Old receptors are removed by endocytosis and degraded by lysosomes.

In myasthenia gravis, the immune system produces antibodies to the ACh receptor. The antigen–antibody complex has a half-life of only 2 days, leading to a progressive loss of receptors and shrinkage of junctional folds (Figure 12.4.1).

The disease usually begins in the second or third decade in women and in the sixth or seventh decade in men. Muscles most affected are those supplied by cranial nerves, usually in a descending sequence. The symptoms and signs are those of variable *weakness*, expressed by inability to maintain contractions: the eyelids tend to droop, the extrinsic ocular muscles are unable to sustain the gaze, the face tends to sag, and the jaw tends to need support. Chewing may be difficult, and swallowing may pose a threat of fluid or food inhalation—sometimes with fatal effect. Respiratory muscle weakness may also precipitate pulmonary infection. Proximal limb muscles are affected late if at all. Patients who survive the first year are likely to improve progressively.

That the weakness is not caused by nerve paralysis is easily verified by the ability to *commence* a movement; all that is required is a minute's rest beforehand.

Confirmation of the diagnosis can be obtained by means of needle electromyography. The abductor pollicis brevis is suitable, although not overtly affected. Nerve conduction velocity is normal, as is ACh release, but, when the nerve is stimulated at a rate of three per second, the compound motor unit action potentials rapidly dwindle to nothing. The spikes return to normal amplitude following injection of either edrophonium or neostigmine, which prolong the binding time of ACh with the surviving receptors.

Anticholinesterase antibody can be detected in the blood of 80–90% of patients. The antibody originates in the thymus gland, which is usually hyperplastic and contains a lymphoid tumor in 12% of patients. Removal of the thymus (if enlarged) may be beneficial if symptoms cannot be otherwise controlled.

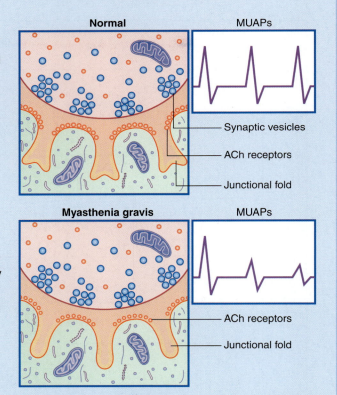

Figure CP 12.4.1 Normal and myasthenic motor end plates and compound motor unit action potentials (MUAPs) compared. Note widening of synaptic cleft in myasthenia gravis, together with reduction of acetylcholine (ACh) receptors and of junctional folds. Note that the nerve terminal itself is not affected.

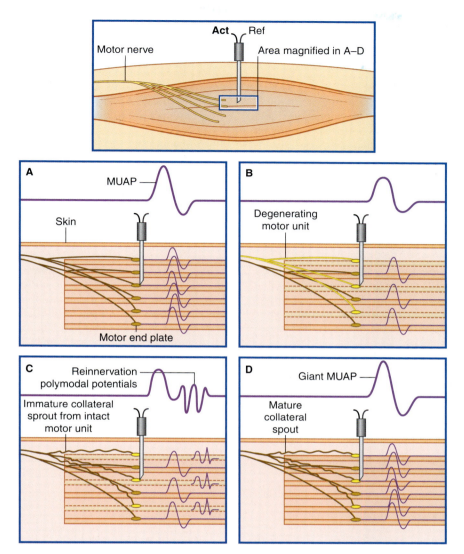

Figure 12.12 Reinnervation of motor end plates. In this composite figure, the topmost diagram represents a muscle about to be activated and its depolarization waves recorded through a needle electrode. **(A)** The two axons at upper left belong to different anterior horn cells; three motor end plates from each unit are shown. **(B)** One anterior horn cell is degenerating. The motor unit action potential (MUAP) has become smaller because of reduced summation. **(C)** Early reinnervation is taking place by collateral sprouts from the intact motor neuron. Depolarization waveforms of these muscle fibers are small, and their appearance is delayed; their MUAPs are characteristically *polyphasic*, showing multiple positive and negative phases. **(D)** Several weeks later, the reinnervated muscle fibers give normal electromyographic responses. All six are now synchronously depolarized, yielding giant motor unit potentials.

Core Information

Nerve conduction studies

The functional state of the peripheral nervous system can be assessed by nerve conduction studies and by electromyography. For motor nerve conduction studies, a stimulating electrode is placed on the skin overlying the selected nerve, and a recording electrode over the midregion of a muscle of supply. The normal onscreen waveform produced is that of compound motor action potentials produced by summation of individual muscle fiber depolarization potentials. The stimulus–response latency (time lapse) is recorded from two separate sites along the same nerve, enabling the motor nerve conduction velocity to be ascertained by subtracting one latency from the other. Preferred nerve trunks in the upper limb are the median and ulnar, in the lower limb the deep peroneal. For sensory nerve conduction studies, antidromic stimulation of a cutaneous nerve is accompanied by proximal recording at two sites over the parent nerve trunk; subtraction yields the sensory nerve conduction velocity.

Spinal nerve roots are assessed by activating muscle spindle reflex arcs at appropriate levels.

Peripheral neuropathies can be classified in accordance with cause, anatomic location, pathology, and time course. A wide variety of nerve entrapment syndromes are encountered in clinical practice.

Electromyography

The recording electrode is inside a needle inserted into a muscle. Motor unit action potentials (MUAPs) normally appear on-screen with any slight voluntary contraction. Each MUAP represents the summated action potentials of the muscle fibers of one motor unit. Overlap of motor units allows several to be sampled simultaneously. Abnormal electromyographic waveforms may be associated with motor nerve injury, acute neuropathy (e.g. Guillain–Barré syndrome), chronic neuropathy, motor neuron disease, or myopathy (e.g. myasthenia gravis). Denervation of muscle from any cause gives rise initially to fibrillation potentials, later to fasciculation potentials, and later again to giant MUAPs indicative of reinnervation by collateral sprouts from neighboring intact motor units.

REFERENCES

Adams RD, Victor M. Diseases of the peripheral nerves. In: Principles of neurology. 5th edn. New York: McGraw-Hill; 1993:1117–1169.

Dyck JD, Thomas PK. Peripheral neuropathy. 4th edn. Philadelphia: Saunders; 2005.

Hallett M, Chokroverty S. Magnetic stimulation in clinical neurophysiology. 2nd edn. Philadelphia: Saunders; 2005.

Lederman RJ. Nerve conduction studies. In: Levin, KH, Lüders HO, eds. Comprehensive clinical neurophysiology. Philadelphia: Saunders; 2000:89–111.

Levin KH. Needle electrode examination. In: Levin, KH, Lüders HO, eds. Comprehensive clinical neurophysiology. Philadelphia: Saunders; 2000:89–111.

Mills KR. The basics of electromyography. J Neurol Neurosurg Psychiatry 2005; 76 (suppl II):ii32–ii35.

Wilbourn AJ, Ferrante MA. Clinical electromyography. In: Joynt RJ, Grigs RC, eds. Clinical neurology, vol 1. New York: Lippincott, Williams & Wilkins; 1998:1–32.

Autonomic nervous system and visceral afferents

STUDY GUIDELINES

1 Resolve the paradox that despite an outflow restricted to 14 or 15 ventral roots, all 31 spinal nerve trunks acquire sympathetic fibers.

2 Appreciate that the sympathetic ganglia along the abdominal aorta are activated by preganglionic fibers, as is the adrenal medulla.

3 Pay special attention to the autonomic innervation of the eye, both here and in Chapter 23.

4 Appreciate that the four parasympathetic ganglia in the head are functionally similar to intramural ganglia elsewhere.

5 The pelvic ganglia are mixed autonomic ganglia.

6 Realize that the preganglionic neurons of both divisions are cholinergic, and that the target receptors in all the autonomic ganglia are nicotinic.

7 Note that, at tissue level, synapses are replaced by looser 'junctions' that permit diffusion of transmitter to outlying receptors.

8 You should focus on four kinds of junctional receptors of the sympathetic system, and on four actions initiated by muscarinic receptors in the parasympathetic system.

9 Learn from Clinical Panel 13.2 how pharmacologists intercept the recycling and degradation sequence at sympathetic nerve endings. The same principles apply to central nervous system drug therapy, notably in psychiatric disorders.

10 Follow Clinical Panel 13.3 to contrast the effects of cholinergic and anticholinergic drugs.

11 Visceral afferents utilize autonomic pathways to gain access to the nervous system. They are especially important in the context of thoracic and abdominal pain.

COMPONENTS OF THE AUTONOMIC NERVOUS SYSTEM

The autonomic ('self-regulating') nervous system is distributed to the peripheral tissues and organs by way of outlying autonomic ganglia. Controlling centers in the hypothalamus and brainstem send central autonomic fibers to synapse on preganglionic neurons located in the gray matter of the brainstem and spinal cord. From these neurons, *preganglionic fibers* (mostly myelinated) project out of the central nervous system (CNS) to synapse on multipolar neurons in the autonomic ganglia. Unmyelinated *postganglionic fibers* emerge and form terminal networks in the target tissues.

Both anatomically and functionally, the autonomic system is composed of sympathetic and parasympathetic divisions.

SYMPATHETIC NERVOUS SYSTEM

The sympathetic system is so called because it acts in sympathy with the emotions. In association with rage or fear, the sympathetic system prepares the body for 'fight or flight': the heart rate is increased, the pupils dilate, and the skin sweats. Blood is diverted from the skin and intestinal tract to the skeletal muscles, and the sphincters of the alimentary and urinary tracts are closed.

The sympathetic outflow from the nervous system is *thoracolumbar*, the preganglionic neurons being located in the lateral gray horn of the spinal cord at thoracic and upper two (or three) lumbar segmental levels. From these neurons, preganglionic fibers emerge in the corresponding anterior nerve roots and enter the paravertebral sympathetic chain. The fibers do one of four things (Figure 13.1).

1 Some fibers synapse in the nearest ganglion. Postganglionic fibers enter spinal nerves T1–L2 and supply blood vessels and sweat glands in the territory of these nerves.

2 Some fibers *ascend* the sympathetic chain and synapse in the superior or middle cervical ganglion, or in the stellate ganglion. (The stellate consists of the fused inferior cervical and first thoracic ganglia; it lies in front of the neck of the first rib.) Postganglionic fibers

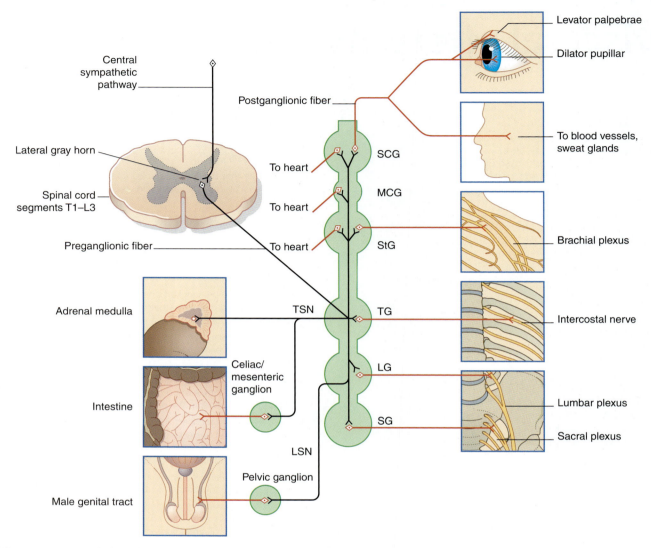

Figure 13.1 General plan of the sympathetic system. Ganglionic neurons and postganglionic fibers are shown in *red*. LG, lumbar ganglia; LSN, lumbar splanchnic nerve; MCG, middle cervical ganglion; SCG, superior cervical ganglion; SG, sacral ganglia; StG, stellate ganglion; TG, thoracic ganglia; TSN, thoracic splanchnic nerve.

supply the head, neck, and upper limbs, also the heart. Of particular importance is the supply to the dilator muscle of the pupil (Clinical Panel 13.1).

3 Some fibers *descend* to synapse in lumbar or sacral ganglia of the sympathetic chain. Postganglionic fibers enter the lumbosacral plexus for distribution to the blood vessels and skin of the lower limbs.

4 Some fibers *traverse* the chain and emerge as the (preganglionic) thoracic and lumbar splanchnic nerves. The thoracic splanchnic nerves (usually called, simply, the splanchnic nerves) pass through the lower eight thoracic ganglia, pierce the diaphragm, and synapse within the abdomen in the celiac and mesenteric prevertebral ganglia, and in renal ganglia. Postganglionic fibers accompany branches of the aorta to reach the gastrointestinal tract, liver, pancreas, and kidneys. Lumbar splanchnic nerves pass through the upper three lumbar ganglia and meet in front of the

bifurcation of the abdominal aorta. They enter the pelvis as the hypogastric nerves before ending in pelvic ganglia, from which the genitourinary tract is supplied.

The medulla of the adrenal gland is the homolog of a sympathetic ganglion, being derived from the neural crest. It receives a direct input from fibers of the thoracic splanchnic nerve of its own side (see later).

The sympathetic system exerts tonic (continuous) constrictor activity on blood vessels in the limbs. In order to improve the blood flow to the hands or feet, impulse traffic along the sympathetic system can be interrupted surgically (Clinical Panel 13.1).

PARASYMPATHETIC NERVOUS SYSTEM

The parasympathetic system generally has the effect of counterbalancing the sympathetic system. It adapts the

Clinical Panel 13.1   Sympathetic interruption

Stellate block

Injection of local anesthetic around the stellate ganglion—*stellate block*—is a procedure used in order to test the effects of sympathetic interruption on blood flow to the hand. Both pre- and postganglionic fibers are inactivated, producing sympathetic paralysis in the head and neck on that side, as well as in the upper limb. A successful stellate block is demonstrated by (a) a warm, dry hand; (b) *Horner syndrome*, which consists of a constricted pupil owing to unopposed action of the pupillary constrictor; and (c) *ptosis* (drooping) of the upper eyelid owing to paralysis of smooth muscle fibers contained in the levator muscle of the upper eyelid (Figure CP 13.1.1).

Dominance of the right stellate ganglion in control of the heart rate is shown by the marked slowing of the

Figure CP 13.1.1 Horner syndrome, patient's right side. Note the moderate ptosis of the eyelid, and the moderate miosis (pupillary constriction). The affected pupil reacts to light but recovers very slowly.

pulse following a right, but not a left, stellate block. (See also Box 13.1.)

Functional sympathectomy of the upper limb may be carried out by cutting the sympathetic chain below the stellate ganglion. This is not an anatomic sympathectomy, because the ganglionic supply to the limb from the middle cervical and stellate ganglia remains intact. It is a functional one, because the ganglionic neurons for the limb are deprived of tonic sympathetic drive. Horner syndrome is avoided by making the cut at the level of the second rib: the preganglionic fibers for the head and neck enter the stellate direct from the first thoracic spinal nerve.

Two indications for interruption of the sympathetic supply to one or both upper limbs are painful blanching of the fingers in cold weather (*Raynaud phenomenon*), and *hyperhidrosis* (excessive sweating) of the hands—usually an embarrassing affliction of teenage girls.

The sympathetic supply to the eye is considered further in Chapter 21.

Lumbar sympathectomy

In order to improve blood flow in the lower limb, the preganglionic nerve supply may be interrupted by cutting the upper end of the lumbar sympathetic chain. The usual procedure is to remove the second and third lumbar sympathetic ganglia. In male patients, bilateral lumbar sympathectomy may result in persistent, painful erections (*priapism*) because of interruption of a pathway that maintains the resting, flaccid state of the penis.

---

eyes for close-up viewing, slows the heart, promotes secretion of salivary and intestinal juices, and accelerates intestinal peristalsis. A notable instance of *concerted* sympathetic and parasympathetic activity occurs during sexual intercourse (Box 13.4).

The parasympathetic outflow from the CNS is *craniosacral* (Figure 13.2). Preganglionic fibers emerge from the brainstem in four cranial nerves—the oculomotor, facial, glossopharyngeal, and vagus—and from sacral segments of the spinal cord.

Cranial parasympathetic system

Preganglionic parasympathetic fibers emerge in four cranial nerves (Figure 13.3).

1 In the oculomotor nerve, to synapse in the **ciliary ganglion**. Postganglionic fibers innervate the sphincter of the pupil and the ciliary muscle. Both muscles act to produce the *accommodation reflex*.

2 In the facial nerve, to synapse in the **pterygopalatine** ganglion, which innervates the lacrimal and nasal glands; and in the **submandibular ganglion**, which innervates the submandibular and sublingual glands.

3 In the glossopharyngeal nerve, to synapse in the **otic ganglion**, which innervates the parotid gland.

4 In the vagus nerve, to synapse in **mural** ('on the wall') or **intramural** ('in the wall') ganglia of heart, lungs, lower esophagus, stomach, pancreas, gall bladder, small intestine, and ascending and transverse parts of the colon.

Sacral parasympathetic system

The sacral segments of the spinal cord occupy the conus medullaris (conus terminalis) at the lower extremity of the spinal cord, behind the body of the first lumbar vertebra. From the lateral gray matter of segments S2, S3, and S4, preganglionic fibers descend in the cauda equina within ventral nerve roots. On emerging from the pelvic sacral foramina, the fibers separate out as the pelvic splanchnic nerves. Some fibers of the left and right pelvic splanchnic nerves synapse on ganglion cells in the wall of the distal colon and rectum. The rest synapse in **pelvic ganglia**, close to the pelvic sympathetic ganglia already mentioned. Postganglionic parasympathetic fibers supply the detrusor muscle of the bladder; also the tunica media of the internal

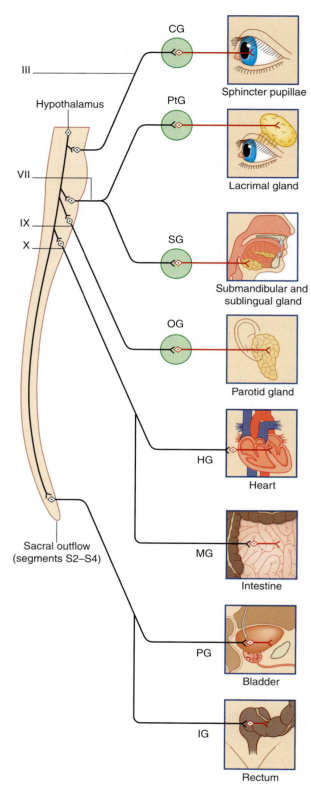

Figure 13.2 General plan of the parasympathetic system. Ganglionic neurons and postganglionic fibers are shown in *red*. CG, ciliary ganglion; HG, heart ganglia; IG, intramural ganglia; MG, myenteric ganglia; OG, optic ganglion; PtG, pterygopalatine ganglion; SG, submandibular ganglion.

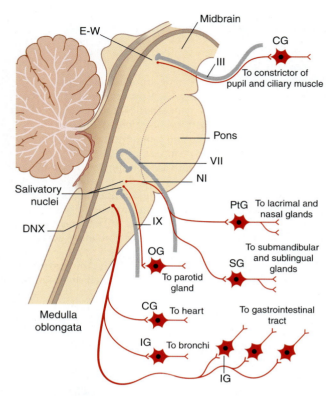

Figure 13.3 Cranial parasympathetic system. DNX, dorsal nucleus of vagus; E–W, Edinger–Westphal nucleus. Other abbreviations as in Figure 13.2.

pudendal artery and of its branches to the cavernous tissue of the penis or clitoris (see later).

NEUROTRANSMISSION IN THE AUTONOMIC NERVOUS SYSTEM

Ganglionic transmission

The preganglionic neurons of the sympathetic and parasympathetic systems are *cholinergic*: the neurons liberate acetylcholine (ACh) on to the ganglion cells at axodendritic synapses (Figure 13.4). The receptors on the ganglion cells are nicotinic, so named because the excitatory effect can be imitated by locally applied nicotine.

Junctional transmission

Postganglionic fibers of the sympathetic and parasympathetic systems form *neuroeffector junctions* with target tissues (Figure 13.4). Transmitter substances are liberated from innumerable varicosities strung along the course of the nerve fibers.

The chief transmitter at sympathetic neuroeffector junctions is *norepinephrine (noradrenaline)*, which is liberated from dense-cored vesicles. The postganglionic sympathetic system in general is described as *adrenergic*. An exception to the adrenergic rule is the *cholinergic* sympathetic supply to the eccrine sweat glands over the body surface.

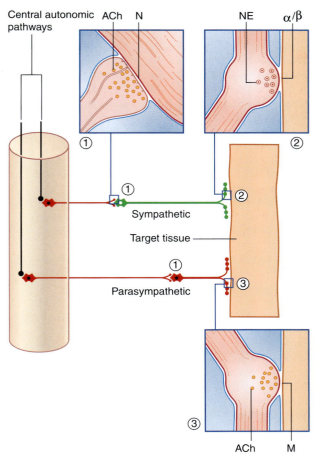

Figure 13.4 Autonomic transmitters and receptors. Ganglionic neurons and postganglionic fibers are shown in *red*. ACh, acetylcholine; M, muscarinic receptors; N, nicotinic receptors; NE, norepinephrine.

The chief transmitter at parasympathetic neuroeffector junctions is ACh. The postganglionic parasympathetic system in general is *cholinergic*.

Junctional receptors

The physiologic effects of autonomic stimulation depend on the nature of the *postjunctional receptors* inserted by target cells into their own plasma membranes. In addition, transmitter release is influenced by *prejunctional receptors* in the axolemmal membrane of the nerve terminals.

Sympathetic junctional receptors (adrenoceptors) (Figure 13.5)
Two kinds of α adrenoceptor and two kinds of β adrenoceptor have been identified for norepinephrine.

1 *Postjunctional α_1 adrenoceptors* initiate contraction of smooth muscle in peripheral small arteries and large arterioles, the dilator pupillae, the sphincters of the alimentary tract and bladder neck, and the vas deferens.

2 *Prejunctional α_2 adrenoceptors* are present on parasympathetic as well as on sympathetic terminals.

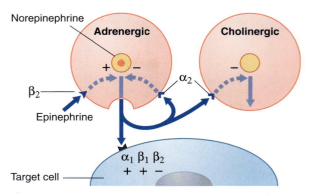

Figure 13.5 Adrenergic activity at a neuroeffector junction. Release of norepinephrine is promoted by epinephrine and inhibited by prejunctional α_2 receptors, which also inhibit transmitter release from neighboring parasympathetic varicosities.

They inhibit transmitter release in both cases. On sympathetic terminals, they are called *autoreceptors*.

3 *Postjunctional β_1 adrenoceptors* increase pacemaker activity in the heart and increase the force of ventricular contraction (Box 13.1). In response to a severe fall of blood pressure, sympathetic activation of β_1 receptors on the juxtaglomerular cells of the kidney causes secretion of renin. Renin initiates production of the powerful vasoconstrictor angiotensin II.

4 *β_2 Receptors* respond to circulating epinephrine (adrenaline) (Figure 13.6) in addition to locally released norepinephrine.

Postjunctional β_2 receptors relax smooth muscle, notably in the tracheobronchial tree and in the accommodatory muscles of the eye. Some postjunctional β_2 receptors are on the surface of hepatocytes in the liver, where they initiate glycogen breakdown to provide glucose for immediate energy needs.

Prejunctional β_2 receptors on adrenergic terminals promote release of norepinephrine.

Most of the norepinephrine liberated at sympathetic terminals is retrieved by an *amine uptake pump*. Some is degraded after uptake by a mitochondrial enzyme, *monoamine oxidase*.

The effects of drugs on the sympathetic system are considered in Clinical Panel 13.2.

Parasympathetic junctional receptors
Parasympathetic junctional receptors are called *muscarinic*, because they can be mimicked by application of the drug muscarine (Figure 13.7). Parasympathetic stimulation produces the following muscarinic effects.

- Slowing of the heart in response to vagal stimulation, and diminished force of ventricular contraction (Box 13.1).

- Contraction of smooth muscle, with the following effects: intestinal peristalsis (Box 13.2), bladder emptying, accommodation of the eye for 'near vision'.

- Glandular secretion.

Box 13.1   Innervation of the heart

The preganglionic sympathetic supply to the heart arises from the lateral gray horn of cord segments T1–5. The fibers synapse in all three cervical and in the uppermost five thoracic ganglia of the sympathetic chain. Postganglionic adrenergic fibers are distributed to the specialized myocardial cells of nodal and conducting tissues, to the general myocardium (of the left ventricle in particular), and to the coronary arteries.

Experimental evidence indicates that the preganglionic parasympathetic supply originates in neurons occupying the caudal ventrolateral medulla oblongata. The fibers descend within the trunk of the vagus and synapse within mural ganglia on the posterior walls of the atria and in the posterior atrioventricular groove (Figure Box 13.1.1). Postganglionic cholinergic fibers supply the same tissues as those of the sympathetic, although the direct supply to ventricles and coronary arteries is slight.

There is a high level of autonomic interaction where innervation is dense, notably within nodal tissue, in the modes shown in Figures 13.5 and 13.7. Many sympathetic nerve endings also release *neuropeptide Y*, which binds to a specific receptor on cholinergic terminals with adjuvant inhibitory effect on ACh release.

Many parasympathetic endings corelease vasoactive intestinal polypeptide (VIP), which attenuates release of ACh by binding with VIP-specific inhibitory autoreceptors on the endings that release it.

An abundance of non-adrenergic, non-cholinergic neurons modulate the activity of parasympathetic ganglion cells. Also found are scattered adrenergic neurons whose preganglionic supply traverses the sympathetic chain, and bipolar local circuit neurons.

Autoregulation of myocardial performance by the intramural ganglionic networks of the normal heart is sufficient to withstand the total extrinsic denervation involved in a cardiac transplant.

A fourth set of neurons is afferent in nature. Unipolar somas in the inferior ganglion of the vagus provide stretch-sensitive nerve endings close to the endocardium—notably in the right atrium, where distension produces reflex slowing of the heart rate by way of a central pathway to the dorsal vagal nucleus via the solitary nucleus (Ch. 24).

Some unipolar somas in spinal dorsal root ganglia send peripheral processes to form chemosensitive endings in the myocardium. Metabolites released by ischemic myocardial cells in response to coronary artery occlusion generate impulse trains that travel along the central processes of these cells to reach the posterior gray horn via *anterior* nerve roots. The central processes synapse on projection cells of the lateral spinothalamic tract, with consequent perception of referred pain (see main text). A prominent transmitter in the nociceptive neurons is substance P, which is released *at both ends simultaneously*: In the gray matter, this peptide is excitatory to spinothalamic projection cells, and in the ischemic tissue it activates specific excitatory receptors on cholinergic endings, thus slowing the heart.

The cardiac pacemaker (sinuatrial node) is on the right side of the body and mainly innervated by the two right-sided sets of autonomic neurons. The atrioventricular node is on the left side and receives a corresponding preponderance.

The sinuatrial node is highly responsive to emotional states having their seat of origin in the right, 'emotional' hemisphere (Ch. 34). The descending pathways concerned are largely ipsilateral and polysynaptic, prior to reaching the lower autonomic nervous system centers of medulla and cord. *Sympathetic* overactivity, in response to 'approach' emotions of sexual or combative nature, may cause the heart to 'miss a beat' (extrasystole) or the 'pulse to race' (tachycardia). *Parasympathetic* overactivity, in response to 'withdraw' (aversive) emotions, usually of olfactory or visual origin, may cause bradycardia—or even cardiac arrest.

The atrioventricular node and Purkinje fibers concordantly increase or reduce the speed of transference of action potentials to the ventricles.

Ventricular contractility, and *synchrony* throughout the ventricular myocardium, is increased by raised sympathetic activity. Both are diminished by the

● Mural parasympathetic ganglia

Figure Box 13.1.1 Disposition of mural cardiac parasympathetic ganglia. (The assistance of Professor Andrew J. Armour, Centre de Recherche de l'Hôpital du Sacre-Coeur de Montréal, Quebec, Canada, is gratefully appreciated.)

Box 13.1 *Continued*

parasympathetic, in this case mainly by autonomic interaction; the scarce cholinergic fibers terminate mainly 'on top' of adrenergic ones without any direct influence on the myocardium.

The coronary arterial tree possesses a considerable degree of autoregulation based on release of myocardial cellular metabolites. However, adrenoceptors are also important. The arterioles (<120 μm in diameter) are rich in β_2 receptors responsive to neural norepinephrine at the commencement of exercise, and to circulating epinephrine when exercise gets under way. The arteries (>120 μm) contain α_1 receptors exerting a restraining effect, directing blood to the *subendocardial* ventricular myocardium, which is vulnerable on two counts: it is the most distal coronary territory; and it is the most compressed myocardial component during systole, receiving blood only during diastole.

Cholinergic coronary nerve endings are scarce, but they have a significant dilator effect on the main arteries—precisely those most at risk of atherosclerosis! It transpires that released acetylcholine acts *indirectly*, by causing release of the potent dilator nitric oxide from the vascular endothelium. Progressive devitalization of the endothelium by underlying atherosclerotic plaques leads to more or less complete failure of beneficial nitric oxide production.

REFERENCES

Ardell JL. Structure and function of mammalian intrinsic cardiac neurons. In: Armour JA, Ardell JL, eds. Neurocardiology. New York: Oxford University Press; 1994:95–114.

Armour JA, Murphy DA, Yuan B-X, et al. Gross and microscopic anatomy of the human intrinsic cardiac nervous system. Anat Rec 1997; 247:289–298.

Armour JA. Peripheral autonomic neuronal interactions in cardiac regulation. In: Armour JA, Ardell JL, eds. Neurocardiology. New York: Oxford University Press; 1994:219–244.

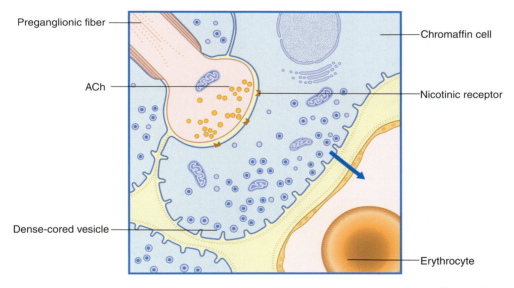

Figure 13.6 Chromaffin cell of the adrenal medulla receiving a synaptic contact from a preganglionic fiber of the thoracic splanchnic nerve. Acetylcholine (ACh)-activated nicotinic receptors. Eighty percent of the cells contain large-cored vesicles (represented here) and secrete epinephrine; the arrow indicates release into the capillary bed. Twenty percent contain small, dense-cored vesicles and secrete norepinephrine.

In addition to the above postjunctional effects, prejunctional muscarinic receptors located on sympathetic varicosities inhibit release of norepinephrine (Figure 13.6).

The effects of *drugs* on the parasympathetic system are considered in Clinical Panel 13.3. Drugs having muscarinic effects are described as *cholinergic*. Drugs that prevent access of ACh to junctional receptors are *anticholinergic*.

A major consideration in the use of drugs either to imitate or to suppress sympathetic or parasympathetic activity is the existence of α, β, and muscarinic receptors in the *central* nervous system. In psychiatric practice, in particular, drugs are often chosen for their action at their central rather than peripheral receptors.

Other types of neuron

Non-adrenergic, non-cholinergic (NANC) neurons are found in both divisions of the autonomic system. In sympathetic

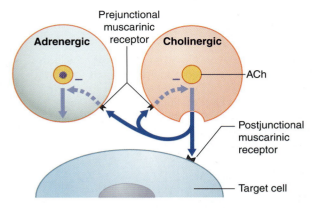

Figure 13.7 Cholinergic activity at a neuroeffector junction. Release of excess acetylcholine (ACh) is inhibited by prejunctional muscarinic receptors, which also inhibit transmitter release from neighboring sympathetic varicosities.

ganglia, small internuncial neurons liberate *dopamine*— a precursor of norepinephrine. Some of the dopamine is secreted into capillaries, the rest binds with dopamine receptors on the main (adrenergic) neurons and exerts a mild inhibitory effect.

The NANC neurons are especially numerous among the ganglion cells in the wall of the alimentary tract, and in the pelvic ganglia. More than 50 different *peptide* substances have been identified, either singly or in various combinations, in these neurons. For the most part, they act as *modulators*, acting either pre- or postjunctionally to influence the duration of action of classic transmitters. Some are *cotransmitters*, for example released together with ACh.

Vasoactive intestinal polypeptide (VIP) is a cotransmitter in the cholinergic supply to the salivary glands and to sweat glands. VIP is a powerful vasodilator and conveniently opens the local vascular bed (through specific VIP receptors on arterioles) just when the muscarinic ACh receptors are raising glandular metabolism.

Nitric oxide is well established as a transmitter in the parasympathetic system. It is a powerful smooth muscle relaxant.

REGIONAL AUTONOMIC INNERVATION

Box 13.1 describes the autonomic innervation of the heart, Box 13.2 the enteric nervous system, Box 13.3 lower-level bladder controls, and Box 13.4 the functional innervation of the genital tract.

Innervation of the genital tract

The *nervi erigentes* ('erectile nerves') are postganglionic pelvic splanchnic nerve fibers supplying the smooth muscle of the internal pudendal arteries and of the trabecular erectile tissue of the phallus in both sexes. The nervi are activated by central parasympathetic neurons following psychic stimulation of the anterior hypothalamus; and/or through a spinal reflex arc in response to direct genital stimulation. Activation produces smooth muscle relaxation, with flooding of the cavernous spaces. In females, a transmural exudate into the vagina acts as a lubricant for the penis. In the male, the bulbourethral glands lubricates the urethra to facilitate passage of semen.

INTERACTION OF THE AUTONOMIC AND IMMUNE SYSTEMS

The lymphatic tissues of the thymus, lymph nodes, and spleen are richly supplied with adrenergic nerve fibers. So,

---

Clinical Panel 13.2   Drugs and the sympathetic system

Considerable scope is offered for pharmacologic interference at sympathetic nerve endings. Drugs that cross the blood–brain barrier (Ch. 5) may exert their effects on central rather than peripheral adrenoceptors. Potential sites of drug action are numbered in Figure CP 13.2.1.

1 Norepinephrine is loosely bound to a protein in the dense-cored vesicles. It can be unbound by specific drugs, whereupon it diffuses into the axoplasm and is degraded by monoamine oxidase (MAO).

2 Exocytosis into the synaptic cleft can be accelerated. Amphetamine, for example, exerts its stimulant effect by flooding the extracellular space with expelled norepinephrine and dopamine.

3 α or β receptors can be selectively either stimulated or blocked. As was mentioned in Chapter 7, a

receptor can be likened to a lock, and a drug that operates the lock is an *agonist*. A drug that 'jams' the lock without operating it is a *blocker*. 'Beta agonists' are used to relax the bronchial musculature in asthmatic patients. Cardioselective 'beta-blockers' are used to limit access of norepinephrine to α_1 receptors.

4 The amine uptake mechanism can be blocked in the central nervous system by the tricyclic antidepressant drugs or by cocaine. As a result, norepinephrine accumulates in the brain extracellular fluid.

5 Some antidepressant drugs increase the norepinephrine content of synaptic vesicles by inhibiting MAO, which normally degrades some of the transmitter after retrieval.

Clinical Panel 13.2 *Continued*

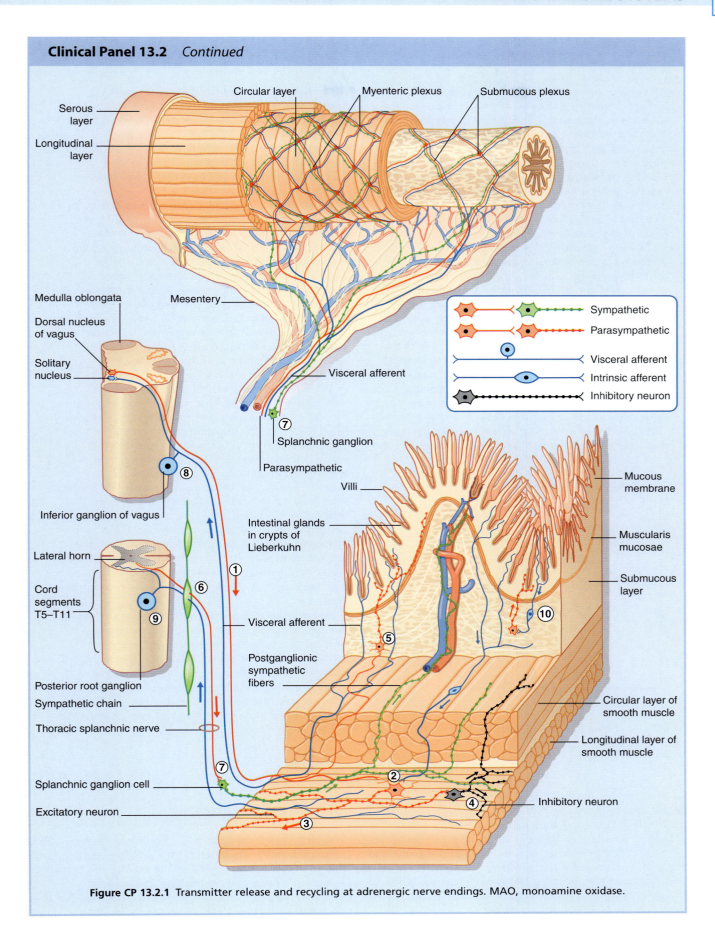

Figure legend:
- Sympathetic
- Parasympathetic
- Visceral afferent
- Intrinsic afferent
- Inhibitory neuron

Labels:
- Serous layer
- Longitudinal layer
- Circular layer
- Myenteric plexus
- Submucous plexus
- Medulla oblongata
- Dorsal nucleus of vagus
- Solitary nucleus
- Mesentery
- Visceral afferent
- Inferior ganglion of vagus
- Splanchnic ganglion
- Parasympathetic
- Villi
- Mucous membrane
- Intestinal glands in crypts of Lieberkuhn
- Lateral horn
- Cord segments T5–T11
- Muscularis mucosae
- Submucous layer
- Visceral afferent
- Postganglionic sympathetic fibers
- Posterior root ganglion
- Sympathetic chain
- Circular layer of smooth muscle
- Longitudinal layer of smooth muscle
- Thoracic splanchnic nerve
- Splanchnic ganglion cell
- Inhibitory neuron
- Excitatory neuron

Figure CP 13.2.1 Transmitter release and recycling at adrenergic nerve endings. MAO, monoamine oxidase.

Box 13.2 Enteric nervous system

The enteric nervous system (ENS) (Figure Box 13.2.1) extends from the midregion of the esophagus all the way to the anal canal. Throughout the length of this tube, it controls peristaltic activity, glandular secretion, and water and ion transfer. In addition, the ENS supplies the pancreas, liver, and gall bladder. The number of intrinsic neurons in the wall of the gastrointestinal tract has been reckoned about the same as in the entire spinal cord. The ENS is sometimes referred to as the 'gut brain' on account of its size and relative functional independence.

The intrinsic neurons of the gut are mainly deployed in two intramural plexuses, namely the **myenteric plexus** (*of Auerbach*) between the longitudinal and circular layers of smooth muscle, and the smaller **submucous plexus** (*of Meissner*). The principal 'drivers' of the muscle and glands belong to the parasympathetic division of the autonomic system. The dorsal (motor) nucleus of the vagus provides the *preganglionic parasympathetic* supply (1) to all parts with the exception of the distal colon and rectum, which receive their preganglionic supply from the *pelvic splanchnic nerves* (having parent neurons in the intermediolateral cell column of cord segments S2–S4). The drivers throughout are *intramural ganglion cells* located in both intramural plexuses. The beaded

postganglionic fibers of the myenteric plexus (2) initiate peristaltic waves by simultaneously causing the gut to contract in their own location (3) and to relax distally by activating inhibitory neurons (4). Parasympathetic ganglion cells in the wall of the gall bladder cause expulsion of bile. Those in the submucous plexus (5), and in the pancreas, cause glandular secretion.

Peristaltic activity persists even after total extrinsic denervation, because of the intrinsic circuitry and the spontaneous excitability of 'pacemaker' patches of smooth muscle (notably in stomach and duodenum).

The *preganglionic sympathetic* nerve supply originates in lateral horn cells of cord segments T5–11. The fibers traverse the paravertebral sympathetic chain (6) without synapsing here and terminate in the prevertebral, splanchnic ganglia (7) within the abdomen (celiac, superior and inferior mesenteric). Their beaded postganglionic fibers supply the smooth muscle of the intestine and of blood vessels, which they relax via β_2 receptors.

Visceral afferents reaching the central nervous system have their unipolar somas in a nodose ganglion of the vagus (8) and in posterior root ganglia at spinal levels T5–11 (9). The spinal afferents reach the posterior gray horn via *anterior* nerve roots. These *ventral root afferents* are of special clinical importance, because they include first-order *nociceptive* afferents that synapse centrally on lateral spinothalamic projection cells providing the principal 'pain pathway' to the brain.

Intrinsic visceral afferent neurons are in the form of bipolar neurons (12). Some participate in local reflex arcs within the myenteric or submucous plexus. Others (not shown) project as far as the splanchnic ganglia with the potential of exerting more widespread reflex effects.

Transmitters and *modulators* are numerous among the enteric ganglion cells. The principal excitatory transmitter is acetylcholine, with substance P cotransmitted as a modulator. The principal inhibitory transmitters are nitric oxide, γ-aminobutyric acid, and vasoactive intestinal polypeptide. Large numbers of different peptides have been revealed by means of histochemistry. More often than not, two or more are present within individual cells.

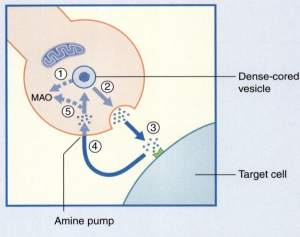

Figure Box 13.2.1 Enteric nervous system.

too, is the bone marrow. Adrenoceptors have been found on T cells, B cells, and macrophages.

During acute psychologic stress, raised levels of circulating norepinephrine may induce lymphatic tissue to respond by increasing the number of natural killer cells and cytotoxic lymphocytes. The consequent reduction of immune response to pathogens results in increased susceptibility to infections.

VISCERAL AFFERENTS

Afferents from thoracic and abdominal viscera utilize autonomic pathways to reach the CNS. They participate in important reflexes involved in the control of circulation, respiration, digestion, micturition, and coition.

Visceral activities are not normally perceived, but they do reach conscious levels in a variety of disease states.

Clinical Panel 13.3　Drugs and the parasympathetic system

Possible peripheral effects of cholinergic and anticholinergic drugs are listed in Figure CP 13.3.1. Some success has been achieved in the search for organ- or tissue-specific drugs. For example, the contribution of the vagus nerve to acid secretion in the stomach involves activation of a muscarinic receptor (M_1) that is distinct from the receptor type (M_2) found in the heart or on smooth muscle. An M_1-receptor blocker is available for patients suffering from peptic ulcer, for the specific purpose of reducing gastric acidity.

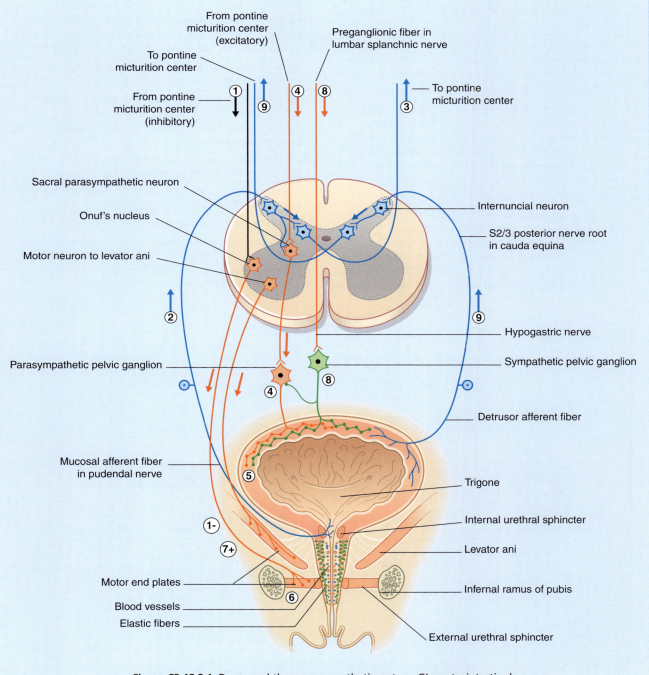

Figure CP 13.3.1 Drugs and the parasympathetic system. GI, gastrointestinal.

Box 13.3 Lower-level bladder controls

The female bladder is selected for this description, and also for higher-level bladder controls in Chapter 24.

Relevant anatomic details
- The smooth muscle of the detrusor in the body (corpus) of the bladder is an interwoven meshwork of fasciculi, and functions as a unit.

- The bladder neck is surrounded by two layers of longitudinal smooth muscle enclosing a layer of circular muscle, constituting the **internal urethral sphincter**.

- The outer longitudinal fibers descend within the mucous membrane of the urethra. When these fibers contract (along with the rest of the detrusor), they shorten and widen the urethral canal.

- The resting urethral canal is kept closed by a rich encircling web of elastic fibers, a highly vascular mucous membrane, a thin circular layer of smooth muscle, and the striated **external urethral sphincter**. Urologists tend to use the term *rhabdosphincter* in order to emphasize the *striated* nature of the external sphincter.

- The external urethral sphincter is richly endowed with slow-twitch, fatigue-resistant muscle fibers. It comes into play when abdominal pressure is raised either briefly (e.g. during a cough or sneeze) or for longer (e.g. while a heavy load is being carried). The cell group innervating this rhabdosphincter is the *nucleus of Onuf*

in the anterior gray horn at spinal cord levels S2 and S3. Most of the axons travel in the pudendal nerve.

The micturition cycle (Figure Box 13.3.1)
1 Immediately prior to the act of micturition, the anterior horn motor neurons to the levator ani and other muscles of the pelvic floor are inhibited by axons descending from the micturition center in the pons (Ch. 24). The neck of the bladder descends passively, and urine trickles into the urethra.

2 Mucosal fibers of the pudendal nerve, sensory to epithelium of trigone and urethra, discharge impulses to the posterior gray horn of cord segments S2–4.

3 From sacral cord, second-order sensory neurons discharge to the pontine micturition center.

4 Sacral parasympathetic neurons serving the bladder are simultaneously activated by the pontine micturition center and by neurons in the posterior horn at segmental levels S2–4.

5 The detrusor responds to postganglionic stimulation by contracting uniformly to expel the urine.

6 The rhabdosphincter, slave to Onuf, contracts to expel urine from the urethral canal.

7 Levator ani contracts to resume its supportive role.

8 Bladder filling recommences, while the bladder wall is rendered *compliant* by tonic inhibitory β_2 action of the sympathetic system on the detrusor muscle and by α_2 receptors on parasympathetic terminals.

Cholinergic drugs					
Pupillary constriction Near vision	Salivation	Constriction Secretion	Slowing	Gastric secretion increased Colic Diarrhea	Voiding of urine
Eye	**Salivary glands**	**Bronchi**	**Heart**	**GI tract**	**Bladder**
Pupillary dilatation Far vision	Dry mouth	Relaxation Sticky dry	Acceleration	Gastric secretion reduced Constipation	Retention of urine

Anticholinergic drugs

Figure Box 13.3.1 Lower-level bladder controls. (The assistance of Dr. Mary Pat FitzGerald, Department of Gynecology, Loyola University School of Medicine, Chicago, is gratefully appreciated.)

Box 13.3 *Continued*

9 When the bladder is half full, the stretch receptor afferents from the detrusor inform higher-level neurons in the brainstem, as described in Box 21.2.

Notes on urinary incontinence

Urinary incontinence afflicts about 30% of the female population at one or more periods of their lives. The two chief causes are detrusor instability and stress incontinence.

Detrusor instability (urge incontinence) is characterized by spontaneous expulsion of urine during the filling phase of the micturition cycle despite conscious attempts to inhibit it. There is as yet no general agreement as to the nature of this problem. Suggestions include (a) the development of 'pacemaker' patches of detrusor smooth muscle (by analogy with those of the intestine); (b) weakness of the bladder neck, allowing urine to escape and excite pudendal nerve endings with a consequent reflex detrusor response; and (c) overreaction of the detrusor to acetylcholine (ACh) being secreted in small amounts by the pelvic splanchnic nerve. In parous women and elderly nullipara, degeneration of pelvic splanchnic fibers may result in *up-regulation* of ACh receptors in the detrusor, whereby the muscle cells have inserted too many additional ACh receptors into their own plasmalemmas.

Parous women are also prone to *stress incontinence*. The underlying problem is *inherent weakness of the pelvic floor*, designed, as it is, to permit passage of the mature fetus. The pelvic floor weakness is often increased during parturition, by injury of the pudendal nerve resulting from its compression against the ischium.

Stress incontinence and urge incontinence also attend advancing age. One or other afflicts about 50% of women in institutional care. The primary problem in the elderly seems to be the progressive loss of striated muscle fibers in the urethral wall and pelvic floor.

In both sexes, disorders of brain function may also be responsible for incontinence (Ch. 31).

The use of serotonin or norepinephrine reuptake inhibitors for treatment of stress incontinence is discussed in Chapter 34.

REFERENCES

FitzGerald MP, Mueller E. Physiology of the lower urinary tract. Clin Obstet Gynecol 2004; 47:18–27.

Holstege G, Mouton LJ. Central nervous system control of micturition. Int Rev Neurobiol 2003; 56:123–1002.

Visceral pain is of immense importance in the context of clinical diagnosis.

Visceral pain

There are three fundamental types of visceral pain.

1 *Pure visceral pain*, felt in the region of the affected organ.
2 *Visceral referred pain*, projected subjectively into the territory of the corresponding somatic nerves.
3 *Viscerosomatic pain*, caused by spread of disease to somatic structures.

Pure visceral pain

Pure visceral pain is characteristically vague and deep-seated. It is often accompanied by sweating or nausea. It is experienced as the initial pain in association with inflammation and/or ulceration in the alimentary tract; with obstruction of the intestine, bile duct, or ureter; or when the capsule of a solid organ (liver, kidney, pancreas) is stretched by underlying disease. In marked contrast, the viscera are completely insensitive to cutting or burning.

Visceral referred pain

As its severity increases, visceral pain is 'referred' to somatic structures innervated from the same segmental levels of the spinal cord. For example, the pain of myocardial ischemia is referred to the chest wall ('angina pectoris'), pains of biliary or intestinal origin are referred to the anterior abdominal wall, and labor pains are referred to the sacral area of the back.

According to the generally accepted 'convergence–projection' theory of referred pain, the brain falsely interprets the source of noxious stimulation because visceral and somatic nociceptors have some spinothalamic neurons in common; in previous experience, these neurons habitually signaled somatic pain.

Viscerosomatic pain

The parietal serous membranes (pleura and peritoneum) receive a rich sensory supply from the overlying intercostal nerves, and they are exquisitely sensitive to acute inflammatory exudates. The extension of an inflammatory process to the surface of stomach, intestine, appendix, or gall bladder gives rise to a severe, steady pain in the abdominal wall directly overlying the inflamed organ. With the onset of acute peritonitis, the abdominal wall is 'splinted' by the muscles in a protective reflex.

Tenderness

Tenderness is *pain elicited by palpation*. In the abdomen, it is sought by pressing the hand and fingers against the abdominal wall. The clinician is, in effect, clothing the finger

Box 13.4 Functional innervation of male genital tract (Figure Box 13.4.1)

1 *Erection.* Psychic stimulation of the central *parasympathetic* pathway activates selected preganglionic neurons (P) to pelvic ganglia supplying parasympathetic fibers to the internal pudendal artery, where muscarinic and vasoactive intestinal polypeptide receptors cause the artery to relax, allowing blood to distend the penile cavernous tissue spaces. Cholinergic fibers also cause the relaxant transmitter nitric oxide to be released from the lining epithelium of the cavernous spaces.

2 *Secretion.* Parasympathetic ganglia in the walls of the prostate and seminal vesicles are stimulated to cause glandular secretion (via muscarinic receptors on the acini). These secretions contribute 80% of total semen volume.

3 *Emission.* Psychic stimulation of the central *sympathetic* pathway activates preganglionic neurons to pelvic ganglia supplying fibers to α_1 receptors on the smooth muscle of vas deferens, seminal vesicles, prostate, and internal urethral

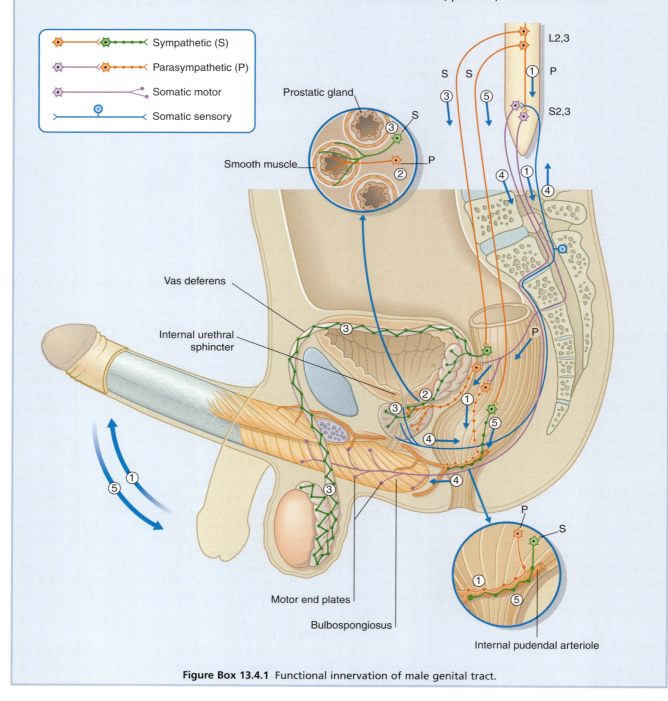

Figure Box 13.4.1 Functional innervation of male genital tract.

Box 13.4   *Continued*

(*preprostatic*) sphincter. Sperm and glandular contents are expelled into the urethra while the sphincter prevents backfire into the bladder. Simultaneous activation of bladder β_2 receptors prevents detrusor contraction.

4 *Ejaculation*. Entry of semen into the urethra activates somatic afferent nerve endings provided by the pudendal nerve. Through a reflex arc at S2–4 segmental levels, somatic motor fibers in the pudendal nerve cause rhythmic contractions of the bulbospongiosus muscles to ejaculate ('throw out') the semen into the vagina.

5 *Detumescence*. Selected central sympathetic fibers activate preganglionic neurons to pelvic sympathetic ganglia supplying fibers to α_1 receptors on pudendal arterioles at points of entry into the cavernous spaces. Arteriolar constriction results in detumescence.

The sympathetic system and psychogenic impotence
In addition to the prevertebral supply of the vas deferens mentioned above (3), a second, 'paravertebral' sympathetic pathway relays in the sacral sympathetic chain to supply the trabecular tissue, rich in β_2 adrenoreceptors. The resting, flaccid state of the penis depends on tonic activity in this pathway. In this context, the corpus cavernosum resembles a well-muscled artery. For erection to take place, the sympathetic supply must be switched off while the parasympathetic, relaxant supply is switched on. Both events may be coordinated at the level of the hypothalamus. Failure to switch off tonic sympathetic activity is regarded as the commonest immediate cause of *psychogenic impotence*, defined as impotence in the presence of intact anatomic pathways. Damage to reflex arcs, for example by spinal cord injury (Ch. 16), may cause *reflexive impotence*.

REFERENCES

Drossman DA, Camilleri M, Mayer E, et al. AGA technical review on irritable bowel syndrome. Gastroenterology 2002; 123:2108–2131.

Mach T. The brain–gut axis in irritable bowel syndrome—clinical aspects. Med Sci Monit 2004; 10:RA125–RA131.

Mulak A, Bonaz B. Irritable bowel syndrome: a model of the brain–gut interactions. Med Sci Monit 2004; 10:RA55–RA62.

pads with the patient's parietal peritoneum and using this to seek out an inflamed organ. If the organ is mobile, like the appendix, 'shifting tenderness' may be elicited if the patient is willing to roll from one side to the other.

Pain and the mind

Although visceral pain has well-established causative mechanisms (inflammation, spasm of smooth muscle, ischemia, distension), thoracic or abdominal pain may be experienced in the complete absence of visceral disease. Pain that recurs or persists over a long period (months), and is not accounted for by standard investigational procedures, is more likely to have a *psychologic* rather than a physical explanation. This is not to deny that the pain is real, but to imply that it originates within the brain itself. An example is the abused child whose abdominal pains represent a cry for help. In adults, recurrent and rather ill-defined pains are a common manifestation of *major depression* (see Ch. 25).

Irritable bowel syndrome is a very common disorder usually arising in the third or fourth decade. In this syndrome, there *is* evidence of abnormality at intestinal cellular level, but alterations of bowel behavior appear to be heightened by a *disorder of the brain–gut axis* (Clinical Panel 13.4).

Note on vascular afferents

Two *vascular* sets of unipolar neurons are customarily included in descriptions of the visceral afferent system. One supplies the carotid sinus and aortic arch with stretch receptors involved in the maintenance of the systemic blood pressure (Ch. 24); the other supplies the carotid body with chemoreceptors and is involved in respiratory control (Ch. 24). There is a progressive tendency to acknowledge *all* vascular afferents as being visceral, because those on peripheral blood vessels are morphologically and functionally the same as those serving the heart. They all contain substance P, are 'silent' in health, and subserve pain in the presence of disease or injury—as witness, the 'dragging' leg pains accompanying varicose veins, or the stab of pain when a clumsily inserted antecubital venipuncture needle strikes the brachial artery. The pathway to the posterior nerve roots is still uncertain, but it appears that (to an approximation) perivascular fibers above elbow and knee send impulses by the sympathetic route (but in the reverse direction), and that more peripheral perivascular fibers send messages in company with cutaneous nerves (and in the same direction). The notion of visceral afferents running in cutaneous nerves is reminiscent of their same service with respect to nerve fibers terminating in Golgi tendon organs at wrist and ankle.

Clinical Panel 13.4 Irritable bowel syndrome

Irritable bowel syndrome (IBS) is considered to be the most prevalent of all gastrointestinal tract disorders, affecting as much as 20% of the population in most countries. The exact incidence is uncertain because of the absence of a specific test; diagnosis is based on a constellation of symptoms associated with a suggestive psychosocial history. The disorder is three times more common in women, and onset is most frequent during the third and fourth decades.

The typical clinical picture is one of recurrent abdominal pain relieved by defecation. Some patients may have less than three bowel movements per week, others more than three per day. Both groups experience bloating (feeling of abdominal distension). Sensitivity to visceral sensations may have been triggered by a previous infectious or food allergy gastroenteritis. The typical psychologic profile is one of introspection and hypochondriasis accompanied by feelings of mental stress and anxiety. Sometimes, a history of childhood physical or sexual abuse may be elicited.

The overall situation is generally accepted as one of *dysfunction of the brain–gut axis*.

Figure 34.9 shows the position of the *emotional nociceptive area* within the cingulate gyrus. This area is activated by aversive (unpleasant) painful stimulation of any body part, as revealed by positron emission tomography. In IBS patient volunteers, it is activated by balloon distension of the distal colon at a lower balloon volume than in healthy controls. Heightened sensitivity to intestinal events seems to some extent *centrally* rather than peripherally generated. It is now believed that the preganglionic neurons of the parasympathetic system synapse mainly on *internuncial* neurons in the intestinal wall, rather than on the 'traditional' postganglionic motor neurons shown in Figure Box 13.2.1. The central drive of the parasympathetic may be an expression of stress, and because internuncial neurons may activate nociceptive afferents as well as motor neurons, heightened sensitivity may be maintained or even increased through this feedback loop.

At peripheral level, biopsies taken from ileum and colon indicate that heightened sensitivity may be the outcome of an immune response generated by earlier gastrointestinal infection or food allergy, as evidenced by proliferation of enterochromaffin cells in the wall of intestinal crypts, and/or mast cells in the lamina propria (Figure CP 13.4.1). The peptide granules of

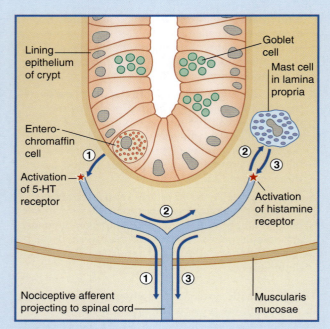

Figure CP 13.4.1 Activation of a nociceptive neuron in the wall of the colon. **(1)** Serotonin liberated by enterochromaffin cells has activated a nociceptive neuron projecting to posterior horn of spinal cord. **(2)** Antidromic impulses liberate substance P, which in turn releases histamine from mast cells. **(3)** Histamine reinforces the effect of serotonin.

enterochromaffin (chromate-staining) cells collectively contain more serotonin than does the entire brain. Serotonin liberated in response to intestinal distension has a double effect: it activates 5-HT$_3$ receptors on smooth muscle cells, thereby promoting peristaltic contractions; and it activates nociceptors on nearby visceral afferents, thereby causing mast cells to liberate histamine that in turn may potentiate the local effect of serotonin.

Following investigations to rule out organic disease, reassurance alone may be sufficient to restore equilibrium, although many patients benefit from psychotherapy. Drug treatments are essentially symptomatic, for example 5-HT$_3$-receptor antagonists or M$_3$-receptor anticholinergics for diarrhea, 5-HT$_3$-receptor agonists or cholinergics for constipation.

Core Information

The autonomic nervous system contains three neuron chains of effector neurons: central neurons project from hypothalamus–brainstem to brainstem–spinal cord preganglionic neurons. These send preganglionic fibers to autonomic ganglion cells, which in turn send postganglionic fibers to target tissues.

Sympathetic preganglionic outflow to the sympathetic chain of ganglia is thoracolumbar. Some fibers synapse in nearest ganglia. Some ascend to the superior cervical, middle cervical, or stellate ganglion, whence postganglionic fibers innervate head, neck, upper limbs, and heart. Some descend to synapse in lumbar or sacral ganglia, whence postganglionic fibers enter the lumbosacral plexus to supply lower limb vessels. Some pass through the chain and synapse instead in central abdominal ganglia (for the supply of gastrointestinal and genitourinary tracts) or in the adrenal medulla.

Parasympathetic preganglionic outflow is craniosacral. Cranial nerve distributions are oculomotor nerve via ciliary ganglion to sphincter pupillae and ciliaris; facial nerve via pterygopalatine ganglion to lacrimal and nasal glands; facial nerve via submandibular ganglion to submandibular and sublingual glands; glossopharyngeal nerve via otic ganglion to parotid gland; vagus nerve via ganglia on or in walls of heart, bronchi, and alimentary tract to muscle tissue and glands. Sacral nerves 2–4 deliver preganglionic fibers to intramural ganglia of distal colon and rectum, and to pelvic ganglia for supply of bladder and internal pudendal artery.

All preganglionic neurons are cholinergic. They activate nicotinic receptors in the ganglia. All postganglionic fibers end at neuroeffector junctions. In the sympathetic system, these are generally adrenergic, liberating norepinephrine that may activate postjunctional α_1 adrenoceptors on smooth muscle, prejunctional α_2 adrenoceptors on local nerve endings, postjunctional β_1 on cardiac muscle, or postjunctional β_2 that are more responsive to epinephrine. Epinephrine is liberated by adrenomedullary chromaffin cells, and resultant activation of β_2 receptors on smooth muscle causes relaxation.

Parasympathetic postganglionic fibers are cholinergic. The cholinoceptive receptors on cardiac and smooth muscle and glands are muscarinic.

Visceral afferents
Nociceptive afferents from thoracic and abdominal viscera and from blood vessels use autonomic pathways to reach the central nervous system. Pure visceral pain is vague and deep-seated. Visceral referred pain is experienced in somatic structures innervated from the same segmental levels. Viscerosomatic pain arises from chemical or thermal irritation of one of the serous membranes; the pain is severe and steady, and accompanied by protective contraction of body wall muscles.

REFERENCES

Andersson KF, Wagner D. Physiology of penile erection. Physiol Rev 1995; 75:191–236.
Cardozo L. Detrusor instability. In: Shaw RW, Soutter WP, Stanton SL, eds. Gynaecology. 2nd edn. New York: Churchill Livingstone; 1997:739–752.
Coupland RB. The natural history of the chromaffin cell. Arch Histol Cytol 1989; 52:331–341.
de Grout WC, Booth AM. Autonomic system to urinary bladder and sexual organs. In: Dyck PJ, Thomas PK, eds. Peripheral neuropathy, 3rd edn. Philadelphia: Saunders; 1993:198–207.
Dixon JS, Jen PYP, Gosling JA. Structure and autonomic innervation of the human vas deferens: a review. Microsc Res Tech 1998; 42:423–432.
Drummond PD. Autonomic innervation of the face. In: Gandevia SC, Burke D, Anthony M, eds. Science and practice of clinical neurology. Cambridge: Cambridge University Press; 1993:223–242.
Feigl EO. Neural control of coronary blood flow. In: Armour JA, Ardell JL, eds. Neurocardiology. New York: Oxford University Press; 1994:139–164.
Gai WP, Blessing WW. Human brainstem preganglionic parasympathetic neurons localized by markers for nitric oxide synthesis. Brain 1996; 119:1145–1152.
Hopkins DA, Bieger D, de Vente J, et al. Vagal efferent projections: viscerotopy, neurochemistry and effects of vagotomy. In: Holstege G, Bandler R, Saper CB, eds. The emotional motor system. Amsterdam: Elsevier; 1996:79–96.
Kinder MV, Bastiaanssen BHC, Janknegt RA, et al. Neuronal circuitry of the lower urinary tract. Anat Embryol 1995; 192:195–209.
Leonard BE. Fundamentals of psychopharmacology. 2nd edn. Chichester: Wiley; 1997.
Martinotti E. Adrenergic subtypes on vascular smooth muscle. Pharmacol Res 1991; 24:297–306.
McLeod JG. Disorders of the autonomic system. In: Gandevia SC, Burke D, Anthony M, eds. Science and practice in clinical neurology. Cambridge: Cambridge University Press; 1993:205–222.
McMahon SB, Dmitrieva N, Kolzenburg M. Visceral pain. Br J Anaesth 1995; 75:132–144.
Paintal AS. The visceral sensations—some basic mechanisms. Prog Brain Res 1986; 67:3–18.
Panuncio AL, De La Pena S, Gualco G, et al. Adrenergic innervation in reactive human lymph nodes. J Anat 1999; 194:143–146.
Perkin GD, Murray-Lyon I. Neurology and the gastrointestinal system. Neurol Neurosurg Psychiatry 1998; 65:291–300.
Perna FM, Schneiderman N, LaPierre A. Psychological stress, exercise and immunity. Int J Sports Med 1997; 18(S1):78–83.
Procacci P, Zoppi M, Maresca M. Clinical approach to visceral sensation. Prog Brain Res 1986; 67:21–28.
Randall WC. Efferent sympathetic innervation of the heart. In: Armour JA, Ardell JL, eds. Neurocardiology. New York: Oxford University Press; 1994:77–94.
Schott GD. Visceral afferents: their contribution to 'sympathetic-dependent' pain. Brain 1994; 117:397–413.

van Leishout JJ, Wieling W, Karemaker JM, et al. The vasovagal response. Clin Sci 1991; 81:575–586.

Wittling W, Block A, Genzel S, et al. Hemisphere asymmetry in parasympathetic control of the heart. Neuropsychologia 1998; 36:461–468.

Wittling W, Block A, Schweiger E, et al. Hemisphere asymmetry in sympathetic control of the human myocardium. Brain Cogn 1998; 38:17–35.

Nerve roots

STUDY GUIDELINES

1 During embryonic development, some immature neurons send out ventral roots, while others project along the marginal zone to form fiber tracts. Sensory fibers enter the dorsal horn from the neural crest.

2 In the mature vertebral canal, note disparities of levels, for example a collapsed T11 vertebra would crush cord segment L1.

3 Recall that ventral roots S2–4 of cauda equina contain preganglionic parasympathetic fibers vital for bladder and bowel control, and that the corresponding posterior roots contain visceral afferents vital for reflexes.

4 Do remember the extradural venous plexus and the harm it can do.

5 It is clinically important to appreciate that a sense of numbness or tingling in the fingers in later life may result from compression of posterior nerve roots!

6 For the commonest and lowest two levels of disk prolapse, the next spinal nerve is the one likely to be caught.

7 Make certain that you understand the anatomy of lumbar puncture (spinal tap).

DEVELOPMENT OF THE SPINAL CORD

Cellular differentiation

The neural tube of the embryo consists of a pseudostratified epithelium surrounding the neural canal (Figure 14.1A). Dorsal to the sulcus limitans, the epithelium forms the **alar plate**; ventral to the sulcus it forms the **basal plate**.

The neuroepithelium contains germinal cells that synthesize DNA before retracting to the innermost, **ventricular zone**, where they divide. The daughter nuclei move outward, synthesize fresh DNA, then retreat and divide again. After several such cycles, postmitotic cells round up in the **intermediate zone**. Some of the postmitotic cells are immature neurons; the rest are **glioblasts** that, after further division, become astrocytes or oligodendrocytes. Some of the glioblasts form an ependymal lining for the neural canal.

The microglial cells of the central nervous system (CNS) are derived from basophil cells of the blood.

Enlargement of the intermediate zone of the alar plate creates the dorsal horn of gray matter. The dorsal horn receives central processes of dorsal root ganglion cells (Figure 14.1B). As explained in Chapter 1, the ganglion cells derive from the neural crest.

Partial occlusion of the neural canal by the developing dorsal gray horn gives rise to the dorsal median septum and to the definitive central canal of the cord (Figure 14.1C).

Enlargement of the intermediate zone of the basal plate creates the ventral gray horn and the ventral median fissure (Figure 14.1C). Axons emerge from the ventral horn and form the ventral nerve roots.

In the outermost, **marginal zone** of the cord, axons run to and from spinal cord and brain.

Neural cord

The neural tube reaches caudally only to the level of the second lumbar mesodermal somites. Following closure of the posterior neuropore, the ectoderm and mesoderm at the level of the more caudal lumbar and sacral somites blend to form a **neural cord**. This ribbon of cells becomes canalized and links up with the neural tube; it forms the lower end of the spinal cord.

Ascent of the spinal cord (Figure 14.2)

The spinal cord occupies the full length of the vertebral canal until the end of the 12th postconceptual week. The sixth to eighth are marked by regression of the caudal end of the neural tube to become a neuroglial thread, the **filum terminale**.

After the 12th week, the vertebral column grows rapidly and drags the cord upward. The tip of the cord is at second or third lumbar level at the time of birth. The adult level (first or second lumbar) is attained 3 weeks later.

In consequence of greater ascent of the lower part of the cord compared with the upper part, the spinal nerve roots show an increasing disparity between their segmental levels of attachment to the cord and the corresponding vertebral levels (Figure 14.3).

Neural arches

During the fifth week, the mesenchymal vertebrae surrounding the notochord give rise to **neural arches** for protection of the spinal cord (Figure 14.4). The arches are initially *bifid* (split). Later, they fuse in the midline and form the vertebral spines.

Conditions where the two halves of the neural arches have failed to unite are collectively known as *spina bifida* (Clinical Panel 14.1).

ADULT ANATOMY

The spinal cord and nerve roots are sheathed by pia mater and float in cerebrospinal fluid contained in the

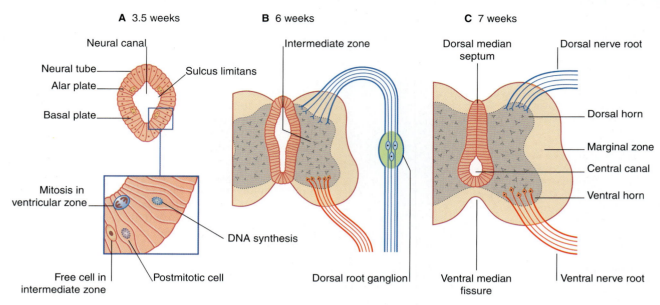

A 3.5 weeks   **B** 6 weeks   **C** 7 weeks

Neural canal
Neural tube
Alar plate
Basal plate
Sulcus limitans
Mitosis in ventricular zone
DNA synthesis
Free cell in intermediate zone
Postmitotic cell

Intermediate zone
Dorsal root ganglion

Dorsal median septum
Dorsal nerve root
Dorsal horn
Marginal zone
Central canal
Ventral horn
Ventral median fissure
Ventral nerve root

Figure 14.1 (A–C) Cellular differentiation in the embryonic spinal cord.

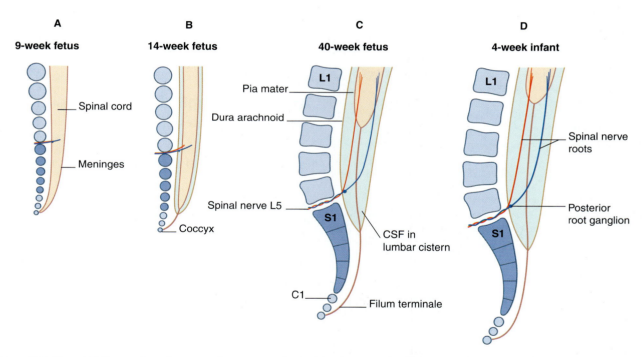

A
9-week fetus

B
14-week fetus

C
40-week fetus

D
4-week infant

Spinal cord
Meninges

Coccyx

Pia mater
Dura arachnoid
Spinal nerve L5
CSF in lumbar cistern
Filum terminale

Spinal nerve roots
Posterior root ganglion

Figure 14.2 (A and B) Regression of coccygeal segments of spinal cord creates the filum terminale. (C and D) Ascent of spinal cord. CSF, cerebrospinal fluid. (*Note*: Recent evidence indicates that, as represented here, the number of embryonic coccygeal vertebrae does not exceed three or four.)

subarachnoid space. The pial **denticulate ligament** pierces the arachnoid and anchors the cord to the dura mater on each side. Outside the dura is the **extradural (epidural) venous plexus** (Figure 14.5), which harvests the vertebral red marrow and empties into the segmental veins (deep cervical, intercostal, lumbar, sacral). These veins are without valves, and reflux of blood from the territory of segmental veins is a *notorious* cause of cancer spread from prostate, lung, breast, and thyroid gland. For example, nerve root compression from collapse of an invaded vertebra may be the presenting sign of cancer in one of these organs.

The respective anterior and posterior nerve roots join at the intervertebral foramina, where the posterior root

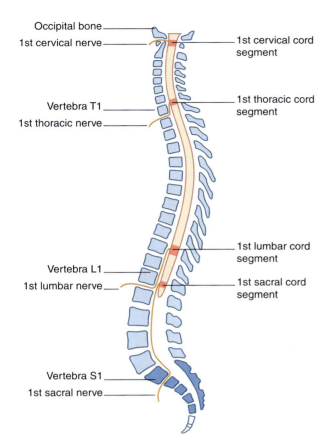

Figure 14.3 Segmental and vertebral levels compared. Spinal nerves 1–7 emerge above the corresponding vertebrae; the remaining spinal nerves emerge below.

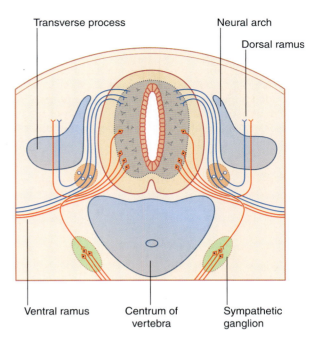

Figure 14.4 Normal, bifid stage of neural arch development in an embryo of 8 weeks.

ganglia are located (Figure 14.5). The arachnoid mater blends with the perineurium of the spinal nerve, and the dura mater blends with the epineurium. The nerve roots carry extensions of the subarachnoid space into the intervertebral foramina.

Below cord level, nerve roots seeking the lower lumbar and sacral intervertebral foramina constitute the **cauda equina** ('horse's tail'). The cauda equina floats in the lumbar subarachnoid cistern (Figures 14.6–14.8), which reaches to the level of the second sacral vertebra. At its upper end, the cauda comprises nerve roots L3–S5 of both sides—*a total of 32 roots* (excluding the insignificant coccygeal roots).

In the center of the cauda equina is the unimportant filum terminale, which pierces the meninges to become attached to the coccyx.

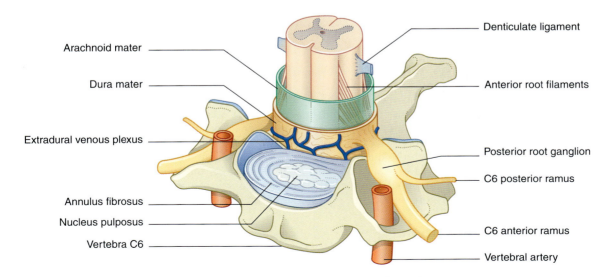

Figure 14.5 Relationships of the sixth cervical spinal nerve.

Clinical Panel 14.1 Spina bifida

Among the more common congenital malformations of the central nervous system are several conditions included under the general heading *spina bifida*. The 'bifid' effect is produced by failure of union of the two halves of the neural arches, usually in the lumbosacral region (Figure CP 14.1.1).

Spina bifida occulta **(A)** is usually symptom-free, being detected incidentally in lumbosacral radiographs.

In *spina bifida cystica*, a meningeal cyst protrudes through the vertebral defect. In 10% of these cases, the cyst is a *meningocele* containing no nervous elements **(B)**. In 90%, unfortunately, the cyst is a *meningomyelocele*, containing either spinal cord or cauda equina **(C)**; the lower limbs, bladder, and rectum are paralyzed, as in the case illustrated in Figure CP 14.1.2, and meningitis is likely to supervene sooner or later. To make matters worse, an *Arnold–Chiari malformation* (Ch. 4) is almost always present as well.

The most severe form of spina bifida is *myelocele* **(D)**, where the neural folds have remained open and cerebrospinal fluid leaks on to the surrounding skin. The clinical outlook is very poor.

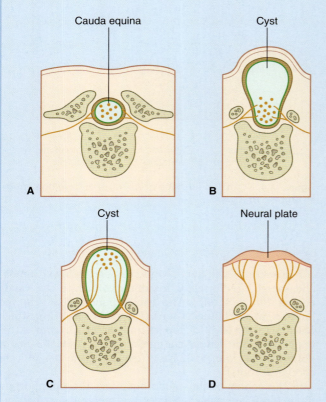

Figure CP 14.1.1 Varieties of spina bifida. **(A)** Spina bifida occulta. **(B)** Meningocele. **(C)** Meningomyelocele. **(D)** Myelocele.

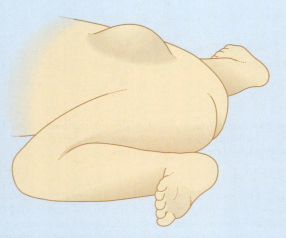

Figure CP 14.1.2 Lumbar meningomyelocele (from a photograph). The 'frog leg' posture is characteristic of combined femoral and sciatic nerve paralysis, with preservation of hip flexion by the iliopsoas.

DISTRIBUTION OF SPINAL NERVES

Each spinal nerve gives off a *recurrent* branch that provides mechanoreceptors and pain receptors for the dura mater, posterior longitudinal ligament, and intervertebral disk. The synovial *facet* joints between successive articular processes are each supplied by the nearest three spinal nerves. Pain caused by injury or disease of any of the above structures is referred to the cutaneous territory of the corresponding posterior rami (Figure 14.9).

Segmental sensory distribution: the dermatomes

A **dermatome** is the strip of skin supplied by an individual spinal nerve. The dermatomes are orderly in the embryo

(Figure 14.10), but they are distorted by outgrowth of the limbs (Figure 14.11). Spinal nerves C5–T1 are drawn into the upper limb, so that C4 dermatome abuts T2 at the level of the sternal angle. Nerves L2–S2 are drawn into the lower limb, so that L2 abuts S3 dermatome over the buttock. Maps like those in Figure 14.11 fail to portray *overlap* in the cutaneous distribution of successive dorsal nerve roots. On the trunk, for example, the skin over an intercostal space is supplied by the nerves immediately above and below in addition to the proper nerve.

Segmental motor distribution

In the limbs, the individual muscles are supplied by more than one spinal nerve because of interchange in the brachial and lumbosacral plexuses. The segmental supply

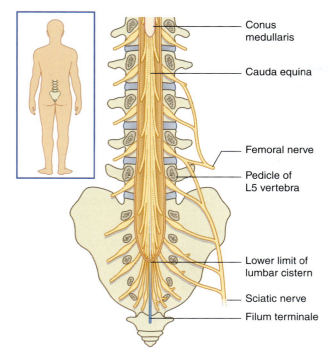

Conus medullaris

Cauda equina

Femoral nerve

Pedicle of L5 vertebra

Lower limit of lumbar cistern

Sciatic nerve

Filum terminale

Figure 14.6 The cauda equina in the lumbar cistern. Contributions to the femoral and sciatic nerves are shown on the right side.

of the limbs is expressed in terms of *movements* in Figure 14.12.

Segmental sensory inputs and segmental motor outputs are combined during execution of *withdrawal* or *avoidance reflexes* (Box 14.1). (The prevalent term, *flexor reflex*, is too limited; e.g. a stimulus applied to the lateral surface of a limb may elicit adduction.)

Nerve root compression syndromes

Nerve root compression within the vertebral canal is most frequent where the spine is most mobile, namely at lower cervical and lower lumbar levels (Clinical Panel 14.2). The effects of root compression may be expressed in five different ways.

1 *Pain* perceived in the muscles supplied by the corresponding spinal nerve(s).
2 *Paresthesia* (numbness or tingling) along the respective dermatome(s).
3 *Cutaneous sensory loss*—more likely if two successive dermatomes are involved, because of overlap.
4 *Motor weakness*.
5 *Loss of a tendon reflex* if the segmental level is appropriate (Table 14.1).

Note: Peripheral nerve entrapment syndromes are considered in Chapter 12.

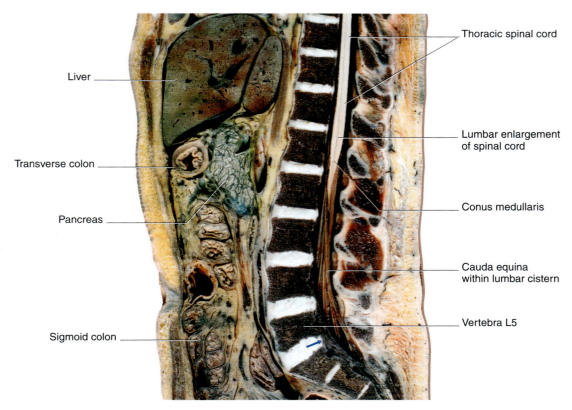

Liver

Transverse colon

Pancreas

Sigmoid colon

Thoracic spinal cord

Lumbar enlargement of spinal cord

Conus medullaris

Cauda equina within lumbar cistern

Vertebra L5

Figure 14.7 Midline sagittal section of embalmed cadaver, displaying thoracic, lumbar, and sacral (cauda equina) spinal cord and cauda equina. *Arrow* indicates most frequent intervertebral disk to prolapse.

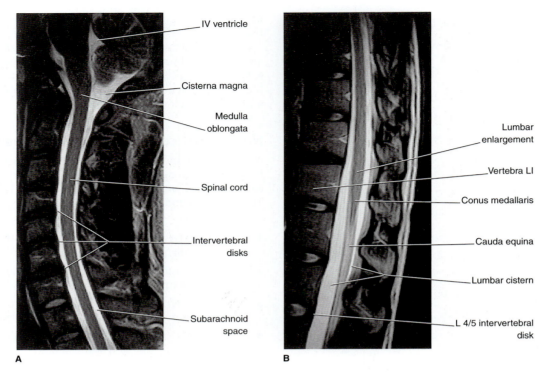

Figure 14.8 Sagittal MRI scans of the vertebral canal, weighted so as to enhance cerebrospinal fluid. **(A)** Brainstem, cerebellum, and cervical spinal cord are outlined. **(B)** Lumbosacral spinal cord and cauda are outlined. (From a series kindly provided by Professor J. Paul Finn, Director, Magnetic Resonance Research, Department of Radiology, David Geffen School of Medicine at UCLA, California.)

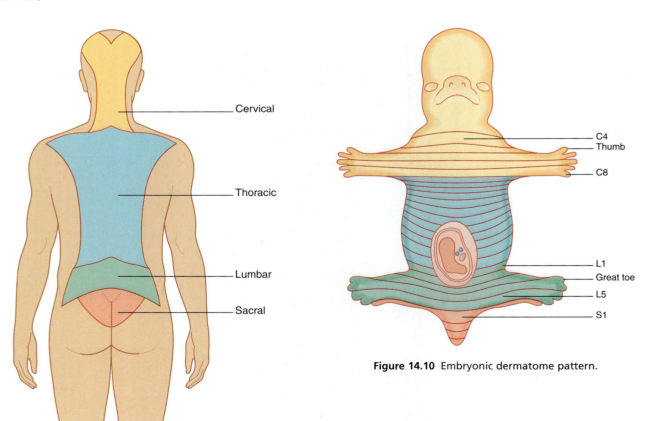

Figure 14.9 Cutaneous distribution of posterior rami of spinal nerves.

Figure 14.10 Embryonic dermatome pattern.

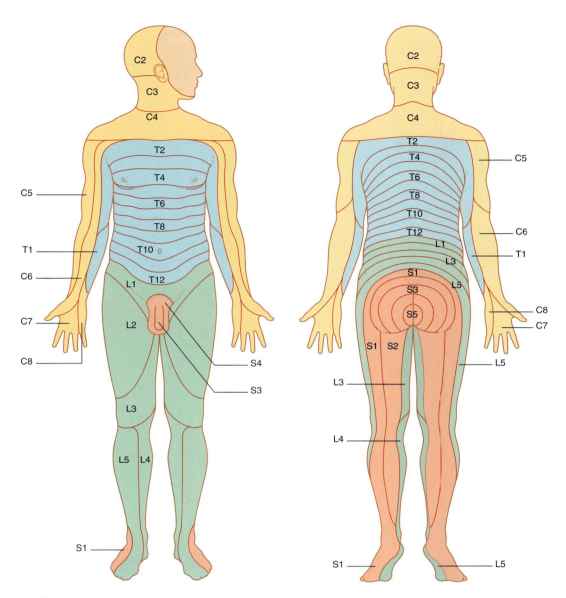

Figure 14.11 Adult dermatome pattern.

Lumbar puncture (spinal tap)

This procedure involves removing a sample of cerebro-spinal fluid from the lumbar cistern. This procedure should not be performed if there is any reason to suspect the presence of raised intracranial pressure. (It *is* performed occasionally when there is some uncertainty, but only under immediate neurosurgical cover.)

Table 14.1 Segmental levels of tendon reflexes	
Segmental level	**Reflex**
C5, C6	Biceps
	Brachioradialis ('supinator reflex')
C7	Triceps
L3, L4	Knee jerk
S1	Ankle jerk

Anesthetic procedures

A so-called *spinal anesthetic* is often given in preference to a general anesthetic prior to surgical procedures on the prostate in the elderly. A local anesthetic is injected into the lumbar cistern in order to block impulse conduction in the lumbar and sacral nerve roots. Care is taken that the anesthetic does not reach a high level in the subarachnoid space, for fear of paralyzing the intercostal and phrenic nerve root fibers serving respiration.

Anesthesia and childbirth

In skilled hands, pain-free labor can be assured by block-ing the lumbar and sacral nerve roots extradurally. For *epidural anesthesia*, local anesthetic is carefully intro-duced into the extradural space by the lumbar route. For *caudal anesthesia* (rarely performed), the extradural space is approached in an upward direction, through the sacral

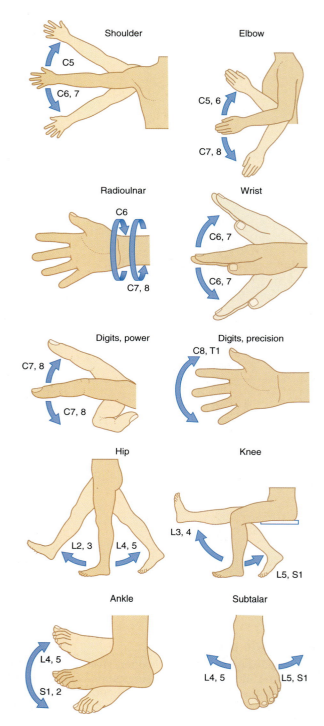

Figure 14.12 Segmental control of limb movements. (After Rosse and Clawson 1980, with permission of Harper & Row, and Last 1973.)

hiatus. In both procedures, the anesthetic diffuses through the dural sheath of the nerve roots where they leave the vertebral canal. Labor may be prolonged because of interruption of excitatory reflex arcs linking perineum to uterus through the lower end of the spinal cord. However, avoidance of general anesthesia is valuable in allowing immediate bonding to take place between mother and child.

Box 14.1   Lower limb withdrawal reflex

Figure Box 14.1.1 depicts a lower limb *withdrawal reflex* with *crossed extensor thrust*. **(A)** The right foot is about to enter the stance phase of locomotion. **(B)** Contact with a sharp object initiates a withdrawal reflex, together with the crossed extensor response required to support the entire body weight.

Sequence of events

1 Plantar nociceptors send impulse trains along tibial–sciatic afferent fibers (a) having parent posterior root ganglion somas within the L5–S1 intervertebral foramen. The impulses ascend cauda equina (b) and enter segment L5 of spinal cord. Some impulses are dispatched up and down Lissauer's tract (c) to activate segments L2–4 and S1.

2 In all five segments, primary nociceptive afferents excite *flexor reflex internuncials* in the base of the posterior horn (2a). Several internuncials may be interposed, in series, between entering afferents and target motor neurons. Axons of medially placed internuncials cross the midline in the gray commissure, allowing impulse trains to activate contralateral internuncials (2b).

3 On the stimulated side, α and γ motor neurons in cord segments L3–S1 contract iliopsoas (a), hamstrings (b), and ankle dorsiflexors (d). At the same time (not shown here), ipsilateral 1a inhibitory internuncials are recruited to silence the antigravity motor neurons.

4 On the contralateral side, α and γ motor neurons in cord segments L2–5 contract gluteus maximus (not visible here) and quadriceps femoris (c).

Note: Not shown in the figure are *lateral spinothalamic tract* relay neurons (see Ch. 15). These neurons receive inputs from nociceptive afferent fibers in Lissauer's tract, and they relay impulses to brain sites able to decode the location and nature of the initial stimulus (Box 14.1).

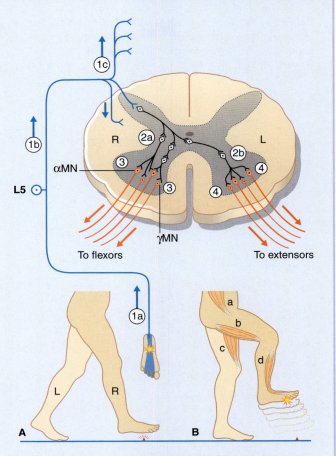

Figure Box 14.1.1 Withdrawal reflex. MN, motor neuron.

Clinical Panel 14.2   Nerve root compression

Cervical roots

The intervertebral disks and synovial joints of the neck are subject to degenerative disease (*cervical spondylosis*) in 50% of 50-year-olds and in 70% of 70-year-olds. Although any or all of the joints may deteriorate, problems are most frequent in relation to vertebra C6, which provides the fulcrum for flexion–extension movements of the neck. Spinal nerve C6 (above) or C7 (below) may be pinched by extruded disk material or by bony outgrowths (*osteophytes*) beside the synovial joints (Figure CP 14.2.1). Sensory, motor, and reflex disturbances may result in accordance with the data in Table 14.1 and in Figures 14.10 and 14.11.

Lumbosacral roots

One important cause of low back pain is a *prolapsed intervertebral disk (herniated nucleus pulposus)*. Fully 95% of all disk prolapses occur immediately above or below the last lumbar vertebra. The typical herniation is *posterolateral*, with compression of the nerve roots passing to the *next* intervertebral foramen (Figure CP 14.2.2).

Symptoms include backache caused by rupture of the annulus fibrosus, and pain in the buttock, thigh, or leg caused by pressure on posterior root fibers contributing to the sciatic nerve. The pain is increased by stretching the affected root, for example by having the straightened leg raised by the examiner.

An L4–5 disk prolapse produces pain or paresthesia over the L5 dermatome. Motor weakness may be detected during dorsiflexion of the great toe (later, of all toes and of the ankle), and during eversion of the foot. Abduction of the hip may also be weak; this movement is tested with the patient lying on one side.

With an L5–S1 prolapse (the commonest of all) (Figure CP 14.2.3), symptoms are felt in the back of the leg or sole of foot (S1 dermatome). Plantar flexion may be weak, and the ankle jerk reduced or absent.

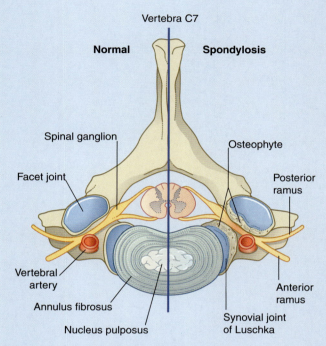

Figure CP 14.2.1 Spondylosis on the right side of vertebra C7. Osteophytes are pinching C7 spinal nerve trunk.

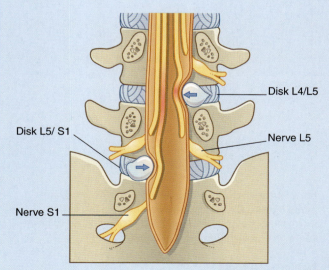

Figure CP 14.2.2 Nerves compressed (arrows) by posterolateral prolapse of the two lowest intervertebral disks.

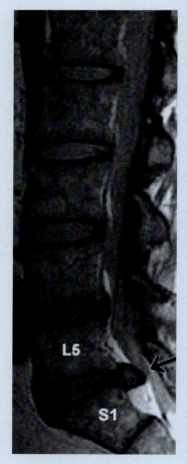

Figure CP 14.2.3 Sagittal MRI revealing a prolapsed L5–S1 intervertebral disk pressing against the cauda equina (*arrow*). (Kindly provided by Professor Robert D. Zimmerman, Department of Radiology, Weil Medical College of Cornell University, New York.)

Core Information

The neuroepithelium of the embryonic cord undergoes mitotic activity in the inner, ventricular zone. Daughter cells move into the intermediate zone and become either neuroblasts or glioblasts. The developing dorsal horn receives central processes of neural crest–derived spinal ganglion cells. The ventral horn issues axons that form ventral nerve roots. The outer, marginal zone contains the axons of developing nerve pathways. The caudal end of the cord develops separately, from the caudal cell mass, which links up with the neural tube. After the 12th week, rapid growth of the vertebral column drags the cord up the vertebral canal; the lower tip of the cord is at L2–3 level at birth and at L1–2 level 3 weeks later. The result is a progressive disparity between segmental levels of nerve root attachment to the cord and intervertebral levels of exit of spinal nerves. The neural arches are dorsal projections of vertebral mesenchyme; the initial bifid arrangement is normally lost by fusion of the projections to form spines.

The mature cord and nerve roots are sheathed by pia mater and float in the subarachnoid space, anchored to dura by the denticulate ligament. The extradural space contains valveless veins that drain vertebral bone marrow into segmental veins and provide potential avenues for spread of cancer cells. Below the level of the cord, the cauda equina comprises paired nerve roots L3–S5 of both sides. As it emerges from the intervertebral foramen (occupied by the posterior root ganglion), each spinal nerve gives a recurrent branch supplying ligaments and dura mater.

Segmental sensory distribution is shown by the regular dermatomal pattern of skin innervation by the posterior roots (via the mixed peripheral nerves). Segmental motor supply is expressed in the form of movements performed by specific muscle groups. Nerve root compression, for example by a prolapsed disk, may be expressed segmentally by muscle pain, dermatomal paresthesia, cutaneous sensory loss, motor weakness, or loss of a tendon reflex.

Lumbar puncture (spinal tap) is performed by passing a careful needle between spines at L3–4 or L4–5—but not if raised intracranial pressure is suspected. A spinal anesthetic is given by injecting local anesthetic into the lumbar cistern. An epidural anesthetic is given into the lumbar epidural space. A caudal anesthetic is given through the sacral hiatus.

REFERENCES

Adams CBT, Logue V. Studies in cervical spondylitic myelopathy. 1. Movement of the cervical roots, dura, and cord, and their relation to the course of extrathecal roots. Brain 1971; 94:557–568.

Auteroche P. Innervation of the zygapophyseal joints of the lumbar spine. Anat Clin 1983; 5:18–28.

Barson AJ, Logue V. The vertebral level of termination of the spinal cord during normal and abnormal development. J Anat 1970; 106:489–497.

Bogduk N. Spinal pain: backache and neck pain. In: Gandevia SC, Burke D, Anthony M, eds. Science and practice in clinical neurology. Cambridge: Cambridge University Press; 1993:39–60.

Groen DJ, Baljet R, Drukker J. Nerves and nerve plexuses of the human vertebral column. Am J Anat 1990; 188:282–296.

Holsheimer J, den Boer JA, Strujik JJ, et al. MR assessment of the normal position of the spinal cord in the spinal canal. Am J Neuroradiol 1994; 16:951–959.

Postacchini F, Rauschning W. Anatomy. In: Postacchini F, ed. Lumbar disc herniation. Berlin: Springer-Verlag; 1999:18–58.

Rosse C, Clawson DK. The musculoskeletal system in health and disease. Hagerstown: Harper & Row; 1980.

Russell EJ. Cervical disc disease. Radiology 1990; 187:314–325.

Sunderland S. Meningeal–dural relationships in the intervertebral foramen. J Neurosurg 1974; 40:756–763.

Spinal cord: ascending pathways

STUDY GUIDELINES

1 Confirm that the mature spinal cord is not segmented internally.

2 Confirm that anterior horn cells take the form of columns rather than laminae.

3 Appreciate that 'non-conscious sensation' simply means that the ascending afferent impulse activity concerned does not generate any kind of perception.

4 Remember that conscious proprioception is more sensitive than either vision or the vestibular labyrinth in telling us when we are going off balance.

5 Get used to the idea that muscles tell us more than joints do about the position of our limbs in space.

6 It is clinically important to appreciate that one of the two 'conscious' pathways crosses the midline at all levels of the spinal cord, whereas the other crosses all at once, within the brainstem.

7 Appreciate the meaning of the term *dissociated sensory loss*.

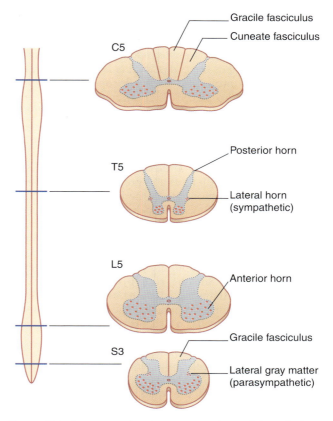

Figure 15.1 Representative transverse sections of the spinal cord.

GENERAL FEATURES

The arrangement of gray and white matter at different levels of the spinal cord is shown in Figure 15.1. The cervical and lumbosacral enlargements are produced by expansions of the gray matter required to service innervation of the limbs. White matter is most abundant in the upper reaches of the cord, which contain the sensory and motor pathways serving all four limbs. In the posterior funiculus, for example, the gracile fasciculus carries information from the lower limb and is present at cervical as well as lumbosacral segmental levels, whereas the cuneate fasciculus carries information from the upper limb and is not seen at lumbar level.

Although it is convenient to refer to different levels of the spinal cord in terms of numbered segments corresponding to the sites of attachment of the paired nerve roots, the cord shows no evidence of segmentation internally. The nuclear groups seen in transverse sections are in reality cell columns, most of them spanning several segments (Figure 15.2).

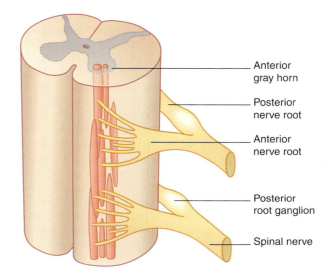

Figure 15.2 Two segments of the spinal cord, showing cell columns in the anterior gray horn.

Types of spinal neuron

The smallest neurons (soma diameters 5–20 µm) are *propriospinal*, being entirely contained within the cord. Some are confined within a single segment; others span two or more segments by way of the neighboring **propriospinal tract**. Many of the smallest neurons participate in spinal reflexes. Others are intermediate cell stations interposed between fiber tracts descending from the brain and motor neurons projecting to the locomotor apparatus. Others again are so placed as to influence sensory transmission from lower to higher levels of the central nervous system (CNS).

Medium-sized neurons (soma diameters 20–50 µm) are found in all parts of the gray matter except the substantia gelatinosa. Most are *relay (projection) cells* receiving inputs from posterior root afferents and projecting their axons to the brain. The projections are in the form of *tracts*, a tract being defined as a functionally homogeneous group of fibers. As will be seen, the term *tract* is often used loosely, because many projections originally thought to be 'pure' contain more than one functional class of fiber.

The largest neurons of all are the **alpha motor neurons** (soma 50–100 µm) for the supply of skeletal muscles. Scattered among them are small, **gamma motor neurons** supplying muscle spindles. In the medial part of the anterior horn are **Renshaw cells**, which exert tonic inhibition on alpha motor neurons.

Spinal reflex arcs originating in muscle spindles and tendon organs have been described in Chapter 10, and the withdrawal reflex in Chapter 14.

In thick sections of the spinal cord, the nerve cells exhibit a laminar (layered) arrangement. True lamination is confined to the posterior horn (Figure 15.3), but 10 laminae of Rexed have been defined in the gray matter as a whole in order to correlate findings from animal research in different laboratories.

Spinal ganglia

The spinal or posterior root ganglia are located in the intervertebral foramina, where the anterior and posterior roots come together to form the spinal nerves. Thoracic ganglia contain about 50 000 unipolar neurons, and those serving the limbs contain about 100 000. The individual ganglion cells are invested with modified Schwann cells called **satellite cells** (Figure 15.4). The common stem axon of each cell bifurcates, sending a centrifugal process into one or other ramus of the spinal nerve (or into the recurrent branch) and a centripetal ('center-seeking') process into the spinal cord. Following stimulation of the peripheral sensory receptors, trains of nerve impulses traverse the point of bifurcation without interruption, although the cell body is also depolarized. The initial segment of the stem axon does not normally generate impulses, but it may do so if the adjacent part of the posterior root is compressed, for example by a prolapsed intervertebral disk.

Traditionally, the centripetal axons of all spinal ganglion cells have been thought to enter posterior nerve roots. It is now known that many visceral afferents (in particular) enter the cord by way of ventral roots and work their way to the posterior gray horn (Ch. 13). This feature accounts for the frequent failure of *posterior rhizotomy* (surgical

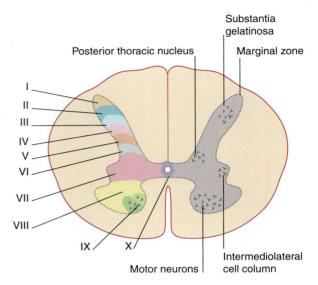

Figure 15.3 Laminae (I–X) and named cell groups at midthoracic level.

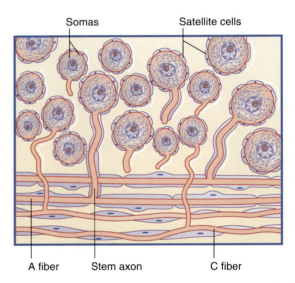

Figure 15.4 Posterior root ganglion. In bottom of figure, note T-shaped bifurcation of stem fibers.

section of posterior roots) to relieve pain originating from intraabdominal cancer.

Central terminations of posterior root afferents (Figure 15.5)

In the *dorsal root entry zone* close to the surface of the cord, the afferent fibers become segregated into medial and lateral streams. The medial stream comprises medium and large fibers that divide within the posterior funiculus into ascending and descending branches. The branches swing into the posterior gray horn, and synapse in laminae II, III, and IV. The largest ascending fibers run all the way to the posterior column nuclei (gracilis–cuneatus) in the medulla oblongata. These long fibers form the bulk of the gracile and cuneate fasciculi.

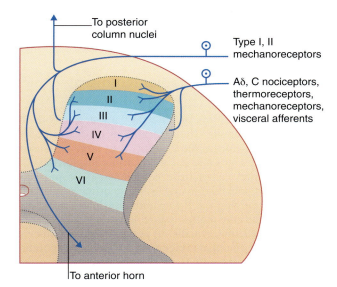

Figure 15.5 Targets of primary afferent neurons in the posterior gray horn.

The lateral stream comprises small (Aδ and C) fibers that, on entry, divide into short ascending and descending branches within the **posterolateral tract** of Lissauer. They synapse on neurons in the marginal zone (lamina I) and in the substantia gelatinosa (lamina II); some fibers synapse on dendrites of cells belonging to laminae III–V.

ASCENDING SENSORY PATHWAYS

Categories of sensation

In accordance with the flow chart in Figure 15.6, neurologists speak of two kinds of sensation: conscious and non-conscious (unconscious). Conscious sensations are perceived at the level of the cerebral cortex. Non-conscious sensations are not perceived; they have reference to the cerebellum (see later).

Conscious sensations

There are two kinds of conscious sensation: *exteroceptive* and *proprioceptive*. Exteroceptive sensations come from the

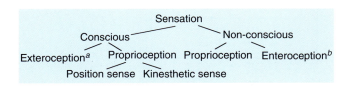

Figure 15.6 Categories of sensation. [a]Exteroceptors can be categorized as *telereceptors* receiving from a distance (retina and cochlea) and somatic receptors on the body surface (touch, pain, etc.). [b]Enteroceptors (*Gr. enteron*, 'gut') are strictly a subdivision of *interoceptors*, a term signifying all of the viscera. In pathologic states, they may produce conscious visceral–viscerosomatic sensations.

external world; they impinge either on somatic receptors on the body surface or on telereceptors serving vision and hearing. Somatic sensations include touch, pressure, heat, cold, and pain.

Conscious proprioceptive sensations arise within the body. The receptors concerned are those of the locomotor system (muscles, joints, bones) and of the vestibular labyrinth. The pathways to the cerebral cortex form the substrate for *position sense* when the body is stationary, and for *kinesthetic sense* during movement.

Non-conscious sensations

These also are of two kinds. *Non-conscious proprioception* is the term used to describe afferent information reaching the cerebellum through the spinocerebellar pathways. This information is essential for smooth motor coordination. Second, *interoception* is a little-used term referring to unconscious afferent signals involved in visceral reflexes.

Sensory testing

Routine assessment of *somatic exteroceptive sensation* includes tests for the following.

- Touch, by grazing the skin with the finger tip or a cotton swab.
- Pain, by applying the point of a pin.
- Thermal sense, by applying warm or cold test tubes to the skin.

In alert and cooperative patients, active and passive tests of conscious proprioception can be performed. Active tests examine the patient's ability to execute set-piece activities with the eyes closed.

- In the erect position, stand still, and then 'toe the line' without swaying.
- In the seated position, bring the index finger to the nose from the extended position of the arm (finger-to-nose test).
- In the recumbent position, place the heel of the foot on the opposite knee (heel-to-knee test).

Passive tests of conscious proprioception include the following.

- *Joint sense.* The clinician grasps the thumb or great toe by the sides and moves it while asking the patient to name the direction of movement ('up' or 'down'). Joint sense is mediated in part by articular receptors but mainly by passive stretching of neuromuscular spindles. (If the nerves supplying a joint are anesthetized, or if the joint is completely replaced by a prosthesis, joint sense is only slightly impaired. Alternatively, activation of spindles by means of a vibrator creates the illusion of movement when the relevant joint is stationary.)
- *Vibration sense.* The clinician assesses the patient's ability to detect the vibrations of a tuning fork applied to the radial styloid process or to the shaft of the tibia.

SOMATIC SENSORY PATHWAYS

Two major pathways are involved in somatic sensory perception. They are the *posterior column–medial lemniscal pathway* and the *spinothalamic pathway*. They have the following features in common (Figure 15.7).

- Both comprise first-order, second-order, and third-order sets of sensory neurons.
- The somas of the first-order neurons, or *primary afferents*, occupy posterior root ganglia.
- The somas of the second-order neurons occupy CNS gray matter on the same side as the first-order neurons.
- The second-order axons *cross the midline* and then ascend to terminate in the thalamus.
- The third-order neurons project from the thalamus to the somatic sensory cortex.
- Both pathways are *somatotopic*: an orderly map of body parts can be identified experimentally in the gray matter at each of the three loci of fiber termination.

- Synaptic transmission from primary to secondary neurons, and from secondary to tertiary, can be modulated (inhibited or enhanced) by other neurons.

Posterior column–medial lemniscal pathway

The first-order afferents include the largest somas in the posterior root ganglia. Their peripheral processes collectively receive information from the largest sensory receptors: Meissner's and Pacinian corpuscles, Ruffini endings and Merkel cell–neurite complexes, neuromuscular spindles, and Golgi tendon organs. The centripetal processes from cells supplying the lower limb and lower trunk give branches to the gray matter before ascending as the **gracile fasciculus (fasciculus gracilis)** to reach the gracile nucleus in the medulla oblongata (Figure 15.8). The corresponding collaterals from the upper limb and upper trunk run in the **cuneate fasciculus (fasciculus cuneatus)** to reach the cuneate nucleus.

The second-order afferents commence in the posterior column nuclei, namely the **nucleus gracilis** and **nucleus cuneatus**. They pass ventrally in the tegmentum of the

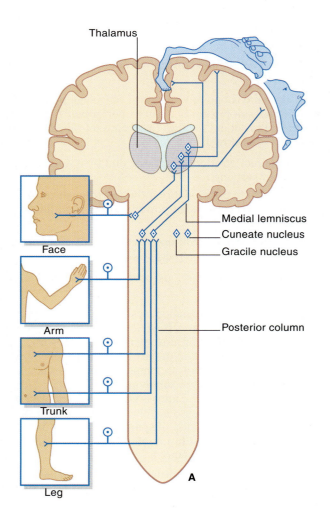

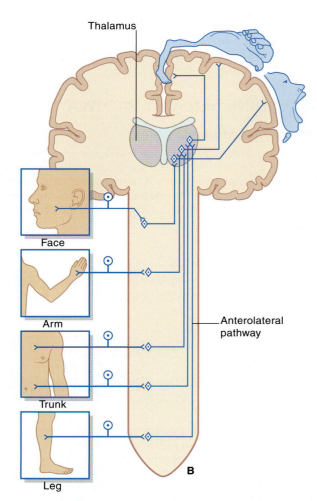

Figure 15.7 Basic plans of **(A)** posterior column–medial lemniscal pathway; **(B)** spinothalamic pathway.

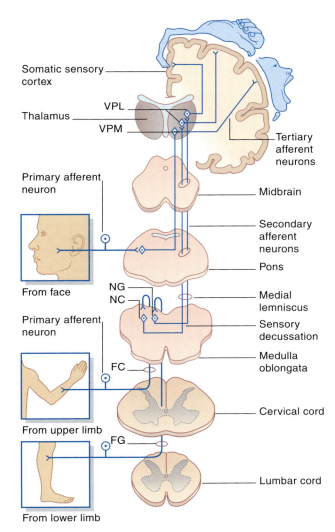

that we are constantly aware of the position of body parts both at rest and during movement. Without this informational background, the execution of movements is severely impaired.

In humans, disturbance of posterior column function is most often observed in association with demyelinating diseases such as multiple sclerosis. The classic symptom is known as *sensory ataxia*. This term signifies a movement disorder resulting from sensory impairment, in contrast to *cerebellar ataxia*, in which a movement disorder results from a lesion within the motor system. The patient with a severe sensory ataxia can stand unsupported only with the feet well apart and with the gaze directed downward to include the feet. The gait is broad-based, with a stamping action that maximizes any conscious proprioceptive function that remains (Figure 15.9).

Sensory testing in posterior column disease reveals severe swaying when the patient stands with the feet together and the eyes closed. This is *Romberg's sign*. (Inability to 'toe the line' with the eyes closed is *tandem Romberg's sign*.) The finger-to-nose and/or heel-to-knee tests may reveal loss of kinesthetic sense. Joint sense and vibration sense may also be impaired.

Note: Romberg's sign may also be elicited in patients suffering from vestibular disorders (Ch. 19); in cerebellar disorders, there may be instability of station whether the eyes are open or closed (Ch. 25).

Tactile, painful, and thermal sensations are preserved, but there is impairment of tactile discrimination. The patient has difficulty in discriminating between single and paired stimuli applied to the skin (*two-point discrimination test*), in identifying numbers traced on to the skin by the examiner's finger, and in distinguishing between objects of similar shape but of different textures.

A difficulty in assigning specific functional deficits to posterior column disease is the rarity of pathology affecting the posterior funiculi *alone*. In particular, the posterior part of the lateral funiculus is likely to be involved as

Figure 15.8 The posterior column–medial lemniscal pathway. FC, fasciculus cuneatus; FG, fasciculus gracilis; NC, nucleus cuneatus; NG, nucleus gracilis; VPL, VPM, ventral posterior lateral, ventral posterior medial nuclei of thalamus.

medulla oblongata before intersecting their opposite numbers in the great **sensory decussation**. Having crossed the midline, the fibers turn rostrally in the **medial lemniscus**.

The medial lemniscus diverges from the midline as it ascends through the tegmentum of the pons and midbrain. It terminates in the lateral part of the ventral posterior nucleus of the thalamus (**ventral posterolateral nucleus**).

Terminating in the medial part of the same nucleus (**ventral posteromedial nucleus**) is the *trigeminal lemniscus*, which serves the head region.

The third-order afferents project from the thalamus to the somatic sensory cortex (see Ch. 27 for details).

Functions

The chief functions of the posterior column–medial lemniscal pathway are those of *conscious proprioception* and *discriminative touch*. Together, these attributes provide the parietal lobe with an instantaneous body image so

Figure 15.9 The 'stamp and stick' gait of sensory ataxia.

well. Postmortem findings from patients having different degrees of pathology in the posterior and lateral funiculi suggest that kinesthetic sense from the lower limb may be mediated in part by fibers that leave the gracile fasciculus at thoracic level and relay rostrally in the posterior part of the lateral funiculus.

Spinothalamic pathway

The spinothalamic pathway consists of second-order sensory neurons projecting from laminae I–II, IV–V of the posterior gray horn to the contralateral thalamus (Figure 15.10). The cells of origin receive excitatory and inhibitory synapses from neurons of the substantia gelatinosa; these have important 'gating' (modulatory) effects on sensory transmission, as explained in Chapter 24.

The axons of the spinothalamic pathway cross the midline in the anterior commissure at all segmental levels. Having crossed, they run upward in the anterolateral part of the cord. This *anterolateral pathway* (as it is usually called) is divisible into an **anterior spinothalamic tract** located in the anterior funiculus and a **lateral spinothalamic tract** located in the lateral funiculus. The two tracts merge in the brainstem as the **spinal lemniscus**. The spinal lemniscus is joined by trigeminal afferents from the head region, and it accompanies the medial lemniscus to the ventral posterior nucleus of the thalamus, terminating immediately behind the medial lemniscus. Third-order sensory neurons project from the thalamus to the somatic sensory cortex (Ch. 27).

Functions

The functions of the spinothalamic pathway have been elucidated by the procedure known as *cordotomy*, whereby the spinothalamic pathway is interrupted on one or both sides for the relief of intractable pain. For a *percutaneous cordotomy*, the patient is sedated and a needle is passed between the atlas and the axis, into the subarachnoid space. Under radiologic guidance, the needle tip is advanced into the anterolateral region of the cord.

A stimulating electrode is passed through the needle. If the placement is correct, a mild current will elicit paresthesia (tingling) on the opposite side of the body. The anterolateral pathway is then destroyed electrolytically. Afterward, the patient is insensitive to *pinprick*, *heat*, or *cold* applied to the opposite side (Figure 15.11). Sensitivity to *touch* is reduced. The effect commences several segments below the level of the procedure because of the oblique passage of spinothalamic fibers across the white commissure.

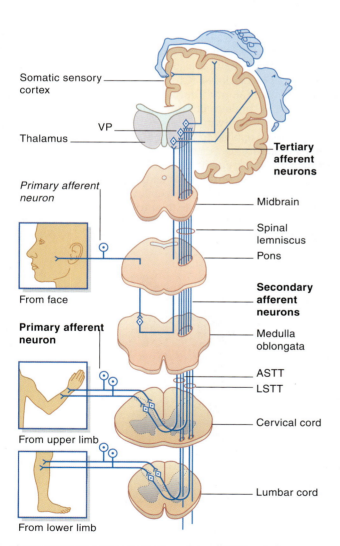

Figure 15.10 The spinothalamic pathway. ASTT, anterior spinothalamic tract; LSTT, lateral spinothalamic tract; VP, ventral posterior nucleus of thalamus.

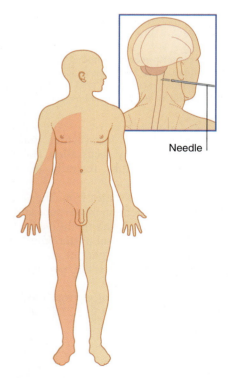

Figure 15.11 Usual extent of analgesia (shaded) following cordotomy at C1–2 segmental level.

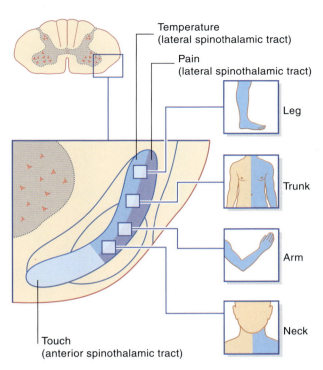

Figure 15.12 Sensory modalities in spinothalamic pathway at upper cervical level.

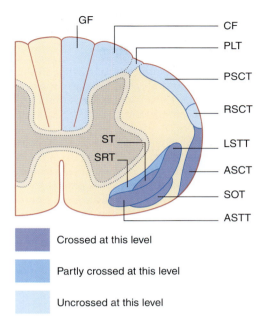

Figure 15.13 Ascending pathways at upper cervical level. ASCT, anterior spinocerebellar tract; ASTT, anterior spinothalamic tract; CF, cuneate fasciculus; GF, gracile fasciculus; LSTT, lateral spinothalamic tract; PLT, posterolateral tract; PSCT, posterior spinocerebellar tract; RSCT, rostral spinocerebellar tract; SOT, spinoolivary tract; SRT, spinoreticular tract; ST, spinotectal tract.

Cordotomy is sometimes performed for patients terminally ill with cancer. It is not used for benign conditions, because the analgesic (pain-relieving) effect wears off after about a year. This functional recovery may be due to nociceptive transmission either in uncrossed fibers of the spinoreticular system (see later) or in C-fiber collaterals sent to the posterior column nuclei by some axons of the lateral root entry stream.

The internal anatomy of the human spinothalamic pathway has been worked out from postoperative sensory testing and is shown in Figure 15.12. The picture is one of *modality segregation*. The lateral spinothalamic tract mediates noxious and thermal sensations separately, and the anterior spinothalamic tract mediates touch. The lateral tract is somatotopically arranged, the neck being represented at the front and the leg at the back. The anterior tract is also somatotopic.

A rare but classic condition illustrating dissociated sensory loss is illustrated in Clinical Panel 15.1.

Spinoreticular tracts

The spinoreticular tracts are the phylogenetically oldest somatosensory pathways. The reticular formation of the brainstem has scant regard for the midline, being often bilaterally distributed in terms of its ascending and descending connections. Spinoreticular fibers originate in laminae V–VII and accompany the spinothalamic pathway as far as the brainstem (Figure 15.13). Postmortem studies of nerve fiber degeneration following cordotomy procedures indicate that at least half of the spinoreticular fibers may be uncrossed. Accurate estimations based on axonal

degeneration are difficult, because some spinothalamic fibers give off collaterals to the reticular formation as they pass by.

The spinoreticular tracts terminate at all levels of the brainstem, and they are not somatotopically arranged. Impulse traffic is continued rostrally to the thalamus in the *ascending reticular activating system* (Ch. 24). Briefly, the spinoreticular system has two interrelated functions.

1 To arouse the cerebral cortex, i.e. to induce or maintain the waking state.
2 To report to the limbic cortex of the anterior cingulate gyrus about the nature of a stimulus. The emotional response may be pleasurable (e.g. to stroking) or aversive (e.g. to pinprick).

In summary, the phylogenetically old, 'paleospinothalamic' pathways through the reticular formation are concerned with the arousal and affective (emotional) aspects of somatic sensory stimuli. In contrast, the direct, 'neospinothalamic' pathway is analytic, encoding information about modality, intensity, and location.

Spinocerebellar pathways

Four fiber tracts run from the spinal cord to the cerebellum. They are:

1 posterior spinocerebellar
2 cuneocerebellar
3 anterior spinocerebellar
4 rostral spinocerebellar.

Syringomyelia is a disorder of uncertain etiology, characterized by development of a *syrinx* (fusiform cyst) in or beside the central canal, usually in the cervical region (Figure CP 15.1.1). Initial symptoms arise from obliteration of spinothalamic fibers decussating in the white commissure.

The early clinical picture is one of *dissociated sensory loss*: sensitivity is lost to painful and thermal stimuli, whereas sensitivity to touch is retained because the posterior column–medial lemniscal pathway is preserved. Typically, the patient develops ulcers on the fingers arising from painless cuts and burns. The joints of the elbow, wrist, and hand may become disorganized over time, or even dislocated, owing to loss of warning sensation from stretched joint capsules. Progressive expansion of the syrinx may compromise conduction in the long ascending and descending pathways.

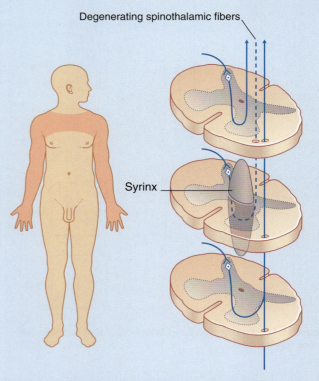

Figure CP 15.1.1 Syringomyelia. Shading shows distribution of analgesia.

The first two are principally concerned with non-conscious proprioception. The second two report continuously about the state of play among the internuncial neurons of the spinal cord.

Non-conscious proprioception

Non-conscious proprioception is served by the posterior spinocerebellar tract for the lower limb and lower trunk, and by the cuneocerebellar tract for the upper limb and upper trunk. Both are uncrossed, in keeping with the known control by each cerebellar hemisphere of its own side of the body.

The posterior spinocerebellar tract originates in the **posterior thoracic nucleus** (formerly, *dorsal nucleus* or *Clarke's column*) in lamina VII at the base of the posterior gray horn (Figure 15.3). The nucleus extends from T1 through L1 segmental levels, and the primary afferents from the lower limb enter the gracile fasciculus to reach it (Figure 15.14). It receives primary afferents of all kinds from the muscles and joints, including an intense input from muscle spindle primaries. It also receives collaterals from cutaneous sensory neurons. The fibers of the posterior spinocerebellar tract are the largest in the entire CNS, measuring 20 μm in external diameter. Very fast conduction is required to keep the cerebellum informed about ongoing movements. The tract ascends close to the surface of the cord (Figure 15.14) and enters the inferior cerebellar peduncle.

The **cuneocerebellar tract** comes from the **accessory cuneate nucleus**, which lies above and outside the cuneate nucleus. The primary afferent inputs are of the same nature as those for the posterior thoracic nucleus; they reach it through the cuneate fasciculus. The cuneocerebellar tract enters the inferior cerebellar peduncle.

Information from reflex arcs

Two tracts originate in the intermediate gray matter of the cord. Although they receive some primary afferents of a similar nature to those already mentioned, their main function is to monitor the state of activity of spinal reflex arcs.

From the lower half of the cord, the pathway is the **anterior spinocerebellar tract** (Figure 15.14). The component fibers cross initially and run close to the surface as far as the midbrain. They then turn into the *superior* cerebellar peduncle and recross within the cerebellar white matter.

From the upper half of the cord, the **rostral spinocerebellar tract** ascends without crossing and enters (mainly) the inferior cerebellar peduncle.

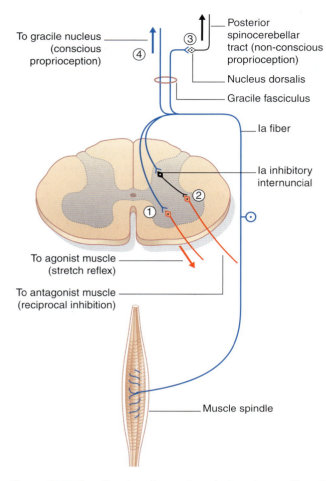

To gracile nucleus (conscious proprioception)

③ ④

Posterior spinocerebellar tract (non-conscious proprioception)

Nucleus dorsalis

Gracile fasciculus

Ia fiber

Ia inhibitory internuncial

② ①

To agonist muscle (stretch reflex)

To antagonist muscle (reciprocal inhibition)

Muscle spindle

Figure 15.14 Functional anatomy of a spindle primary afferent from the lower limb. **(1)** Stretch reflex; **(2)** Ia internuncial serving reciprocal inhibition; **(3)** non-conscious proprioception; **(4)** kinesthesia.

OTHER ASCENDING PATHWAYS

The **spinotectal tract** runs alongside the spinothalamic pathway (Figure 15.13), which it resembles in its origin and functional composition. It ends in the superior colliculus, where it joins crossed visual inputs involved in turning the eyes, head, trunk toward sources of sensory stimulation (*visuospinal reflex*).

The **spinoolivary tract** sends tactile information to the **inferior olivary nucleus** in the medulla oblongata. The inferior olivary nucleus has an important function in *motor learning* through its action on the contralateral cerebellar cortex (Ch. 25). Spinoolivary discharge can modify cerebellar activity in response to environmental change, for example while climbing a surprisingly steep stairway. This feature is called *motor adaptation*. On the other hand, learning to perform routine motor programs automatically is a function of the basal ganglia (Ch. 33).

A *spinocervical tract* is well developed in the cat, where the spinothalamic pathways are small. It seems to be vestigial or absent in humans.

Core Information

The unipolar neurons of the spinal ganglia are first-order (primary) sensory neurons. They send a centrifugal process to peripheral tissues and a centripetal process into the cord. They serve all categories of somatic and visceral sensation, conscious and non-conscious.

Conscious proprioception and discriminative touch are served by large centripetal processes that ascend to the posterior column nuclei in the medulla, where second-order neurons project via the sensory decussation to the contralateral thalamus; third-order neurons project to the somatic sensory cortex.

Painful, thermal, and more crude tactile sensations are served by fine processes that enter Lissauer's tract and end in the posterior gray horn; second-order neurons project across the midline at all segmental levels, coalescing as the spinothalamic tract (anterior and lateral), which is similarly relayed by the thalamus.

The spinoreticular tract projects to the brainstem reticular formation of both sides; it has an arousal function and is concerned with qualitative aspects of stimuli.

First-order neurons serving unconscious proprioception from the lower body end in the posterior thoracic nucleus, for relay to the ipsilateral cerebellum by the posterior spinocerebellar tract; from the upper body, they run via cuneate fasciculus to accessory cuneate nucleus for relay by the cuneocerebellar tract.

Information about activity in spinal reflex arcs is relayed (partly crossed) by the anterior and rostral spinocerebellar tracts.

The spinotectal tract (tactile function, crossed) runs to the superior colliculus for integration with visual data. The spinoolivary tract runs to the inferior olivary nucleus.

REFERENCES

Cervero F. Dorsal horn neurons and their sensory inputs. In: Yaksh TL, ed. Spinal afferent processing. New York: Plenum Press; 1986:197–217.

Coggeshall RE. Unmyelinated primary afferent fibers in the dorsal column, a possible alternate ascending pathway for noxious information. In: Dimitrijivic S, et al. eds. Recent achievements in restorative neurology 3: altered sensation and pain. Basel: Karger; 1990:158–141.

Dykes RW. Parallel processing of somatosensory information: a theory. Brain Res Rev 1983; 6:47–116.

Nathan PW, Smith MC, Cook AW. Sensory effects in man of lesions of the posterior columns and of some other afferent pathways. Brain 1986; 109:1003–1041.

Price DD, Greenspan JD, Dubner R. Neurons involved in the exteroceptive function of pain. Pain 2003; 106:216–219.

Smith MC, Deacon P. Topographical anatomy of the posterior columns of the spinal cord in man. Brain 1984; 107:671–698.

Willis WD. Ascending somatosensory systems. In: Yaksh TL, ed. Spinal afferent processing. New York: Plenum Press; 1985:243–274.

Spinal cord: descending pathways

STUDY GUIDELINES

1 Each of the tracts descending the spinal cord is strategically placed for access to its particular set of motor neurons, in accordance with the layout in Figure 16.7.

2 Note the six different kinds of target neurons selected by the lateral corticospinal tract (LCST).

3 Note that the reticulospinal tracts are concerned with automatic movements and with postural fixation.

4 The Clinical Panels dealing with upper and lower motor neuron disease and spinal cord injury are especially important.

ANATOMY OF THE ANTERIOR GRAY HORN

Cell columns

Each of the columns of motor neurons in the anterior gray horn supplies a group of muscles having similar functions. The individual muscles are supplied from cell groups (nuclei) within the columns. Axial (trunk) muscles are supplied from medially placed columns, proximal limb segment muscles from the midregion, and distal limb segment muscles from lateral columns (Figure 16.1). Columns supplying extensor muscles lie anterior to columns supplying flexors, hence the presence of ventromedial and dorsomedial columns for the trunk, and ventrolateral and dorsolateral columns for the limbs. A *retrodorsolateral nucleus* is devoted to the intrinsic muscles of the hand and foot. An isolated, *central nucleus* supplies the diaphragm.

The segmental levels of the six somatomotor cell columns are listed in Table 16.1. The autonomic nervous system is represented by the intermediolateral cell column.

Table 16.1 The somatomotor cell columns

Cell column	Muscle(s)
Ventromedial (all segments)	Erector spinae
Dorsomedial (T1–L2)	Intercostals, abdominals
Ventrolateral (C5–8, L2–S2)	Arm, thigh
Dorsolateral (C6–8, L3–S3)	Forearm, leg
Retrodorsolateral (C8, T1, S1–2)	Hand, foot
Central (C3–5)	Diaphragm

Cell types

Large, α (alpha) motor neurons supply the extrafusal fibers of the skeletal muscles. Interspersed among them are small, γ (gamma) motor neurons supplying the intrafusal fibers of neuromuscular spindles.

Tonic and phasic motor neurons

The α motor neurons have large dendritic trees receiving some 10 000 excitatory boutons from propriospinal neurons and from supraspinal pathways descending from the cerebral cortex and brainstem. (The term *supraspinal* refers to any pathway descending to the cord from a higher level.) The somas of α motor neurons receive some 5000 inhibitory boutons, mostly from propriospinal sources.

Two principal types of α motor neuron are recognized: tonic and phasic. Tonic α motor neurons innervate slow, oxidative–glycolytic muscle fibers; they are readily depolarized and have relatively slowly conducting axons with small spike amplitudes. Phasic α motor neurons innervate squads of fast, oxidative and fast, oxidative–glycolytic muscle fibers. The phasic neurons are larger, have higher thresholds, and have rapidly conducting axons with large spike amplitudes.

Tonic neurons are usually the first recruits when voluntary movements are initiated, even if the movement is to be fast.

Renshaw cells

The axons of the α motor neurons give off recurrent branches that form excitatory, cholinergic synapses on inhibitory internuncial neurons called *Renshaw cells* in the medial part of the anterior horn. The Renshaw cells form inhibitory, *glycinergic* synapses on the α motor neurons. This is a classic example of *negative feedback*, or *recurrent inhibition*, through which the discharges of α motor neurons are self-limiting (cf. Clinical Panel 8.1).

Segmental-level inputs to α motor neurons

At each segmental level, α motor neurons receive powerful excitatory and inhibitory inputs. Note that any inhibitory

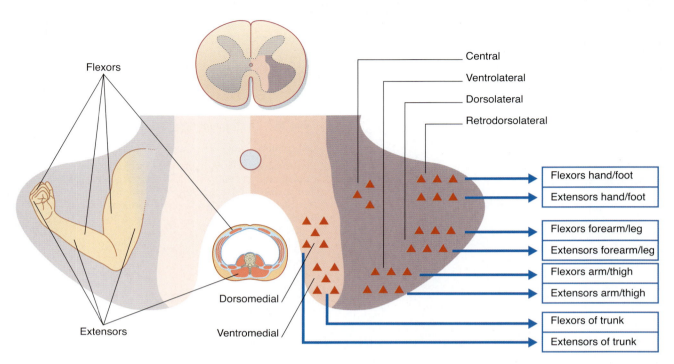

Flexors

Central
Ventrolateral
Dorsolateral
Retrodorsolateral

Flexors hand/foot
Extensors hand/foot

Flexors forearm/leg
Extensors forearm/leg

Flexors arm/thigh
Extensors arm/thigh

Flexors of trunk
Extensors of trunk

Dorsomedial

Extensors

Ventromedial

Figure 16.1 Cell columns in the anterior gray horn of the spinal cord: somatotopic organization.

effect produced by activity in dorsal nerve root fibers requires interpolation of inhibitory internuncials, because all primary afferent neurons are excitatory in nature.

Segmental-level inputs to a flexor α motor neuron include the following.

- Type Ia and type II afferents from spindles in the flexor muscles provide the afferent limb of the monosynaptic stretch reflex (e.g. the biceps reflex).
- Type Ia afferents from spindles in extensor muscles exert reciprocal inhibition on the flexor motor neurons via Ia inhibitory internuncials.
- Type Ib afferents from Golgi tendon organs in the flexor muscles exert autogenetic inhibition on the flexor motor neurons.
- Type Ib afferents from Golgi tendon organs in extensor muscles exert reciprocal excitation of flexors via excitatory internuncials.
- Afferents from the flexor aspect of relevant synovial joints are stimulated when the capsule becomes taut in extension. They initiate an articular protective reflex, as described in Chapter 10.
- In execution of the withdrawal reflex described in Chapter 14, large numbers of excitatory, 'flexor reflex' internuncials are activated over several spinal segments on the same side as the stimulus, as well as inhibitory internuncials supplying motor neurons to antagonist muscles.
- Renshaw cells.

A reciprocal list can be drawn up for extensor motor neurons, with substitution of extensor thrust inputs for flexor reflex internuncials.

DESCENDING MOTOR PATHWAYS

Important pathways descending to the spinal cord are:

- corticospinal (pyramidal)
- reticulospinal (extrapyramidal)
- vestibulospinal
- tectospinal
- raphespinal
- aminergic
- autonomic.

Corticospinal tract

The corticospinal tract is the great voluntary motor pathway. About 50% of its fibers take their origin from the primary motor cortex in the precentral gyrus. Other sources include the supplementary motor area on the medial side of the hemisphere, the premotor cortex on the lateral side, the somatic sensory cortex, the parietal lobe, and the cingulate gyrus (Figure 16.2). The contributions from the two sensory areas mentioned terminate in sensory nuclei of the brainstem and spinal cord, where they modulate sensory transmission.

The corticospinal tract descends through the corona radiata and internal capsule to reach the brainstem. It continues through the crus of the midbrain and the basilar pons to reach the medulla oblongata (Figure 16.3). Here it forms the **pyramid** (hence the synonym *pyramidal tract*).

During its descent through the brainstem, the corticospinal tract gives off fibers that activate motor cranial nerve nuclei, notably those serving the muscles of the face, jaw, and tongue. These fibers are called *corticonuclear* (Figure 16.4). (The term *corticobulbar* is sometimes used, but *bulb* is open to different interpretations.)

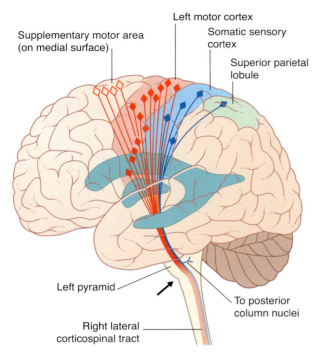

Figure 16.2 Pyramidal tract visualized from the left side. The supplementary motor area is on the medial surface of the hemisphere. *Arrow* indicates level of pyramidal decussation. Non-motor neurons are shown in *blue*.

Just above the spinomedullary junction (Figure 16.5):

- About 80% of the fibers cross the midline in the **pyramidal decussation**.
- These fibers descend on the contralateral side of the spinal cord as the **lateral corticospinal tract** (crossed corticospinal tract).

- About 10% enter the **anterior corticospinal tract**, which occupies the anterior funiculus at cervical and upper thoracic levels. These fibers cross in the white commissure and supply motor neurons serving deep muscles in the neck.
- About 10% of the pyramidal fibers enter the LCST on the same side.

The corticospinal tract contains about one million nerve fibers. The average conduction velocity is 60 m/s, indicating an average fiber diameter of 10 μm ('rule of six' in Ch. 6). About 3% of the fibers are extra large (up to 20 μm); they arise from **giant neurons** (**cells of Betz**), located mainly in the leg area of the motor cortex (Ch. 26). All corticospinal fibers are excitatory and appear to use glutamate as their transmitter substance.

Targets of the lateral corticospinal tract

Distal limb motor neurons

In the anterior gray horn, LCST axons synapse on the dendrites of α and γ motor neurons supplying limb muscles, notably in the upper limb. A unique property of these *corticomotoneuronal fibers* of LCST is that of *fractionation*, whereby small groups of neurons can be selectively activated. This is most obvious in the case of the index finger, which can be flexed or extended quite independently, although three of its long tendons arise from muscle bellies devoted to all four fingers. Fractionation is essential for the execution of skilled movements such as buttoning a coat or tying shoe laces. Skilled movements are lost, and seldom recover completely, following damage to the corticomotoneuronal system anywhere from motor cortex to spinal cord.

Although fractionation is a manifestly important function, even simple voluntary movements, such as flexion of the elbow or abduction of the shoulder, are initiated by corticomotoneuronal fibers.

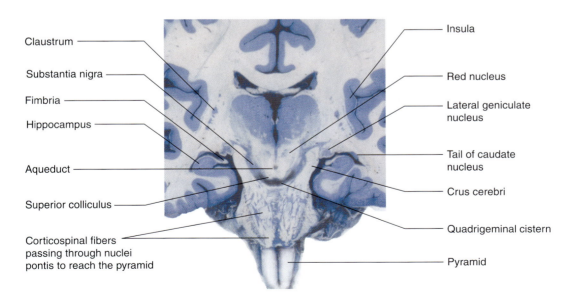

Figure 16.3 Coronal section of embalmed brain, following treatment with copper sulfate (Mulligan's stain), showing unstained corticospinal fibers displacing nuclei pontis en route to the pyramid. (Illustration kindly provided by Professor David Yew, University of Hong Kong, Hong Kong.)

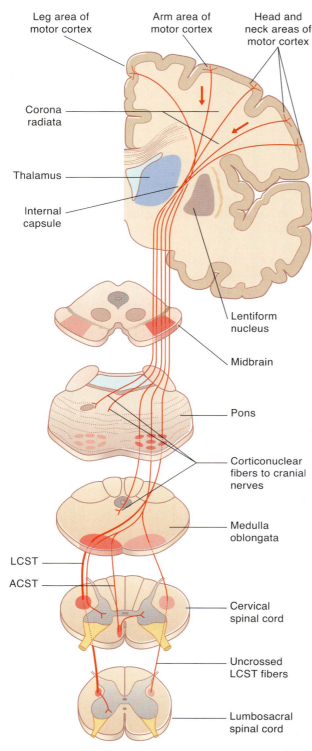

Figure 16.4 The pyramidal tract. ACST, anterior corticospinal tract; LCST, lateral corticospinal tract. *Note*: Only the motor components are shown; the parietal lobe components are omitted.

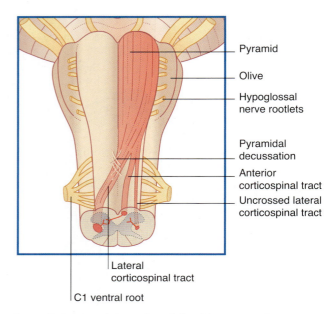

Figure 16.5 Ventral view of medulla oblongata and upper spinal cord, showing the three spinal projections of the left pyramid.

As mentioned already in Chapter 10, the α and γ motor neurons are coactivated by the LCST during a given movement, so that spindles in the prime movers are signaling active stretch while those in the antagonists are signaling passive stretch.

Renshaw cells
The number of possible functions served by LCST synapses on Renshaw cells is large, because some of the cells synapse mainly on Ia inhibitory internuncials, and others on other Renshaw cells. Probably the most important function is to permit *cocontraction* of prime movers and their antagonists, in order to fix one or more joints, for example when a chopping or shoveling action is required of the hand. Cocontraction is achieved by inactivation of Ia inhibitory internucials by Renshaw cells.

Excitatory internuncials
In the intermediate gray matter and the base of the anterior horn, motor neurons supplying axial (vertebral) and proximal limb muscles are recruited mainly indirectly by the LCST, by way of excitatory internuncials.

Ia inhibitory internuncials
Also located in the intermediate gray matter are the Ia inhibitory internuncials, and these are the *first* neurons to be activated by the LCST during voluntary movements. Activity of the Ia internuncials causes the antagonist muscles to relax before the prime movers (agonists) contract. In addition, it renders the antagonists' motor neurons refractory to stimulation by spindle afferents passively stretched by the movement. The sequence of events is shown in Figure 16.6 and its caption for voluntary flexion of the knee.

(*Note on terminologies*: During quiet standing, the knees are 'locked' in slight hyperextension and the

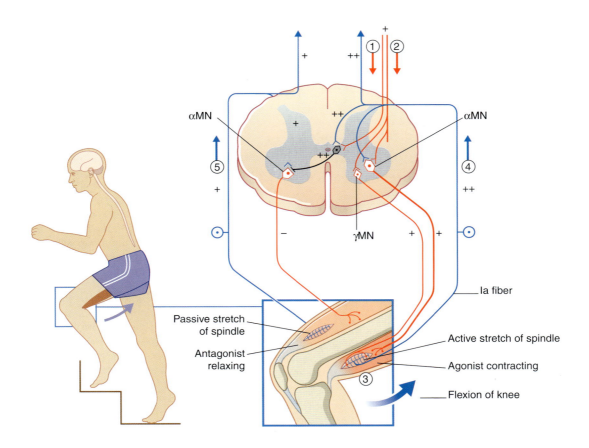

αMN

αMN

γMN

la fiber

Passive stretch of spindle

Active stretch of spindle

Antagonist relaxing

Agonist contracting

Flexion of knee

Figure 16.6 Sequence of events in a voluntary movement (flexion of the knee). **(1)** Activation of Ia internuncials to inhibit antagonist α motor neurons; **(2)** activation of agonist α and γ motor neurons; **(3)** activation of extrafusal and intrafusal muscle fibers; **(4)** feedback from actively stretched spindles increases excitation of agonist α motor neurons and inhibition of antagonist α motor neuron; **(5)** Ia fibers from passively stretched antagonist spindles find the respective α motor neurons refractory. MN, motor neuron. *Note*: The sequence γ motor neuron—Ia fiber—α motor neuron is known as the *gamma loop*.

quadriceps is inactive, as indicated by the patellae being 'loose'. Any tendency of one or both knees to go into flexion is counteracted by a twitch of quadriceps in response to passive stretching of dozens of muscle spindles there. Because the flexion movement is resisted in this way, the reflex concerned is called a *resistance reflex*. During voluntary flexion of the knee, on the other hand, the movement is helped along in the manner described in the caption to Figure 16.6, through an *assistance reflex*. The *change of sign*, from negative to positive, is called *reflex reversal*.)

Presynaptic inhibitory neurons serving

the stretch reflex

Consider a sprinter. At each stride, gravity pulls the body out of the air on to a knee extended by the quadriceps muscle. At the moment of impact, all the muscle spindles in the contracted quadriceps are thrown into active stretch. The obvious danger is that the quadriceps may rupture. Golgi tendon endings (Ch. 10) offer some protection through autogenetic inhibition, but the main protection seems to be through presynaptic inhibition by the LCST of spindle afferents close to their contact points with motor neurons. At the same time, preservation of the ankle jerk

is advantageous in this situation, giving immediate recruitment of calf motor neurons for the next takeoff. The extent of suppression of the stretch reflex by the LCST in fact appears to depend on the particular motor program being executed.

Presynaptic inhibition of first-order afferents

In the posterior gray horn, there is some suppression of sensory transmission into the spinothalamic pathway during voluntary movement. This is brought about by activation of inhibitory internuncials synapsing on primary afferent nerve terminals.

Modulation is more subtle at the level of the gracile and cuneate nuclei, where pyramidal tract fibers (after crossing) are capable of either enhancing sensory transmission during slow, exploratory movements, or reducing it during rapid movements.

Upper and lower motor neurons

In the context of disease, clinicians refer to the corticospinal (and corticonuclear) neurons as upper motor neurons (Clinical Panel 16.1), and those of the brain stem and spinal cord as lower motor neurons (Clinical Panel 16.2).

Clinical Panel 16.1   Upper motor neuron disease

Upper motor neuron disease is a clinical term used to denote interruption of the corticospinal tract somewhere along its course. If the lesion occurs above the level of the pyramidal decussation, the signs will be detected on the opposite side of the body; if it occurs below the decussation, the signs will be detected on the same side.

Sudden interruption of the corticospinal tract is characterized by the following features.

1 The affected limb(s) show an initial flaccid (floppy) paralysis with loss of tendon reflexes. Normal muscle tone—defined as the resistance to passive movement (e.g. flexion–extension of the knee by the examiner)—is lost.

2 After several days or weeks, some return of voluntary motor function can be expected. At the same time, muscle tone increases progressively. The typical long-term effect on muscle tone is one of *spasticity*, with abnormally brisk reflexes (*hyperreflexia*). Classically, spasticity in the leg is 'clasp knife' in character: after initial strong resistance to passive flexion of the knee, the joint gives way.

3 *Clonus* can often be elicited at the ankle or wrist. It consists of rhythmic contraction of the flexor muscles 5–10 times per second in response to sudden passive dorsiflexion.

4 *Babinski sign (extensor plantar response)* consists of dorsiflexion of the great toe and fanning of the other toes in response to a scraping stimulus applied to the sole of the foot. The normal response is flexion of the toes (Figure CP 16.1.1).

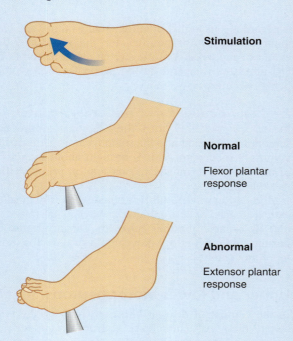

Stimulation

Normal

Flexor plantar response

Abnormal

Extensor plantar response

Figure CP 16.1.1 The plantar reflex.

5 The *abdominal reflexes* are absent on the affected side. A normal abdominal reflex consists of brief contraction of the abdominal muscles when the overlying skin is scraped.

The above features are most commonly observed after a vascular *stroke* interrupting the corticospinal tract on one side of the cerebrum or brainstem. The usual picture here is one of initial flaccid *hemiplegia* ('half-paralysis'), followed by a permanent spastic *hemiparesis* ('half-weakness'). As illustrated in Clinical Panel 35.3, the spasticity following a stroke characteristically affects the antigravity muscles. In the lower limb, these are the extensors of the knee and the plantar flexors of the foot; in the upper limb, they are the flexors of the elbow and of the wrist and fingers. Following complete transection of the spinal cord, on the other hand, there may be a *paraplegia in flexion* of the lower limbs, owing to concurrent interruption of the vestibulospinal tract (Clinical Panel 16.3).

The 'positive' signs listed under **2**, **3**, and **4** cannot be explained on the basis of interruption of the corticospinal tract alone. In the rare cases in which the human pyramid has been transected surgically, spasticity and hyperreflexia have not been prominent later on, although a Babinski sign has been present.

Spasticity and hyperreflexia are largely explained by the fact that stretch reflexes in spastic muscle groups are hyperactive. Electromyography (EMG) records of spastic muscles show enhanced motor unit activity in response to relatively slow rates of stretch, for example slow passive elbow extension. However, this is not the sole basis of explanation. In patients with spastic hemiparesis, the ankle flexors show increased tone (resistance to passive dorsiflexion) even with very slow rates of stretch—too slow to elicit any EMG response. The resistance takes several weeks to become pronounced. It is called *passive stiffness* and may be caused by progressive accumulation of collagen within the muscles affected. In addition, biochemical changes within paretic muscle lead to increasing change of fast-twitch to slow-twitch fibers, accounting for progressively greater difficulty in execution of rapid movements.

Why are motor neurons hyperexcitable?

In paraplegic patients, spasticity and hyperreflexia are often accompanied by increased cutaneomuscular reflex excitability, through polysynaptic propriospinal pathways. Pulling on a pair of trousers may be enough to produce spasms of the hip and knee flexors, sometimes accompanied by autonomic effects (sweating, hypertension, emptying of the bladder). Where the requisite technical facilities exist, the situation can be dramatically improved by perfusion of the lumbar cerebrospinal fluid cistern with minute amounts of *baclofen*, a γ-aminobutyric acid (GABA)-mimetic (imitative) drug. The first inference is that the drug diffuses through the pia–glial membrane of the

Clinical Panel 16.1 *Continued*

spinal cord, activates GABA receptors located on the surface of primary afferent nerve terminals, and dampens impulse traffic by means of presynaptic inhibition. The second inference is that the resident population of GABA neurons in the substantia gelatinosa has fallen silent in these cases through loss of tonic supraspinal 'drive'. The normal source of supraspinal drive seems to derive in part from the corticospinal tract, and in part from corticoreticulospinal fibers that reach the spinal cord via the tegmentum of the brainstem rather than via the pyramids.

Figure CP 16.1.2 shows the distribution of inhibitory nerve endings derived from Renshaw cells. Not alone do they normally have a tonic breaking action on α and γ motor neurons at their own segmental level, they also tonically inhibit heteronymous motor neurons (i.e. those serving other muscle groups). For example, they act simultaneously on motor neurons controlling knee and ankle movements, as part of the executive arm of central motor programs regulating successive muscle engagements and disengagements during locomotion. Locomotion is controlled by reticulospinal rather than corticospinal neurons, and any reduction in reticulospinal drive will render motor neurons hyperexcitable, and accounts for the frequent occurrence of ill-timed contractions produced by heteronymous motor neurons.

How do voluntary movements recover?
Multiple explanations are discussed in the final Clinical Panel in the final chapter.

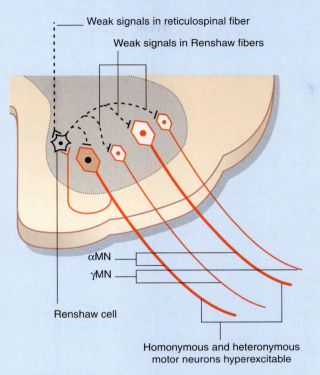

Figure CP 16.1.2 Impaired Renshaw cell activity in spasticity. MN, motor neuron.

Reticulospinal tracts

The reticulospinal tracts originate in the reticular formation of the pons and medulla oblongata. They are partially crossed.

The **pontine reticulospinal tract** descends ipsilaterally in the anterior funiculus, and the **medullary reticulospinal tract** descends, partly crossed, in the lateral funiculus (Figure 16.7). Both tracts act, via internuncials shared with the corticospinal tract, on motor neurons supplying axial (trunk) and proximal limb muscles. Information from animal experiments indicates that the pontine reticulospinal tract acts on extensor motor neurons, and the medullary reticulospinal tract on flexor motor neurons. Both pathways exert reciprocal inhibition.

The reticulospinal system is involved in two different kinds of motor behavior: *locomotion* and *postural control*.

Locomotion

Walking and running are rhythmic events involving all four limbs. Movements of the two sides are reciprocal with respect to flexor and extensor contractions and relaxations. In lower animals, locomotion is regulated by a hierarchic system in which the lowest members are internuncial neurons on both sides at cervical and lumbosacral levels, activating the flexors and extensors of the individual limbs. They are called *pattern generators*. Coordinating the pattern generators for the individual limbs is a further generator situated in the intermediate gray matter at the upper end of the spinal cord; it is capable of initiating rhythmic movements after section of the neuraxis at the spinomedullary junction. Locomotion is initiated from a *locomotor center* located in the lower midbrain of humans, in the pons in laboratory animals. In anesthetized cats, electrical stimulation of the locomotor center with pulses of increasing frequency produces walking movements, then trotting, and finally galloping.

Although the basic locomotor patterns are inbuilt, they are modulated by sensory feedback from the terrain. Overall control of the motor output resides in the premotor cortex, which has direct projections to the brainstem neurons that give rise to the reticulospinal tracts. The tracts are used to steer the animal as it walks or runs, and to override the spinal generators, for example in scaling a wall.

Human locomotion is less 'spinal' than that of quadrupeds. However, the general neuroanatomic framework

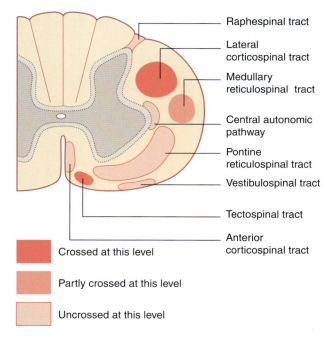

- Raphespinal tract
- Lateral corticospinal tract
- Medullary reticulospinal tract
- Central autonomic pathway
- Pontine reticulospinal tract
- Vestibulospinal tract
- Tectospinal tract
- Anterior corticospinal tract

[red] Crossed at this level

[medium] Partly crossed at this level

[light] Uncrossed at this level

Figure 16.7 Descending pathways at upper cervical level. *Notes*: The anterior corticospinal tract crosses completely at cervical and upper thoracic levels to engage anterior horn cells. Some 10% of lateral corticospinal tract fibers descend ipsilaterally.

has been conserved during higher evolution, and the basic physiology seems to be in place as well. In particular, a bilaterally organized motor system controlling proximal and axial muscles *must* exist to account for the return of near-perfect locomotor function following removal of an entire cerebral hemisphere during childhood or adoescence. Such people never recover manual skill on the contralateral side, and this reinforces the belief among physical therapists that two distinct pathways are involved in motor control: *pyramidal* and *'extrapyramidal'*. The latter term denotes the reticulospinal pathway and its controls upstream in the cerebral cortex and basal ganglia.

Higher-level locomotor controls are described in Chapter 24.

Posture

Definitions of *posture* vary with the context in which the term is used. In the general context of standing, sitting, and recumbency, posture may be defined as *the position held between movements*. In the local context of a single hand or foot, the term signifies *postural fixation*—the immobilization of proximal limb joints by cocontraction of the surrounding muscles, leaving the distal limb parts free to do voluntary business. As will be noted in Chapter 29, there is reason to believe that the human premotor cortex is programmed to select appropriate proximal muscle groups by way of the reticulospinal tracts, to set the stage for any particular movement of the hand or foot.

Clinical Panel 16.2   Lower motor neuron disease

Disease of lower motor neurons may be caused by a variety of infectious agents—notably the virus of poliomyelitis. The term *motor neuron disease*, or *MND*, is used to describe a symptom complex characterized by degeneration of upper and lower motor neurons in late middle age.

During the first year or two, lower motor neurons alone may be involved, especially in the upper limbs. This phase is called *progressive muscular atrophy*, and it has the following manifestations.

1 *Weakness* of the muscles affected, together with

2 *Wasting*. The wasting is not merely a disuse atrophy but results from loss of a trophic (nourishing) factor produced by motor neurons and conveyed to muscle by axonal transport.

3 *Loss of tendon reflexes* (areflexia) in the wasted muscles.

4 *Fasciculations*, which are visible twitchings of small groups of muscle fibers in the early stage of wasting. They arise from spontaneous discharge of motor neurons with activation of motor units, as described in the context of electromyography in Clinical Panel 12.2.

5 *Fibrillations*, which are minute contractions detectable only by needle electromyography, also described in Clinical Panel 12.2.

Sooner or later, signs of upper MND appear. The lower limbs become weak, with increased muscle tone and brisk reflexes. This condition is called *amyotrophic lateral sclerosis*. Motor cranial nerve nuclei in the pons and medulla oblongata may be involved from the start (*progressive bulbar palsy*, Ch. 18) or only terminally. Death, from respiratory complications, usually occurs within 5 years of onset.

The search for etiologic clues is intense. Damage to motor neurons by free radicals has long been suspected, and it is of interest that mutation of a free-radical scavenging enzyme has been detected in some of the 10% of patients who inherit MND in an autosomal dominant mode. Because it is known that retrograde transport of neurotrophins is essential for long-term neuronal survival, recent research has focused on retrograde transport of signaling endosomes. In particular, failure of the dynein retrograde motor (Ch. 6) has been implicated in the development of several neurodegenerative disorders.

The interpolation of internuncial neurons between the two main motor pathways acting on motor neurons serving axial and proximal limb muscles means that either pathway may be in command for a particular movement sequence—the extrapyramidal (reticulospinal) pathway for routine tasks such as walking along a clear path, the pyramidal pathway for tasks requiring close attention, for example picking one's way along a path strewn with rubble.

Tectospinal tract

The tectospinal tract is a crossed pathway descending from the tectum of the midbrain to the medial part of the anterior gray horn at cervical and upper thoracic levels. It is strategically placed for access to axial motor neurons (Figure 16.7).

This tract is an important motor pathway in the reptilian brain, being responsible for orienting the head–trunk toward sources of visual stimulation (superior colliculus) or auditory stimulation (inferior colliculus). It is likely to have similar automatic functions in humans.

Vestibulospinal tract

The vestibulospinal tract is an important uncrossed pathway whereby the tone of appropriate antigravity muscles is automatically increased when the head is tilted to one side. It descends in the anterior funiculus (Figure 16.7), and its function is to keep the center of gravity between the feet. It originates in the vestibular nucleus in the medulla oblongata. (*Note*: As explained in Ch. 17, there are in fact two vestibulospinal tracts on each side. The unqualified term refers to the lateral vestibulospinal tract.)

Raphespinal tract

The raphespinal tract originates in and beside the raphe nucleus situated in the midline in the medulla oblongata. It descends on both sides within the posterolateral tract of Lissauer. Its function is to modulate sensory transmission between first- and second-order neurons in the posterior gray horn—particularly with respect to pain (see Ch. 24).

Aminergic pathways

Aminergic pathways descend from specialized cell groups in the pons and medulla oblongata (Ch. 24). The principal neurotransmitters involved are *noradrenaline* (*norepinephrine*) and *serotonin*, both of which are classed as biogenic amines. The aminergic pathways descend in the outer parts of the anterior and lateral funiculi, and are distributed widely in the spinal gray matter. In general terms, they have inhibitory effects on sensory neurons and facilitatory effects on motor neurons.

Central autonomic pathways

Central sympathetic and parasympathetic fibers descend beside the intermediate gray matter (Figure 16.7). They originate in part from autonomic control centers in the hypothalamus and in part from several nuclear groups in the brainstem. They terminate in the intermediolateral cell columns that give rise to the preganglionic sympathetic and parasympathetic fibers of the peripheral autonomic system.

The central sympathetic pathway is required for normal *baroreceptor reflex* activity. If the spinal cord is crushed in a neck injury, the patient loses consciousness if raised from the recumbent position within the first week or so, because a fall of blood pressure in the carotid sinus on sitting up normally causes a compensatory increase in sympathetic activity in order to maintain blood flow to the brain.

The central parasympathetic pathway is required for normal bladder (and rectal) function. The fibers concerned originate in the reticular formation, mainly at the level of the pons (Ch. 24). The pontine micturition center has a tonic inhibitory action on the sacral parasympathetic system. Severe injury to the spinal cord or cauda equina results in reflex voiding when the bladder is only half full (Clinical Panel 16.3).

Note on the rubrospinal tract

The rubrospinal tract is an important motor pathway in cats and dogs, where it arises in the contralateral red nucleus and descends in front of the corticospinal tract. In monkeys, this tract is small and in humans it is quite negligible.

BLOOD SUPPLY OF THE SPINAL CORD

Arteries

Close to the foramen magnum, the two vertebral arteries give off **anterior** and **posterior spinal** branches. The anterior branches fuse to form a single **anterior spinal artery** in front of the anterior median fissure (Figure 16.8). Branches are given alternately to the left and right sides of the spinal cord. The posterior spinal arteries descend along the line of attachment of the dorsal nerve roots on each side.

The three spinal arteries are boosted by several **radiculospinal branches** from the vertebral arteries and from intercostal arteries. They are distinguishable from the small **radicular arteries** that enter every intervertebral foramen to nourish the nerve roots. The largest radiculospinal artery is the **artery of Adamkiewicz**, which arises from a lower intercostal artery or upper lumbar artery on the left side and supplies the lumbar enlargement and conus medullaris.

Vascular disorders of the spinal cord are quite rare. As part of a generalized atherosclerosis, a branch of the anterior spinal artery may become occluded, causing necrosis of the anterior half of the cord on one side. The clinical picture has some resemblance to a one-sided amyotrophic lateral sclerosis owing to destruction of anterior horn motor neurons and diminished function in the LCST on the same side. However, arterial disease should be suspected here, because of the relatively abrupt onset of symptoms and because concurrent damage to the spinothalamic pathway produces loss of pain and of thermal sense on the opposite side, below the level of the lesion.

Clinical Panel 16.3   Spinal cord injury

In the industrialized world, automobile accidents are the commonest cause of spinal cord injury. More than half of the victims are between 16 and 30 years old, and the cervical cord is most commonly affected. Injury at thoracic or lumbar segmental level results in *paraplegia* (paralysis of lower limbs). Injury at cervical level causes *tetraplegia (quadriplegia)*, in which the extent of upper limb paralysis depends on the number or level of cervical segments involved.

Spinal shock
The following features are found below the segmental level of the injury in the first few days following a complete cord transection.

- Paralysis of movement. The limbs are flaccid and tendon reflexes are absent.
- Anesthesia (loss of all forms of sensation).
- Paralysis of the bladder and rectum.

Spinal shock is currently attributed to a generalized hyperpolarization of spinal neurons below the level of the lesion, perhaps because of large-scale release of the inhibitory transmitter glycine. In addition, the patient develops *postural hypotension* when raised from the recumbent position, owing to interruption of the baroreceptor reflex. Wearing an abdominal binder may be sufficient to compensate for the lost reflex.

Return of spinal function
Several days or weeks later, reflex functions of the cord become progressively restored, and 'upper motor neuron signs' appear. Muscle tone becomes excessive (spastic). Tendon reflexes become abnormally brisk. A Babinski sign can be elicited on both sides. Ankle clonus is commonly seen when a patient's leg is lifted into contact with the footplate of a wheelchair.

If extensor spasticity in the lower limbs is dominant, the patient develops *paraplegia in extension*; if flexor spasticity is dominant, *paraplegia in flexion*. An extended posture may permit *spinal standing*; it is promoted by appropriate passive placement of the limbs, and it is the rule following cord injury that is either incomplete or low. A flexed posture is promoted by repetitive mass flexor reflexes involving the ankles, knees, and hips; mass reflexes can follow any cutaneous stimulation of the legs if the flexor reflex internuncial neurons of the cord are already sensitized by afferent discharges from a pressure sore or from an infected bladder.

The condition of the bladder is of great importance because of the twin dangers of infection and formation of bladder stones. For the initial, *atonic* bladder, a sterile catheter is inserted in order to ensure unobstructed drainage. Later, the bladder becomes *automatic*, emptying itself every 4–6 h through a reflex arc involving the sacral autonomic center in the conus medullaris.

In animals, much of the damage done to the cord by injury has been shown to be secondary to local shifts in electrolyte concentrations, and to vascular changes including arterial spasm and venous thrombosis. Modest success is being achieved in counteracting these effects. Another line of experimental research is to implant *embryonic* spinal gray matter at the site of injury. These grafts often survive and establish local synaptic connections, but the goal of functional recovery has not been attained.

Considerable interest has been aroused by observations in several spinal rehabilitation centers, to the effect that patients with complete cord transections can be trained to activate spinal locomotor generators, as described in Chapter 24.

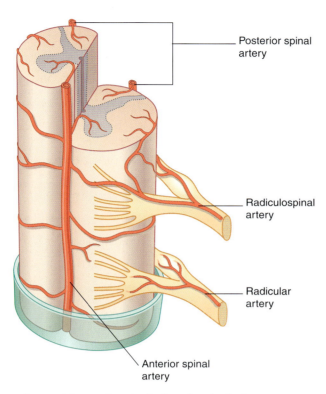

Posterior spinal
artery

Radiculospinal
artery

Radicular
artery

Anterior spinal
artery

Figure 16.8 Arteries of spinal cord and spinal nerve roots.

The artery of Adamkiewicz has to be borne in mind by the vascular surgeon attempting to deal with an abdominal aortic aneurysm. If a clamp is placed across the aorta and the artery happens to arise below that level, the patient is at risk of postoperative paraplegia with incontinence!

Veins

The venous drainage of the cord is by means of anterior and posterior spinal veins, which drain outward along the nerve roots. Any obstruction to the venous outflow is liable to produce edema of the cord, with progressive loss of function.

Core Information

Fibers of the corticospinal tract (CST) governing voluntary movement originate in motor, premotor, and supplementary motor areas of the cerebral cortex; fibers governing sensory transmission during movement originate in the parietal lobe. The CST includes corticonuclear fibers innervating motor cranial nerve nuclei. The lateral CST innervates anterior horn cells supplying trunk and limb muscles; 80% of these fibers cross in the pyramidal decussation and enter the lateral CST, 10% descend ipsilaterally in the anterior CST prior to crossing at lower levels, and 10% remain entirely ipsilateral. Lateral CST targets include alpha and gamma motor neurons, Ia inhibitory internuncials, and Renshaw cells.

Clinically, the CST is the upper motor neuron. Damage (e.g. in hemiplegia from stroke) is characterized by initial flaccid paralysis, later by spasticity, brisk reflexes, clonus, and Babinski sign. Lower motor neuron (anterior horn cell) disease is characterized by muscle weakness, wasting, fasciculation, and loss of related segmental reflexes. Spinal cord transection is characterized by initial flaccid paraplegia or tetraplegia with areflexia, atonic bladder, and (permanent) anesthesia below the segmental level

involved; later, by spasticity, hyperreflexia, clonus, Babinski sign, and automatic bladder.

Reticulospinal tracts are activated by the premotor cortex. For locomotion, they originate in a midbrain locomotor center and travel to pattern generators in the cord. For postural fixation, they originate in pons and medulla and supply motor neurons via internuncials.

The tectospinal tract descends (crossed) from colliculi to anterior horn; it operates to direct the gaze toward visual, auditory, tactile stimuli. The (lateral) vestibulospinal tract (uncrossed) increases antigravity tone on the side to which the head is tilted. The raphespinal tract descends from the medullary raphe nucleus to the posterior horn via Lissauer's tract; it modulates sensory transmission, especially for pain.

A central sympathetic pathway from hypothalamus–brain stem to the lateral horn includes the efferent limb of the baroreflex. A central parasympathetic pathway activates the bladder and rectum.

The cord receives spinal branches from the vertebral arteries, boosted by radiculospinal arteries at segmental levels. Venous drainage is into segmental veins.

REFERENCES

Burne A, Lippold OCJ. Reflex inhibition following electrical stimulation over muscle tendons in man. Brain 1996; 119:1107–1115.

Busches A, El Manira A. Sensory pathways and their modulation in the control of locomotion. Curr Opin Neurobiol 1998; 8:733–739.

Crone C, Nielson J. Central control of disynaptic inhibition in humans. Acta Physiol Scand 1994; 162:351–363.

Davidoff RA. The pyramidal tract. Neurology 1990; 40:332–339.

Dietz V. Spastic gait disorder. In: Bronstein AM, Brandt T, Woollacott M, eds. Clinical disorders of balance posture and gait. London: Arnold; 1996:1–17.

Halezparast M, Klocke R, Ruhrberg C, et al. Mutations in dynein link motor neuron degeneration to defects in retrograde transport. Science 2003; 300:808–812.

Jeanmonod D. Neuroanatomical bases of spasticity. In: Sindou M, Abbott R, Keravel Y, eds. Neurosurgery for spasticity. New York: Springer-Verlag; 1991:3–15.

Katz R, Pierrot-Deseilligny E. Recurrent inhibition in man. Prog Neurobiol 1998; 57:325–355.

Levin ML, Feldman AG. The role of stretch reflex threshold regulation in normal and impaired motor control. Brain Res 1994; 657:23–30.

Levy LM. Brain γ-aminobutyric acid changes in stiff-person syndrome. Arch Neurol 2005; 62:970–974.

Martaens de Nordhout A, Rapisarda G, Bogacz D, et al. Corticomotoneural synaptic connections in man. Brain 1999; 122:1627–1640.

Mazzocchio R, Rossi A. Involvement of spinal recurrent inhibition in spasticity. Brain 1997; 120:991–1003.

Meinck HM, Benecke R, Kuster S, et al. Cutaneomuscular (flexor) reflex organization in normal man and in patients with motor disorders. In: Desmedt JE, ed. Motor control systems in health and disease. New York: Raven Press; 1983:787–796.

Mitz AR, Winstein C. The motor system. 1, Lower centers. In: Cohen M, ed. Neuroscience for rehabilitation. Philadelphia: Lippincott; 1993:151–185.

Nacimiento W, Noth J. What, if anything, is spinal shock? Arch Neurol 1999; 53:1033–1035.

Nathan PW, Smith M, Deacon P. Vestibulospinal, reticulospinal and descending propriospinal nerve fibers in man. Brain 1996; 119:1809–1833.

Nathan PW, Smith MC. The rubrospinal and central tegmental tracts in man. Brain 1982; 105:223–269.

Schoenen J, Faull RLM. Spinal cord: cytoarchitectural, dendroarchitectural, and myeloarchitectural organization. In: Paxinos G, ed. The human nervous system. San Diego: Academic Press; 1990:19–54.

Ugawa Y, Uesaka Y, Terao Y, et al. Magnetic stimulation of corticospinal pathways at the foramen magnum level in humans. Ann Neurol 1994; 36:618–624.

Brainstem

STUDY GUIDELINES
1 This chapter is largely to do with identification of structures in transverse sections of the brainstem. A separate study guide is provided for the sections.
2 Four brainstem decussations should recall those described in Box 3.1.
3 Note that, in magnetic resonance images, brainstem orientation is the reverse of the anatomic convention.

GENERAL ARRANGEMENT OF CRANIAL NERVE NUCLEI

In the thoracic region of the developing spinal cord, four distinct cell columns can be identified in the gray matter on each side (Figure 17.1A, B). In the basal plate, the *general somatic efferent column* supplies the striated muscles of the trunk and limbs. The *general visceral efferent column* contains preganglionic neurons of the autonomic system. In the alar plate, the *general visceral afferent column* receives afferents from thoracic and abdominal organs. A *general somatic afferent* column receives afferents from the body wall.

In the brainstem, these four cell columns can be identified, but they are fragmented, and not all contribute to each cranial nerve. Their connections are as follows.

- *General somatic efferent column.* Supplies the striated musculature of the orbit (via the oculomotor, trochlear, and abducens nerves) and tongue (via the hypoglossal nerve).
- *General visceral efferent column.* Gives rise to the cranial parasympathetic system introduced in Chapter 13. The target ganglia are the ciliary, pterygopalatine, otic, and submandibular ganglia in the head, and the vagal ganglia in the thorax and abdomen.

- *General visceral afferent column.* Receives from the visceral territory of the glossopharyngeal and vagus nerves.
- *General somatic afferent column.* Receives from skin and mucous membranes, mainly in trigeminal nerve territory whose most important components are the skin and mucous membranes of the oronasofacial region, and the dura mater.

Three additional cell columns (Figure 17.1C, D) serve branchial arch tissues and the inner ear, as follows.

- *Special visceral (branchial) efferent column.* To branchial arch musculature of face, jaws, palate, larynx, and pharynx (via facial, trigeminal, glossopharyngeal, and vagus nerves). These striated muscles have visceral functions in relation to food and air intake (hence, *visceral*).
- *Special visceral afferent column.* Receives from taste buds located in the endoderm lining the branchial arches.
- *Special sense afferent column.* Receives vestibular (balance) and cochlear (hearing) from the inner ear.

Figure 17.2 shows the position of the various nuclei in a dorsal view of the brainstem.

In this chapter, details of the internal anatomy of the brainstem accompany nine representative transverse sections and their captions. Connections (direct or indirect) with the *right* cerebral hemisphere have been highlighted in accordance with information to be provided.

BACKGROUND INFORMATION

As stated earlier, exteroceptive and conscious proprioceptive information are transferred (by spinothalamic and dorsal column–medial lemniscal pathways, respectively) from left trunk and limbs to right cerebral hemisphere. It was also explained that corticospinal fibers of the pyramidal tract arising from motor areas of the cerebral cortex supply contralateral anterior horn cells and give a small ipsilateral supply of similar nature, and that those arising from the parietal lobe project to the contralateral posterior gray horn.

The same arrangement holds good for the brainstem. The pyramidal tract fibers terminating in the brainstem are *corticonuclear*. As shown in Figure 17.3, their distribution is predominantly contralateral to somatic and branchial motor nuclei, and entirely contralateral to the somatic sensory nuclei.

Absent from this figure are the three pairs of motor ocular nuclei. Why? Because these do not receive a direct corticonuclear supply. Instead, their predominantly contralateral supply synapses on adjacent cell groups known as *gaze centers* having the function of synchronizing conjugate (conjoint parallel) movements of the eyes.

For a basic understanding of neural relationships in the brainstem, it is also essential to appreciate hemisphere

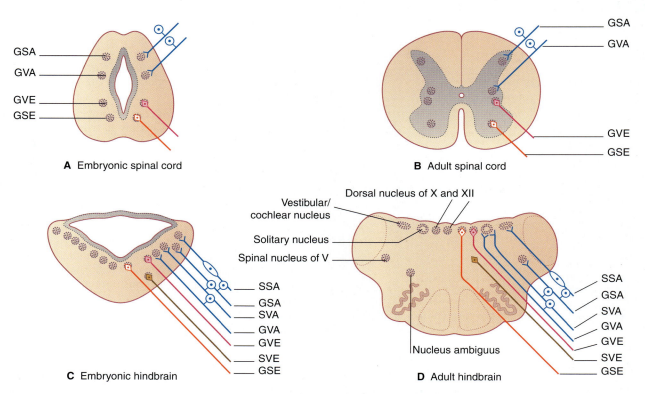

Figure 17.1 Cell columns of the spinal cord and brainstem. **(A)** Embryonic spinal cord. **(B)** Adult spinal cord. **(C)** Embryonic hindbrain. **(D)** Adult hindbrain. *Afferent cell columns*: GSA, general somatic afferent; GVA, general visceral afferent; SSA, special somatic afferent; SVA, special visceral afferent. *Efferent cell columns*: GSE, general somatic efferent; GVE, general visceral efferent; SVE, special visceral efferent.

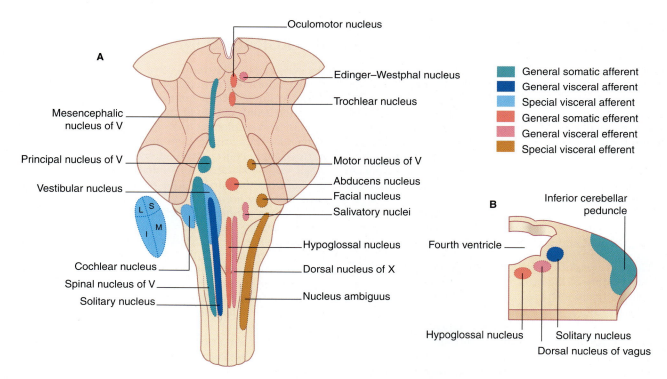

Figure 17.2 Posterior view of adult brainstem, showing position of cranial nerve cell columns. L, S, I, M, lateral, superior, inferior, medial vestibular nuclei.

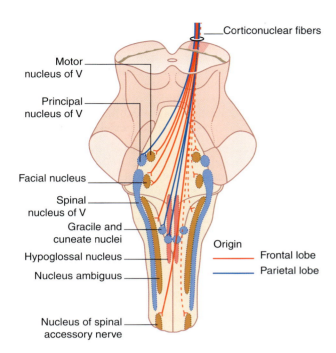

Figure 17.3 Posterior view of brainstem, showing distribution of corticonuclear fibers from the right cerebral cortex.

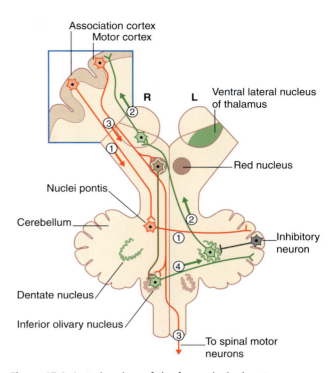

Figure 17.4 Anterior view of the four principal motor decussations of the brainstem. Pathways are numbered in accordance with their sequence of activation in voluntary movements. **(1)** corticopontocerebellar; **(2)** dentatothalamocortical; **(3)** corticospinal; **(4)** olivocerebellar. Also shown is the rubroolivary connection.

linkages to the inferior olivary nucleus and to the cerebellum (Figure 17.4).

The general layout of the **reticular formation** (Figure 17.5) is borrowed from a figure in Chapter 24 that is devoted to this topic. It may be consulted when reading under this heading in successive descriptions.

Figure 17.6 depicts the main components of the **medial longitudinal fasciculus (MLF)**. This fiber bundle extends the entire length of the brainstem, changing its fiber composition at different levels. This figure, too, may be consulted during study of the brainstem sections to be described, following inspection of C1 segment of the spinal cord.

Study guide

The presentation departs from the traditional method, which is to describe photographs or diagrams at successive levels in ascending order without highlights. In the present approach:

1 The various nuclei and pathways are highlighted and labeled on the side having primary affiliation with the right cerebral hemisphere.

2 The nuclei and pathways are color-coded by systems, for example red for motor, blue for sensory, green for connections of cerebellum and reticular formation.

3 Highlighting together with color-coding makes it possible to study individual systems in vertical, 'multiple window' mode. The descriptive text related to the brainstem sections enables a logical sequence of study whereby afferent pathways can be followed from below upward to thalamic level (commencing with Figure 17.10), and efferent pathways can be followed from above downward (commencing with

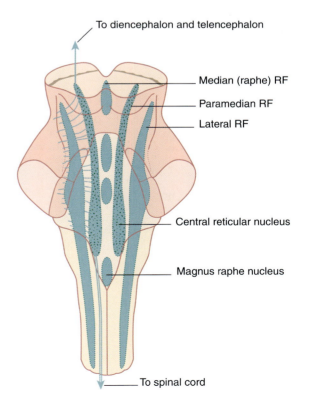

Figure 17.5 Layout of the reticular formation (RF).

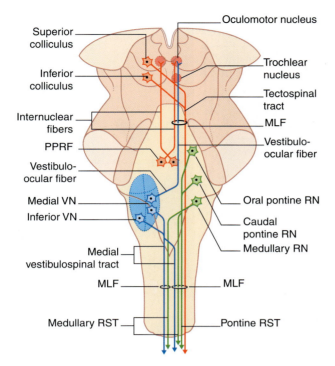

Figure 17.6 Main fiber composition of the medial longitudinal fasciculus (MLF). PPRF, paramedian pontine reticular formation; RN, reticular nucleus; RST, reticulospinal tract; VN, vestibular nucleus.

Figure 17.19). This accounts for the *First figure* and *Last figure* headings in relation to level C1. It must be emphasized that, following study in the vertical mode, a horizontal approach must be undertaken, the location of the various systems to be identified at each level. This is because occlusion of a small artery of supply to the brainstem may affect function in a patch that may include several distinct nuclei or pathways.

At each level, miniature replicas of the diagrams in Figure 17.7 are inserted to assist left–right orientation.

Special note: Readers unfamiliar with the internal anatomy of the brainstem may be disconcerted by the amount of new information contained in the series of sections to be described. It may be reassuring to know that *all* the information will come up again in later chapters. Therefore a sensible approach could be to undertake an initial browse through the sections and to recheck the location of individual items during later reading.

Overview of three pathways in the brainstem

Figure 17.8 shows the *posterior column–medial lemniscal* and *anterolateral pathways* already described in Chapter 15. Recall that the latter comprises the lateral spinothalamic tract serving pain and temperature, and the anterior spinothalamic tract serving touch. Within the brainstem, the two are combined as the **spinal lemniscus**.

The corticospinal tract, treated in Chapter 16, is shown in Figure 17.9. Also included are corticonuclear projections to the facial and hypoglossal nuclei.

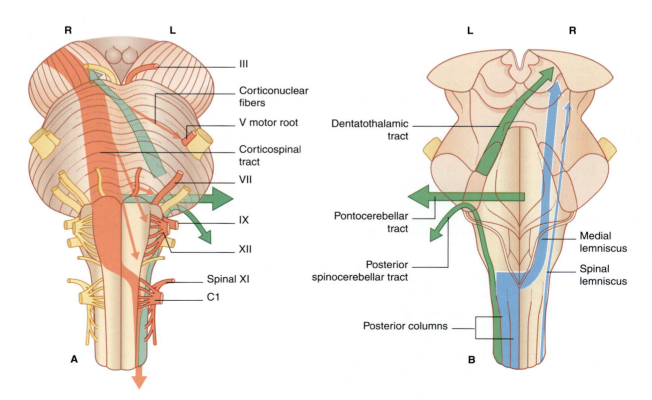

Figure 17.7 (A) Anterior and **(B)** posterior view of brainstem, showing disposition of some major pathways.

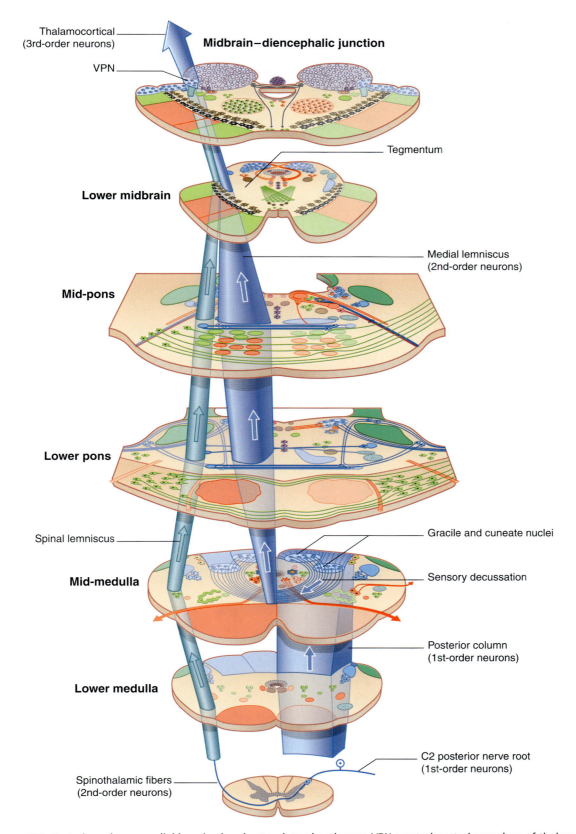

Thalamocortical
(3rd-order neurons)

VPN

Midbrain–diencephalic junction

Tegmentum

Lower midbrain

Medial lemniscus
(2nd-order neurons)

Mid-pons

Lower pons

Spinal lemniscus

Gracile and cuneate nuclei

Mid-medulla

Sensory decussation

Posterior column
(1st-order neurons)

Lower medulla

C2 posterior nerve root
(1st-order neurons)

Spinothalamic fibers
(2nd-order neurons)

Figure 17.8 Posterior column–medial lemniscal and anterolateral pathways. VPN, ventral posterior nucleus of thalamus.

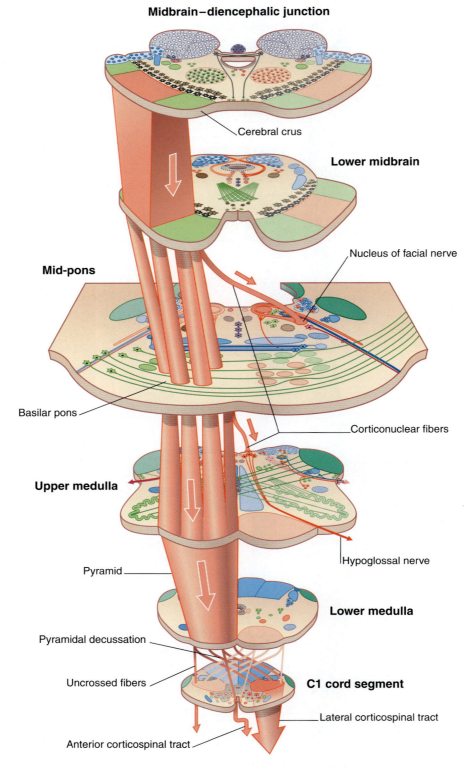

Midbrain–diencephalic junction

Cerebral crus

Lower midbrain

Nucleus of facial nerve

Mid-pons

Basilar pons

Corticonuclear fibers

Upper medulla

Hypoglossal nerve

Pyramid

Lower medulla

Pyramidal decussation

Uncrossed fibers

C1 cord segment

Lateral corticospinal tract

Anterior corticospinal tract

Figure 17.9 Corticospinal tract; two corticonuclear projections.

C1 SEGMENT OF SPINAL CORD
(Figure 17.10)

Landmarks

Immediately below the pyramidal decussation, the lateral corticospinal tract (6) is crossing to occupy the lateral funiculus of the spinal cord.

First figure of ascending sequence

Afferent nuclei and pathways

The positions of the **gracile** (1) and **cuneate** (2) **fasciculi**, and of the **anterolateral pathway** (16) are unchanged from lower levels (Ch. 15).

The inner three laminae of the posterior gray horn are compressed. The outer three laminae constitute the **spinal nucleus of the trigeminal nerve** (3). The **spinal tract of the trigeminal nerve** (4) is at this level conveying posterior root axons from nerves C2 and C3 to the spinal nucleus. (C1 usually lacks a posterior root.)

Cerebellar connections (ascending)

Of the four spinocerebellar pathways (posterior, anterior and rostral spinocerebellar, and cuneocerebellar), only the posterior spinocerebellar tract (5) is shown in this series. Its position is unchanged from lower levels.

Final figure of descending sequence

Efferent nuclei and pathways

The medial part of the anterior horn of gray matter gives rise to the anterior root of nerve C1 (11). The lateral part gives rise to the uppermost root (7) of the spinal accessory nerve.

The highlighted lateral corticospinal tract (6) has not quite completed its journey from the right pyramid of the medulla to the left lateral funiculus of the spinal cord. Fibers of the anterior corticospinal tract (17) will cross in the anterior commissure to supply deep muscles in the left side of the neck. The highlighted tectospinal tract (12) arose in the right midbrain tectum (Figure 17.20). Both medial vestibulospinal tracts (14) are highlighted, because each medial vestibular nucleus sends axons into both tracts.

Reticular formation

The pontine reticulospinal tract (10) is descending ipsilaterally to supply motor neurons innervating antigravity muscles. The medullary reticulospinal tract (8), partly crossed, will supply flexor motor neurons.

The raphespinal tract (17) arises in the midline of the upper medulla oblongata, descends bilaterally within the posterolateral tract of Lissauer, and terminates in the posterior gray horn at all segmental levels.

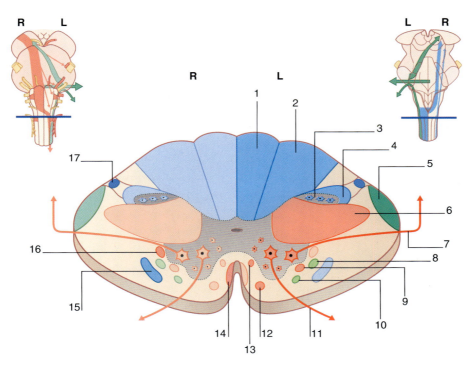

Figure 17.10 C1 segment of spinal cord. (After Noback CR et al. 1996, with permission of Williams & Wilkins.)

Autonomic system

The central autonomic pathway (17) conveys sympathetic and parasympathetic fibers from the ipsilateral hypothalamus to the lateral gray horn of the cord at appropriate levels.

SPINOMEDULLARY JUNCTION
(Figure 17.11)

Landmarks

The central canal and central gray matter (1) occupy the midregion. The remainder of the spinal gray matter is broken into discrete cranial nerve nuclei, as will be seen at all higher levels. Ventral to (1) is (19), the **medial longitudinal fasciculus**, whose contents at this level are shown in Figure 17.6.

Afferent nuclei and pathways

Dorsally, the left **gracile fasciculus** (2) and **cuneate fasciculus** (4) are highlighted. They subserve conscious proprioception and discriminative touch for the left limbs and trunk.

The right spinal lemniscus (16) is highlighted. This is the brainstem continuum of the anterolateral pathway (anterior and lateral spinothalamic tracts). Included within the spinal lemniscus is the small, crossed, spinotectal tract

that accompanies it all the way to the tectal plate of the midbrain.

The spinal tract and nucleus (5) of the trigeminal nerve are seen again.

Cerebellar connections (ascending)

The position of the posterior spinocerebellar tract (6) is unchanged.

Efferent nuclei and pathways

In the ventral region, fibers of the right pyramidal tract (16) are passing through the **pyramidal decussation** (12). Some have emerged from the decussation and are proceeding to the lateral funiculus of the spinal cord as the **lateral (crossed) corticospinal tract** (10). Also seen are fibers entering the ipsilateral **anterior corticospinal tract** (11) to cross lower down, and a bundle (14) representing the 10% of fibers that remain uncrossed.

Dorsal to the pyramids are the lateral and medial vestibulospinal tracts (9).

Reticular formation

Somas of the paramedian (3) and lateral (7) columns of the reticular formation are seen, also the pontine and medullary reticulospinal tracts (8) and the raphespinal tracts (19).

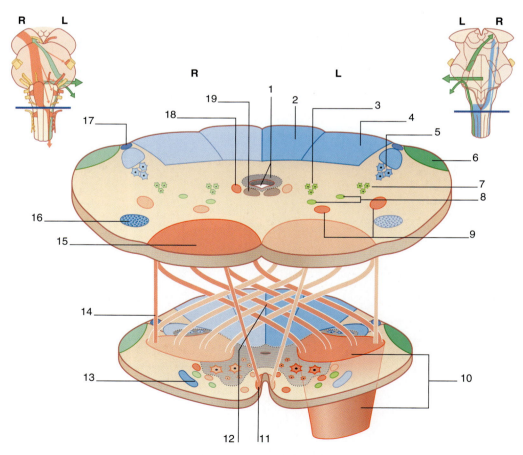

Figure 17.11 Spinomedullary junction.

Autonomic system

Central autonomic fibers (18) are descending from the hypothalamus to autonomic cell stations in the spinal cord.

MIDDLE OF MEDULLA OBLONGATA
(Figure 17.12)

Landmarks

The central canal and central gray matter (1) are beginning to move dorsally.

Afferent nuclei and pathways

A striking feature at this level is the great **sensory decussation** (18). The fibers concerned arise in the **gracile nucleus** (4) and **cuneate nucleus** (10), having received their inputs from the respective fasciculi (3, 6). Dorsal to these nuclei are some fascicular fibers passing ventrally to enter them. The sensory decussation is formed by the decussation of **internal arcuate fibers** (16) emerging from the nuclei and sweeping around the central gray matter. The highlighted left set, having crossed the midline, turn upward immediately, as the right **medial lemniscus** (18). Each lemniscus belongs to the second-order pathway from the periphery to the contralateral thalamus. (The third-order pathway extends from thalamus to somatic sensory cortex.)

Another second-order somatic sensory pathway here is the spinal lemniscus (22) already encountered. The spinal tract and nucleus of the trigeminal nerve (9) are also seen again.

The solitary tract and nucleus (5) will be considered with the next section.

Cerebellar connections (ascending)

The posterior spinocerebellar tract is unchanged. Fibers (not shown) separate from the cuneate fasciculus and synapse in the **accessory cuneate nucleus** (7), which projects cuneocerebellar fibers into the inferior cerebellar peduncle just above this level. This pathway was seen in Chapter 15 to be the upper limb counterpart of the posterior spinocerebellar tract.

The inferior olivary nucleus (20) will be considered with the next section.

Efferent nuclei and pathways

Two pairs of nuclei are associated with cranial nerves. The *somatic efferent* **hypoglossal nucleus** (2) sends the **hypoglossal nerve** (15) to the muscles on its own side of the tongue. The *branchial efferent* nucleus ambiguus (14) sends the **cranial accessory nerve** (11) for distribution (via the vagus nerve) to muscles of the palate, larynx, and pharynx. Lateral to the nucleus ambiguus is the lateral vestibulospinal tract (12).

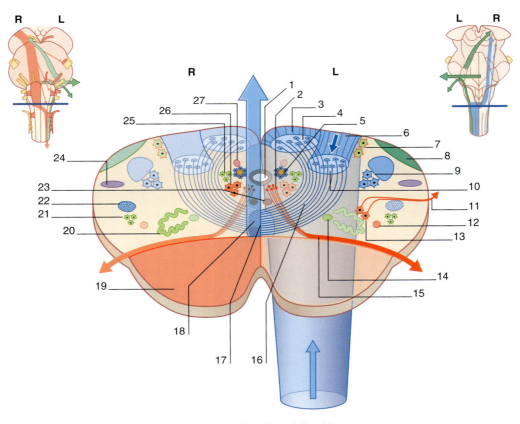

Figure 17.12 Middle of medulla oblongata.

Reticular formation

Close to the arch of internal arcuate fibers is the paramedian reticular formation (25). The somas seen send the medullary reticulospinal tract the full length of the spinal cord (partly crossed) to participate in withdrawal reflexes (Ch. 14).

Embedded in the lateral reticular formation (21) are cells important for control of the sympathetic nervous system. Variously designated as the *superficial ventro-lateral area*, *vasomotor center*, and *vasopressor center*, it contains numerous noradrenergic and adrenergic neurons involved in the control of arterial blood pressure (Ch. 24).

The pontine reticulospinal tract (17) is lateral to the hypoglossal nerve.

The raphespinal tract (24) is moving toward the surface.

Autonomic system

The dorsal nucleus of vagus (26) will be considered with the next section. Dorsal to the sensory decussation is the MLF (23). Autonomic fibers (27) continue their descent from the hypothalamus in the dorsal longitudinal fasciculus (DLF).

UPPER PART OF MEDULLA OBLONGATA   (Figure 17.13)

Landmarks

The central canal has opened into the fourth ventricle (1). The central gray matter in the floor of the ventricle contains cranial nerve nuclei.

Afferent nuclei and pathways

The **medial** (3) and **inferior** (4) **vestibular nuclei** are prominent. These are two of four nuclei in receipt of incoming terminals from the vestibular division of the eighth cranial nerve.

Afferents in the **glossopharyngeal nerve** (8) include general visceral afferent fibers from the oropharynx, terminating in the **solitary nucleus** (5). Some of these fibers provide the afferent limb of the swallowing reflex. Others provide the afferent limb of the baroreceptor reflex arc (Ch. 24). Also seen are glossopharyngeal fibers entering the spinal tract and nucleus of the trigeminal nerve (7); these are activated by inflammation of the oropharynx.

On the right side of the medulla are the medial lemniscus (16) and spinal lemniscus (19).

Cerebellar connections (ascending)

A striking feature at this level is the wrinkled **inferior olivary nucleus** (18). The principal cells of the inferior olivary nucleus, and of the small **accessory olivary nuclei** (18), give rise to the **olivocerebellar tract** (marked by eight arrows), which intersects with its opposite number before entering the opposite **inferior cerebellar peduncle** (6). The right nuclei and tract are highlighted because the main input to their nucleus is from the ipsilateral red nucleus (Figure 17.4). Here the rubroolivary fibers (22) are fanning prior to termination.

Efferent nuclei and pathways

Motor neurons highlighted in the medial and inferior vestibular nuclei give rise to the **medial vestibulospinal**

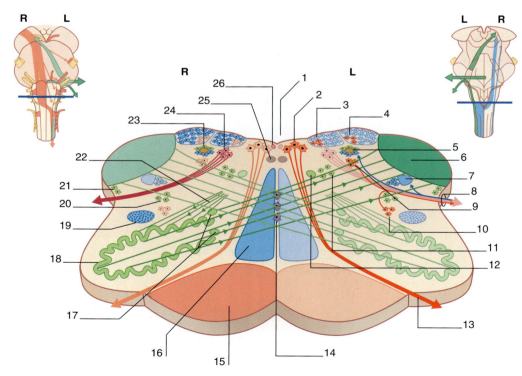

Figure 17.13 Upper medulla oblongata.

tract, which enters the MLF (25) for ultimate distribution to anterior horn cells in the cervical spinal cord (see head-righting reflexes in Ch. 19). The medial vestibular nucleus also sends fibers to the contralateral motor ocular nuclei (Figure 17.6) to participate in the vestibuloocular reflex (Ch. 19).

In the ventral tegmentum, the nucleus ambiguus (10) sends branchial efferent fibers into the glossopharyngeal nerve to innervate the stylopharyngeus. Fibers from the hypoglossal nucleus (2) are emerging in a rootlet of the hypoglossal nerve (13).

Reticular formation

In the midline is the **magnus raphe nucleus** (14), which sends streams of serotonergic fibers into both raphespinal tracts. They seek the posterolateral tract of Lissauer, extend the full length of the cord, and are significant in relation to 'gate control' of pain (Ch. 24).

Cells of the paramedian reticular formation at this level (11) give rise to the (partly crossed) medullary reticulospinal tract seen at lower levels.

The pontine reticulospinal tract (12) will be considered with the pontomedullary junction.

The **retroambiguus nucleus** (20) is the *expiratory center*. The *inspiratory center* is a group of reticular neurons in the solitary nucleus (23).

The *chemoreceptive area* (21) contains neurons sensitive to the carbonic acid levels in the overlying cerebrospinal fluid.

Autonomic system

Ventral to the medial vestibular nucleus is the **dorsal (motor) nucleus of the vagus** (24, sometimes referred to as the *DMX*). This is the *general visceral efferent* motor nucleus supplying preganglionic fibers to autonomic ganglia in the walls of the thoracic and abdominal viscera (Ch. 13). Information from animal experiments indicates that vagal neurons serving heart and lungs may be lodged in the nucleus ambiguus.

The small, *general visceral efferent* **inferior salivatory nucleus** (9) provides the preganglionic parasympathetic secretomotor supply for the parotid gland (Ch. 22).

The slender **dorsal longitudinal fasciculus** (26) is at its level of termination. DLF fibers travel from the limbic lobe and hypothalamus to central autonomic nuclei in the medulla and spinal cord.

PONTOMEDULLARY JUNCTION
(Figure 17.14)

Landmarks

The tegmentum of the pons extends from the floor of the fourth ventricle to the posterior set of transverse fibers (23). The transverse fibers and pyramidal tract (22) occupy the basilar pons.

Afferent nuclei and pathways

The **cochlear** and **vestibular** divisions of the eighth cranial nerve are present. Both comprise *bipolar* neurons whose

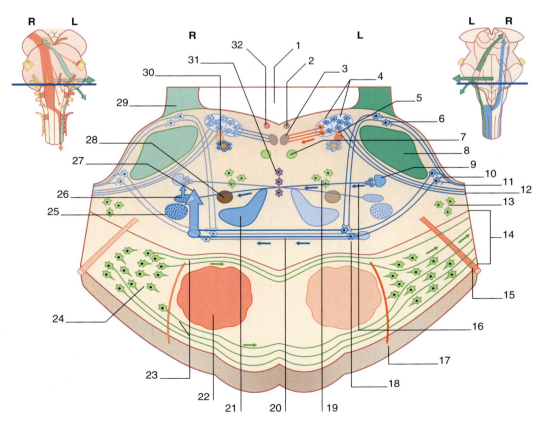

Figure 17.14 Pontomedullary junction.

somas are housed in sensory ganglia within the temporal bone (Chs 19 and 20). The cochlear nerve (11) terminates by synapsing on the *special sense afferent* **dorsal** (6) and **ventral** (10) **cochlear nuclei**, respectively located on the dorsal and lateral aspects of the inferior cerebellar peduncle (8). Many second-order afferents follow the general plan of sensory pathways and project across the midline—in this instance within the **trapezoid body** (20)—and run upward in the **lateral lemniscus** (27). Others synapse in the ipsilateral **superior olivary nucleus** (18) to be relayed either to the opposite lateral lemniscus or into the *ipsilateral* one. The presence of ipsilateral fibers in these ensures bilateral representation of hearing at higher levels (Ch. 20).

The bipolar neurons of the vestibular nerve (12) synapse in the special sense afferent **lateral** and **superior vestibular nuclei** (4), as well as on those of the medial and inferior vestibular nuclei previously identified.

The medial lemniscus (21) is taking up a more lateral position. Lateral to the lateral lemniscus is the **trigemino-thalamic tract** (26). This tract comprises axons that originate in the spinal nucleus of the trigeminal nerve and cross (17) to accompany the spinal lemniscus. It is the homolog, for the head and upper neck regions, of the lateral spinothalamic tract, signaling painful and/or thermal stimulation of peripheral nerve endings. The spinal nucleus sends fibers across to the tract at all brainstem levels of that nucleus.

Also seen are the spinal lemniscus (25) and the solitary tract and nucleus (30).

Cerebellar connections (ascending)

The inferior cerebellar peduncle (8) contains the olivocerebellar tract seen in the previous section, also the posterior spinocerebellar tract.

The **superior cerebellar peduncle** (29) will be seen again at higher levels.

Efferent nuclei and pathways

The pyramidal tract (22) occupies the midregion of the basilar pons on each side.

The **lateral vestibulospinal tract** (5) is given off by the lateral vestibular nucleus and descends in the anterior funiculus of the cord to motor neurons supplying anti-gravity muscles.

The emergent fibers of the **facial nerve** (15) and **abducens nerve** (17) are seen in the basilar pons.

Cerebellar connections (descending)

The nature of the nuclei pontis (24) and transverse fibers (23) is the same as in Figure 17.15.

Reticular formation

The **caudal pontine reticular nucleus** (19), together with the oral pontine nucleus at a higher level, gives rise to the pontine reticulospinal tract, which will be seen in the next section. Somas of the lateral reticular nucleus (14) are also seen.

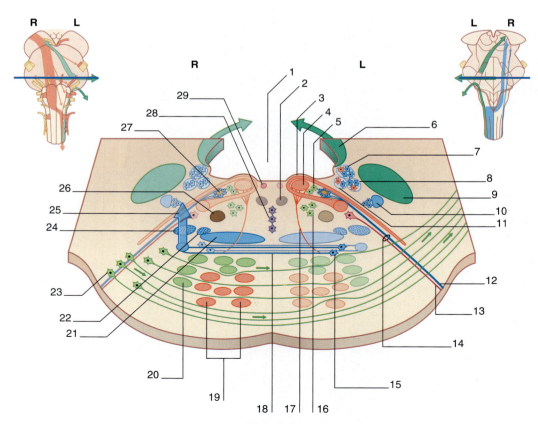

Figure 17.15 Mid-pons.

The **central tegmental tract** (28), containing rubroolivary fibers, continues its descent from the ipsilateral red nucleus to the inferior olivary nucleus.

From the pontine raphe nucleus (31), serotonergic fibers are distributed to pons and cerebellum.

Autonomic system

Fibers of hypothalamic origin (32) are descending in the DLF (2).

MID-PONS (Figure 17.15)

Landmarks

The facial colliculus (3) is in the floor of the fourth ventricle (1). The ventricle is bounded laterally by the superior cerebellar peduncles (6).

Afferent nuclei and pathways

Previously seen are the spinal tract and nucleus of the trigeminal nerve (10), now at their highest level; the medial (21) and spinal (24) lemnisci; and the trigeminothalamic tract (22). The superior olivary nucleus (16) continues to send fibers through the trapezoid body (18) into the contralateral lateral lemniscus (25).

Gustatory (taste) fibers (12) leave the nervus intermedius to enter the uppermost part of the solitary tract and nucleus.

Cerebellar connections (ascending)

The inferior cerebellar peduncle (9) is embedded in cerebellar white matter. The superior cerebellar peduncles (6) are converging (arrows).

Efferent nuclei and pathways

The facial colliculus is created by the *somatic efferent* **abducens nucleus** (4) from which the **abducens nerve** (17) passes through the tegmentum. From the *branchial efferent* **facial nucleus**, the **facial nerve** (3) loops mediolaterally around the abducens nucleus before traveling forward through the tegmentum.

The pyramidal tract (19) is parceled into fascicles by transverse fibers.

Cerebellar connections (descending)

Corticopontine fibers (20) continue to synapse on pontine nuclei (23), which send pontocerebellar tract fibers through the contralateral middle cerebellar peduncle.

Reticular formation

Close to the abducens nucleus is the **paramedian pontine reticular formation** (5) or *pontine gaze center* (Ch. 23), which enables *saccades* (glances) toward the same side.

In the medial tegmentum are cells of the paramedian reticular formation (17).

Autonomic system

The **superior salivatory nucleus** is sending secretomotor fibers (14) alongside the gustatory fibers identified earlier. Together, the two sets constitute the **nervus intermedius** (14), which will accompany the facial nerve into the internal acoustic meatus.

The DLF (29) contains autonomic fibers descending from the hypothalamus.

UPPER PONS (Figure 17.16)

Landmarks

The fourth ventricle (1) is being narrowed by convergence of the superior cerebellar peduncles (5). The DLF (2) lies in its floor.

Afferent nuclei and pathways

Previously encountered are the medial (17), spinal (18), and lateral (19) lemnisci, also the trigeminothalamic tract (20).

The **sensory root** (10) of the trigeminal nerve (11) terminates in the **principal** (*pontine*) **nucleus** (8), which serves a tactile function for the face, mouth, and other areas supplied by the sensory root. The nucleus projects second-order fibers across the midline to join those of the trigeminothalamic tract. Above this level, the crossed pair are called the **trigeminal lemniscus**.

Dorsal to the principal sensory nucleus lies the **mesencephalic tract** (7) of the trigeminal nerve. The parent nucleus is in the midbrain.

Cerebellar connections (ascending)

The superior cerebellar peduncles (5) again form the side walls of the fourth ventricle. They are composed mainly of a projection from the dentate nucleus of cerebellum to the contralateral thalamus.

Efferent nuclei and pathways

The *branchial efferent* **motor nucleus** of the **trigeminal nerve** (9) sends the **motor root** through tegmentum and basilar pons to its point of emergence from the brainstem. Its main targets are the muscles of mastication (chewing).

The trigeminal nerve is used to indicate the demarcation line between pons proper and middle cerebellar peduncle.

The tectospinal tract has been incorporated into the MLF (3).

The pyramidal tract (17) has been parceled into fascicles by transverse fibers (13).

Cerebellar connections (descending)

The corticopontine fibers descended from the crus cerebri synapse on millions of somas that make up the **nuclei pontis** (18). The nuclei end their axons to the contralateral cerebellar cortex via **transverse fibers** (13) that enter the cerebellum through the middle cerebellar peduncle.

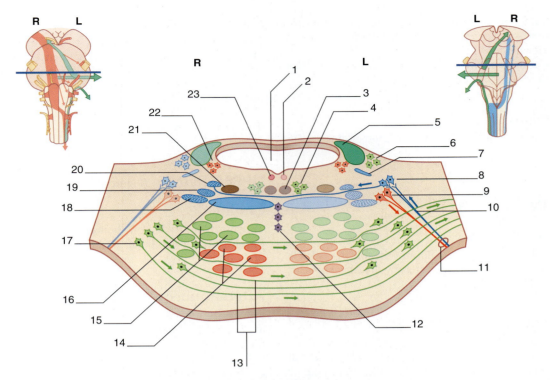

Figure 17.16 Upper pons.

Reticular formation

In the upper, lateral part of the fourth ventricle on each side is the **cerulean nucleus** (22). It contains the largest group of noradrenergic neurons in the brain. It distributes long, fine, beaded axons to all parts of the cerebellar and cerebral cortex (see Ch. 24). Immediately lateral to the superior cerebellar peduncle is the **pedunculopontine nucleus** (6), part of the locomotor control center (Ch. 24).

The DLF contains autonomic fibers (23) descending from the ipsilateral hypothalamus.

LOWER MIDBRAIN (Figure 17.17)

Landmarks

The central canal (2) is the **cerebral aqueduct** (*aqueduct of Sylvius*), which in life contains cerebrospinal fluid pouring down from the third ventricle to the fourth. In the lateral part of the **periaqueductal gray matter** (PAG) are *enkephalinergic* neurons (3) projecting to the magnus raphe nucleus of the medulla, with a significant role in the suppression of pain sensation (see Ch. 24).

Afferent nuclei and pathways

In the most ventral tegmentum are the **compact** (10) and **reticular** (11) parts of the **substantia nigra**. The compact part, comprising pigmented *dopaminergic neurons*, is the source of the *nigrostriatal pathway* to the corpus striatum. The nigrostriatal pathway loses both pigment and cells progressively, in those unfortunates bound for *Parkinson's*

disease (Ch. 33). The reticular part comprises GABAergic neurons.

The **ventral tegmental nuclei** (14) contain *mesocortical* dopaminergic neurons projecting to the frontal cortex, and *mesolimbic* dopaminergic neurons projecting to the nucleus accumbens (a ventral part of the corpus striatum). Both elements are clinically significant (Ch. 34).

In the lateral tegmentum are the medial (20), spinal (22), and trigeminal (23) lemnisci. The lateral lemniscus (24) is terminating in the **inferior colliculus** (25), the subcortical center for hearing.

The **mesencephalic nucleus** of the trigeminal nerve (4) is anatomically unique in the central nervous system: it is entirely composed of *unipolar neurons* that remained within the pons and midbrain at the time (third week of embryonic life) when homologous precursors elsewhere moved outside, within the neural crest (Ch. 1), to form posterior root ganglion cells. The nucleus and tract serve proprioception in the trigeminal territory.

Cerebellar connections (ascending)

The most prominent feature at this level is the **decussation of the superior cerebellar peduncles** (16). The highlighted fibers originated mainly in the dentate nucleus of the left cerebellar hemisphere (Figure 17.4).

Efferent nuclei and pathways

From the *somatic efferent* **trochlear nucleus** (6), the **trochlear nerve** skirts PAG and decussates (1) prior to emergence on the dorsum of the brainstem.

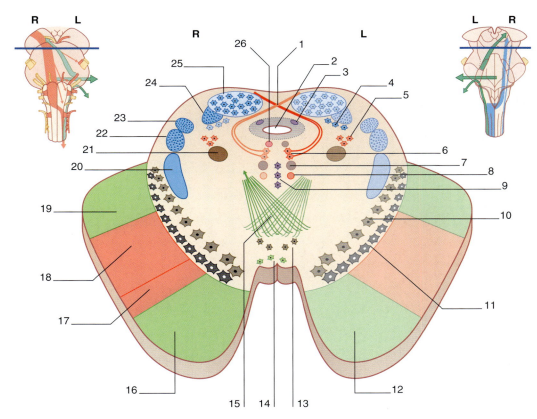

Figure 17.17 Lower midbrain.

The corticonuclear (17) and corticospinal (18) fibers are unchanged.

Cerebellar connections (descending)

The two sets of corticopontine fibers (17,19) are unchanged. The red nucleus is giving rise to rubroolivary fibers that descend within the central tegmental tract (21) to synapse in the ipsilateral inferior olivary nucleus.

Reticular formation

The interpeduncular nucleus (17) occupies the most ventral tegmentum.

Autonomic system

The DLF (26) contains ipsilateral descending autonomic fibers.

UPPER MIDBRAIN   (Figure 17.18)

Landmarks

The cerebral aqueduct and PAG (1) are unchanged, as are the compact and reticular parts of the substantia nigra (9, 8).

Afferent nuclei and pathways

The medial (21), trigeminal (22), and spinal (23) lemnisci continue to move dorsally as they near the thalamus. Fibers leaving the spinal lemniscus are fellow travelers that have

reached their station. They have emerged as the small **spinotectal tract** (24) to enter the **superior colliculus** (25).

Cerebellar connections (ascending)

Dentatothalamic fibers (20) of the crossed superior cerebellar peduncle have bypassed the red nucleus. Some dentate fibers (not shown here) enter the red nucleus; their significance is discussed in Chapter 25. *Rubrothalamic* fibers (19) ascend in the central tegmental tract.

Efferent nuclei and pathways

Close to the midline is the *somatic efferent* **oculomotor nucleus** (2). Vestibuloocular and internuclear fibers leave the MLF (5) and enter the nucleus.

The **oculomotor nerves** traverse the red nuclei prior to emergence in the interpeduncular fossa (11).

The **tectospinal tract** crosses the midline before descending (7). This tract belongs to the efferent limb of the *spinovisual reflex*, which turns the head toward something glimpsed. Behaviors such as these are better called *responses* because reaction is optional.

Unchanged from Figure 17.19 are corticopontine (15, 18), corticonuclear (17), and corticospinal (18) fibers.

Reticular formation

The interpeduncular nucleus (13) is unchanged. The **midbrain raphe nucleus** (14) sends an enormous *serotonergic*

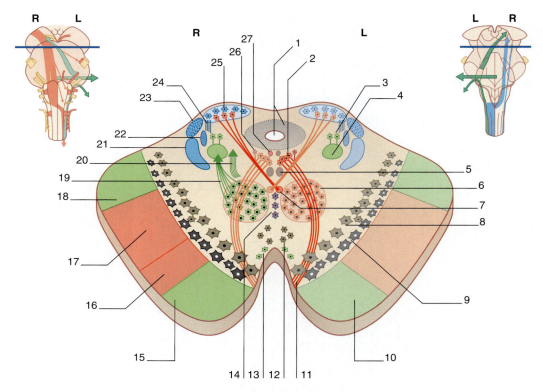

Figure 17.18 Upper midbrain.

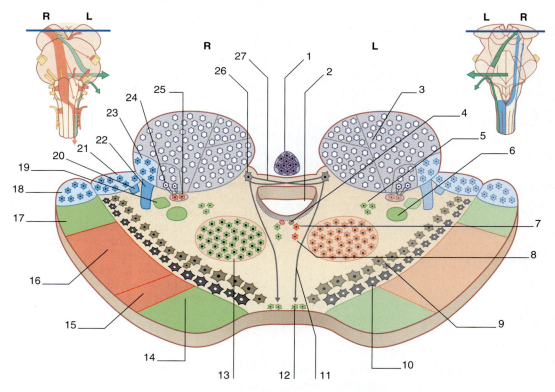

Figure 17.19 Midbrain–thalamic junction.

projection to both cerebral hemispheres via the medial forebrain bundle (Ch. 24).

The **cuneiform nucleus** (3) projects upward within the central tegmental tract (4) to participate in *arousal* mechanisms at thalamic level (Ch. 24).

Autonomic system

Lateral to the oculomotor nucleus is the tiny general visceral efferent **anteromedial** (*Edinger–Westphal*) **nucleus** (26), which contributes parasympathetic fibers to the oculomotor nerve. The DLF (27) contains autonomic fibers descending to lower levels.

MIDBRAIN–THALAMIC JUNCTION
(Figure 17.19)

Landmarks

The bulky **thalamus** (3) lies above the upper extremity of the midbrain tegmentum. In the midline is the overhanging **pineal gland** (1). The central canal is opening into the **third ventricle** (2).

The **medial forebrain bundle** (6) contains fibers passing to and from the hypothalamic area. It merges with the central tegmental tract.

The compact and reticular parts of the substantia nigra (10, 9) are unchanged.

Sensory nuclei and pathways

Three thalamic nuclei are highlighted. The **ventral posterior nucleus** (23) is receiving the medial, spinal, and trigeminal lemnisci. The **medial geniculate body** (21) is receiving the **inferior brachium** (20) from the inferior colliculus. Most lateral is the **lateral geniculate body** (18), which receives the **superior brachium** (not shown) from the superior colliculus.

Cerebellar connections (ascending)

In the lateral tegmentum, the dentatothalamic tract (19) is aiming for the ventral lateral nucleus of the thalamus, which lies anterior to the ventral posterior nucleus. From there, a final projection will reach the motor cortex and will coordinate ongoing movements.

Efferent nuclei and pathways

The middle three-fifths of the cerebral crus contain the **corticospinal tract** (16) together with **corticonuclear fibers** (15) and corticoreticular fibers (not shown) concerned with automatic movements (Ch. 29).

The corticonuclear fibers are of two kinds. One kind arises in cortical areas anterior to the central sulcus. They innervate contralateral motor cranial nerve nuclei of pons and medulla directly (Figure 17.3) and the contralateral oculomotor and trochlear nerve nuclei indirectly, via gaze centers (see later). The second kind of corticonuclear fibers arises in the parietal lobe. Their main targets are the contralateral gracile and cuneate nuclei, which may be either facilitated or inhibited (Ch. 16).

The corticospinal fibers are also of two kinds, one destined for anterior horn motor neurons, the other for the posterior gray horn.

The **subthalamic nucleus** (25) is of great current interest in connection with Parkinson's disease (Ch. 33). At this level, it is receiving inputs from the compact part of substantia nigra and from the **centromedian nucleus** (24) of the thalamus. It projects to the lentiform nucleus.

Cerebellar connections (descending)

The association areas of the cerebral cortex send massive projections to cell stations in the ipsilateral pons. *Frontopontine* fibers (14) travel in the medial part of the crus cerebri. *Parieto-*, *occipito-*, and *temporopontine* fibers occupy the lateral part (17).

The right red nucleus is highlighted in green because it receives inputs from the right motor cortex and from the left cerebellum. It gives rise to an ipsilateral rubroolivary projection (Figure 17.4).

Reticular formation

The **habenular nucleus** (26) projects the **fasciculus retroflexus** (11) on to the **interpeduncular nucleus** (12). The habenular nuclei exchange fibers through the **habenular commissure** (27).

Close to the midline, two nodules of the reticular formation are the *centers for upward* (7) and *downward* (8) *gaze*. They receive a crossed projection from an *eye field* in the frontal lobe and they synapse bilaterally on the oculomotor and trochlear nuclei. Rarely, a *pinealoma* (pineal gland tumor) may signal its presence by causing *paralysis of upward gaze*.

In the dorsal tegmentum is the **pretectal nucleus** (5). Each of these nuclei supplies *both* Edinger–Westphal nuclei for bilateral execution of the pupillary light reflex (Ch. 23).

Autonomic system

Central autonomic fibers descend ipsilaterally in the **DLF** (4) to reach autonomic nuclei at lower levels.

Epilog

Figure 17.20 offers an unlabeled overview of the brainstem sections. Gentle skimmers may opt to photocopy lightly and to crayon pathways.

ORIENTATION OF BRAINSTEM SLICES IN MAGNETIC RESONANCE IMAGES

Figure 17.21 shows brainstem slices in magnetic resonance images. Their orientation is the opposite of those in the preceding sections. In photographs and drawings, the convention is to represent anterior structures below. As already mentioned in Chapter 2, in magnetic resonance imaging scans, anterior structures are represented above.

Figure 17.20 Brainstem review.

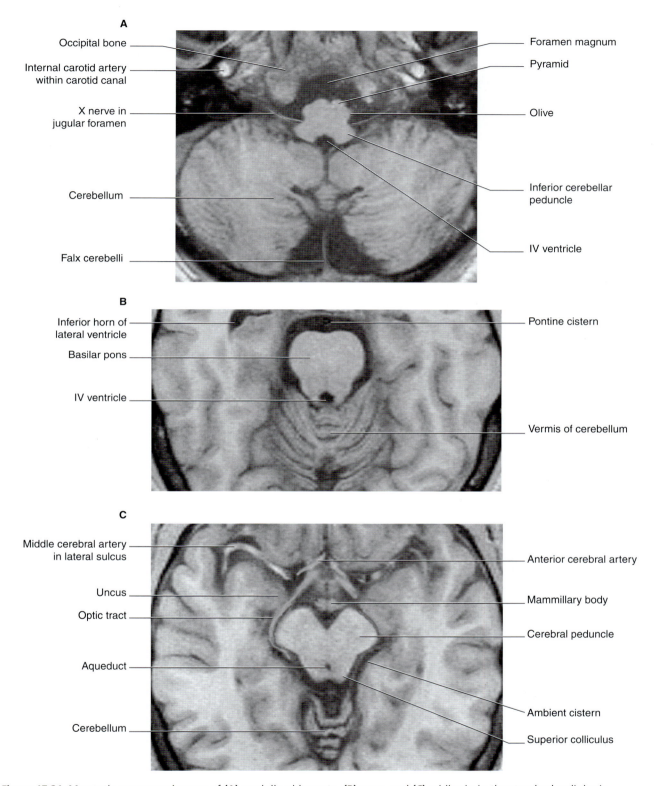

A

Occipital bone

Internal carotid artery
within carotid canal

X nerve in
jugular foramen

Cerebellum

Falx cerebelli

Foramen magnum

Pyramid

Olive

Inferior cerebellar
peduncle

IV ventricle

B

Inferior horn of
lateral ventricle

Basilar pons

IV ventricle

Pontine cistern

Vermis of cerebellum

C

Middle cerebral artery
in lateral sulcus

Uncus

Optic tract

Aqueduct

Cerebellum

Anterior cerebral artery

Mammillary body

Cerebral peduncle

Ambient cistern

Superior colliculus

Figure 17.21 Magnetic resonance images of **(A)** medulla oblongata, **(B)** pons, and **(C)** midbrain in the standard radiologic orientation. (From a series kindly provided by Professor J. Paul Finn, Director, Magnetic Resonance Research, Department of Radiology, David Geffen School of Medicine at UCLA, California.)

Core Information

Cell columns

Cranial nerve cell columns and their representations are as follows.

- *General somatic efferent*, represented in the medulla by the hypoglossal nucleus, in the pons by the abducens nucleus, and in the midbrain by the oculomotor and trochlear nuclei.
- *Special visceral efferent*, supplying muscles of branchial arch origin, represented in the medulla by the nucleus ambiguus and in the pons by trigeminal and facial motor nuclei.
- *General visceral efferent*, represented in the medulla by dorsal nucleus of the vagus and in the medulla and in the pons by the salivatory nucleus.
- *General visceral afferent*, represented in the medulla by the inferior solitary nucleus.
- *Special visceral afferent*, represented in the pons by the superior solitary nucleus.
- *General somatic afferent*, represented by trigeminal sensory nuclei: spinal in the medulla, principal sensory in the pons, and mesencephalic in the midbrain.
- *Special sense afferent*, represented at the pontomedullary junction by the cochlear and vestibular nuclei.

Ascending pathways

The gracile and cuneate nuclei send internal arcuate fibers across the midline to form the medial lemniscus, which goes through pons and midbrain to reach the thalamus. The spinal trigeminal nucleus sends fibers across the midline to form the trigeminothalamic tract. The posterior spinocerebellar and cuneocerebellar tracts send their fibers into the ipsilateral inferior cerebellar peduncle, where they mingle with olivocerebellar fibers crossing from the inferior and accessory olivary nuclei. The (crossed) anterior spinocerebellar tract enters the superior cerebellar peduncle; its fibers cross a second time within the cerebellar white matter.

- The spinal lemniscus is formed of the anterior and lateral spinothalamic tracts. It is accompanied first by trigeminothalamic fibers, later by the lateral lemniscus and by fibers crossing from the principal trigeminal nucleus completing the trigeminal lemniscus.
- The cochlear nuclei project fibers across the trapezoid body to form the lateral lemniscus, which ascends to the inferior colliculus. Some fibers synapse instead in a superior olivary nuclear relay to the ipsilateral inferior colliculus. Third-order neurons of the inferior colliculus project via inferior brachium to the medial geniculate body. The medial and superior vestibular nuclei send fibers to the oculomotor nucleus to execute the vestibuloocular reflex.

- The upper part of the central tegmental tract contains fibers of the ascending reticular activating system.
- In the ventral tegmentum of the midbrain are the pigmented (compact) substantia nigra giving rise to the nigrostriatal pathway, and the ventral tegmental nucleus giving rise to the mesocortical and mesolimbic pathways. The non-pigmented (reticular) nigral neurons are inhibitory.

Descending pathways other than reticulospinal

Corticonuclear fibers from motor areas of the cerebral cortex are distributed preferentially to contralateral motor cranial nerve nuclei excepting the ocular motor slaves of gaze centers. Corticonuclear fibers from sensory areas synapse in contralateral trigeminal and posterior column nuclei and posterior gray horn of spinal cord.

- Prior to the initiation of a voluntary movement on the left side of the body, the left cerebellar hemisphere is notified by discharges from association areas of the right cerebral cortex, along the corticopontocerebellar pathway. The left cerebellum responds via the dentatothalamocortical pathway, to the right primary motor cortex. Then the right pyramidal tract discharges and on its way down notifies the cerebellum a second time by activating the right red nucleus, which in turn activates the right olivocerebellar tract.
- Corticospinal fibers pass through the middle three-fifths of the cerebral crus and through the basilar pons (where they are segregated into bundles by transverse fibers), finally creating the pyramid of the medulla before four-fifths enter the pyramidal decussation.
- The lateral vestibular nucleus gives rise to the lateral vestibulospinal tract having an antigravity function. The medial and inferior vestibular nuclei give rise to the medial vestibulospinal tract involved in head-righting reflexes. The tectospinal tract belongs to the spinovisual reflex arc. The dorsal longitudinal fasciculus contains ipsilateral central autonomic fibers.
- A sleep-related pathway from the septal area reaches the interpeduncular nucleus by way of the habenular nucleus and fasciculus retroflexus.

Reticular formation

In the uppermost midbrain are the upward and downward gaze centers. (The lateral gaze centers adjoin the abducens nucleus in the pons.) The midbrain also contains an up-going, 'arousal' projection from the cuneiform nucleus, a down-going, pain-suppressant projection from the periaqueductal gray matter, and a locomotor generator, the pedunculopontine nucleus.

Core Information *(Continued)*

The pons contains the noradrenergic, cerulean nucleus, also the oral and caudal pontine reticular nuclei, which send ipsilateral pontine reticulospinal tracts to extensor motor neurons. The medullary reticulospinal tract is partly crossed and supplies flexor motor neurons. Three respiratory reticular nuclei also occupy the medulla.

REFERENCES

Literature references to individual brainstem nuclei and pathways are to be found in the relevant chapters.

REFERENCES

Kautcherov Y, Huang X-F, Halliday G, et al. Organization of human brainstem nuclei. In: Paxinos G, Mai JK, eds. The human nervous system. 2nd edn. Amsterdam: Elsevier; 2004:267–320.

Noback CR et al. The human nervous system: structure and function. Baltimore: Williams & Wilkins; 1996.

STUDY GUIDELINES

Comments on the last four cranial nerves in ascending order:

1 The hypoglossal nerve is straightforward: it is motor to the tongue. The spinal accessory nerve is straightforward: it is motor to the sternomastoid and trapezius.

2 The cranial accessory nerve supplies the intrinsic muscles of larynx and pharynx, and the levator palati. It is *distributed* by the vagus.

3 The vagus nerve proper is the principal preganglionic parasympathetic nerve. It is also the principal visceral afferent nerve.

4 Main features of the glossopharyngeal are:
(a) it provides the afferent limb of the gag reflex
(b) it tells us when we have an inflamed throat
(c) it signals bitterness
(d) its carotid branch carries afferents from the carotid sinus monitoring blood pressure and from the carotid body monitoring blood gases
(e) it gives a clinically significant branch to the middle ear.

HYPOGLOSSAL NERVE

The hypoglossal nerve (cranial nerve XII) contains somatic efferent fibers for the supply of the extrinsic and intrinsic muscles of the tongue. Its nucleus lies close to the midline in the floor of the fourth ventricle and extends almost the full length of the medulla (Figure 18.1). The nerve emerges as a series of rootlets in the interval between the pyramid and the olive. It crosses the subarachnoid space and leaves the skull through the hypoglossal canal. Just below the skull, it lies close to the vagus and spinal accessory nerves (Figure 18.2). It descends on the carotid sheath to the level of the angle of the mandible, then passes forward on the surface of the hyoglossus muscle where it gives off its terminal branches.

Afferents from about 100 muscle spindles in the same side of the tongue hitchhike along the distal part of the nerve before leaving to enter the upper end of the spinal cord via the cervical plexus.

Phylogenetic note

In reptiles, the lingual muscles, the geniohyoid muscle, and the infrahyoid muscles develop together from the uppermost mesodermal somites. The somatic efferent neurons supplying this *hypobranchial muscle sheet* form a continuous ribbon of cells extending from lower medulla to the third cervical spinal segment. In mammals, the hypoglossal nucleus is located more rostrally, and its rootlets emerge separately from the cervical rootlets. However, the caudal limit of the hypoglossal nucleus remains linked to the cervical motor cell column by the *supraspinal nucleus*, from which the thyrohyoid muscle is supplied via

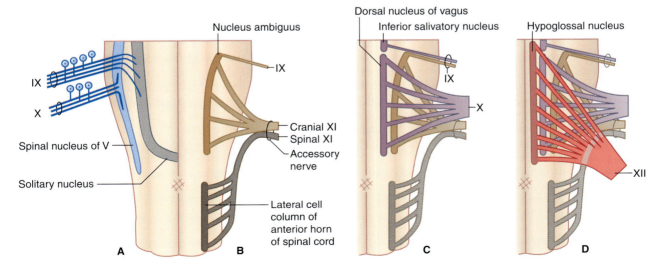

Figure 18.1 (A) Sensory nuclei (left) and motor nuclei (right) serving cranial nerves IX–XII. **(B)** The special visceral efferent cell column giving a contribution to the glossopharyngeal nerve and forming the cranial accessory nerve. **(C)** The general visceral efferent cell column contributing to the glossopharyngeal and vagus nerves. **(D)** The somatic efferent cell column giving rise to the hypoglossal nerve.

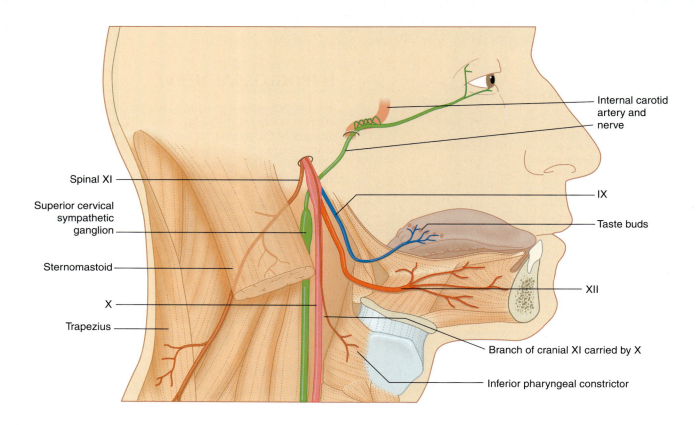

Figure 18.2 Semischematic illustration of the lowest four cranial nerves and the internal carotid branch of the superior cervical ganglion.

the first cervical ventral root. In rodents, some of the intrinsic muscle fibers of the tongue receive their motor supply indirectly, from axons that leave the most caudal cells of the hypoglossal nucleus and emerge in the first cervical nerve to join the hypoglossal nerve trunk in the neck. Whether this arrangement holds for primates is not yet known.

Supranuclear supply to the hypoglossal nucleus

The hypoglossal nucleus receives inputs from the reticular formation, whereby it is recruited for stereotyped motor routines in eating and swallowing. For delicate functions, including articulation, most of the fibers from the motor cortex cross over in the upper part of the pyramidal decussation; some remain uncrossed and supply the ipsilateral hypoglossal nucleus.

Supranuclear, nuclear, and infranuclear lesions of the hypoglossal nerve are described together with lesions of the accessory nerve (see Clinical Panels 18.1–18.3).

SPINAL ACCESSORY NERVE

The spinal accessory nerve (cranial nerve XI) is a purely motor nerve attached to the uppermost five segments of the spinal cord. The nucleus of origin is a column of α and γ motor neurons in the basolateral anterior gray horn.

The nerve runs upward in the subarachnoid space, behind the denticulate ligament. It enters the cranial cavity through the foramen magnum and leaves it again through the jugular foramen. While in the jugular foramen, it shares a dural sheath with the cranial accessory nerve, but there is no exchange of fibers (Figure 18.3). On leaving the cranium, it crosses the transverse process of the atlas and enters the sternomastoid muscle, in company with twigs from roots C2 and C3 of the cervical plexus. It emerges from the posterior border of the sternomastoid and crosses the posterior triangle of the neck to reach the trapezius. It pierces the trapezius in company with twigs from roots C3 and C4 of the cervical plexus. In the posterior triangle, the nerve is vulnerable, being embedded in prevertebral fascia and covered only by investing cervical fascia and skin.

The spinal accessory nerve provides the extrafusal and intrafusal motor supply to the sternomastoid and trapezius. The branches from the cervical plexus are proprioceptive in function to the sternomastoid and to the craniocervical part of the trapezius. The thoracic part of the trapezius, which arises from the spines of all the thoracic vertebrae, receives its proprioceptive innervation from the posterior rami of the thoracic spinal nerves. Some of the afferents supplying muscle spindles in the thoracic trapezius do not meet up with the fusimotor supply before reaching the spindles. This is the only instance, *in any muscle known*, where the fusimotor and afferent fibers to some spindles travel by completely independent routes.

Clinical Panel 18.1   Supranuclear lesions of the IX, X, and XI cranial nerves

Supranuclear lesions of all three are commonly seen following vascular strokes damaging the pyramidal tract in the cerebrum or brainstem.

Effects of unilateral supranuclear lesions

1 The supranuclear supply to the hypoglossal nucleus is mainly crossed. The usual picture following a hemiplegic stroke is as follows: during the first few hours or days, the tongue, when protruded, deviates toward the paralyzed side because of the stronger pull of the healthy genioglossus. Later, the tongue does not deviate on protrusion. However, normal hypoglossal nerve function on the affected side is *not* restored. Electrophysiologic testing has revealed that tongue movement, in response to electrical stimulation of the crossed monosynaptic corticonuclear supply to the hypoglossal nucleus, is both delayed and weaker than normal. This, together with comparable deficiency in the corticonuclear supply to the facial nerve (which includes a motor supply to the lips), accounts for the *dysarthria* (slurred speech) that persists after a hemiplegic stroke.

2 Damage to the supranuclear supply to the nucleus ambiguus may cause temporary interference with phonation and swallowing.

3 On testing the power of trapezius by asking the patient to shrug the shoulders against resistance, the muscle on the affected side is relatively weak. This accords with expectation. But on testing sternomastoid (SM) by asking the patient to turn the head against resistance applied to the side of the jaw, the SM on the *unaffected* side appears to be relatively weak. Given that electrical stimulation applied to the supranuclear supply for SM has shown

that the crossed supply is strong and monosynaptic and that the uncrossed is weak and disynaptic, there appears to be a 'sternomastoid paradox'. However, the most parsimonious explanation is that the prime mover for the 'no' headshake is not the contralateral SM but the *ipsilateral inferior oblique*, a muscle within the suboccipital triangle passing from spine of axis to transverse process of atlas. Supplementary ipsilateral muscles include splenius capitis and longissimus capitis. All three are typical spinal muscles and would be expected to share in the general muscle weakness on the affected side.

During the head rotation test, the functionally intact contralateral (healthy side) SM does contract strongly. However, the head rotators also have a *tilting* action at the atlantooccipital joint. The laterally placed insertion of SM has strong leverage potential and is well placed to counter the tilting action of the four ipsilateral muscles inserting on to the skull.

Effects of bilateral supranuclear lesions

The supranuclear supply to the hypoglossal nucleus and nucleus ambiguus may be compromised *bilaterally* by thrombotic episodes in the brainstem in patients suffering from arteriosclerosis of the vertebrobasilar arterial system. The motor nuclei of the trigeminal nerve (to the masticatory muscles) and of the facial nerve (to the facial muscles) may also be affected. The characteristic picture, known as *pseudobulbar palsy*, is that of an elderly patient who has spastic (tightened) oral and pharyngeal musculature, with consequent difficulty with speech articulation, chewing, and swallowing. The gait is slow and shuffling because of involvement of corticospinal fibers descending to lower limb motor neurons.

Clinical Panel 18.2   Nuclear lesions of the X, XI, and XII cranial nerves

Lesions of the hypoglossal nucleus and nucleus ambiguus occur together in *progressive bulbar palsy*, a variant of progressive muscular atrophy (Ch. 16) in which the cranial motor nuclei of the pons and medulla are attacked at the outset. The patient quickly becomes distressed by a multitude of problems: difficulty in chewing and articulation (mandibular and facial nerve

nuclei), and difficulty in swallowing and phonation (hypoglossal and cranial accessory nuclei).

Unilateral lesions at nuclear level may be caused by occlusion of the vertebral artery or of one of its branches (see *Lateral medullary syndrome* in Ch. 19). The distribution of motor weakness is the same as for infranuclear lesions (see Clinical Panel 18.3).

Clinical Panel 18.3   Infranuclear lesions of the lowest four cranial nerves

Jugular foramen syndrome

The last four cranial nerves, and the internal carotid (sympathetic) nerve nearby, are at risk of entrapment by a tumor spreading along the base of the skull. The tumor may be a primary one in the nasopharynx, or a metastatic one within lymph nodes of the upper cervical chain. In the second case, the primary tumor may be in an air sinus or in the tongue, larynx, or pharynx. In either case, a mass can usually be felt behind the ramus of the mandible. The symptomatology varies with the number of nerves caught up in the tumor, and the degree to which they are compromised.

Symptoms

- Pain in or behind the ear, attributable to irritation of the auricular branches of the IX and X nerves. Whenever an adult complains of constant pain in one ear, without evidence of middle ear disease, a cancer of the pharynx must be suspected.
- Headache, from irritation of the meningeal branch of the vagus.
- Hoarseness, owing to paralysis of laryngomotor fibers.
- Dysphagia (difficulty in swallowing) owing to paralysis of pharyngomotor fibers.

Signs (Figure CP 18.3.1)

- Horner syndrome (ptosis of the upper eyelid, with some pupillary constriction) from interruption of the sympathetic internal carotid nerve.
- Infranuclear paralysis of the hypoglossal nerve, with wasting of the affected side of the tongue and deviation of the tongue to the affected side on protrusion.

- When the patient is asked to say 'Aahh', the uvula is pulled away from the affected side by the unopposed healthy levator palati.
- Sensory loss in the oropharynx on the affected side.
- On laryngoscopic examination, inability to adduct the vocal cord to the midline.
- Interruption of the spinal accessory nerve produces weakness and wasting of the sternomastoid and trapezius.

A jugular foramen syndrome may also be caused by invasion of the jugular foramen from above, for instance by a tumor extending from the cerebellopontine angle (Ch. 20). In this case, the sympathetic and spinal accessory nerves will be out of reach and unaffected.

Isolated lesion of the spinal accessory nerve

The surface marking for the spinal accessory nerve in the posterior triangle of the neck is a line drawn from the posterior border of the sternomastoid one-third of the way down to the anterior border of the trapezius two-thirds of the way down. It may be injured in this part of its course by a stab wound, or during a surgical procedure for removal of cancerous lymph nodes. The trapezius is selectively paralyzed, whereupon the scapula and clavicle sag noticeably because trapezius normally helps to carry the upper limb. Shrugging of the shoulder is weakened because the levator scapulae must work alone. Progressive atrophy of the muscle leads to characteristic scalloping of the contour of the neck (Figure CP 18.3.2).

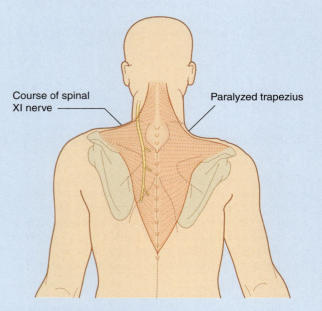

Drooping of soft palate

Anesthesia of oropharynx

Wasting of the tongue

Laryngeal mirror   Inability to adduct the vocal fold

Course of spinal XI nerve

Paralyzed trapezius

Figure CP 18.3.1 Left-sided jugular foramen syndrome. A laryngeal mirror is being used to inspect the vocal folds during an attempt to cough.

Figure CP 18.3.2 Visible effects of right-sided spinal XI paralysis: scalloping of the neck and drooping of the shoulder.

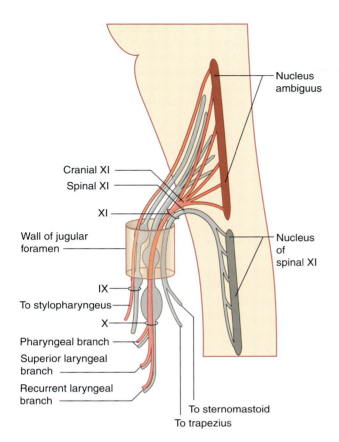

Figure 18.3 Course and distribution (in *red*) of special visceral efferent fibers derived from the nucleus ambiguus.

GLOSSOPHARYNGEAL, VAGUS, AND CRANIAL ACCESSORY NERVES

Especially relevant to nerves IX, X, and cranial XI are the solitary nucleus and the nucleus ambiguus. The solitary nucleus extends from the lower border of the pons to the level of the gracile nucleus. Its lower end merges with its opposite number in the midline, hence the term **commissural nucleus** for the lower part of the solitary nucleus.

Anatomically, the nucleus is divisible into eight parts. Functionally, four *regions* have been clarified (Figure 18.4).

1 The uppermost region is the **gustatory nucleus**, which receives primary afferents supplying taste buds in the tongue and palate.

2 The lateral midregion is the **dorsal respiratory nucleus** (see Ch. 22).

3 The medial midregion is the **baroreceptor nucleus**, which receives the primary afferents supplying blood pressure detectors in the carotid sinus and aortic arch.

4 The most caudal region, including the commissural nucleus, is the major **visceral afferent nucleus** of the brainstem. It receives primary afferents supplying the alimentary tract and respiratory tract.

From the nucleus ambiguus, *special visceral efferent* fibers supply the constrictor muscles of the pharynx, the stylopharyngeus, levator palati, intrinsic muscles of the larynx, and (via the recurrent laryngeal nerve) the striated muscle of the upper one-third of the esophagus.

Glossopharyngeal nerve

The glossopharyngeal nerve is almost exclusively sensory. It carries no less than five different kinds of afferent fibers traveling to five separate afferent nuclei in the brainstem. The largest of its peripheral territories is the oropharynx, which is bounded in front by the back of the tongue, hence the name for the nerve.

The glossopharyngeal rootlets are attached behind the upper part of the olive. The nerve accompanies the vagus through the anterior compartment of the jugular foramen (the posterior compartment contains the bulb of the internal jugular vein). Within the foramen, the nerve shows small superior and inferior ganglia; these contain unipolar sensory neurons.

Immediately below the skull, the glossopharyngeal is in the company of three other nerves (Figure 18.2): the vagus, the spinal accessory, and the internal carotid (sympathetic) branch of the superior cervical ganglion. Together with the stylopharyngeus, it slips between the

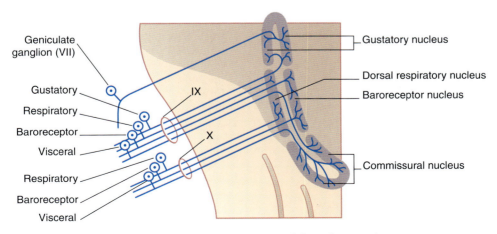

Figure 18.4 Functional composition of the solitary nucleus.

superior and middle constrictor muscles to reach the mucous membrane of the oropharynx.

Functional divisions and branches

- Before emerging from the jugular foramen, the IX nerve gives off a *tympanic branch*, which ramifies on the tympanic membrane and is a potential source of referred pain (see later). The central processes of the tympanic branch synapse in the spinal nucleus of the trigeminal nerve (Figure 18.1).

- Some fibers of the tympanic branch are parasympathetic fibers originating in the inferior salivary nucleus. They pierce the roof (tegmen tympani) of the middle ear as the lesser petrosal nerve, leave the skull through the foramen ovale, and synapse in the otic ganglion. Postganglionic fibers supply secretomotor fibers to the parotid gland (Figure 13.3).

- The branchial efferent supply to the stylopharyngeus comes from the nucleus ambiguus.

- Branches serving 'common sensation' (touch) supply the mucous membranes bounding the oropharynx (throat), including the posterior one-third of the tongue. The neurons synapse centrally in the commissural nucleus. The glossopharyngeal branches provide the afferent limb of the *gag reflex*— contraction of the pharyngeal constrictors in response to stroking the wall of the oropharynx. (The gag reflex is unpleasant because of accompanying nausea. To test the integrity of the IX nerve, it is usually sufficient to test sensation on the pharyngeal wall.) Generalized stimulation of the oropharynx elicits a complete swallowing reflex, through a linkage between the commissural nucleus and a specific swallowing center nearby (Ch. 24).

- Gustatory neurons supply the taste buds contained in the circumvallate papillae of the tongue; they terminate centrally in the gustatory nucleus (Figure 18.4).

- An important *carotid branch* descends to the bifurcation of the common carotid artery. This branch contains two different sets of afferent fibers. One set ramifies in the wall of the carotid sinus (at the commencement of the internal carotid artery), terminating in *stretch receptors* responsive to systolic blood pressure; these *baroreceptor neurons* terminate centrally in the medial part of the nucleus solitarius (Figure 18.4).

- The second set of afferents in the carotid branch supplies glomus cells in the carotid body. These nerve endings are *chemoreceptors* monitoring the carbon dioxide and oxygen levels in the blood. The central terminals enter the dorsal respiratory nucleus (Figure 18.4).

Vagus and cranial accessory nerves

The vagus is the main parasympathetic nerve. Its preganglionic component has a huge territory, which includes the heart, the lungs, and the alimentary tract from esophagus through transverse colon (Ch. 13). At the same time, the vagus is the largest visceral afferent nerve; afferents outnumber parasympathetic motor fibers by four to one. Overall, the vagus contains the same seven fiber classes as the glossopharyngeal, and they will be listed in the same order.

The rootlets of the vagus and cranial accessory nerves are in series with the glossopharyngeal, and the three nerves travel together into the jugular foramen. At this point, the cranial accessory nerve shares a dural sheath with the spinal accessory, but there is no exchange of fibers (Figure 18.3). Just below the foramen, the cranial accessory is incorporated into the vagus. The vagus itself shows a small, jugular (superior) and a large, nodose (inferior) ganglion; both are sensory.

Functional divisions and branches

- An *auricular branch* supplies skin lining the outer ear canal, and a *meningeal branch* ramifies in the posterior cranial fossa. Both branches have their cell bodies in the jugular ganglion; the central processes enter the spinal trigeminal nucleus.

- The parasympathetic neurons for the alimentary tracts from lower esophagus to transverse colon originate from the dorsal nucleus of the vagus. As already mentioned in Chapter 17, parasympathetic neurons serving heart and lungs are embedded in the nucleus ambiguus in laboratory animals and are presumed to be similarly placed in humans.

- Special visceral efferent neurons of the nucleus ambiguus constitute the motor elements in the pharyngeal and laryngeal branches of the vagus. They supply the pharyngeal and laryngeal muscles already noted, and levator palati. They also supply the striated musculature of the upper third of the esophagus.

- General visceral afferent fibers from the heart and from the respiratory and alimentary tracts have their cell bodies in the nodose ganglion and synapse centrally in the commissural nucleus. They serve important reflexes including the *Bainbridge reflex* (cardiac acceleration brought about by distension of the right atrium), the *cough reflex* (stimulation of a coughing center [Ch. 24] by irritation of the tracheobronchial tree), and the *Hering–Breuer reflex* (inhibition of the dorsal respiratory center by pulmonary stretch receptors). In addition, afferent information from the stomach (in particular) is forwarded to the hypothalamus and influences feeding behavior (Ch. 26).

- A few taste buds on the epiglottis report to the gustatory nucleus.

- *Baroreceptors* in the aortic arch are supplied.

- *Chemoreceptors* in the tiny aortic bodies are supplied; these supplement the corresponding receptors at the carotid bifurcation.

Supranuclear, nuclear, and infranuclear lesions of the IX, X, and XI nerves are described in the Clinical Panels.

Core Information

Hypoglossal nerve
XII contains somatic efferent neurons supplying extrinsic and intrinsic muscles of the tongue. Its nucleus is close to midline and is innervated by reticular neurons for automatic or reflex movements and by (mainly crossed) corticonuclear neurons for speech articulation. XII emerges beside the pyramid, exits the hypoglossal canal, and descends on the carotid sheath, where it collects cervical proprioceptive fibers for the supply of lingual muscle spindles. Supranuclear paralysis of XII is characterized by temporary deviation to the paralyzed side on protrusion. Nuclear–infranuclear paralysis is characterized by wasting and fasciculation of the tongue, as well as deviation.

Spinal accessory nerve
Spinal XI is purely motor. From motor neurons of spinal segments C1–5, the axons enter the foramen magnum and exit the jugular foramen; they pierce and supply sternomastoid, then pass deep to trapezius and supply it. Proprioceptive connections are received from cervical and thoracic spinal nerves. Supranuclear lesions are characterized by weakness of the contralateral trapezius and contralateral head rotators; nuclear–infranuclear lesions by ipsilateral wasting of the two muscles and drooping of the scapula.

Glossopharyngeal nerve
IX emerges behind the olive and exits the jugular foramen, where it shows two unipolar cell ganglia and gives off a tympanic branch that is partly sensory to the middle ear, partly parasympathetic to the parotid gland via the otic ganglion. IX then passes between superior and middle constrictors to gain the oropharynx, where it supplies sensation to that mucous membrane including the posterior third of tongue (hence the name), and taste fibers to the circumvallate papillae. A carotid branch supplies the carotid sinus and carotid body.

Vagus and cranial accessory nerves
X and cranial XI rootlets emerge behind the olive and unite in the jugular foramen. Cranial XI fibers arise in nucleus ambiguus and utilize laryngeal and pharyngeal branches of X to supply the intrinsic muscles of larynx and pharynx, and levator palati.

Preganglionic X fibers travel to intramural ganglia in the walls of heart, bronchi, and alimentary tract. Visceral afferents from these regions, and from larynx and pharynx, have unipolar cell bodies in the nodose ganglion and project to the commissural nucleus.

REFERENCES

Andresen MC. Nucleus tractus solitarii—gateway to neural circulatory control. Ann Rev Physiol 1994; 56:93–117.

Dampney RAL. Functional organization of the central pathways regulating the cardiovascular system. Physiol Rev 1994; 74:323–364.

FitzGerald MJT, Comerford PT, Tuffery AR. Sources of innervation of the neuromuscular spindles in sternomastoid and trapezius. J Anat 1982; 144:184–190.

FitzGerald MJT, Sachithanandan SR. The structure and source of lingual proprioceptors in the monkey. J Anat 1979; 128:523–552.

FitzGerald MJT. Sternomastoid paradox. Eur J Neurosci 2001; 12(suppl 11):157; Clin Anat 2001; 15:330–331.

Mtui EP, Anwar M, Reis DJ, et al. Medullary visceral reflex circuits: local afferents to nucleus tractus solitarii synthesize catecholamines and project to thoracic spinal cord. J Comp Neurol 1995; 351:5–26.

Saxina PR. Serotonin receptors: subtypes, functional responses and therapeutic relevance. Pharmacol Ther 1995; 66:339–368.

Thompson PD, Thickbroom GW, Mastaglia FL. Corticomotor representation of the sternocleidomastoid muscle. Brain 1997; 120:245–255.

Urban PP, Hopf HC, Fleischer S, et al. Impaired corticobulbar tract function in dysarthria due to hemispheric stroke. Brain 1997; 120:1077–1084.

Vestibular nerve

CHAPTER SUMMARY
Vestibular system
    Static labyrinth: anatomy and actions
    Kinetic labyrinth: anatomy and action
CLINICAL PANELS
Vestibular disorders
Lateral medullary syndrome

STUDY GUIDELINES
1 The static labyrinth is primarily concerned with maintenance of balance when the head is off-center.
2 The dynamic labyrinth enables us to maintain our gaze on a particular object while the head is moving.
3 Clinically, the lateral semicircular is the canal of choice for testing labyrinthine function.

The **vestibulocochlear nerve** is primarily composed of the centrally directed axons of bipolar neurons housed in the petrous temporal bone (Figure 19.1). The peripheral processes are applied to neuroepithelial cells in the vestibular labyrinth and cochlea. The nerve enters the brainstem at the junctional region of pons and medulla oblongata. The functional anatomy of the vestibular division of the nerve is described in this chapter.

VESTIBULAR SYSTEM

The **bony labyrinth** of the inner ear is a very dense shell containing **perilymph**, which resembles extracellular fluid in general. The perilymph provides a water jacket for the **membranous labyrinth**, which encloses the sense organs of balance and of hearing. The sense organs are bathed in **endolymph**. The endolymph resembles intracellular fluid, being potassium-rich and sodium-poor.

The vestibular labyrinth comprises the **utricle**, the **saccule**, and three **semicircular ducts** (Figure 19.2). The utricle and saccule contain a 3×2 mm^2 **macula**. Each semicircular duct contains an **ampulla** at one end, and the ampulla houses a **crista**. (It should be pointed out that clinicians commonly speak of 'canals' where 'ducts' would be strictly more appropriate.)

The two maculae are the sensory end organs of the *static labyrinth*, which signals head position. The three cristae are the end organs of the *kinetic* or *dynamic labyrinth*, which signals head movement.

The bipolar cells of the **vestibular ganglion** occupy the internal acoustic meatus. Their peripheral processes are applied to the five sensory end organs. Their central processes, which constitute the **vestibular nerve**, cross the subarachnoid space and synapse in the vestibular nucleus previously seen in Figures 17.14 and 17.15.

Static labyrinth: anatomy and actions

The position and structure of the maculae are shown in Figure 19.3. The utricular macula is relatively horizontal; the **saccular macula** is relatively vertical. The cuboidal cells lining the membranous labyrinth become columnar supporting cells in the maculae. Among the supporting cells are so-called **hair cells**, to which vestibular nerve endings are applied. Some hair cells are almost completely

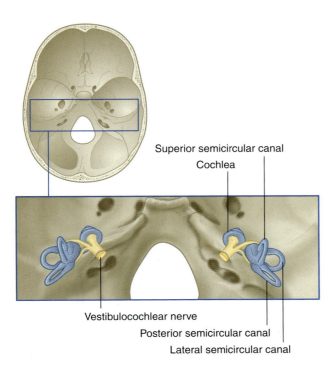

Superior semicircular canal
Cochlea

Vestibulocochlear nerve
Posterior semicircular canal
Lateral semicircular canal

Figure 19.1 Bony labyrinth, viewed from above.

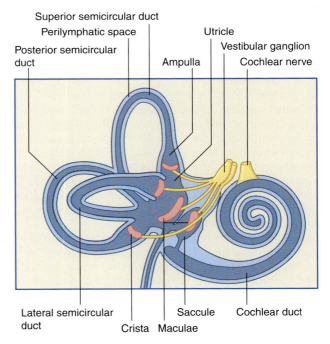

Superior semicircular duct
Perilymphatic space
Posterior semicircular duct
Ampulla
Utricle
Vestibular ganglion
Cochlear nerve

Lateral semicircular duct
Crista Maculae
Saccule
Cochlear duct

Figure 19.2 Locations of the five vestibular sense organs.

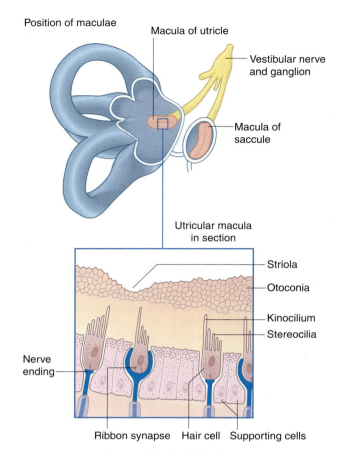

Position of maculae
Macula of utricle
Vestibular nerve and ganglion
Macula of saccule
Utricular macula in section
Striola
Otoconia
Kinocilium
Stereocilia
Nerve ending
Ribbon synapse   Hair cell   Supporting cells

Figure 19.3 Static labyrinth.

enclosed by large nerve endings, whereas others (phylogenetically older) receive only small contacts. At the cell bases are **ribbon synapses**, the synaptic vesicles being lined up along synaptic bars. Projecting from the free surface of each hair cell are about 100 **stereocilia** and, close to the cell margin, a single, long **kinocilium**. The hair cells discharge continuously, the resting rate being about 100 Hz.

The cilia are embedded in a gelatinous matrix containing protein-bound calcium carbonate crystals called **otoconia** ('ear sand'). (The term *otoliths*, when used, refers to the larger 'ear stones' of reptiles.) The otoconia exert gravitational drag on the hair cells. Whenever kinocilia are dragged away from stereocilia, depolarization is facilitated. The macula has a central groove (**striola**), and the hair cell orientations have a mirror arrangement in relation to the groove. Electrical activity of hair cells is facilitated on one side of the groove by a given gravitational vector, and disfacilitated on the other side.

The maculae also respond to linear acceleration of the head in the horizontal plane (e.g. during walking) or in the vertical (gravitational) plane. Also, when the tilted head is stationary in a flexed or extended position, the facilitated half of the utricular macula discharges intensely in both ears. The saccular ones are more responsive when the head is held to the side.

The primary function of the static labyrinth is to signal the vestibular position of the head relative to the trunk. In response

to this signal, the vestibular nucleus initiates compensatory movements, with the effect of maintaining the center of gravity between the feet (in standing) or just in front of the feet (during locomotion), and of keeping the head horizontal. These effects are mediated by the vestibulospinal tracts.

The **lateral vestibulospinal tract**, seen earlier in sections of medulla oblongata in Chapter 17, arises from large neurons in the **lateral vestibular nucleus** (of *Deiters*). The fibers descend in the anterior funiculus on the same side of the spinal cord and synapse on extensor (antigravity) motor neurons. Both α and γ motor neurons are excited, and a significant part of the increased muscle tone is exerted by way of the gamma loop (Ch. 16). During standing, the tract is tonically active on both sides of the spinal cord. During walking, activity is selective for the quadriceps motor neurons of the leading leg; this commences following heel strike and continues during the stance phase (when the other leg is off the ground). Deiters's nucleus is somatotopically organized, and the functionally appropriate neurons are selected by the flocculonodular lobe of the cerebellum. The flocculonodular lobe (Ch. 25) has two-way connections with all four vestibular nuclei.

Antigravity action is triggered mainly from the horizontal macula of the utricle. The vertical macula of the saccule, on the other hand, is maximally activated by a *free fall*. The shearing effect on the macula produces a powerful extensor thrust in anticipation of a hard landing.

A small, **medial vestibulospinal tract** arises in the medial and inferior vestibular nuclei (Figure 17.8). It descends bilaterally in the medial longitudinal fasciculus and terminates on excitatory and inhibitory internuncials in the cervical spinal cord. It operates *head-righting reflexes*, which serve to keep the head—and the gaze—horizontal when the body is craned forward or to one side. Good examples of head-righting reflexes are to be seen around pool tables and in bowling alleys. An added twist can be provided, if required, by torsion of the eyeballs (up to 10°) within the orbital sockets. This *eye-righting reflex* is mediated by axons *ascending* the medial longitudinal fasciculus from the *lateral* vestibular nucleus to reach nuclei controlling the extraocular muscles. Evidence derived from unilateral vestibular destruction (Clinical Panel 19.1) indicates that the horizontal position of the eyes in the upright head is the result of a canceling effect of bilateral tonic activity in these Deiteroocular pathways.

The medial vestibulospinal tract is also activated by the kinetic labyrinth.

The static labyrinth contributes to the sense of position. The sense of position of the body in space is normally provided by three sensory systems: the visual system, the conscious proprioceptive system, and the vestibular system. Deprived of one of the three, the individual can stand and walk by using information provided by the other two. Following loss of vision, for example, the subject can get about, although the constraints imposed by blindness are known to all. Following loss of conscious proprioception instead, the subject uses vision as a substitute for proprioceptive sense, and is disabled by closure of the eyes (sensory ataxia, Ch. 15). If the static labyrinths alone are active, closure of the eyes may lead to a heavy fall.

Clinical Panel 19.1   Vestibular disorders

Unilateral vestibular disease

Acute failure of one vestibular labyrinth may follow spread of disease from the middle ear or thrombosis of the labyrinthine artery. A common cause of unilateral vestibular symptoms in the elderly is a *transient ischemic attack* involving the vertebrobasilar arterial system. Transient ischemic attacks usually last 15 min or less and leave no residual neurologic deficit.

The effects of unilateral vestibular disease are well demonstrated when the vestibular system is inactivated surgically, either during removal of an acoustic neuroma (Ch. 22) or as a last resort in treating paroxysmal attacks of vertigo. During the immediate postoperative period, the patient shows triple effects of loss of tonic input from the static labyrinth.

- Loss of function in the Deiteroocular pathway on one side leads to about 10° of torsion of both eyeballs toward that side. The patient's perception of the horizontal shows a corresponding tilt, so that reaching movements become inaccurate.

- The head tilts to the same side, matching the gaze with the tilted horizon.

- The patient tends to fall to the same side, because the Deiterospinal tract no longer compensates for tilting of the head.

Because function continues in the normal lateral semicircular canal, there is a nystagmus to the normal side.

Bilateral vestibular disease

Following total loss of static labyrinthine function, visual guidance becomes important, and the patient dare not walk out of doors after twilight. By day, any distraction causing the patient to look overhead may result in a heavy fall. Loss of kinetic labyrinthine function makes it impossible to fix the gaze on an object while the head is moving. During walking, the scene bobs up and down as if it were being viewed through a hand-held camera.

Kinetic labyrinth: anatomy and action

Basic features of macular epithelium are repeated in the three cristae. Again, there are supporting cells, and hair cells to which vestibular nerve endings are applied. The kinocilia of the hair cells are long, penetrating into a gelatinous projection called the **cupula** (Figure 19.4). The cupula is bonded to the opposite wall of the ampulla.

The cristae are sensitive to angular acceleration of the labyrinths. Angular acceleration occurs during rotary 'yes' and 'no' movements of the head. The endolymph tends to lag behind because of its inertia, and the cupula balloons like a sail when thrust against it. The disposition of the kinocilia is uniform across each crista, and is such that the *lateral* ampullary crista is facilitated by cupular displacement *toward* the utricle; the *superior* and *posterior* cristae are facilitated by cupular displacement *away from* the utricle. In practical terms, the right lateral ampulla is activated by turning the head to the right, both superior ampullae are activated by flexion of the head, and both posterior ampullae by extension of the head.

Afferents from the cristae terminate in the medial and superior vestibular nuclei. As with the macular afferents, there are two-way connections with the flocculonodular lobe of the cerebellum.

The function of the kinetic labyrinth is to provide information for compensatory movements of the eyes in response to movement of the head. *Vestibuloocular reflexes* operate to maintain the gaze on a selected target. A simple example is our ability to gaze at the period (full stop) at the end of a sentence, while moving the head about. The two eyes move *conjugately*, i.e. in parallel.

The horizontal vestibuloocular reflex response to a rightward turn of the head is depicted in Figures 19.4 and 19.5, and described in their captions. Appropriate point-to-point connections also exist between the vestibular nuclei and gaze centers in the midbrain for similar reflexes in the vertical plane.

In order to control the vestibuloocular reflexes, the cerebellum is informed about the initial position of the head in relation to the trunk. This information is provided by a great wealth of muscle spindles in the deep muscles surrounding the cervical vertebral column. The spindle afferents enter the rostral spinocerebellar tract and relay in the accessory cuneate nucleus on each side (Figure 17.12).

Nystagmus

A horizontal vestibuloocular reflex can be elicited artificially by warming or cooling the endolymph in the semicircular canals. In routine tests of vestibular function, advantage is taken of the proximity of the lateral semicircular canal to the middle ear. The canal is angled at 30° to the horizontal plane. Tilting the head back by 60° brings the canal into the vertical plane, with the ampulla uppermost. In the *warm caloric test*, water at 44°C is then instilled into the ear. The air in the middle ear is heated, and heat transfer to the lateral canal produces convection currents within the endolymph. Whether through displacement of the cupula or by some other mechanism, the crista of the warmer lateral ampulla becomes more active than its opposite number. The result is a slow drift of the eyes away from the stimulated side. *It is as if the head had been turned to the side being tested.* The drift is followed by a recovery phase in which the eyes snap back to the resting position. These slow and fast phases are repeated several times per second. This is *vestibular nystagmus*. The direction of the nystagmus is named in accordance with the fast phase because of the obvious 'beat'. A warm

A

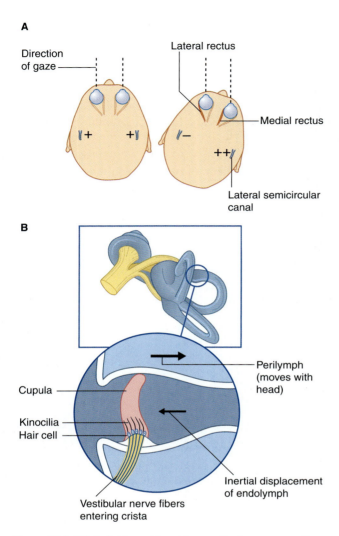

Direction of gaze

Lateral rectus

Medial rectus

Lateral semicircular canal

B

Perilymph (moves with head)

Cupula

Kinocilia
Hair cell

Inertial displacement of endolymph

Vestibular nerve fibers entering crista

Figure 19.4 (A) A rightward head turn activates nerve endings in the right lateral semicircular canal, resulting in contraction of the left lateral and right medial rectus muscles. **(B)** The nerve endings in the cupula are excited by passive displacement of the cupula toward the ampulla. Impulse traffic increases along parent bipolar neurons whose central fibers excite the medial and superior vestibular nuclei.

caloric test applied to the right ear should produce a right-beating nystagmus ('nystagmus to the right').

Subjectively, nystagmus is accompanied by vertigo—a sense of rotation of self in relation to the external world, or vice versa.

Unilateral and bilateral vestibular syndromes are considered in Clinical Panel 19.1. A vascular syndrome involving the vestibular system in the medulla oblongata is described in Clinical Panel 19.2.

Vestibulocortical connections

Second-order sensory neurons project from the vestibular nucleus mainly to the *ipsilateral* thalamus. The fibers relay via the ventral posterior nucleus to the *parietoinsular vestibular cortex (PIVC)* and to the adjacent region of the superior temporal gyrus, as shown in Figure 19.6. How-

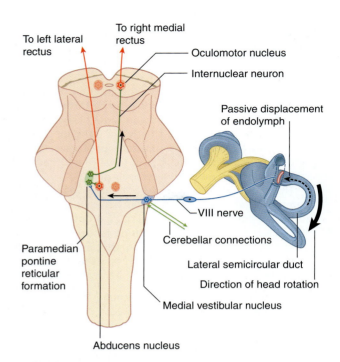

To left lateral rectus

To right medial rectus

Oculomotor nucleus

Internuclear neuron

Passive displacement of endolymph

VIII nerve

Cerebellar connections

Lateral semicircular duct

Direction of head rotation

Medial vestibular nucleus

Paramedian pontine reticular formation

Abducens nucleus

Figure 19.5 Under cerebellar guidance, the right medial vestibular nucleus responds to a rightward head turn by sending impulses to the contralateral paramedian reticular formation (PPRF, Figure 17.15). The PPRF selects abducens motor neurons supplying the left lateral rectus, and sends internuclear fibers up the right medial longitudinal fasciculus to the right oculomotor nucleus, where they seek out motor neurons serving the right medial rectus. (Not shown is the superior vestibular nucleus, which sends ipsilateral fibers having the function of inhibiting motor neurons to the two antagonist recti.)

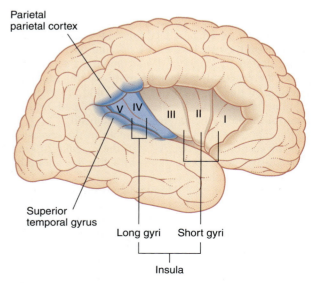

Parietal parietal cortex

V IV III II I

Superior temporal gyrus

Long gyri   Short gyri

Insula

Figure 19.6 Parietoinsular vestibular cortex (blue). I–III, short insular gyri; IV, V, long insular gyri (IV and V are commonly fused as a single, Y-shaped long insular gyrus).

ever, the PIVC cannot be named as a primary sensory area, because it is in fact multisensory, receiving visual and tactile inputs as well as vestibular. (By analogy, positron emission tomography studies of tactile sensation show activity throughout most of the parietal lobe, but we know from other sources that the postcentral gyrus is the primary area, being the takeoff point for analysis in the posterior parietal cortex.)

The relationship of PIVC activity to hemisphere dominance is discussed in Chapter 32.

Clinical Panel 19.2   Lateral medullary syndrome

Thrombosis of the vertebral or posterior inferior cerebellar artery may produce an infarct (area of necrosis) in the lateral part of the medulla. The clinical picture depends on the extent to which the related nuclei and pathways are damaged. Brainstem pathology must always be suspected when a cranial nerve lesion on one side is accompanied by 'upper motor neuron signs' on the other side—so-called *alternating* or *crossed hemiplegia*.

Lateral medullary syndrome (Figure CP 19.2.1)
1 Damage to the vestibular nucleus leads to vertigo (often with initial vomiting), together with the symptoms of unilateral disconnection of the labyrinth described in Clinical Panel 19.1.
2 Interruption of posterior and rostral spinocerebellar fibers may produce signs of cerebellar ataxia in the ipsilateral limbs. Cerebellar ataxia is a prominent feature if blood flow is interrupted in the posterior inferior cerebellar artery.
3 Damage to the spinal tract of the trigeminal nerve interrupts fine primary afferent fibers descending the brainstem from the trigeminal ganglion (Ch. 20). These fibers are functionally equivalent to those of Lissauer's tract in the spinal cord (Ch. 15). The result of interruption is loss of pain and thermal senses from the face on the same side.
4 Interruption of the central sympathetic pathway to the spinal cord produces a complete Horner syndrome (ptosis, miosis, anhidrosis).

5 Damage to the nucleus ambiguus causes hoarseness, and sometimes difficulty in swallowing.
6 The only *contralateral* sign is loss of pain and temperature sense in the trunk and limbs, resulting from damage to the lateral spinothalamic tract. There is no motor weakness, because the corticospinal tract is spared.

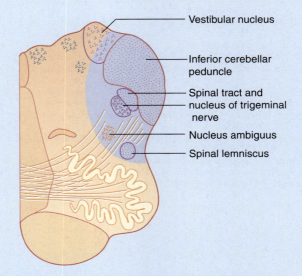

Vestibular nucleus
Inferior cerebellar peduncle
Spinal tract and nucleus of trigeminal nerve
Nucleus ambiguus
Spinal lemniscus

Figure CP 19.2.1 Lateral medullary infarct (shaded).

Core Information

The static labyrinth comprises the maculae in utricle and saccule. The dynamic labyrinth comprises the semicircular ducts and their cristae. Vestibular bipolar neurons supply all five and synapse in the vestibular nucleus, which is controlled by the flocculonodular lobe of cerebellum. The static labyrinth functions to control balance, via the lateral vestibulospinal tract, by increasing antigravity tone on the side to which the head is tilted. This system is in partnership with proprioceptors and retina in maintaining upright posture. In the absence of good vision, or at night outdoors, a fall is likely if the system has been compromised.

The dynamic labyrinth operates vestibuloocular reflexes so as to keep the gaze on target during rotatory movements of the head. For sideways rotation, the main projection is from medial vestibular nucleus to contralateral paramedian reticular formation, which activates VI neurons supplying the lateral rectus muscle and internuclear neurons projecting via the medial longitudinal fasciculus to the contralateral medial rectus. Clinically, this pathway can be activated by the caloric test, which normally elicits nystagmus.

REFERENCES

Anniko M. Functional morphology of the vestibular system. In: Jahn AF, Santos-Sacchi J, eds. Physiology of the ear. New York: Raven Press; 1988:457–475.

Brandt T, Dieterich M. Postural imbalance in peripheral and central vestibular disorders. In: Bronstein AM, Brandt T, Woollacott M, eds. Clinical disorders of posture and gait. London: Arnold; 1996:131–146.

Dieterich M, Bense S, Lutz S, et al. Dominance for vestibular cortical function in the non-dominant hemisphere. Cereb Cortex 2003; 13:994–1007.

Elliott LL. Functional brain imaging and hearing. J Acoust Soc Am 1994; 96:1397–1408.

Fitzpatrick R, McCloskey DI. Proprioceptive, visual and vestibular thresholds for the perception of sway during standing in humans. J Physiol 1994; 478:173–196.

Hart CW, McKinley PA, Peterson BW. Compensation following acute unilateral total loss of peripheral vestibular function. In: Graham MD, Kemink JL, eds. The vestibular system: neurologic and clinical research. New York: Raven Press; 1987:197–192.

Markham CH. Vestibular control of muscular tone and posture. Can J Neurol Sci 1987; 14:493–496.

Paulesu E, Frackowiak RSJ, Bottini G. Maps of somatosensory systems. In: Frackowiak RSJ, Friston KJ, Frith CD, et al, eds. Human brain function. San Diego: Academic Press; 1997:219–231.

Spoendlin H. Neural anatomy of the inner ear. In: Jahn AF, Santos-Sacchi J, eds. Physiology of the ear. New York: Raven Press; 1988:201–219.

Cochlear nerve

STUDY GUIDELINES
1 Vibrations created by sound waves cross the vestibular membrane to reach the organ of Corti, where cochlear nerve fibers terminate.
2 The central auditory pathway is partly ipsilateral, to the extent that damage on one side does not compromise hearing unduly.
3 Nuclei strung along the brainstem part of the pathway serve to magnify tiny differences in relative intensities of sounds entering the two ears.

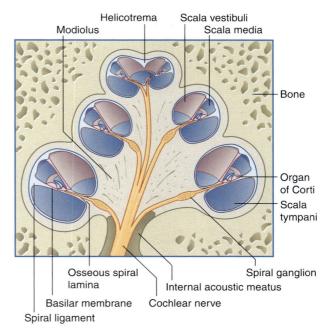

Figure 20.1 The cochlea in section.

AUDITORY SYSTEM

The auditory system comprises the cochlea, the cochlear nerve, and the central auditory pathway from the cochlear nucleus in the brainstem to the cortex of the temporal lobe. The central auditory pathway is more elaborate than the somatosensory or visual pathway. This is because the same sounds are detected by both ears. In order to signal the location of a sound, a very complex neuronal network is in place, with numerous connections (mainly inhibitory) between the two central pathways in order to magnify minute differences in intensity and timing of sounds that exist during normal, binaural hearing.

The cochlea

The main features of cochlear structure are seen in Figures 20.1 and 20.2. The cochlea is pictured as though it were upright, but in life it lies on its side, as shown earlier in Figure 19.1. The central bony pillar of the cochlea (the **modiolus**) is in the axis of the internal acoustic meatus. Projecting from the modiolus, like the flange of a screw, is the **osseous spiral lamina**. The **basilar membrane** is attached to the tip of this lamina; it reaches across the cavity of the bony cochlea to become attached to the **spiral ligament** on the outer wall. The osseous spiral lamina and spiral ligament become progressively smaller as one ascends the two and one-half turns of the cochlea, and the fibers of the basal lamina become progressively longer.

The basal lamina and its attachments divide the cochlear chamber into upper and lower compartments. These are the **scala vestibuli** and the **scala tympani**, respectively, and they are filled with perilymph. They communicate at the apex of the cochlea, through the **helicotrema**. A third compartment, the **scala media** (**cochlear duct**), lies above

the basilar membrane and is filled with endolymph. It is separated from the scala vestibuli by the delicate **vestibular membrane**.

Sitting on the basilar membrane is the **spiral organ** (*organ of Corti*). The principal sensory receptor epithelium consists of a single row of **inner hair cells**, each one having up to 20 large afferent nerve endings applied to it. The hair cells rest on **supporting cells**, and there are ancillary cells as well. The organ of Corti contains a central tunnel, filled with perilymph diffusing through the basilar membrane. On the outer side of the tunnel are several rows of **outer hair cells**, attended by supporting and ancillary cells.

All the hair cells are surmounted by **stereocilia**. Unlike the vestibular hair cells, they have no kinocilium in the adult state. The stereocilia of the outer hair cells are embedded in the overlying tectorial membrane. Those of the inner hair cells lie immediately below the membrane.

The outer hair cells are contractile (in tissue culture), and they have substantial efferent nerve endings (Figure 20.2). In theory at least, oscillatory movements of outer hair cells could influence the sensitivity of the inner hair cells through effects on the tectorial or basilar membrane.

Sound transduction

The vibrations of the tympanic membrane in response to sound waves are transmitted along the ossicular chain. The footplate of the stapes fits snugly into the oval window, and vibrations of the stapes are converted to pressure waves in the scala vestibuli. The pressure waves are transmitted through the vestibular membrane to reach the basilar membrane. High-frequency pressure waves,

239

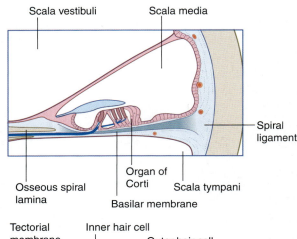

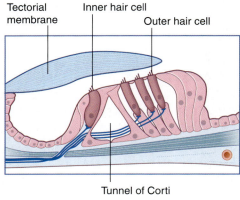

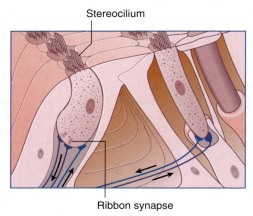

Figure 20.2 Organ of Corti at three levels of magnification. Arrows indicate directions of impulse traffic.

Cochlear nerve

The bulk of the cochlear nerve consists of the myelinated central processes of some 30 000 large bipolar neurons of the spiral ganglion. Unmyelinated fibers come from small ganglion cells supplying dendrites to the outer hair cells. (Motor fibers do not travel in the cochlear nerve trunk.) The cochlear nerve traverses the subarachnoid space in company with the vestibular and facial nerves, and it enters the brainstem at the pontomedullary junction.

Central auditory pathways

The general plan of the central auditory pathway from the left cochlear nerve to the cerebral cortex is shown in Figure 20.3. The first cell station is the **cochlear nucleus**, where all cochlear nerve fibers terminate on entry to the brainstem. From here, some second-order fibers project all the way to the opposite **inferior colliculus** by way of the **trapezoid body** and **lateral lemniscus**. The **inferior brachium** links the inferior colliculus to the **medial geniculate body**, which projects to the **primary auditory cortex** in the temporal lobe.

A small but important purely ipsilateral relay passes from the superior olivary nucleus to the higher auditory centers.

Functional anatomy (Figure 20.4)
Cochlear nucleus
The cochlear nucleus comprises dorsal and ventral nuclei on the surface of the inferior cerebellar peduncle, as was shown in Figure 17.12. Many incoming fibers of the cochlear nerve bifurcate and enter both nuclei. The cells in both are tonotopically arranged.

created by high-pitched sounds, cause the short fibers of the basilar membrane in the basal turn of the cochlea to resonate and absorb their energy. Low-frequency waves produce resonance in the apical turn where the fibers are longest. The basilar membrane is therefore *tonotopic* in its fiber sequence. Not surprisingly, the inner hair cells have a similar tonotopic sequence. In response to local resonance, the cells become depolarized and liberate excitatory transmitter substance from synaptic ribbons (Figure 20.2).

The nerve fibers supplying the hair cells are the peripheral processes of the bipolar spiral ganglion cells lodged in the base of the osseous spiral lamina.

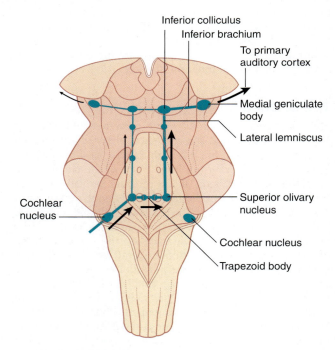

Figure 20.3 Dorsal view of brainstem, showing basic plan of central auditory pathways. The strand linking the two inferior colliculi is the collicular commissure.

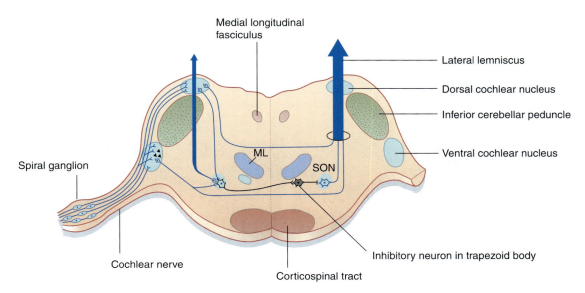

Figure 20.4 Transverse section of lower end of pons, showing central connections of cochlear nerve. ML, medial lemniscus; SON, superior olivary nucleus.

Responses of many cells in the ventral nucleus are called primary-like, because their frequency (firing rate) resembles that of primary afferents. Most of the output neurons project to the nearby superior olivary nucleus.

The cells of the dorsal nucleus are heterogeneous. At least six different cell types have been characterized by their morphology and electrical behavior. Most of the output neurons project to the contralateral inferior colliculus. Individually, they exhibit an extremely narrow range of tonal responses, being 'focused' by collateral inhibition.

Superior olivary nucleus

The superior olivary complex of nuclei is relatively small in the human brain. It contains *binaural neurons* affected by inputs from both ears. Ipsilateral inputs are excitatory to the binaural neurons, whereas contralateral inputs are inhibitory. The inhibitory effect is mediated by inter-nuncial neurons in the **nucleus of the trapezoid body**.

The superior olivary nucleus is responsive to differences in intensity and timing between sounds entering both ears simultaneously. On the side ipsilateral to a sound, stimulation of cochlea and nucleus is earlier and more intense than on the contralateral side. By exaggerating these differences through crossed inhibition, the superior olivary nucleus helps to indicate the spatial direction of incoming sounds. At the same time, the excited nucleus projects to the inferior colliculus of both sides, giving rise to binaural responses in the neurons of the inferior colliculus and beyond.

Lateral lemniscus

Fibers of the lateral lemniscus arise from the dorsal and ventral cochlear nuclei and from the superior olivary nuclei—in each case, mainly contralaterally. The tract terminates in the **central nucleus of the inferior colliculus**. Nuclei within the lateral lemniscus participate in reflex arcs (see later).

Inferior colliculus

Spatial information from the superior olivary nucleus, intensity information from the ventral cochlear nucleus, and pitch information from the dorsal cochlear nucleus are integrated in the inferior colliculus. The main (central) part of the nucleus is laminated in a tonotopic manner. Within each tonal lamina, cells differ in their responses: some have a characteristic 'tuning curve' (they respond only to a particular tone), some fire spontaneously but are inhibited by sound, and some respond only to a moving source of sound.

In addition to projecting to the medial geniculate body (nucleus), the inferior colliculus exerts inhibitory effects on its opposite number through the **collicular commissure** (Figure 20.3). It also contributes to the tectospinal tract.

Medial geniculate body

The medial geniculate body is the specific thalamic nucleus for hearing. The main (ventral) nucleus is laminated and tonotopic, and its large (magnocellular) principal neurons project as the **auditory radiation** to the primary auditory cortex (Figure 20.5).

Primary auditory cortex

The upper surface of the temporal lobe shows two or more **transverse temporal gyri**. The anterior one (the **gyrus of Heschl**) contains the **primary auditory cortex** (Figure 20.6). Tonotopic arrangement is preserved in Heschl's gyrus, its posterior part being responsive to high tones and its anterior part to low tones. The auditory cortex responds to auditory stimuli within the *contralateral sound field*. In cats, destruction of a patch of primary cortex on one side produces a *sigoma* or 'deaf spot' in the contralateral sound field. In humans, ablation of the superior temporal gyrus (in the course of tumor removal) does not cause deafness, but it significantly reduces ability to judge the direction and distance of a source of sound.

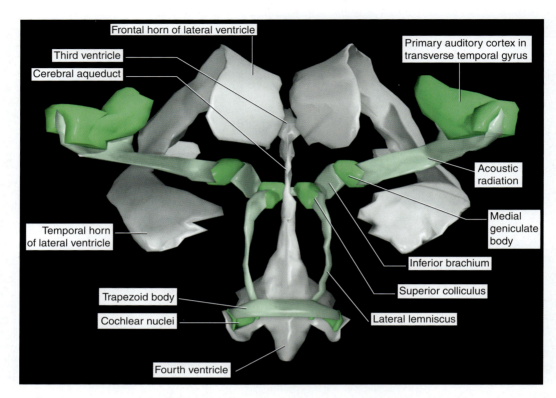

Figure 20.5 Graphic reconstruction of the central auditory pathways from a postmortem brain. (From Kretschmann and Weinrich 1998, with permission of Thieme and the authors.)

Brainstem auditory evoked potentials are described in Chapter 31.

Brainstem acoustic reflexes

Collateral branches emerging from the lateral lemniscus form the internuncial linkage for certain reflex arcs.

- Fibers entering the motor nuclei of the trigeminal and facial nerves link up with motor neurons supplying the tensor tympani and stapedius, respectively. These muscles exert a damping action on the ossicles of the middle ear. The tensor tympani is activated by the subject's own voice, the stapedius by external sounds.

- Fibers entering the reticular formation have an important arousal effect, as exemplified by the alarm clock. Sudden loud sounds cause the subject to flinch; this is the 'startle response', mediated by outputs from the reticular formation to the spinal cord and to the motor nucleus of the facial nerve.

Descending auditory pathways

A cascade of descending fibers runs from the primary auditory cortex to the medial geniculate nucleus and inferior colliculus, and from the inferior colliculus to the superior olivary nucleus. From here, the **olivocochlear bundle** emerges in the vestibular nerve and carries efferent, cholinergic fibers to the cochlea along with some for the vestibular labyrinth. The cochlear fibers apply large synaptic boutons to outer hair cells, and small boutons to the afferent nerve endings on inner hair cells.

The function of the olivocochlear bundle is uncertain. Experimental evidence indicates an involvement in enhancing detection of faint sounds.

Deafness

Deafness is a widespread problem in the community. About 10% of adults suffer from it in some degree. The cause may lie in the outer, middle, or inner ear, or in the cochlear neural pathway. The two fundamental types of deafness are described in Clinical Panel 20.1.

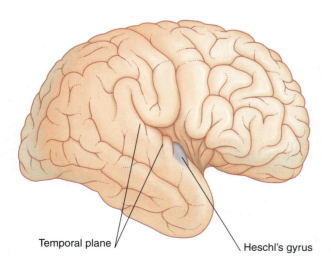

Figure 20.6 Tilted view of the right cerebral hemisphere; the frontal and parietal opercula of the insula have been removed to show the anterior temporal gyrus of Heschl (*blue*).

Clinical Panel 20.1   Two kinds of deafness

All forms of deafness can be grouped into two categories. *Conductive deafness* is caused by disease in the outer ear canal or in the middle ear. *Sensorineural deafness* is caused by disease in the cochlea or in the neural pathway from cochlea to brain.

Common causes of conductive deafness include accumulation of cerumen (wax) in the outer ear, and *otitis media* (inflammation in the middle ear). *Otosclerosis* is a disorder of the oval window in which the capsule of the synovial joint between the footplate of the stapes and the vestibule of the bony labyrinth is progressively replaced by bone. The stapes becomes immobilized, with severe impairment of hearing throughout the full tonal range. Replacement of the stapes by a prosthesis (artificial substitute) often restores normal hearing.

Sensorineural deafness usually originates within the cochlea. The commonest form is the high-frequency hearing loss of the elderly, resulting from deterioration of the organ of Corti in the basal turn. As a result, the elderly have difficulty in distinguishing between high-frequency consonants (d, s, t); vowels, which are low frequency, are quite audible. Therefore the elderly should be addressed distinctly rather than loudly.

Occupational deafness arises from a noisy environment at work. A persistent noise, especially indoors, may eventually lead to degeneration of the organ of Corti in the region corresponding to the particular frequency.

Ototoxic deafness may follow administration of drugs, including streptomycin, neomycin, and quinine.

Infectious deafness may follow more or less complete destruction of the cochlea by the virus of mumps or congenital rubella (German measles).

An important cause of sensorineural deafness in adults is an *acoustic neuroma*. Because the trigeminal and facial nerves may be affected as well as the cochlear and vestibular, this tumor is described in Chapter 22.

Core Information

The bipolar cochlear neurons occupy the osseous spiral lamina of the modiolus. Their peripheral processes supply hair cells in the organ of Corti. Their central processes end in the cochlear nucleus; from here, a polyneuronal pathway leads mainly through the trapezoid body and lateral lemniscus to the inferior colliculus, but there is a significant ipsilateral pathway too. From the inferior colliculus, fibers run to the medial geniculate body and from there to the primary auditory cortex on the upper surface of the temporal lobe of the brain.

Clinically, deafness is of two kinds: conductive, involving disease in the outer or middle ear; and sensorineural, involving disease of the cochlea (usually) or of central auditory pathways. Hearing is seldom significantly compromised by central pathway lesions, because of the bilateral projections to the inferior colliculus and beyond.

REFERENCES

Adams JC. Neuronal morphology in the human cochlear nucleus. Arch Otolaryngol Head Neck Surg. 1986; 112:1253–1261.

Aitkin LM. The auditory system. In: Bjorklund A, Hokfeld T, Swanson LW, eds. Handbook of chemical neuroanatomy, vol 7: integrated systems of the CNS, part II. New York: Elsevier; 1989:165–218.

Corwin JT, Warchol ME. Auditory hair cells: structure, function, development, and regeneration. Ann Rev Neurosci 1991; 14:301–333.

Kretschmann H-J, Weinrich W. Neurofunctional systems: 3D reconstructions with correlated neuroimaging: text and CD-ROM. New York: Thieme; 1998

Phillips DP. Introduction to anatomy and physiology of the central auditory nervous system. In: Jahn AF, Santos-Sacchi J, eds. Physiology of the ear. New York: Raven Press; 1988:407–427.

Trigeminal nerve

STUDY GUIDELINES
1 The motor nucleus supplies the muscles of mastication.
2 The mesencephalic unipolar neurons are proprioceptive.
3 The neurons of the principal sensory nucleus receive sensory inputs from the face and underlying mucous membranes.
4 The spinal nucleus is of special clinical importance because of its huge nociceptive territory.

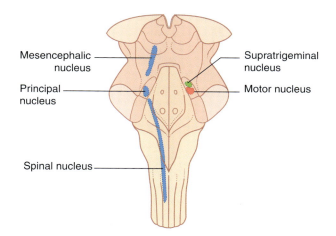

Figure 21.1 Trigeminal nuclei. *Left*, sensory nuclei; *right*, motor nucleus, supratrigeminal nucleus.

TRIGEMINAL NERVE

The trigeminal nerve has a very large sensory territory that includes the skin of the face, the oronasal mucous membranes and the teeth, the dura mater, and major intracranial blood vessels. The nerve is also both motor and sensory to the muscles of mastication. The **motor root** lies medial to the large **sensory root** at the site of attachment to the pons (Figure 17.16). The **trigeminal** (*Gasserian*) **ganglion**, near the apex of the petrous temporal bone, gives rise to the sensory root and consists of unipolar neurons.

Details of the distribution of the ophthalmic, maxillary, and mandibular divisions are available in gross anatomy textbooks. Accurate appreciation of their respective territories on the face is essential if trigeminal neuralgia is to be distinguished from other sources of facial pain (Clinical Panel 21.1).

Motor nucleus (Figures 17.16 and 21.1)

The motor nucleus is the special visceral nucleus supplying the muscles derived from the embryonic mandibular arch. These comprise the four masticatory muscles attached to each half of the mandible (Figure 21.2), tensor tympani and tensor palate, and mylohyoid and anterior belly of digastric. The nucleus occupies the lateral pontine tegmentum. Embedded in its upper pole is a node of the reticular formation, the **supratrigeminal nucleus**, which acts as a pattern generator for masticatory rhythm.

Voluntary control is provided by corticonuclear projections from each motor cortex to both motor nuclei, but mainly the contralateral one (Figure 17.3).

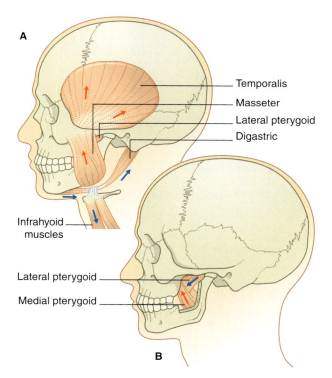

Figure 21.2 **(A)** Masticatory and infrahyoid muscles viewed from the left side. **(B)** Medial view of the pterygoid muscles of the right side. *Red arrows* indicate directions of pull of jaw-closing muscles. *Blue arrows* indicate directions of pull of jaw openers.

Clinical Panel 21.1   Trigeminal neuralgia

Trigeminal neuralgia is an important condition occurring in middle age or later, characterized by attacks of excruciating pain in the territory of one or more divisions of the trigeminal nerve (usually II and/or III). The patient (who is usually more than 60 years old) is able to map out the affected division(s) accurately. Because it must be distinguished from many other causes of facial pain, the clinician should be able to mark out a trigeminal sensory map (Figure CP 21.1.1). Attacks are triggered by everyday sensory stimuli, for example brushing teeth, shaving, and chewing, and the tendency of patients to wince at the onset of attacks accounts for the French term *tic doloureux*.

Episodes of paroxysmal facial pain occurring in young adults should raise a suspicion of multiple sclerosis as the cause. Postmortem histology in such cases has revealed demyelination of the sensory root of the trigeminal nerve where it enters the pons. Demyelination of large sensory fibers receiving tactile signals from skin or mucous membranes in trigeminal territory may cause their exposed axons to come into direct contact with unmyelinated axons serving pain receptors. Animal experiments have shown that this type of contact can initiate *ephaptic transmission* of action potentials between them. It is now widely accepted that the most frequent etiology in later years is *vascular compression*, usually by a 'sagging' posterior cerebral artery in transit around the brainstem. The trigeminal central nervous system–peripheral nervous system transition zone (Ch. 6) is several millimeters lateral to the entry zone into the pons, and postmortem histology has provided evidence of the demyelinating effect of chronic pulsatile compression.

Antiepileptic drugs that exert a blocking effect on sodium and/or calcium channels (e.g. carbamazepine) may suffice to keep ephapsis at bay. For those in whom they do not, surgery is indicated.

A procedure that can be performed under local anesthesia is electrocoagulation of the affected division, through a needle electrode inserted through the foramen rotundum or ovale from below. The intention is to heat the nerve sufficiently to destroy only the finest fibers, in which case analgesia is produced but touch (including the corneal reflex) is preserved.

The final option is to *decompress* the afflicted nerve root through an intracranial approach whereby neighboring vessels are lifted away from it.

A surgical procedure of historical interest is *medullary tractotomy*, whereby the spinal root was sectioned through the dorsolateral surface of the medulla. In successful cases, pain and temperature sensitivity were lost from the face but touch (mediated by the pontine nucleus) was preserved. This procedure was abandoned owing to a high mortality rate associated with compromise of underlying respiratory and cardiovascular centers.

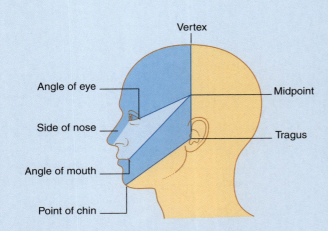

Figure CP 21.1.1 Trigeminal nerve sensory map.

REFERENCE

Love S, Coakham HB. Trigeminal neuralgia: pathology and pathogenesis. Brain 2001; 124:2347–2360.

Sensory nuclei

Three sensory nuclei are associated with the trigeminal nerve: **mesencephalic**, **pontine** (*principal*), and **spinal**.

Mesencephalic nucleus

The mesencephalic nucleus is unique in being the only nucleus in the central nervous system that contains the cell bodies of primary sensory neurons. Their peripheral processes enter the sensory root via the mesencephalic tract of the trigeminal. Some travel in the mandibular division to supply stretch receptors (neuromuscular spindles) in the masticatory muscles. Others travel in the maxillary and mandibular divisions to supply stretch receptors (Ruffini endings) in the suspensory, periodontal ligaments of the teeth.

The central processes of the mesencephalic afferent neurons descend through the pontine tegmentum in the small *tract of Probst*. Most fibers of this tract terminate in the supratrigeminal nucleus; others end in the motor nucleus or in the pontine sensory nucleus; a few travel as far as the dorsal nucleus of the vagus.

Pontine nucleus

The pontine (principal sensory) nucleus (Figure 17.11) is homologous with the posterior column nuclei (gracile and

Facial nerve

STUDY GUIDELINES
1 VII is the most commonly paralyzed of all peripheral nerves, owing to the great length of its canal in the temporal bone, where it is at risk of compression when swollen. Because VII supplies the muscles of facial expression, the effects of peripheral facial nerve paralysis are obvious to all.
2 Learn the distinctions between upper and lower motor neuron lesions of VII.
3 Note that VII participates in several important reflex arcs.

FACIAL NERVE

The **facial nerve** supplies the muscles derived from the second branchial arch. These include the muscles of facial expression, and four others mentioned below. It is accompanied during part of its course by the **nervus intermedius**, which supplies secretomotor fibers to glands in the eye, nose, and mouth, also gustatory fibers to the tongue and palate.

The facial nerve arises from the branchial (special visceral) efferent cell column caudal to the motor nucleus of the trigeminal nerve (Figure 17.2). The **facial nucleus** occupies the lateral region of the tegmentum in the caudal part of the pons (Figures 17.15 and 22.1). Before emerging from the brainstem, it loops, as the **internal genu**, around the abducens nucleus, creating the **facial colliculus** in the floor of the fourth ventricle.

The nerve emerges at the lower border of the pons together with the nervus intermedius. Both nerves cross the subarachnoid space in company with the vestibulocochlear nerve, to the internal acoustic meatus. Above the vestibule of the labyrinth, it enters a 7-shaped bony canal having a backward bend at the **external genu** of the facial nerve. Prior to escaping the canal at the stylomastoid foramen, it supplies the stapedius muscle. On escape, it supplies the posterior belly of the occipitofrontalis, the stylohyoid, and the posterior belly of the digastric. It then turns forward within the substance of the parotid gland while breaking up into the five named branches to the muscles of facial expression (Figure 22.2).

Supranuclear connections

All the cells of the motor nucleus receive a corticonuclear supply from the 'face' area of the contralateral motor cortex. In addition, those to the muscles of the upper face (occipitofrontalis and orbicularis oculi) receive an equal supply from the *ipsilateral* motor cortex. The bilateral

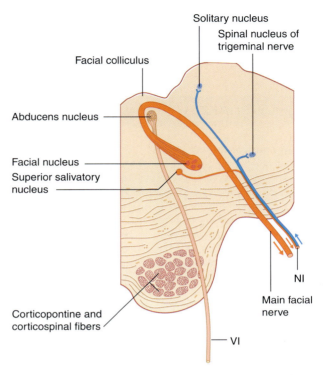

Figure 22.1 Transverse section of the pons, showing the facial nerve and the nervus intermedius (NI).

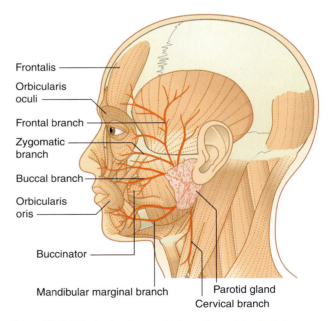

Figure 22.2 Principal extracranial branches of the facial nerve.

supply for the upper facial muscles is reflected in their habitual paired activities in wrinkling the forehead, blinking, and squeezing the eyes closed. The muscles around the mouth, on the other hand, are often activated unilaterally for some expressive purpose. The partial bilateral supply to the facial muscles helps to distinguish a supranuclear from a nuclear or infranuclear lesion of the nerve (Clinical Panel 22.1).

More than any other muscle group, the muscles of facial expression are responsive to emotional states. A limbic contribution to the supranuclear supply is to be expected, and indeed two have been identified. One is the **nucleus accumbens** at the base of the forebrain, identified in Figure 33.1. The nucleus accumbens is a ventral part of the basal ganglia, which in turn influence the motor cortex.

Clinical Panel 22.1 Lesions of the facial nerve

Supranuclear lesions

Much the commonest cause of a supranuclear lesion of the seventh nerve is a vascular stroke, in which corticonuclear and corticospinal fibers are interrupted at or above the level of the internal capsule. The usual effect of a stroke is to produce a contralateral motor weakness of the lower part of the face and of the limbs. (The lower part of the face may appear to recover momentarily when participating in a spontaneous smile, as mentioned earlier.) The upper face escapes because of the bilateral supranuclear supply to the upper part of the facial nucleus.

Nuclear lesions

The main motor nucleus may be involved in thrombosis of one of the pontine branches of the basilar artery. As might be anticipated from the relationships depicted in Figure 22.1, the usual result of such a lesion is an *alternating (crossed) hemiplegia*: complete paralysis of the facial and/or abducens nerve on one side combined with motor weakness of the limbs on the opposite side owing to concomitant involvement of the corticospinal tract.

Infranuclear lesions

Bell's palsy is a common disorder caused by a neuritis (possibly viral in origin) of the facial nerve. The inflammation causes the nerve to swell, and conduction is compromised by the close fit of the nerve in its bony canal in the interval between geniculate ganglion and stylomastoid foramen. There may be some initial pain in the ear, but the condition is otherwise painless.

Facial paralysis is usually complete. On the affected side, the patient is unable to raise the eyebrow, close the eye, or retract the lip. The patient may experience *hyperacusis*: ordinary sounds may be unpleasantly loud owing to loss of the damping action of the stapedius muscle.

The tight-fit segment is usually compromised. Tests may reveal blockage of nervus intermedius fibers, with ipsilateral reduced lacrimal and salivary secretions and loss of taste from the anterior part of the tongue.

Four out of five patients recover completely within a few weeks because the nerve has suffered only a conduction block (*neuropraxia*). In the remainder, the nerve undergoes Wallerian degeneration (Ch. 9); recovery takes about 3 months and is often incomplete. During regeneration, some preganglionic fibers of the nervus intermedius may enter the greater petrosal nerve instead of the chorda tympani, with the result that the lacrimal gland becomes active at mealtimes (so-called 'crocodile tears').

Other causes of infranuclear palsy include a patch of demyelination within the pons in the course of multiple sclerosis (Figure CP 22.1.1), tumors in the cerebellopontine angle (Clinical Panel 22.2), middle ear disease, and tumors of the parotid gland. *Herpes zoster oticus* is a rare but well-recognized viral infection of the geniculate ganglion. Severe pain in one ear precedes a vesicular rash in and around the external acoustic meatus. Swelling of the geniculate ganglion may result in a complete facial palsy (*Ramsay Hunt syndrome*).

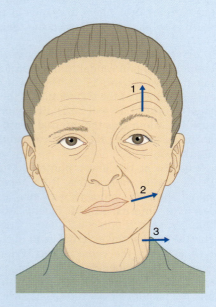

Figure CP 22.1.1 Complete facial nerve paralysis, patient's right side. The patient has been asked to smile and to look upward. To compare the two sides, cover the left and right halves of the photograph alternately with a card. On the left side,
(1) the frontalis muscle has raised the eyebrow;
(2) the buccinator has retracted the lips;
(3) the platysma is in moderate contraction.
On the right side, the lower eyelid is drooping due to paralysis of orbicularis oculi.

That circuit is compromised in Parkinson's disease, which is often characterized by a mask-like facies (Ch. 33). The other occupies the *affective area* of the cingulate gyrus (illustrated in Figure 34.10), an emotionally responsive region in the territory of the anterior cerebral artery. It is active during production of a spontaneous smile, and this is of clinical interest, as explained in Clinical Panel 22.1.

Nuclear connections

Five reflex arcs engaging the facial nucleus are listed in Table 22.1. Most important clinically is the corneal reflex.

Corneal reflex

The usual test is to touch the cornea with a cotton wisp. This should elicit a bilateral blink response. The afferent limb of the reflex is the ophthalmic division of the trigeminal nerve (nasociliary branch). The efferent limb is the facial nerve (branch to palpebral element of orbicularis oculi). Because the reflex can still be elicited following section of the spinal tract of the trigeminal nerve (*tractotomy*, Ch. 21), the ophthalmic afferents evidently synapse in the principal (pontine) nucleus of the trigeminal. Internuncials projecting from each principal nucleus to both facial nuclei complete the reflex arc.

The corneal reflex may be lost following a lesion of either the ophthalmic or facial nerve. A gradual compression of ophthalmic fibers in the sensory root of the trigeminal nerve may damage corneal neurons selectively. For this reason, the corneal reflex must be tested in patients under suspicion of an acoustic neuroma (Clinical Panel 22.2).

Nervus intermedius

Nervus intermedius aligns with the facial nerve distal to the internal genu. It comprises two sets of parasympathetic and two sets of special sense fibers (Figure 22.3).

The *parasympathetic root* of the nerve arises from the **superior salivatory nucleus** in the pons. This is the motor component of the **greater petrosal** and **chorda tympani** nerves. The greater petrosal synapses in the **pterygopalatine ganglion** ('the ganglion of hay fever'), whose postganglionic fibers stimulate the lacrimal and nasal glands. The motor component of chorda tympani synapses in the **submandibular ganglion**, whose postganglionic fibers stimulate the submandibular and sublingual glands.

The *special sense root* of this nerve has unipolar cell bodies in the **geniculate ganglion** of the facial nerve. The

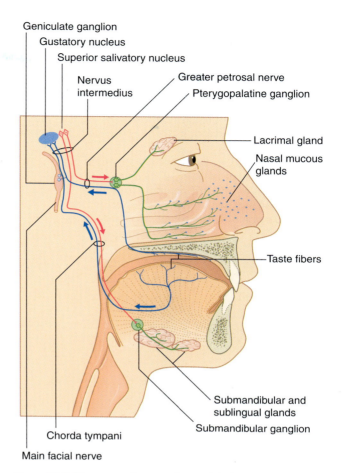

Figure 22.3 The nervus intermedius and its branches. Arrows indicate direction of impulse traffic.

peripheral processes of these ganglion cells supply taste buds in the palate via the great petrosal nerve, and taste buds in the anterior two-thirds of the tongue via the chorda tympani. The central processes enter the gustatory part of the solitary nucleus, which also receives fibers from the glossopharyngeal nerve (Ch. 18). From here, second-order neurons project to the thalamus on the *same* side, for relay to the anterior parts of insula and cingulate cortex.

A few cells of the geniculate ganglion supply skin in and around the external acoustic meatus (Clinical Panel 22.1).

Table 22.1	Brainstem reflexes involving the facial nerve				
	Corneal reflex	**Sucking reflex**	**Blinking to light**	**Blinking to noise**	**Sound attenuation**
Receptor	Cornea	Lips	Retina	Cochlea	Cochlea
Afferent	Ophthalmic nerve	Mandibular nerve	Optic nerve	Cochlear nucleus	Cochlear nucleus
First synapse	Spinal nucleus of trigeminal	Pontine nucleus of trigeminal	Superior colliculus	Inferior colliculus	Superior olivary nucleus
Second synapse	Facial nucleus	Facial nucleus	Facial nucleus	Facial nucleus	Facial nucleus
Muscle	Orbicularis oculi	Orbicularis oculi	Orbicularis oculi	Orbicularis oculi	Stapedius

Clinical Panel 22.2 Syndromes of the cerebellopontine angle

The *cerebellopontine angle* is the recess between the hemisphere of the cerebellum and the lower border of the pons. The petrous temporal bone, laterally, completes a triangle having the V nerve at its upper corner and IX and X at its lower corner, and bisected by VII and VIII.

Several kinds of space-occupying lesions may compromise one or more of the nerves. The most frequent is an *acoustic neuroma* (Figure CP 22.2.1), a slow-growing, benign tumor of Schwann cells (*neurolemmoma*). The tumor originates on the

vestibular nerve within the internal acoustic meatus, but the initial symptoms are more often cochlear than vestibular. *An acoustic neuroma must be suspected in every middle-aged or elderly patient presenting with auditory or vestibular symptoms.* Early diagnosis is important because of the difficulty of removing a large neuroma extending into the posterior cranial fossa, also because the cumulative motor and sensory disturbances may not show significant improvement after surgery.

The following is a fairly typical sequence of symptoms and signs in a case escaping early detection.

- *Tinnitus* is experienced on the affected side, in the form of a high-pitched ringing or fizzing sound.
- *Deafness* on the affected side is slowly progressive over a period of months or years.
- *Vertigo* occurs episodically. Severe vertigo with nystagmus signifies compression of the brainstem.
- *Loss of the corneal reflex* is an early sign of distortion of the V nerve by a tumor emerging from the internal acoustic meatus into the posterior cranial fossa.
- *Weakness of the masticatory muscles* is a later sign of V nerve involvement. The jaw deviates toward the affected side when the mouth is opened, because the normal lateral pterygoid is unopposed. Wasting of the masseter may be detected by palpation.
- *Weakness of the facial musculature* develops as the VII nerve becomes stretched.
- *Anesthesia of the oropharynx* signifies involvement of the IX nerve.
- *Ipsilateral 'cerebellar signs'* in the arm and leg appear when the cerebellum is compressed.
- *'Upper motor neuron signs'* in the limbs signify compression of the brainstem.
- *Signs of raised intracranial pressure* (headache, drowsiness, papilledema) signify obstruction of cerebrospinal fluid circulation either inside or around the brainstem.

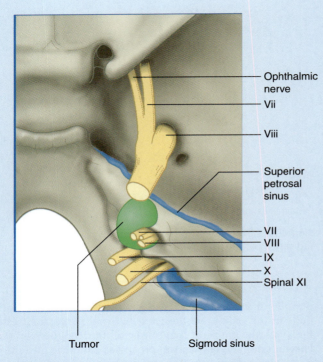

Labels: Ophthalmic nerve — Vii — Viii — Superior petrosal sinus — VII — VIII — IX — X — Spinal XI — Tumor — Sigmoid sinus

Figure CP 22.2.1 An acoustic neuroma invading the right posterior cranial fossa.

Core Information

On leaving its nucleus, the facial nerve whirls around the facial colliculus, emerges at the lower border of pons, and at the internal auditory meatus enters a long bony canal opening at the stylomastoid foramen. It supplies the muscles of facial expression, the occipital portion of occipitofrontalis, the stapedius, and the posterior digastric. The upper half of the facial nucleus receives a bilateral supranuclear supply from the motor cortex; the lower half receives only a contralateral supply.

The nervus intermedius travels in part with VII. The superior salivatory nucleus provides the motor components of greater petrosal (for lacrimal and nasal glands by the pterygopalatine ganglion) and chorda tympani (for submandibular and sublingual glands via the submandibular ganglion). The geniculate ganglion of VII has unipolar neurons receiving taste from the palate via greater petrosal nerve and tongue via chorda tympani. A few unipolar neurons supply skin in and around the external acoustic meatus.

REFERENCES

Fischer U, Hess CW, Rosler KM. Uncrossed corticomuscular projections in humans are abundant to facial muscles of the upper and lower face, but may differ between sexes. J Neurol 2005; 252:21–26.

Lang J. Clinical anatomy of the cerebellopontine angle and internal acoustic meatus. Adv Oto-Rhino-Laryngol 1984; 34:8–24.

Manni JJ, Stennert E. Diagnostic methods in facial nerve pathology. Adv Oto-Rhino-Laryngol 1984; 34:202–223.

Parnes SM. The facial nerve. In: Jahn AF, Santos-Sacchi J, eds. Physiology of the ear. New York: Raven Press; 1988:125–142.

Parsons M. Diagnostic picture test in clinical neurology. London: Wolfe Medical; 1987

Ocular motor nerves

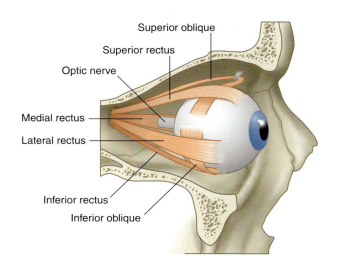

Figure 23.1 Extrinsic ocular muscles.

STUDY GUIDELINES

General
Because of the immense diagnostic and therapeutic importance of ocular innervation, and because of its inherent complexity, neuroophthalmology has become a branch of medicine in its own right.

It is especially important to appreciate the way in which premotor centers are able to operate bilaterally in order to keep the gaze on target, even when the head is moving.

Particular
1 In addition to nerves III, IV, and VI, the sympathetic supply to the eye is reviewed here because it counterbalances the parasympathetic innervation contained in nerve III.

2 The nerve supply to the six muscles that move the eyeball is straightforward. However, nerve III also supplies the elevator of the upper eyelid.

3 The autonomic supply to the eye is straightforward. For the sympathetic, think of someone being startled awake at night. For the parasympathetic, think of someone eyeing food on a fork.

THE NERVES

The ocular motor nerves comprise the **oculomotor** (III cranial), **trochlear** (IV cranial), and **abducens** (VI cranial) **nerves**. They provide the motor nerve supply to the four recti and two oblique muscles controlling movements of the eyeball on each side (Figure 23.1). The oculomotor nerve contains two additional sets of neurons: one to supply the levator of the upper eyelid, the other to control the sphincter of the pupil and the ciliary muscle.

The nuclei serving the extraocular muscles (extrinsic muscles of the eye) belong to the somatic efferent cell column of the brainstem, in line with the nucleus of the hypoglossal nerve. The oculomotor nucleus has an additional, parasympathetic nucleus that belongs to the general visceral efferent cell column.

Oculomotor nerve

The nucleus of the third nerve is at the level of the superior colliculus. It is partly embedded in the periaqueductal gray matter (Figure 23.2A). It is composed of five individual nuclei for the supply of striated muscles, and one parasympathetic nucleus.

The nerve passes through the tegmentum of the midbrain and emerges into the interpeduncular fossa (arachnoid cistern). It crosses the apex of the petrous temporal bone, pierces the dural roof of the cavernous sinus, runs in the lateral wall of the sinus, and breaks into upper and lower divisions within the superior orbital fissure. The upper division supplies the superior rectus and the levator palpebrae superioris; the lower division supplies the inferior and medial recti, and the inferior oblique.

The parasympathetic fibers originate in the **Edinger–Westphal nucleus**. They accompany the main nerve as far as the orbit, then leave the branch to the inferior oblique and synapse in the **ciliary ganglion**. Postganglionic fibers emerge from the ganglion in the **short ciliary nerves**, which pierce the lamina cribrosa ('sieve-like layer') of the sclera and supply the *ciliaris* and *sphincter pupillae* muscles.

Trochlear nerve

The nucleus of the fourth nerve is at the level of the inferior colliculus. The nerve itself is unique in two respects (Figure 23.2B): it is the only nerve to emerge from the back of the brainstem, and it decussates with its opposite number.

The IV nerve winds around the crus of the midbrain and travels through the cavernous sinus in company with the III nerve (Figure 23.3). It passes through the superior orbital fissure and supplies the *superior oblique* muscle.

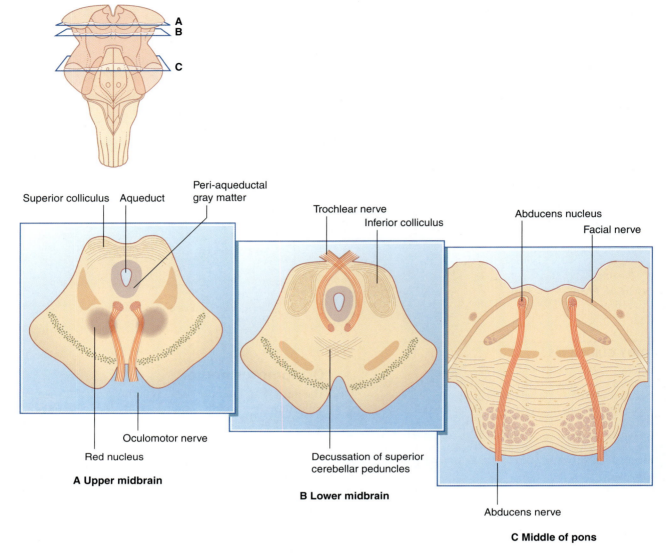

Superior colliculus   Aqueduct   Peri-aqueductal gray matter

Trochlear nerve
Inferior colliculus

Abducens nucleus
Facial nerve

Oculomotor nerve

Red nucleus

A Upper midbrain

Decussation of superior cerebellar peduncles

B Lower midbrain

Abducens nerve

C Middle of pons

Figure 23.2 (A–C) Transverse sections of the brainstem, showing the origins of the ocular motor nerves.

Abducens nerve

The nucleus of the sixth nerve, in the floor of the fourth ventricle, is at the level of the facial colliculus, in the middle of the pons (Figure 23.2C). The nerve descends, to emerge at the lower border of the pons, and runs up the pontine subarachnoid cistern beside the basilar artery. It angles over the apex of the petrous temporal bone and passes through the cavernous sinus beside the internal carotid artery (Figure 23.3). It enters the orbit through the superior orbital fissure and supplies the *lateral rectus* muscle, which abducts the eye.

NERVE ENDINGS

Motor endings

All the ocular motor units are small, containing 5–10 muscle fibers apiece (compared with 1000 or more in the tibialis anterior).

Type A fibers produce the fast twitches required for saccadic movements. *Type B* are slow-twitch and may be used for smooth pursuit. *Type C* show only local contractions beneath the individual plates. Type C fibers may be involved in keeping the visual axes of the two eyes parallel with one another. Because the visual axes diverge following administration of muscle relaxants, keeping them parallel must require continuous muscle action, even during sleep.

Sensory endings

In addition to neuromuscular spindles of standard type, numerous *palisade endings* exist in the form of nerve spirals around individual muscle fibers.

The extraocular muscle proprioceptors are the peripheral terminals of neurons in the mesencephalic nucleus of the trigeminal nerve. In monkeys, some of the central processes of these neurons reach as far caudally as the

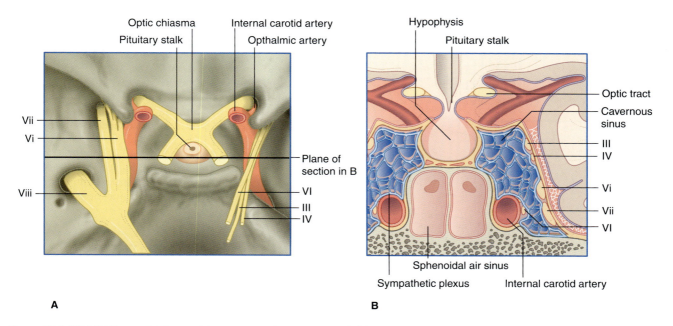

Figure 23.3 (A) Middle cranial fossa with cavernous sinuses removed. (B) Coronal section in the plane of the hypophysis with the cavernous sinuses in place. III, oculomotor nerve; IV, trochlear nerve; VI, abducens nerve; Vi, Vii, Viii, ophthalmic, maxillary, mandibular divisions of trigeminal nerve.

accessory cuneate nucleus in the medulla oblongata. This nucleus also receives proprioceptive terminals from the neck muscles, and it projects both to the ipsilateral cerebellum and to the contralateral superior colliculus. The conjunction of ocular and cervical proprioceptive information presumably assists in the coordination of simultaneous movements of the eyes and head.

PUPILLARY LIGHT REFLEX (Figure 23.4)

Constriction of the pupils in response to light involves four sets of neurons, as follows.

1 The afferent limb commences in the ganglionic layer of the retina, which gives rise to the optic nerve. Fibers leaving the chiasma enter both optic tracts and terminate in the **pretectal nuclei**, situated just rostral to the superior colliculus on each side (Figure 17.19).

2 Each pretectal nucleus is linked by internuncial neurons to both Edinger–Westphal (parasympathetic) nuclei; the contralateral nucleus is reached by way of the **posterior commissure**.

3 Preganglionic parasympathetic fibers enter the oculomotor nerve, leave the branch to the inferior oblique, and synapse in the ciliary ganglion.

4 Postganglionic fibers run in the short ciliary nerves and enter the iris to supply the sphincter pupillae.

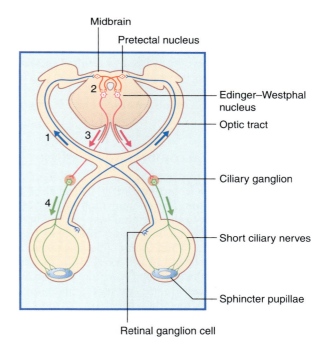

Figure 23.4 Pupillary light reflex. For numbers, see text.

ACCOMMODATION

The near response

When the eyes view an object close up, the ciliary muscle contracts reflexly, thereby relaxing the suspensory liga-ment of the lens (Figure 23.5). Because the lens at rest is somewhat compressed (flattened) by tension exerted on the lens capsule by the suspensory ligament, the lens bulges passively when the ciliary muscle contracts. The thicker lens has the greater refractive power required to bring close-up objects into focus on the retina. The response of the lens is one of *accommodation*.

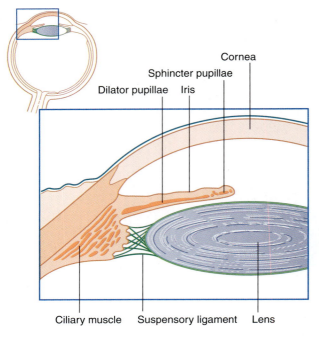

Figure 23.5 Intrinsic muscles of the eye.

The *accommodation reflex*, as understood clinically, involves two additional features. The sphincter pupillae contracts in order to eliminate passage of light through the peripheral, thinner part of the lens. At the same time, the visual axes of the two eyes converge as a result of increased tone in the medial rectus muscles. The convergence is known clinically as *vergence*.

The three features described are also known as the *near response*.

Pathway for the accommodation reflex

In order to execute the near response, a stereoscopic analysis of the object is carried out at the level of the visual association cortex. The afferent limb of the reflex passes from the retina to the occipital lobe via the lateral geniculate body. The efferent limb passes from the occipital lobe to the midbrain, where some fibers activate the Edinger–Westphal nucleus and others activate *vergence* (convergence) cells in the reticular formation. The vergence cells activate the nuclear groups serving the medial recti, with the effect of *fixating* the object on to the fovea centralis of each eye. The (con)vergence response is called the *fixation reflex*.

The far response

Just as the state of the pupil depends on the balance of sympathetic and parasympathetic activity, so does the state of the lens. At rest, both are in midposition. The resting focal length of the lens averages 1 m (with considerable variation between individuals). This is because the ciliary muscle is tonically active. In order to bring a distant object into focus, the ciliary muscle must be inhibited, so that the suspensory ligament becomes taut and the lens flat. The sphincter of the pupil is inhibited as well.

The sympathetic system innervates all the intrinsic muscles. It has a dual mode of action. It causes contraction of the dilator pupillae by way of *alpha* receptors on the muscle fibers, and it causes *relaxation* of the ciliary muscle and pupillary sphincter by way of *beta* receptors. This dual effect constitutes the *far response*, and it is used to focus the eyes on objects at a distance. (*Note*: The unqualified use of *alpha* and *beta* receptors signifies α_1 and β_2, respectively.)

In stressed individuals, heightened sympathetic activity may interfere with the normal process of accommodation. For example, students taking an important written test may have difficulty in bringing the questions into proper focus.

NOTES ON THE SYMPATHETIC PATHWAY TO THE EYE

The great length of the sympathetic pathway to the eye is indicated in Figure 23.6.

1 *Central fibers* descending from the hypothalamus cross to the other side in the midbrain. In the pons and medulla, they are joined by ipsilateral fibers descending from the reticular formation.

2 *Preganglionic fibers* emerge in the first thoracic ventral nerve root, and run up in the sympathetic chain to the superior cervical ganglion.

3 *Postganglionic fibers* run along the external and internal carotid arteries and their branches.

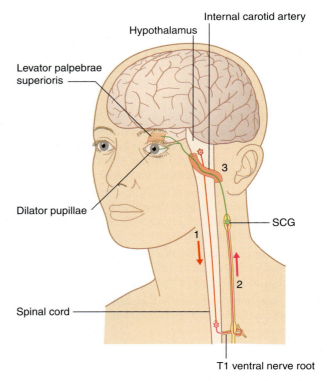

Figure 23.6 Three neuron pathways from the hypothalamus to the eye. Arrows indicate directions of impulse conduction. SCG, superior cervical ganglion. For numbers, see text.

The *external* carotid sympathetic fibers accompany all the branches of the external carotid artery. Those accompanying the facial artery supply the arterioles of the cheek and lips and are particularly responsive to emotional states. Those accompanying the maxillary artery supply the cavernous tissue covering the nasal conchae (turbinate bones).

Two sets of sympathetic fibers accompany the *internal* carotid artery. One set leaves it to join the ophthalmic division of the V nerve in the cavernous sinus, then leaves this in the long and short ciliary nerves to supply the vessels and smooth muscles of the eyeball. The second set forms a plexus around the internal carotid artery and its branches, including the ophthalmic artery. The ophthalmic artery gives off supratrochlear and supraorbital branches, which carry sympathetic fibers to the skin of the forehead and scalp.

Interruption of the postganglionic fibers at the jugular foramen (see *jugular foramen syndrome*, Ch. 18), or in the cavernous sinus, produces anhidrosis (loss of sweating) on the forehead and scalp.

OCULAR PALSIES

The effects of paralysis of the motor nerves to the eye are described in Clinical Panel 23.1.

Clinical Panel 23.1   Ocular palsies

One or more of the three ocular motor nerves may be paralyzed by disease within the brainstem (e.g. multiple sclerosis, vascular occlusion), in the subarachnoid space (e.g. meningitis, aneurysm in the circle of Willis, distortion by an expanding intracranial lesion), or in the cavernous sinus (e.g. thrombosis of the sinus, aneurysm of the internal carotid artery there).

Oculomotor nerve
Complete III nerve palsy
Characteristic signs of complete third nerve paralysis are shown in Figure CP 23.1.1A. They are:

1 complete ptosis of the eyelid (unopposed orbicularis oculi)

2 a fully dilated, non-reactive pupil (unopposed dilator pupilae)

3 a fully abducted eye (unopposed lateral rectus), which is also depressed (unopposed superior oblique).

Partial III nerve palsy
The pupils are *always* monitored when cases of head injury come to medical attention. Rapidly increasing intracranial pressure, resulting from an acute extradural or subdural hematoma (Ch. 4), often compresses the third nerve against the crest of the petrous temporal bone. The parasympathetic fibers are superficially placed and are the first to suffer, and the pupil dilates progressively on the affected side. *Pupillary dilatation is an urgent indication for surgical decompression of the brain.*

Trochlear nerve
The IV nerve is rarely paralyzed alone. The cardinal symptom is diplopia (double vision) on looking down, for example when going down stairs. This happens because the superior oblique normally assists the inferior rectus in pulling the eye downward, especially when the eye is in a medial position.

Abducens nerve
The effect of a *complete* VI nerve paralysis is shown in Figure CP 23.1.1B. The eye is fully adducted by the unopposed pull of the medial rectus.

The abducens has the longest course in the subarachnoid space of any cranial nerve. It also bends sharply over the crest of the petrous temporal bone. A space-occupying lesion affecting *either* cerebral hemisphere may cause compression and paralysis of one abducens nerve.

'Spontaneous' paralysis of the VI nerve may be caused by an arterial aneurysm at the base of the brain or by hardening (atherosclerosis) of the internal carotid artery in the cavernous sinus.

Internuclear ophthalmoplegia
Interruption of the linkage between the abducens nucleus and the contralateral oculomotor nucleus gives

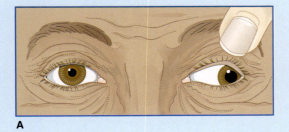

A   B

Figure CP 23.1.1 (A) Complete left III nerve paralysis. The closed eyelid has been raised by the examiner's finger. **(B)** Complete left VI nerve paralysis.

Clinical Panel 23.1 *Continued*

rise to the condition known as *internuclear ophthalmoplegia*. As an example, a lesion of the VI–III connection shown in Figure 23.7 would leave a saccade to the right unaffected, whereas on attempting a saccade to the left, the paralyzed right medial rectus would create a divergent strabismus with accompanying diplopia. Integrity of the nucleus serving the right medial rectus is shown by its normal behavior during the vergence component of the near response.

Below the age of 40, the chief cause of internuclear ophthalmoplegia is a plaque of demyelination associated with multiple sclerosis. Above the age of 60, the chief cause is a patch of tissue necrosis associated with occlusion of a pontine branch of the basilar artery.

Ocular sympathetic supply

Any one of the three sequential sets of neurons depicted in Figure 23.6 may be interrupted by local pathology.

1 The *central* set may be interrupted by a vascular lesion of the pons or medulla oblongata. The usual picture is one of Horner syndrome (ptosis and miosis, as described in Ch. 13) and cranial nerve involvement on one side, together with motor weakness and/or sensory loss in the limbs on the contralateral side. The Horner syndrome is associated with anhidrosis—absence of sweating—in the face and scalp on the same side, together with congestion of the nose (engorged turbinates).

2 The *preganglionic* set is most often interrupted by stony, cancerous deep cervical lymph nodes in the lower part of the neck. A Horner syndrome is associated with anhidrosis of the face and scalp (and nasal congestion) on the same side.

3 The *postganglionic* set accompanying the *external* carotid artery is rarely damaged directly. The set accompanying the internal carotid artery may be interrupted as part of a jugular foramen syndrome (Ch. 18), or by pathology in the cavernous sinus. Horner syndrome is accompanied by anhidrosis of the forehead and anterior scalp (territory of the supraorbital and supratrochlear arteries).

CONTROL OF EYE MOVEMENTS

The eyes normally move as a pair. This *conjugate* movement is of three fundamentally different kinds, as follow.

1 *Scanning.* The eyes flick from one visual target to another in high-speed movements called *saccades*.

2 *Tracking.* In tracking, or *smooth pursuit*, the eyes follow an object of interest across the visual field.

3 *Compensation.* The gaze can be held on an object of interest during movements of the head. This is the *vestibuloocular reflex*, which depends on displacement of endolymph in the kinetic labyrinth (Ch. 19).

Scanning

Four separate *gaze centers* in the brainstem pick out motor neurons appropriate to the direction of movement: leftward, rightward, upward, or downward. The centers are small nodes in the reticular formation. They contain *burst cells*, which discharge at 1000 Hz (impulses/s) and entrain the appropriate motor neurons momentarily at this rate.

The paired centers (left and right) for horizontal saccades are in the paramedian pontine reticular formation (PPRF) (Figure 17.15). Each pulls the eyes to its own side (Figure 23.7). The midbrain contains a bilateral center for upward saccades located in the rostral end of the medial longitudinal fasciculus, at the level of the pretectal nucleus. It is called the **rostral interstitial nucleus**. At the same level but a little ventral to this is a bilateral center for downward gaze (Figure 17.19).

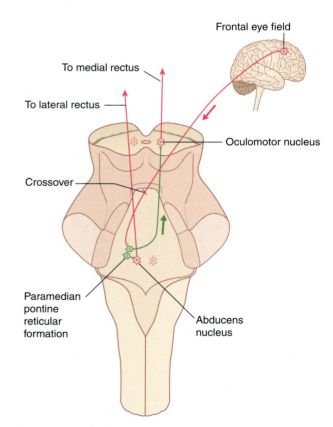

Figure 23.7 Principal pathways involved in a voluntary ocular saccade to the left. The VI–III connection (*green arrow*) is, clinically, internuclear (see Clinical Panel 23.1).

Automatic scanning movements are activated by the superior colliculus, on receipt of visual information from the retina through the medial root of the optic tract. Examples of automatic scanning include the sideward glance toward an object attracting attention in the peripheral visual field, and the saccadic movements used in reading. The tectoreticular projections concerned cross the midline before engaging the gaze centers. Saccadic accuracy is controlled by the midregion (vermis) of the cerebellum, which receives afferents from the superior colliculi and projects to the vestibular nucleus.

Voluntary scanning movements are initiated in the *frontal eye fields*, located at the junction of motor and premotor cortex (Ch. 29). From each frontal eye field, a projection descends in the anterior limb of the internal capsule. Most of the fibers cross over before terminating in the gaze centers.

As explained in Chapter 29, the ipsilateral superior colliculus is activated at the same time, to reinforce the excitation of the appropriate gaze center.

The projection from the frontal eye field is interrupted in about one-third of patients who suffer a stroke involving the internal capsule. The result is *paralysis of contraversive horizontal gaze*. 'Contraversive' refers to an inability to make a voluntary saccade away from the side of the lesion. The gaze paralysis vanishes within a week, even if the hemiplegia remains profound—presumably because of takeover by uncrossed fibers.

The best-known afferents to the frontal eye field come from the parietal cortex, from cells concerned with *visual attention*. In monkeys, some cells in the posterior parietal cortex become active when an object of interest is seen. These cells project to the frontal eye field and are thought to facilitate eye movement in the direction of the object. In humans, neglect of the contralateral visual field is a well-known feature of damage to the posterior parietal lobe, especially on the right side (Ch. 32).

Tracking

The neural mechanisms for tracking must be complex because of the following basic requirements: (a) intact visual pathways to monitor the position of the object throughout the movement, (b) neurons to signal the rate of movement of the object (velocity detectors), (c) neurons to coordinate movements of the eyes and head (neural integrator), and (d) a system to monitor smooth execution of the tracking movement. Monkey and cat experiments indicate the following.

* Object position information is forwarded from the visual cortex to the posterior parietal cortex, and from there to the reticular formation of the pons.

* Velocity detectors are present in the upper part of the pons, apparently receiving information direct from the retina via the medial root of the optic tract.

* Head movement is signaled by the dynamic labyrinth, and is integrated with spatial and velocity information in the **nucleus prepositus hypoglossi**—a node of the reticular formation that is in fact closer to the abducens nucleus than to the hypoglossal nucleus. The nucleus prepositus projects to the PPRF, which controls conjugate eye movements. The pathway to the neck muscles (for turning the head) may involve the superior colliculus.

* Smooth execution of tracking movements is monitored by the flocculus of the cerebellum, which has two-way connections to the vestibular nucleus and pontine reticular formation.

The dynamic labyrinth and cerebellum cooperate to keep the eyes on target during movement of the head, as described in Chapter 19.

Core Information

Oculomotor nerve

Somatic efferent fibers of III arise from the main nucleus at superior collicular level. The nerve passes intact through the cavernous sinus and in two divisions through the superior orbital fissure. The upper division supplies superior rectus and levator palpebrae superioris; the lower division supplies inferior and medial recti and inferior oblique.

Parasympathetic fibers emerge from the Edinger–Westphal nucleus, travel with the main nerve, and synapse in the ciliary ganglion for supply of sphincter pupillae and ciliaris.

Paralysis of III is shown by a dilated pupil, followed by ptosis, and later by a divergent squint in addition.

Trochlear nerve

The nucleus of IV is at inferior collicular level. The nerve crosses the midline before emerging below the inferior colliculus. It passes through the cavernous sinus to supply the superior oblique.

Paralysis is characterized by diplopia on looking down.

Abducens nerve

The nucleus is at the level of the facial colliculus in the pons. The nerve runs in the subarachnoid space from lower border of pons to apex of petrous temporal bone, and passes through cavernous sinus and superior orbital fissure supplies lateral rectus.

Paralysis is characterized by convergent squint with inability to abduct the affected eye.

Sympathetic

Muscles stimulated (via α_1 receptors) are dilator pupillae and levator palpebrae superioris. Paralysis is characterized by ptosis with a constricted pupil (Horner syndrome). Muscles inhibited (via β_2 receptors) are sphincter pupillae and ciliaris.

Parasympathetic

Muscles stimulated are the sphincter pupillae and the ciliaris.

Reflex pathways

For the pupillary light reflex: from retina to pretectal nucleus to both Edinger–Westphal nuclei to ciliary ganglion to sphincter pupillae.

For the accommodation reflex: from retina to lateral geniculate body to occipital cortex to Edinger–Westphal nucleus to ciliary ganglion to ciliaris.

Oculomotor controls

Scanning (saccading) is locally activated by six gaze centers. Clinically most important is the paramedian pontine reticular formation, which operates to pull ipsilateral lateral rectus and contralateral medial rectus conjugately to its own side. Automatic scanning is controlled by the superior colliculi, and voluntary scanning by the frontal eye fields.

Tracking is complex and involves occipital cortex, dynamic labyrinth, cerebellum, superior colliculus, and reticular formation.

REFERENCES

American Academy of Ophthalmology. Principles of ophthalmology. San Francisco: AAO; 1995.

Anderson TJ, Jenkins IH, Brooks DJ, et al. Cortical control of saccades and fixation in man. Brain 1994; 117:1073–1084.

Dean P, Mayhew JEW, Langdon P. Learning and maintaining saccadic accuracy: a model of brain stem-cerebellar interactions. J Cogn Neurosci 1994; 6:117–138.

Fukushima K. The interstitial nucleus of Cajal in the midbrain reticular formation and vertical eye movement. Neurosci Res 1991; 10:159–187.

Keane JR. Internuclear ophthalmoplegia. Arch Neurol 2005; 62:714–717.

Keller EL, Heinen SJ. Generation of smooth pursuit eye movements: neuronal mechanisms and pathways. Neurosci Res 1991; 11:79–107.

Miyazaki S. Location of motoneurons in the oculomotor nucleus and the course of their axons in the oculomotor nerve. Brain Res 1985; 348:57–63.

Moschovakis AK. The neural integrators of the mammalian saccadic system. Front Biosci 1997; 15:D552–D557.

Oda K. Motor innervation and acetylcholine receptor distribution of human extraocular muscle fibers. J Neurol Sci 1986; 74:125–133.

Parkinson D. Further observations on the sympathetic pathways to the pupil. Anat Rec 1988; 232:108–109.

Petit L, Clark VP, Ingeholm J, et al. Dissociation of saccade-related and pursuit-related activation in human frontal eye fields as revealed by fMRI. J Neurophysiol 1997; 77:3386–3390.

Wilhelm H. Neuro-ophthalmology of pupillary function. J Neurol 1998; 245:573–583.

Reticular formation

STUDY GUIDELINES

1 The reticular formation has very diverse functions. Some of its nuclear groups have direct access to motor neurons in the brainstem and spinal cord. Others have direct access to autonomic effector nuclei including cardiovascular controls. Some may act simultaneously on somatic and autonomic nuclei.

2 All the sensory systems contribute afferents to the reticular formation.

3 Ascending projections of the reticular formation to the forebrain, including the ascending reticular activating system, are essential for the state of consciousness.

4 Disturbance of function of aminergic projections from the reticular formation have been correlated with psychiatric states including major depression and schizophrenia.

The reticular formation is phylogenetically a very old neural network, being a prominent feature of the reptilian brainstem. It originated as a slowly conducting, polysynaptic pathway intimately connected with olfactory and limbic regions. The progressive dominance of vision and hearing over olfaction led to lateralization of sensory and motor functions within the tectum of the midbrain. Direct spinotectal and tectospinal tracts bypassed the reticular formation, which was largely relegated to automatic functions. In mammals, the tectum in turn has been relegated to minor status with the emergence of very fast pathways linking the cerebral cortex with the peripheral sensory and motor apparatus.

In the human brain, the reticular formation continues to be of importance in automatic and reflex activities, and it has retained its linkages to the limbic system.

ORGANIZATION

The term *reticular formation* refers only to the polysynaptic network in the brainstem, although the network continues rostrally into the thalamus and hypothalamus, and caudally into the propriospinal network of the spinal cord.

The ground plan is shown in Figure 24.1A. In the midline, the *median reticular formation* comprises a series of *raphe nuclei* (*pron.* 'raffay' and derived from the Greek word for seam). The raphe nuclei are the major source of serotonergic projections throughout the neuraxis (see next section).

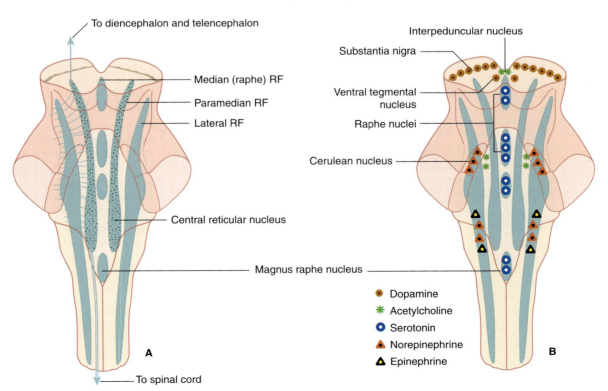

Figure 24.1 Reticular formation (RF). **(A)** Subdivisions. **(B)** Aminergic and cholinergic cell groups.

Next to this is the *paramedian reticular formation*. This part of the network contains *magnocellular* neurons throughout; in the lower pons and upper medulla, some *gigantocellular* neurons also appear, before the network blends with the *central reticular nucleus* of the medulla oblongata.

Outermost is the *lateral, parvocellular* (small-celled) *reticular formation*. Parvocellular dendrites are long and they branch at regular intervals. They have a predominantly transverse orientation, and their interstices are penetrated by long pathways running to the thalamus. The lateral network is mainly afferent in nature. It receives fibers from all the sensory pathways, including the special senses.

- Olfactory fibers are received through the median forebrain bundle, which passes alongside the hypothalamus.
- Visual pathway fibers are received from the superior colliculus.
- Auditory pathway fibers are received from the superior olivary nucleus.
- Vestibular fibers are received from the medial vestibular nucleus.
- Somatic sensory fibers are received from the spinoreticular tracts and from the spinal and principal (pontine) nuclei of the trigeminal nerve.

Most parvocellular axons ramify extensively among the dendrites of the paramedian reticular formation. However, some synapse within the nuclei of cranial nerves and act as pattern generators (see later).

The paramedian reticular formation is a predominantly *efferent* system. The axons are relatively long. Some ascend to synapse in the midbrain reticular formation or in the thalamus. Others have both ascending and descending branches contributing to the polysynaptic network. The magnocellular component receives corticoreticular fibers from the premotor cortex and gives rise to the pontine and medullary reticulospinal tracts.

Aminergic neurons of the brainstem

Embedded in the reticular formation are sets of aminergic neurons (Figure 24.1B). They include one set producing *serotonin* (5-hydroxytryptamine) and three sets producing *catecholamines*, as listed in Table 24.1.

- The *serotonergic neurons* have the largest territorial distribution of any set of central nervous system (CNS) neurons. In general terms, those of the midbrain

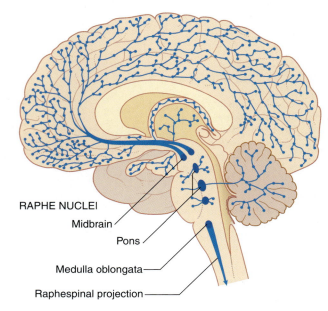

RAPHE NUCLEI
Midbrain
Pons
Medulla oblongata
Raphespinal projection

Figure 24.2 Serotonergic projections from the brainstem midline (raphe).

project rostrally into the cerebral hemispheres, those of the pons ramify in the brainstem and cerebellum, and those of the medulla supply the spinal cord (Figure 24.2). All parts of the CNS gray matter are permeated by serotonin-secreting axonal varicosities. Clinically, enhancement of serotonin activity is part of the treatment for a prevalent condition known as major depression (Ch. 26).

- The *dopaminergic neurons* of the midbrain fall into two groups. At the junction of tegmentum and crus are those of the substantia nigra, which will be considered in Chapter 33. Medial to these, dopaminergic neurons in the **ventral tegmental nuclei** (Figure 24.3) project *mesocortical* fibers to the frontal lobe and *mesolimbic* fibers to the nucleus accumbens in particular (Ch. 34).
- The *noradrenergic neurons* are only marginally less prodigious than the serotonergic ones. About 90% of the somas are pooled in the **cerulean nucleus** (*locus ceruleus*), a 'violet spot' in the floor of the fourth ventricle at the upper end of the pons (Figure 24.4). Neurons of the cerulean nucleus project in *all* directions, as indicated in Figure 24.5.
- *Epinephrine (adrenaline)-secreting neurons* are relatively scarce and are confined to the medulla oblongata. Some project rostrally to the hypothalamus, others project caudally to synapse on preganglionic sympathetic neurons in the spinal cord.

In the cerebral cortex, the ionic and electrical effects of aminergic neuronal activity are quite variable. First, more than one kind of postsynaptic receptor exists for each of the amines. Second, some aminergic neurons liberate a peptide substance also, capable of modulating the transmitter action—usually by prolonging it. Third, the larger cortical neurons receive many thousands of excitatory and inhibitory synapses from local circuit neurons, and they

Table 24.1 Aminergic neurons of the reticular formation	
Transmitter	**Location**
Serotonin	Raphe nuclei of midbrain, pons, medulla
Dopamine	Tegmentum of midbrain
Norepinephrine (noradrenaline)	Midbrain, pons, medulla
Epinephrine (adrenaline)	Medulla

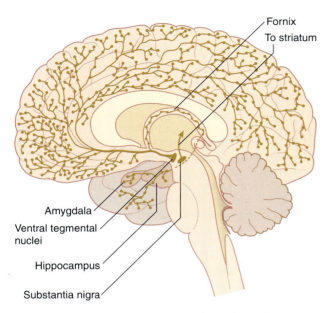

Figure 24.3 Dopaminergic projections from the midbrain.

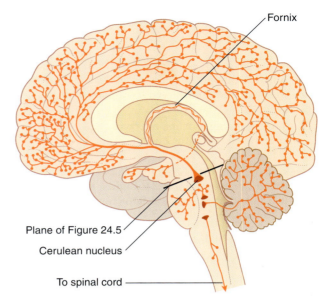

Figure 24.5 Noradrenergic projections from the pons and medulla oblongata

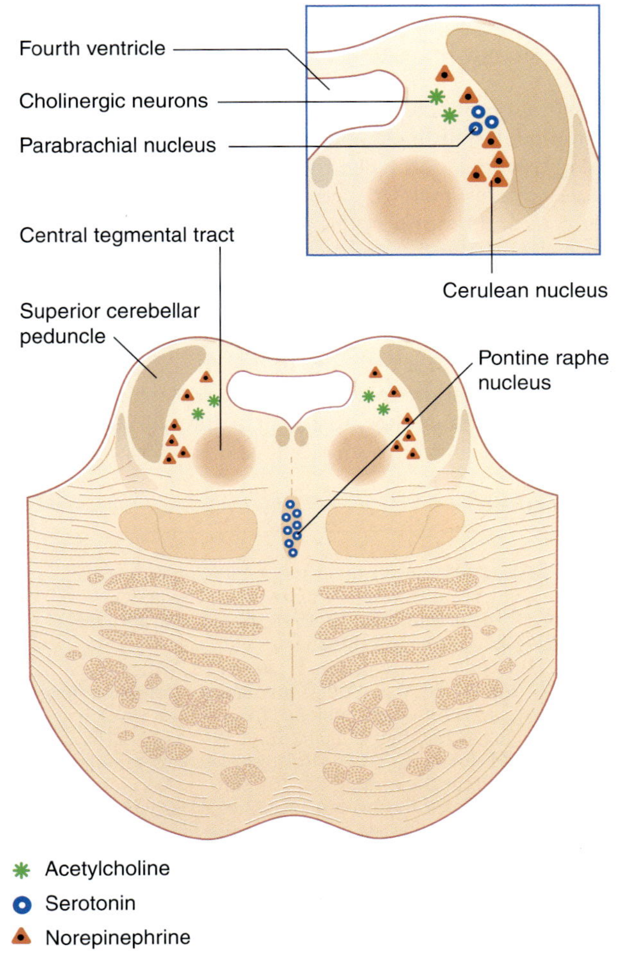

Fourth ventricle

Cholinergic neurons

Parabrachial nucleus

Central tegmental tract

Superior cerebellar peduncle

Cerulean nucleus

Pontine raphe nucleus

✳ Acetylcholine
⊙ Serotonin
▲ Norepinephrine

Figure 24.4 Part of a transverse section through the upper part of the pons, showing elements of the reticular formation.

have numerous different receptors. Activation of a single kind of aminergic receptor may have a large or small effect depending on the existing excitatory state.

Although our understanding of the physiology and pharmacology of the monoamines is far from complete, no one disputes their relevance to a wide range of behavioral functions.

FUNCTIONAL ANATOMY

The range of functions served by different parts of the reticular formation is indicated in Table 24.2.

Table 24.2 Elements of the reticular formation and their perceived functions

Element	Function
Premotor cranial nerve nuclei	Patterned cranial nerve activities
Pontine locomotor center	Pattern generation
Magnocellular nuclei	Posture, locomotion
Salivatory nuclei	Salivary secretion, lacrimation
Pontine micturition center	Bladder control
Medial parabrachial nucleus	Respiratory rhythm
Central reticular nucleus of medulla oblongata	Vital centers (circulation, respiration)
Lateral medullary nucleus	Conveys somatic and visceral information to the cerebellum
Ascending reticular activating system	Arousal
Aminergic neurons	Sleeping and waking, attention and mood, sensory modulation, blood pressure control

Pattern generators

Patterned activities involving cranial nerves include the following.

- Conjugate (in parallel) movements of the eyes locally controlled by premotor nodal points (*gaze centers*) in the midbrain and pons linked to the nuclei of the ocular motor nerves (Ch. 23).
- Rhythmic chewing movements controlled by the supratrigeminal premotor nucleus in the pons (Ch. 21).
- Swallowing, vomiting, coughing, and sneezing, controlled by separate premotor nodal points in the medulla linked to the appropriate cranial nerves and to the respiratory centers.

Locomotor pattern generators are described in Box 24.1. An overview of gait controls is provided in Box 24.2. Higher-level bladder controls are described in Box 24.3.

The salivatory nuclei belong to the parvocellular reticular formation of pons and medulla. They contribute preganglionic parasympathetic fibers to the facial and glossopharyngeal nerves.

Respiratory control

The respiratory cycle is largely regulated by *dorsal* and *ventral respiratory nuclei* located at the upper end of the medulla oblongata on each side. The dorsal respiratory nucleus occupies the midlateral part of the solitary nucleus. The ventral nucleus is dorsal to the nucleus ambiguus (hence the term *retroambiguus nucleus* in Figure 17.11). A third, *medial parabrachial nucleus*, adjacent to the cerulean nucleus, seems to have a pacemaker function governing respiratory rate (cycles per minute). As will be seen in Chapter 34, stimulation of this nucleus by the amygdala, in *anxiety states*, results in characteristic hyperventilation.

Box 24.1   Locomotor pattern generators

From animal experiments, it has long been agreed that lower vertebrates and lower mammals possess *locomotor pattern generators* in the spinal cord, within the gray matter neurologically connected to each of the four limbs. These *spinal generators* comprise electrically oscillating circuits delivering rhythmically entrained signals to flexor and extensor muscle groups. Spinal generator activity is subject to supraspinal commands from a *mesencephalic locomotor area* which in turn obeys commands from motor areas of the cerebral cortex and corpus striatum.

The human *locomotor center* comprises cells of the **pedunculopontine nucleus** (Figure 17.16) and the mesencephalic part of the cerulean nucleus. These nuclei send fibers down the central tegmental tract to the oral and caudal pontine nuclei serving extensor motor neurons and to medullary magnocellular neurons serving flexor motor neurons.

A major focus of spinal rehabilitation is on activation of spinal locomotor reflexes in patients who have experienced injury resulting in partial or complete spinal cord transection. It is now well established that even after complete transection at cervical or thoracic level, a lumbosacral locomotor pattern can be activated by continuous electrical stimulation of the dura mater at lumbar segmental level. The stimulation strongly activates posterior root fibers feeding into the generator in the base of the anterior gray horn. Surface EMG recordings taken from flexor and extensor muscle groups reveal an oscillating pattern of flexor and extensor motor neuron activation, although the pattern is not identical to the normal one. A normal pattern requires that lesion is incomplete with preservation of

some supaspinal projection from the pedunculopontine nucleus.

Generation of actual stepping movements is possible in complete lesions if the individual is supported over a moving treadmill belt while the dura is being stimulated, presumably because of the additional cutaneous and proprioceptive inputs to the generator. Muscle strength and stepping speed improve over a period of weeks but not enough to enable unassisted locomotion within a walking frame.

Current research aims at improving the opportunity for supraspinal motor fibers to 'bridge the gap' by clearing tissue debris from the gap and replacing that tissue with a medium having a matrix that will support regenerating axons both physically and chemically.

REFERENCES

Blessing, W.W. (2004) Lower brainstem regulation of visceral, cardiovascular and respiratory function. In The Human Nervous System, 2nd. Edn. (Paxinos, G.and Mai, J.K., eds.), pp. 465–479. Amsterdam: Elsevier.

Calancie, B., Needham-Shropshire, B., Jacobs, P., Willer, K., Zych, G. and Green, B.A. (1994) Involuntary stepping after chronic spinal cord injury: evidence for a central rhythm generator for locomotion in man. Brain 117: 1143–1159.

Fouad, K. And Pearson, K. (2004) Restoring walking after spinal cord injury. Progr. Neurobiol. 73: 107–126.

Grasso, R., Ivanenko, Y.P., Zago, M., Molinari, M.,Scivoletto, G., Castellano, V., Macellari, V. and Lacquaniti, F. (2004) Distributed plasticity of locomotor pattern generators in spinal cord injured patients. Brain 127: 1029–1034.

Jordan, L.M. (1998) Initiation of locomotion in mammals. Ann. NY Acad. Sci. 860: 8–93.

Box 24.2 Overview of gait controls. (Assistance of Professor Tim O'Brien, Director, Gait Laboratory, Central Remedial Clinic, Dublin is gratefully acknowledged.)

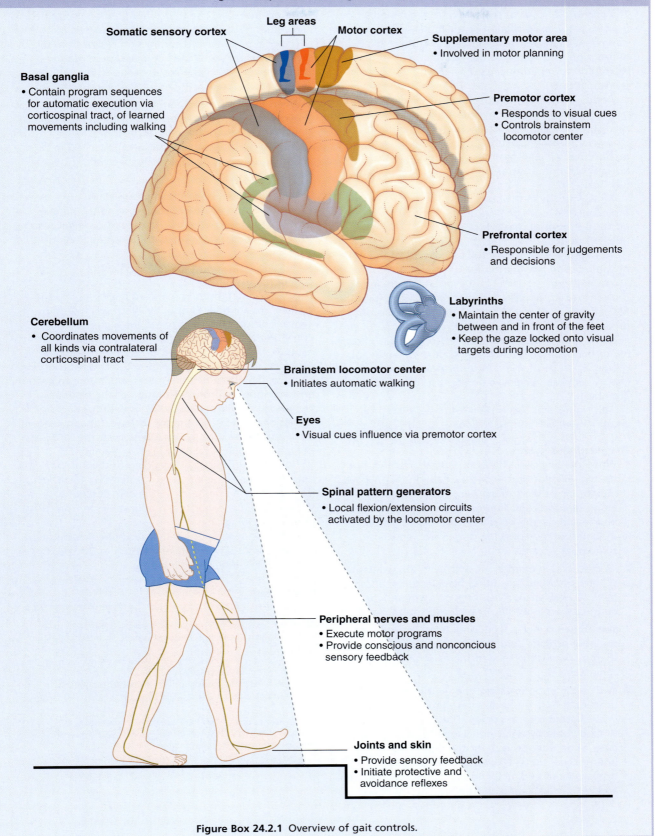

Somatic sensory cortex **Leg areas** **Motor cortex**

Supplementary motor area
• Involved in motor planning

Basal ganglia
• Contain program sequences for automatic execution via corticospinal tract, of learned movements including walking

Premotor cortex
• Responds to visual cues
• Controls brainstem locomotor center

Prefrontal cortex
• Responsible for judgements and decisions

Labyrinths
• Maintain the center of gravity between and in front of the feet
• Keep the gaze locked onto visual targets during locomotion

Cerebellum
• Coordinates movements of all kinds via contralateral corticospinal tract

Brainstem locomotor center
• Initiates automatic walking

Eyes
• Visual cues influence via premotor cortex

Spinal pattern generators
• Local flexion/extension circuits activated by the locomotor center

Peripheral nerves and muscles
• Execute motor programs
• Provide conscious and nonconcious sensory feedback

Joints and skin
• Provide sensory feedback
• Initiate protective and avoidance reflexes

Figure Box 24.2.1 Overview of gait controls.

Box 24.3 Higher-level bladder controls

The *micturition control center* is in the paramedian pontine reticular formation on each side, with interconnections across the midline. Magnocellular neurons project from here all the way to micturition-related parasympathetic neurons in segments S2–4 of the spinal cord.

In cats, stimulation of the micturition control center produces not only a rise in intravesical pressure, but also relaxation of the *rhabdosphincter* (external urethral *striated* sphincter) brought about by simultaneous excitation of GABAergic internuncials synapsing in the nucleus of Onuf in sacral segments of the spinal cord (Ch. 13).

More laterally in the pons is the L (lateral) center projecting to Onuf's nucleus. In this context, the micturition control center is referred to as the M (medial) center.

At higher levels, cells in the lateral part of the *right periaqueductal gray matter* (PAG) receive fibers ascending from the sacral posterior gray horn and project excitatory fibers to the M center. The lateral PAG also receives an excitatory input from the *right* **preoptic nucleus** in the anterior hypothalamus.

Some spinoreticular projections from the sacral cord excite the L center. Others relay via the thalamus to cells in a part of the *right* **anterior cingulate cortex** (ACCx) known to be active during tasks requiring attention.

This right-sided bias is thought to be related to emotional aspects of micturition.

The micturition cycle
1 When the bladder is half-full, vesical afferents from stretch receptors in the detrusor and in the mucous membrane of the trigone relay this information along spinoreticular fibers reaching pons, midbrain, and thalamus. The right L center, lateral PAG, and ACCx respond by glowing on positron emission tomography scans.

2 As described in Chapter 13, activity in the sympathetic system is stepped up so that *bladder compliance* can be increased (via β_2 receptors); parasympathetic neurons are silenced by α_2 neuronal interaction.

3 Spinoreticular fibers synapsing in the L nucleus of the pons activate Onuf's nucleus in the sacral cord, thereby raising the tone of the external urinary sphincter.

4 With completion of filling, there is perception of urgency. If time or place is unsuitable, part of the inferior frontal gyrus comes alive. This area puts the ACCx on hold by reducing its level of activity via association fiber projections to inhibitory internuncials there. Likewise, projections to hypothalamus and midbrain inhibit the preoptic area and PAG by activating appropriate internuncials.

5 A final measure, one that cannot be long sustained, is voluntary contraction of the entire pelvic floor. The command for this contraction is sent from the prefrontal cortex to the perineal representation on the medial side of the motor cortex in the paracentral lobule.

6 When time and place permit, the inferior frontal gyrus releases its three prisoners. The pelvic floor is allowed to sag in the manner described in Chapter 13, and the preoptic area joins PAG in activating the M nucleus while inactivating L via inhibitory internuncials.

The right-sided bias of micturition control is consistent with the clinical observation that, among stroke patients of either sex, urinary incontinence is more commonly associated with right-sided lesions of the brain.

Role of monoamines
The motor and sensory spinal cord nuclei serving the bladder are abundantly supplied with serotonergic neurons descending from the raphe magnus nucleus (MRN) in the medulla oblongata. Animal experiments have shown that local applications of serotonin are excitatory to pelvic sympathetic and sacral anterior horn neurons, inhibitory to pelvic parasympathetic and sacral posterior horn neurons. Distension of the bladder is known to stimulate the MRN (via spinoreticular activation of the PAG. A quick review of the lower-level bladder controls (Box 12.3) suggests that the MRN sets the general tone in favor of bladder filling.

Noradrenergic fibers descending to the anterior gray horn from the cerulean nucleus potentiate the effect of local glutamate release on to the cells of Onuf's nucleus, thereby enhancing rhabdosphincter tone during the filling phase.

A drug showing promise is *duloxetine*, which is both a specific serotonin reuptake inhibitor and a specific norepinephrine reuptake inhibitor.

REFERENCES

Blok FM, Holstege G. The central nervous system control of micturition in cats and humans. Behav Brain Res 1998; 92:121–125.

FitzGerald MP, Mueller E. Physiology of the lower urinary tract. Clin Obstet Gynecol 2004; 47:18–27.

Fowler CJ. Neurological disorders of micturition and their treatment. Brain 1999; 122:1213–1231.

The assistance of Dr. Mary Pat FitzGerald, Department of Urogynecology, Stritch School of Medicine, Loyola University, Chicago, is gratefully acknowledged.

Box 24.3 *Continued*

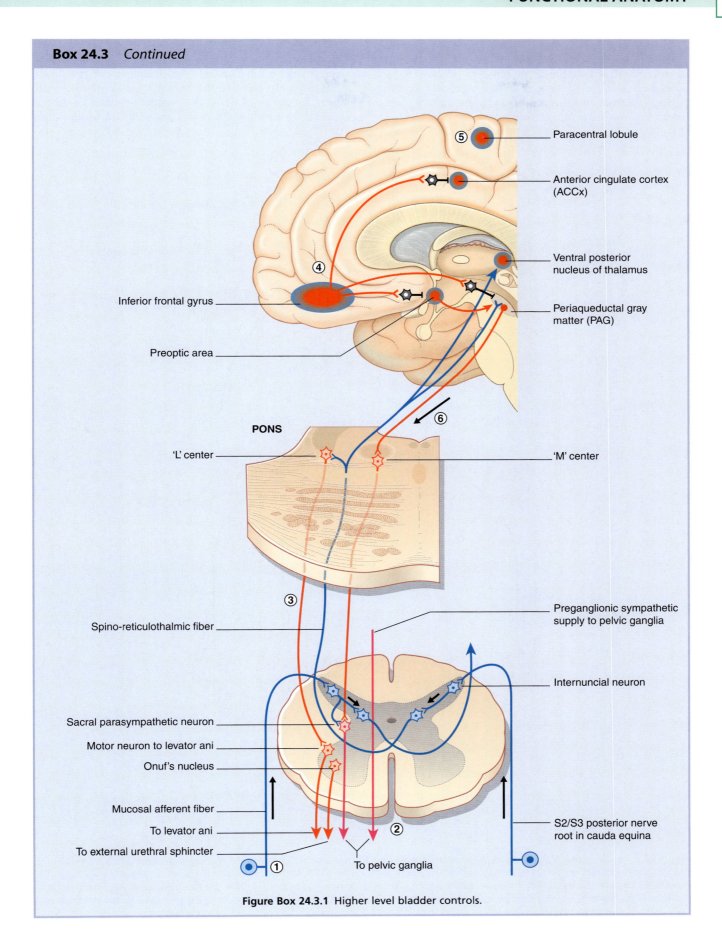

Figure Box 24.3.1 Higher level bladder controls.

The dorsal respiratory nucleus has an inspiratory function. It projects to motor neurons on the opposite side of the spinal cord supplying diaphragm, intercostals, and accessory muscles of inspiration. It receives excitatory projections from chemoreceptors in the medullary chemosensitive area and in the carotid body.

Medullary chemosensitive area

Close to the site of attachment of the glossopharyngeal nerve to the brainstem, the choroid plexus of the fourth ventricle pouts through the lateral aperture of the fourth ventricle (Figure 24.6). At this location, cells of the lateral reticular formation at the medullary surface are exquisitely

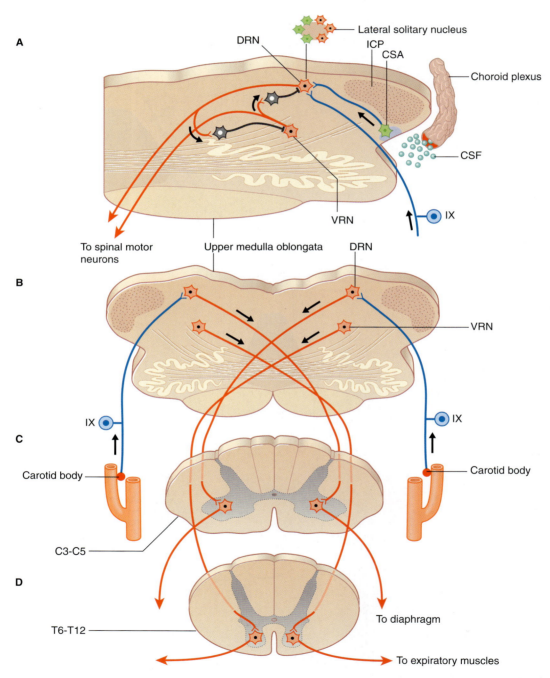

Figure 24.6 Respiratory control systems. All sections are viewed from below and behind. (A) is an enlargement taken from (B). **(A)** Inhibitory interaction between dorsal and ventral respiratory nuclei (DRN, VRN). Chorodial capillaries discharge cerebrospinal fluid (CSF) close to the medullary chemosensitive area (CSA), whence neurons project to DRN. **(B)** The glossopharyngeal nerve (IX) contains chemoreceptive neurons reaching from carotid body to DRN. **(C)** Phrenic motor neurons are activated by the contralateral DRN. **(D)** Muscles of the abdominal wall are activated by the contralateral VRN to produce forced expiration.

sensitive to the H$^+$ ion concentration in the neighboring cerebrospinal fluid. In effect, this *chemosensitive area* samples the $P\text{CO}_2$ level in the blood supplying the brain. Any increase in H$^+$ ions stimulates the dorsal respiratory nucleus through a direct synaptic linkage. (Several other nuclei within the medulla are also chemosensitive.)

Carotid chemoreceptors

The pinhead **carotid body**, close to the stem of the internal carotid artery (Figure 24.6), receives from this artery a twig that ramifies within it. Blood flow through the carotid body is so intense that the arteriovenous $P\text{O}_2$ changes by less than 1% during passage. The chemoreceptors are glomus cells to which branches of the sinus nerve (branch of IX) are applied. The carotid chemoreceptors respond to either a fall in $P\text{O}_2$ or a rise in $P\text{CO}_2$ and cause reflex adjustment of blood gas levels.

Chemoreceptors in the *aortic bodies* (beneath the aortic arch) are relatively insignificant in humans.

The ventral respiratory nucleus is expiratory (in the main). During quiet breathing, it functions as an oscillator, engaged in reciprocal inhibition (via GABAergic internuncials) with the inspiratory center. During forced breathing, it activates anterior horn cells supplying the abdominal muscles required to empty the lungs.

Cardiovascular control

Cardiac output and peripheral arterial resistance are controlled by the neural and endocrine systems. Because of the prevalence of essential hypertension in late middle age, major research efforts are under way to understand the mechanisms of cardiovascular control.

Afferents signaling increased arterial pressure arise in stretch receptors (a multitude of free nerve endings) in the wall of the carotid sinus and aortic arch (Figure 24.7). Known as *baroreceptors*, these afferents project to medially placed cells of the solitary nucleus constituting the *baroreceptor center*. Afferents from the carotid sinus travel in the glossopharyngeal nerve; those from the aortic arch travel in the vagus nerve. The baroreceptor nerves are known as 'buffer nerves', because they act to correct any deviation of the arterial blood pressure from the norm.

Cardiac output and peripheral arterial resistance depend on a balance in the activity of sympathetic and parasympathetic efferents. Two major reflexes, barovagal and barosympathetic, help to lower a raised blood pressure, as detailed in the caption to Figure 24.7.

Sleeping and wakefulness

Electroencephalography (EEG) reveals characteristic patterns in the electrical activity of cerebral cortical neurons that accompany various states of consciousness. The normal waking state is characterized by rapid, low-amplitude waves. The onset of sleep is accompanied by slow, high-amplitude waves, the higher amplitude being due to the synchronized activity of larger numbers of neurons. This type of sleep is called S (synchronized) sleep. It lasts for about 90 min before being replaced by D (desynchronized) sleep, in which the EEG pattern resembles

that of the waking state. Dreams occur during D sleep, and there are rapid eye movements (hence the more usual term, *REM sleep*). Several S and D phases occur during a normal night's sleep, as described in Chapter 30.

Details of brainstem involvement in sleep phenomena are available in psychology texts. Some salient experimental evidence is summarized.

- In animal experiments, destruction of the midbrain raphe neurons, or pharmacologic prevention of serotonin synthesis, results in insomnia lasting for several days.
- Serotonin and norepinephrine (noradrenaline) neuronal activities fluctuate in parallel. Both are most active during attentive wakefulness, sluggish during S sleep, and virtually silent during REM sleep.
- Brainstem serotonin neurons form numerous surface varicosities on the walls of the third ventricle. Serotonin liberated into the cerebrospinal fluid seems to be metabolized by hypothalamic neurons to form a sleep-inducing substance.
- Cholinergic neurons close to the cerulean nucleus are active during REM sleep, and they appear to cause the rapid eye movements by playing on the ocular motor nuclei.

Ascending reticular activating system

This term refers to the participation of reticular formation neurons in activation of the cerebral cortex, as shown by a change in EEG records from high-amplitude, slow waves to low-amplitude, fast waves during spontaneous arousal from sleep. The strongest candidates for such a role are sets of cholinergic neurons close to the cerulean nucleus (Figure 24.3). In addition to supplying the above-mentioned fibers to the ocular motor nuclei, these neurons project to almost all thalamic nuclei, and they have an excitatory effect on thalamic neurons projecting to the cerebral cortex.

The **hypothalamus** is an important control center for various body rhythms, including sleep and wakefulness. A second candidate for cortical activation has been found in the hypothalamus, namely the **tuberomammillary nucleus**. This nucleus contains histaminergic neurons having widespread projections to the cerebral cortex (Ch. 25).

Following arousal, the waking-state EEG pattern seems to be sustained by the continuing discharge of the brainstem and hypothalamic neurons mentioned; also by a third set of neurons, embedded in the basal forebrain immediately above the optic chiasm. The third set occupies the **basal nucleus of Meynert** (Ch. 33) projecting cholinergic axons to most parts of the cerebral cortex.

Sensory modulation: gate control

Sensory transmission from primary to secondary afferent neurons (at the levels of the posterior gray horn and posterior column nuclei) and from secondary to tertiary (at the level of the thalamus) is subject to *gating*. The term *gating* refers to the degree of freedom of synaptic transmission from one set of neurons to the next.

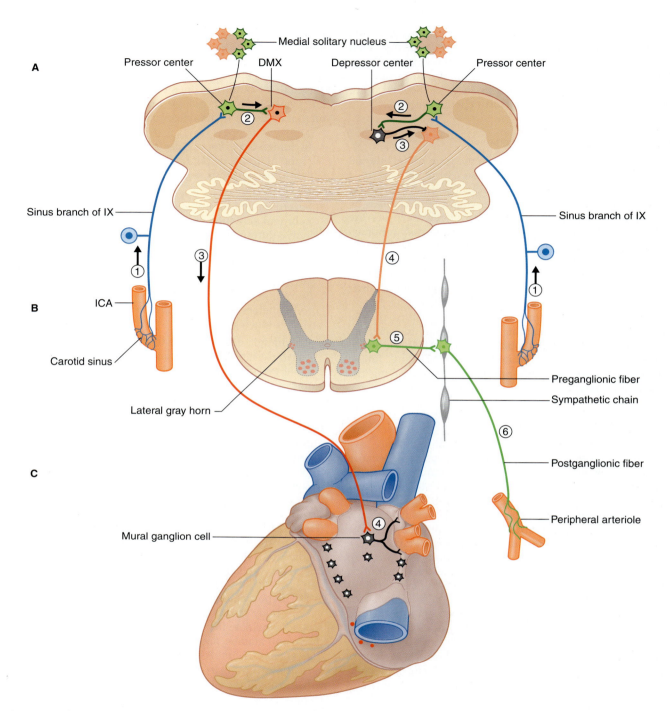

Figure 24.7 **(A)** Upper medulla oblongata; **(B)** Spinal cord segments T1–L3; **(C)** Posterior wall of heart.

Barovagal reflex (left):
(1) Stretch receptors in the carotid sinus excite fibers in the sinus branch of the glossopharyngeal nerve. ICA, internal carotid artery.
(2) Baroreceptor neurons of the solitary nucleus respond by stimulating cardioinhibitory neurons in the dorsal (motor) nucleus of the vagus (DMX). **(3)** Preganglionic, cholinergic parasympathetic vagal fibers synapse on mural ganglion cells on the posterior wall of the heart. **(4)** Postganglionic, cholinergic parasympathetic fibers reduce pacemaker activity, thus reducing the heart rate.

Barosympathetic reflex (right).
(1) Carotid sinus stretch receptor afferents excite baroreceptor medial neurons of the solitary nucleus. **(2)** Baroreceptor neurons respond by exciting the inhibitory neurons of the vasomotor depressor center in the central reticular nucleus of the medulla. **(3)** Adrenergic and noradrenergic neurons of the pressor center in the lateral reticular nucleus (rostral ventrolateral medulla) are inhibited. **(4)** Tonic excitation of the lateral gray horn is reduced. **(5)** and **(6)** Preganglionic and postganglionic sympathetic tone to peripheral arterioles is reduced, thus lowering the peripheral arterial resistance.

Tactile sensory transmission is gated at the level of the posterior column nuclei. Corticospinal neurons projecting from the postcentral gyrus may facilitate or inhibit sensory transmission at this level, as mentioned in Chapter 16.

Nociceptive transmission from the trunk and limbs is gated in the posterior gray horn of the spinal cord. From the head and upper part of the neck, it is gated in the spinal trigeminal nucleus. A key structure in both areas of gray matter is the substantia gelatinosa, which is packed with small excitatory and inhibitory internuncial neurons. The excitatory transmitter is glutamate; the inhibitory one is GABA for some internuncials, enkephalin (an opiate pentapeptide) for others.

Finely myelinated (Aδ) polymodal nociceptive fibers synapse directly on dendrites of relay neurons of the lateral spinothalamic tract and of its trigeminal equivalent. The Aδ fibers signal sharp, well-localized pain. Unmyelinated, C-fiber nociceptive afferents have mainly indirect access to relay cells, via excitatory gelatinosa internuncials. The C fibers signal dull, poorly localized pain. Most of them contain substance P, which may be liberated as a cotransmitter with glutamate.

Segmental antinociception

Large (A category) mechanoreceptive afferents from hair follicles synapse on *anterior* spinothalamic relay cells (and their trigeminal equivalents). They also give off collaterals to inhibitory (mainly GABA) gelatinosa cells that synapse in turn on *lateral* spinothalamic relay cells (Figure 24.8). Some of the internuncials also exert presynaptic inhibition on C-fiber terminals, either by axoaxonic contacts (which are very difficult to find in experimental material), or by dendroaxonic contacts. Gating of the spinothalamic response to C-fiber activity can be induced by stimulating the mechanoreceptive afferents, thereby recruiting inhibitory gelatinosa cells. This simple circuit accounts for the relief afforded by 'rubbing the sore spot'. It also provides a rationale for the use of *transcutaneous electrical nerve stimulation (TENS)* by physical therapists for pain relief in arthritis and other chronically painful conditions. The standard procedure in TENS is to apply a stimulating electrode to the skin at the same segmental level as the source of noxious C-fiber activity, and to deliver a current sufficient to produce a pronounced buzzing sensation.

Supraspinal antinociception

Magnus raphe nucleus (Figure 24.8)
From the magnus raphe nucleus (MRN) in the medulla oblongata, *raphespinal fibers* descend bilaterally within Lissauer's tract and terminate in the substantia gelatinosa at all levels of the spinal cord. In animals, electrical stimulation of the MRN may produce total analgesia throughout the body, with little effect on tactile sensation. Many fibers of the raphespinal tract liberate serotonin, which excites inhibitory internuncials in the posterior gray horn and spinal trigeminal nucleus. The internuncials induce both pre- and postsynaptic inhibition on the relevant relay cells.

There is evidence that noradrenergic projections to the posterior horn from pons and medulla are also involved in

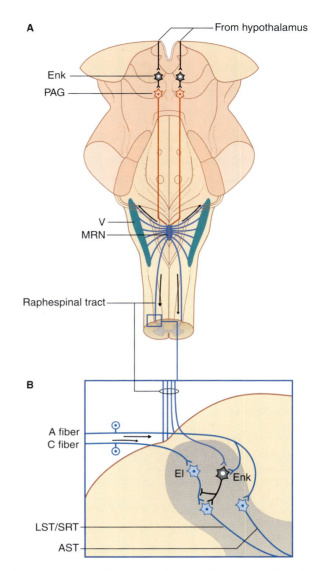

Figure 24.8 Antinociceptive pathways. **(A)** Posterior view of brainstem. **(B)** Right posterior gray horn of spinal cord, viewed from above. The periaqueductal gray matter (PAG) contains an excitatory projection to the midbrain raphe nucleus (MRN), and enkephalinergic internuncials (Enk), which exert tonic inhibition on the projection cells. Inhibitory fibers from the hypothalamus release (*disinhibit*) the excitatory neurons, and MRN responds in turn.

The effects within the spinal nucleus of trigeminal (V) and posterior gray horn are the same: serotonin liberated by MRN neurons excites enkephalinergic internuncials that inhibit nociceptive projection cells.

The nociceptive pathway at cord level is represented by the C-fiber input to an excitatory internuncial (EI), which in turn excites the lateral spinothalamic or spinoreticular projection cell (LST, SRT), unless this is being inhibited by the enkephalinergic internuncial. Rubbing the sore spot sends impulse trains along A fibers inducing the Enk cell to exert presynaptic inhibition on the EI terminal, and postsynaptic inhibition on the LST–SRT projection cell. Passage of purely tactile information into the anterior spinothalamic tract (AST) is not impeded.

supraspinal antinociception, by a direct inhibitory effect on spinothalamic neurons.

- *Diffuse noxious inhibitory controls.* The MRN is not somatotopically arranged, but it does receive inputs from spinoreticular and trigeminoreticular neurons responding to peripheral noxious stimulation. This anatomic connection accounts for what are called difuse noxious inhibitory controls. *Painful stimulation of one part of the body may produce pain relief in all other parts.* The arrangement accounts well for the heterotopic relief of pain in acupuncture (Ch. 31), where needles are used to excite nociceptive afferents in the most superficial musculature rather than in the skin.

- *Stimulus-induced analgesia.* MRN is intensely responsive to stimulation of the periaqueductal gray matter (PAG) of the midbrain. This connection has been used to advantage for patients suffering intractable pain: a fine stimulating electrode can be inserted into PAG and wired so that the patient can control the level of self-stimulation.

- *Stress-induced analgesia.* At rest, the PAG projection to the MRN is under tonic inhibition by inhibitory internuncials present within PAG. The internuncials are themselves inhibited by opioid peptides, notably by β-endorphin released from a small set of hypothalamic neurons projecting to PAG. In life-threatening situations, where injury may be the price to be paid for escape, PAG may be released (disinhibited) by the hypothalamus. This seems to be the mechanism whereby a bullet wound may be scarcely noticed in the heat of battle. (As will be seen in Ch. 34, excitatory neurons in the PAG may also be stimulated directly by the amygdala, located in the anterior temporal lobe, in fearful situations.)

In addition to the segmental and supraspinal controls of nociceptive transmission from primary to secondary afferents, gating occurs within the thalamus (see Ch. 27).

Furthermore, perception of the aversive (unpleasant) quality of pain seems to require participation of the anterior cingulate cortex (Ch. 34), which is rich in opiate receptors.

Core Information

Ground plan
The reticular formation extends the entire length of the brain stem, mainly in three cell columns. The lateral, parvocellular column receives afferents from the sensory components of all cranial and spinal nerves. It projects into the paramedian magnocellular reticular formation, which in turn sends long axons to the brain and spinal cord. The median reticular formation contains serotonergic neurons.

Aminergic neurons
Serotonergic neurons of the raphe nuclei project to all parts of the gray matter of the central nervous system (CNS). Dopaminergic neurons project from substantia nigra to striatum, and from midbrain ventral tegmental nuclei to prefrontal cortex and nucleus accumbens. Noradrenergic neurons of the cerulean nucleus project to all parts of the CNS gray matter. Epinephrine-secreting neurons of the medulla project to hypothalamus and spinal cord.

Pattern generators
Gaze centers in midbrain and pons control conjugate eye movements, a midbrain locomotor area regulates walking, a pontine supratrigeminal nucleus regulates chewing rhythm, and a pontine micturition center controls the bladder. In the medulla oblongata are respiratory, emetic, coughing, and sneezing centers, and pressor and depressor centers for cardiovascular control. In addition, the medullary chemosensitive area contains reticular formation neurons sensitive to H^+ ion levels in the cerebrospinal fluid.

Sleeping and wakefulness are influenced by serotonin and norepinephrine neurons, and by cholinergic neurons in the upper pons. The ascending reticular activating system is a physiologic concept based on brainstem neuronal networks having an arousal effect on the brain, as seen in electroencephalography traces. An important component is a set of pontine cholinergic neurons having an excitatory effect on thalamocortical neurons.

Antinociception
Segmental antinociception is induced by stimulating A fibers from hair follicles. *Supraspinal antinociception* is a function of the medullary magnus raphe nucleus (MRN), which is activated from hypothalamus and midbrain. Serotonin from MRN terminals in substantia gelatinosa of spinal posterior horn or trigeminal nucleus activates enkephalinergic internuncials that inhibit transmission in spinothalamic or trigeminothalamic neurons.

REFERENCES

Bentivoglio M, Steriade M. Brainstem–diencephalic circuits as a structural substrate of the ascending reticular activation concept. In: Mancia M, Marini G, eds. The diencephalon and sleep. New York: Raven Press: 1990:7–29.

Bianchi AL, Denavit-Saube M, Champagnet J. Central control of breathing. Physiol Rev 1995; 75:1–45.

Chalmers J, Pilowski P. Brainstem and bulbospinal systems in the control of blood pressure. J Hypertens 1991; 9:675–694.

Edgerton VR, Roy RR. Paralysis recovery in humans and model systems. Curr Opin Neurobiol 2002; 12:658–667.

Gonzalez C, Almaraz L, Obeso A, et al. Carotid body chemoreceptors: from natural stimuli to sensory discharges. Physiol Rev 1994; 74:829–898.

Rosenfeld JP. Interacting brain components of opiate-activated, descending, pain-inhibitory systems. Neurosci Behav Rev 1994; 18:403–409.

Siddall PJ. Pain mechanisms and management. Clin Exp Pharmacol Physiol 1995; 22:679–688.

Cerebellum

STUDY GUIDELINES

1 The cerebellar cortex has much the same microscopic structure throughout.

2 Afferents to the cerebellar cortex enter from spinal cord and brainstem. They are excitatory, whereas the cortical outputs are inhibitory, and are exclusively from Purkinje cells.

3 The outputs from the cerebellum as a whole are in general excitatory. This is because they originate from deep, excitatory nuclei that are only partially inhibited by Purkinje cells.

4 Of special research interest are the roles played by the inferior olivary and red nuclei in relation to novel motor programs.

5 Clinically, cerebellar disease is characterized by incoordination of movement, with some differences being evident depending on the mediolateral position of the lesion where this is discrete (e.g. a tumor).

Phylogenetically, the initial development of the cerebellum (in fishes) took place in relation to the vestibular labyrinth. With development of quadrupedal locomotion, the anterior lobes (in particular) became richly connected to the spinal cord. Assumption of the erect posture and achievement of a whole new range of physical skills have been accompanied by the appearance of massive linkages between the posterior lobes and the cerebral cortex. In general, cerebellar connections with the labyrinth, spinal cord, and cerebral cortex are arranged such that each cerebellar hemisphere is primarily concerned with the coordination of movements *on its own side*.

The gross anatomy of the cerebellum is described briefly in Chapter 3, where it may be reviewed at this time.

FUNCTIONAL ANATOMY

Phylogenetic and functional aspects can be combined (to an approximation) by dividing the cerebellum into strips, as shown in Figure 25.1. The median strip contains the cortex of the vermis, together with the **fastigial nucleus** in

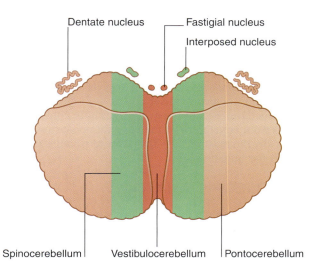

Figure 25.1 Zonation of cerebellum. The central nuclei are represented separately.

the white matter close to the nodule (Figure 25.2). This strip is the *vestibulocerebellum*; it has two-way connections with the vestibular nucleus. It controls the responses of that nucleus to signals from the vestibular labyrinth. The fastigial nucleus also projects to the gaze centers of the brainstem (Ch. 24).

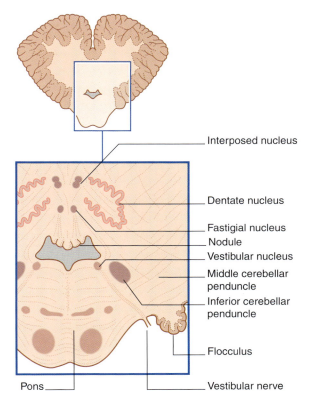

Figure 25.2 Transverse section of lower pons and cerebellum, showing the position of the central and vestibular nuclei.

A paramedian strip, the *spinocerebellum*, includes the paravermal cortex and the **globose** and **emboliform nuclei** (Figure 25.2). The two nuclei are together called the **interposed nucleus**. The spinocerebellum is rich in spinocerebellar connections. It is involved in the control of posture and gait.

The remaining, lateral strip is much the largest and takes in the wrinkled **dentate nucleus** (Figure 25.2). This strip is the *pontocerebellum*, because it receives a massive input from the contralateral nuclei pontis. It is also called the **neocerebellum**, because the nuclei pontis convey information from large areas of the cerebral neocortex (phylogenetically the most recent). The neocerebellum is uniquely large in the human brain.

MICROSCOPIC ANATOMY

The structure of the cerebellar cortex is uniform throughout. From within outward, the cortex comprises granular, piriform, and molecular layers (Figure 25.3).

The **granular layer** contains billions of **granule cells**, whose somas are only 6–8 μm in diameter. Their short dendrites receive so-called **mossy fibers** from all sources except the inferior olivary nucleus. Before reaching the cerebellar cortex, the mossy fibers, which are excitatory in nature, give off collateral branches to the central nuclei.

The axons of the granule cells penetrate to the molecular layer, where they divide in a T-shaped manner to form **parallel fibers**. The parallel fibers run parallel to the axes of the folia. They make excitatory contacts with dendrites of Purkinje cells.

The granular layer also contains Golgi cells (see later).

The **piriform layer** consists of very large **Purkinje cells**. The fan-shaped dendritic trees of the Purkinje cells are the largest dendritic trees in the entire nervous system. The fans are disposed at right angles to the parallel fibers.

The dendritic trees of Purkinje cells are penetrated by huge numbers of parallel fiber axons of granule cells, each one making successive, one per cell, synapses on dendritic spines of about 400 Purkinje cells. Not surprisingly, stimulation of small numbers of granule cells by mossy fibers has a merely facilitatory effect on Purkinje cells. Many thousands of parallel fibers must act simultaneously to bring the membrane potential to firing level.

Each dendritic tree also receives a single **climbing fiber** from the contralateral inferior olivary nucleus. In stark contrast to the one-per-cell synapses of parallel fibers, the olivocerebellar fiber divides at the Purkinje dendritic branch points and makes thousands of synaptic contacts with dendritic spines. A single threshold pulse applied to one climbing fiber is sufficient to elicit a short burst of action potentials from the client Purkinje cell. Climbing fiber effects on Purkinje cells are so powerful that, for some time after they cease firing, the synaptic effectiveness of bundles of parallel fibers is reduced. In this sense, the Purkinje cells *remember* that they have been excited by olivocerebellar fibers.

The axons of the Purkinje cells are the only axons to emerge from the cerebellar cortex. Remarkably, they are entirely inhibitory in their effects. Their principal targets are the central nuclei. They give off collateral branches also, mainly to Golgi cells.

The **molecular layer** is almost entirely taken up with Purkinje dendrites, parallel fibers, supporting neuroglial cells, and blood vessels. However, two sets of inhibitory neurons are also found there, lying in the same plane as the Purkinje cell dendritic trees. Near the cortical surface are small, **stellate cells**, and close to the piriform layer are

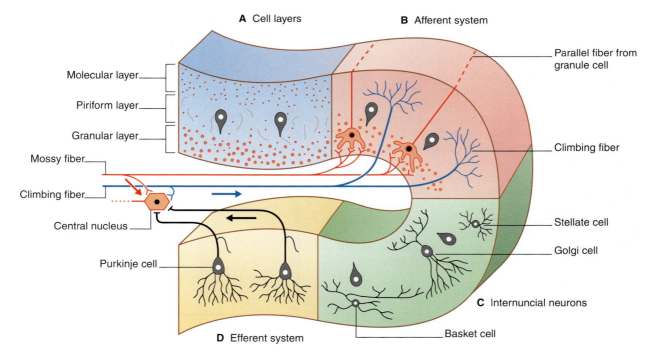

Figure 25.3 Cerebellar cortex. **(A)** Cell layers. **(B)** Afferent system. **(C)** Internuncial neurons. **(D)** Efferent system.

larger, **basket cells**. Both sets are contacted by parallel fibers, and they both synapse on Purkinje cells. The stellate cells synapse on dendritic shafts, whereas the basket cells form a 'basket' of synaptic contacts around the soma, as well as forming axoaxonic synapses on the initial segment of the axon. A single basket cell synapses on some 250 Purkinje cells.

The final cell type in the cortex is the **Golgi cell**, whose dendrites are contacted by parallel fibers and whose axons divide extensively before synapsing on the short dendrites of granule cells. The synaptic ensemble that includes a mossy fiber terminal, granule cell dendrites, and Golgi cell boutons is known as a **glomerulus** (Figure 25.4). See also Figure 25.5.

Spatial effects of mossy fiber activity
(Figure 25.6)

As already noted, cerebellar afferents other than olivocerebellar ones form mossy fiber terminals after giving off excitatory collaterals to one of the deep nuclei. The afferents excite groups of granule cells, which in turn facilitate many hundreds of Purkinje cells. Along most of the beam of excitation, known as a *microzone*, the Purkinje cells begin to fire and to inhibit patches of cells in one of the

deep nuclei. At the same time, weakly facilitated Purkinje cells along the edges of the microzone are shut off by stellate and basket cells. As a result, the beam of excitation is sharply focused. The excitation is terminated by Golgi cell inhibition of the granule cells that initiated it. Powerful excitation will last longer, because highly active Purkinje cells inhibit underlying Golgi cells through their collateral branches.

REPRESENTATION OF BODY PARTS

Representation of body parts in the human cerebellar cortex is currently under investigation by means of functional magnetic resonance imaging (MRI). These investigations, along with some evidence from clinical cases, indicate the presence of somatotopic maps in the anterior and posterior lobes (Figure 25.7).

The maps have been worked out in some detail in laboratory animals during movements. The maps for movement match up with maps of skin, eye, ear, and visceral representation worked out by stimulation of body parts. The expression *fractionated somatotopy* refers to the patchy nature of the representation of body parts. Simple representations like those in Figure 25.7 are likely to be quite inaccurate in view of the vast areas of cortex buried in the fissures.

Figure 25.8, based on positron emission tomography (PET) scans, shows simultaneous activation of the left motor cortex and right cerebellum during repetitive movements of the fingers of the right hand.

See also *The cerebellum and higher brain functions*, later.

AFFERENT PATHWAYS

From the muscles and skin of the trunk and limbs, afferent information travels in the posterior spinocerebellar and the cuneocerebellar tract, and enters the inferior cerebellar peduncle on the same side. Comparable information from the territory served by the trigeminal nerve enters all three cerebellar peduncles.

Afferents from spinal reflex arcs run in the anterior spinocerebellar tract, which reaches the upper pons before looping into the superior cerebellar peduncle.

Special sense (visual, auditory, vestibular) pathways comprise tectocerebellar fibers entering the superior peduncle from the ipsilateral midbrain colliculi, and vestibulocerebellar fibers from the ipsilateral vestibular nucleus.

Two massive pathways enter from the contralateral brainstem. The pontocerebellar tract enters through the middle peduncle, and the olivocerebellar tract enters through the inferior peduncle.

Reticulocerebellar fibers enter the inferior peduncle from the paramedian and lateral reticular nuclei of the medulla oblongata.

Finally, aminergic fibers enter all three peduncles from noradrenergic and serotonergic cell groups in the brainstem. Under experimental conditions, both kinds of neuron appear to facilitate excitatory transmission in mossy and climbing fiber terminals.

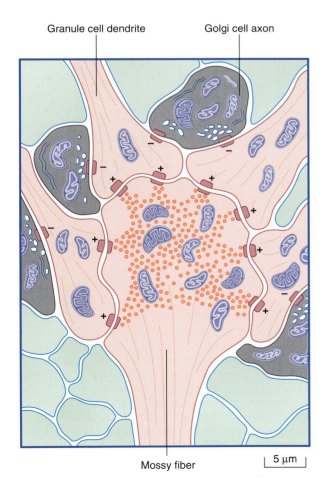

Granule cell dendrite   Golgi cell axon

Mossy fiber   ⌐ 5 μm ⌐

Figure 25.4 A synaptic glomerulus. +, − indicate excitation, inhibition.

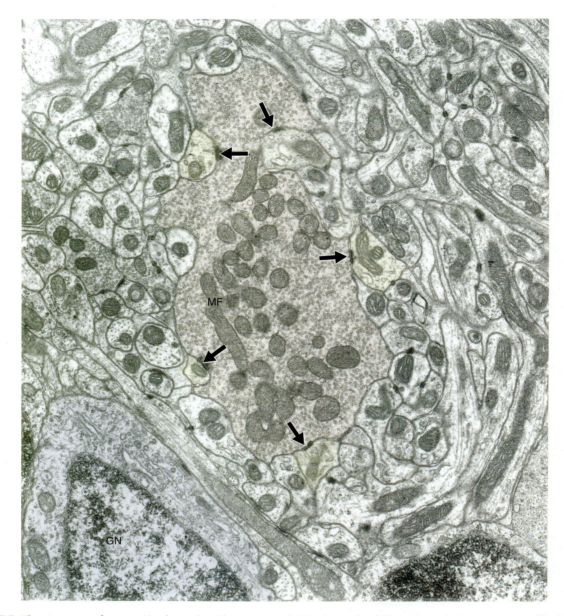

Figure 25.5 Ultrastructure of a synaptic glomeulus. The arrows point to six axodendritic synapses between a mossy fiber (MF) and granule cells. GN, nucleus of granule cell. (Reproduced with permission from Pannese, E. (1994) *Neurocytology*. Fine Structure of Neurons, *Nerve Processes and Neuroglial Cells*. New York: Thieme.)

Olivocerebellar tract

The sensorimotor cortex projects, via corticospinal collaterals, in an orderly, somatotopic manner on to the ipsilateral inferior and accessory olivary nuclei. The order is preserved in the olivary projections on to the body maps in the contralateral cerebellar cortex (from principal nucleus to the posterior map, from the accessory nuclei to the anterior map). Under resting conditions in animal experiments, groups of olivary neurons discharge synchronously at 5–10 Hz (impulses/second). The synchrony is probably due to the observed presence of electrical synapses (gap junctions) between dendrites of neighboring neurons. In the cerebellar cortex, the response of Purkinje cells takes the form of *complex spikes* (multiple action potentials in response to single pulses), because of the spatiotemporal effects of climbing fiber activity along the branches of the dendritic tree.

When a monkey has been trained to perform a motor task, increased discharge of Purkinje cells during task performance takes the form of simple spikes produced by bundles of active parallel fibers. If an unexpected obstacle is introduced into the task (e.g. momentary braking of a lever that the monkey is operating), bursts of complex spikes occur each time the obstacle is encountered. As the animal learns to overcome the obstacle so that the task is completed in the set time, the spike bursts dwindle in number and finally disappear. This is just one of several experimental indicators that the inferior olivary nucleus has a significant *teaching function* in the acquisition of new motor skills.

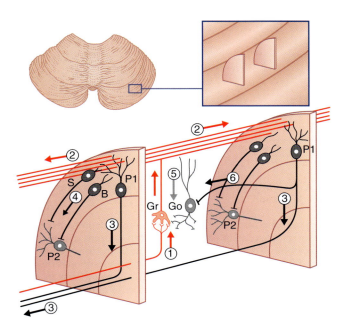

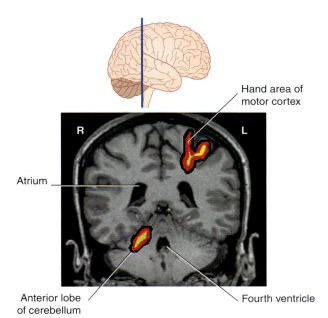

Figure 25.6 Scheme of effects of mossy fiber activity.
1 Mossy fiber stimulating a granule cell (Gr).
2 Beam of parallel fiber activity follows simultaneous activation of many granule cells.
3 Activation of on-line Purkinje cells (P1) results in selective inhibition of neurons within the appropriate central cerebellar nucleus.
4 Activation of stellate (S) and basket cells (B) inhibits off-line Purkinje cells (P2).
5 Golgi cells (Go) terminate granule cell activity.
6 Intense on-line activity can be sustained by inhibition of Golgi cells by Purkinje cells.

Figure 25.8 Representation of functional MRI activity (viewed from behind) in a volunteer executing repetitive movement of the fingers of the right hand. (From a series kindly provided by Professor J. Paul Finn, Director, Magnetic Resonance Research, Department of Radiology, David Geffen School of Medicine at UCLA, California.)

both from cortical fibers descending to the olive and from cerebellar output fibers ascending to the thalamus. Much the largest output from the red nucleus is to the ipsilateral olive, which it appears to inhibit. On detection of a mismatch between a movement intended and a movement organized, the red nucleus could release the appropriate cell groups in the olive until the two are harmonized.

As mentioned in Chapter 15, *motor adaptation* is primarily a function of the cerebellum. Here the cerebellum oversees modification of routine motor programs in response to changes in the environment, for example walking uphill versus walking on the flat. Experimental evidence indicates that prolonged motor adaptation, for example walking over a period of weeks while wearing an ankle plaster cast, is accompanied by long-term potentiation of cerebellothalamic synapses, thereby facilitating the influence of the cerebellum on the motor cortex.

Motor sequence learning, for example learning to walk during infancy, is a function of the basal ganglia (Ch. 33).

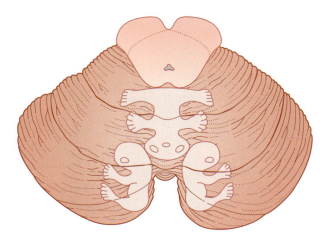

Figure 25.7 Upper surface of cerebellum, showing position of somatotopic maps based on animal experiments.

The olive receives direct ipsilateral projections from the premotor and motor areas of the cerebral cortex, and from the visual association cortex, providing an apparently suitable substrate for its activities. It is also in touch with the outside world through the spinoolivary tract (Ch. 15).

In theory, the red nucleus of the midbrain could function as a *novelty detector*, because it receives collaterals

EFFERENT PATHWAYS
(Figures 25.9 and 25.10)

From the *vestibulocerebellum* (fastigial nucleus), axons project to the vestibular nuclei of both sides, through the inferior cerebellar peduncle. The contralateral projection crosses over within the cerebellar white matter.

Vestibulocerebellar outputs to the medial and superior vestibular nuclei control movements of the eyes through the medial longitudinal fasciculus (Chs 17 and 23). A separate output to the lateral vestibular nucleus (of Deiters) of the same side controls the balancing function of the

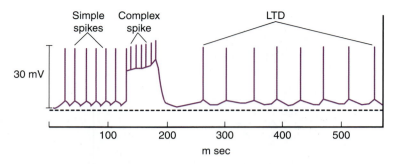

Figure 25.9 Recording from a Purkinje cell dendrite. The complex spike elicited by activation of a climbing fiber has depressed the frequency of the simple spikes elicited by a parallel fiber. LTD, long term depression.

Motor learning is believed to be achieved by means of a phenomenon called *long term depression*. This refer to depression of ongoing parallel fiber activity for up to several hours, following a burst of complex spikes (Figure 25.9). Both neurons concerned are glutamatergic and Purkinje dendrites possess both AMPA and metabotropic receptors. The key molecule in the interaction is the second messenger protein kinase C (PKC), which is activated by parallel fiber activity and mediates protein phosporylation in ion channels. The molecular sequence is as illustrated in Ch. 8 Figure 8. Complex spikes are associated with a large increase in intracellular calcium and this interacts with PKC to diminish the postsynaptic response of the AMPA receptors to glutamate stimulation.

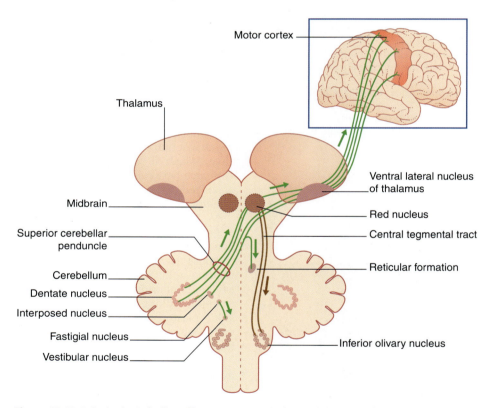

Figure 25.10 Principal cerebellar efferents. Arrows indicate directions of impulse conduction.

vestibulospinal tract. Some Purkinje axons skirt the fastigial nucleus and exert direct tonic inhibition on Deiters' nucleus.

From the interposed nucleus of the *spinocerebellum*, axons emerge in the superior cerebellar peduncle. They terminate mainly in the contralateral reticular formation and red nucleus. Those reaching the pontomedullary reticular formation regulate the functions of the reticulo-

spinal tracts in relation to posture and locomotion. Those ascending to the red nucleus may be involved in motor learning.

From the *neocerebellum*, the massive **dentatorubrothalamic tract** forms the bulk of the superior cerebellar peduncle. It decussates with its opposite number in the lower midbrain and gives collaterals to the red nucleus before synapsing in the ventral lateral nucleus of the thalamus. The onward projection from the thalamus is to the motor cortex.

ANTICIPATORY FUNCTION OF THE CEREBELLUM

The cerebellum has a sophisticated function in relation to *postural stabilization* and *postural fixation*, as indicated by the following examples.

Postural stabilization

Figure 25.11 illustrates anticipatory contraction of the gastrocnemius serving to stabilize a trunk about to receive a displacement impetus produced by contraction of the biceps brachii. In more general terms, displacement of the upper trunk away from the center of gravity by a voluntary movement of the head or upper limb is *anticipated* by the cerebellum. Having read instructions delivered from premotor areas of the frontal lobe (Ch. 29) concerning the *intended* movement, the cerebellum ensures proportionate contractions of postural muscles in a bottom-up manner, from leg to thigh to trunk, in order to keep the center of gravity in the midline between the feet. Damage to the cerebellar vermis affects normal anticipatory activation, via the lateral vestibulospinal tract, of slow-twitch, close-to-the-bone muscle bundles, with consequent failure to counter the effect of gravity displacement produced by movement of any body part (see Clinical Panel 25.1).

Damage to the anterior lobe is associated with failure of the reticulospinal tracts to anticipate the gravitational effects produced by locomotion (see Clinical Panel 25.2).

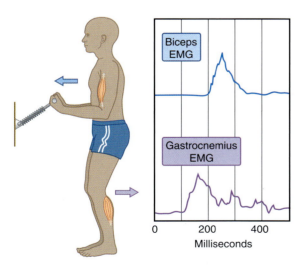

Figure 25.11 *Postural stabilization*. The subject is pulling a stiff spring attached to the wall. Flexion of the elbow during contraction of biceps brachii tends to pull the trunk forward (*arrow*). This movement is prevented by equivalent contraction of the gastrocnemius, exerting downward pressure on the forefoot, which tends to thrust the trunk backward (*arrow*). Simultaneous electromyographic (EMG) recordings show that onset of (automatic) gastrocnemius contraction precedes voluntary biceps contraction by 80 ms. (After Nashner, with permission.)

Postural fixation

Figure 25.12 illustrates an experiment where the subject was instructed to execute sudden wrist extension and to maintain the extended wrist posture for 2 s, while electromyographic records were being taken from prime wrist extensors (extensors carpi radialis longus and brevis) and a prime antagonist (flexor carpi radialis). The readout revealed that the *antagonist* began to contract prior to completion of the movement, and that it played 'shivering ping pong' with the prime mover during the fixation period. The contribution of the antagonist is to prevent spontaneous oscillatory torques (tremors) caused by viscoelastic properties of the muscles. It has been shown that this 'freeze' arrangement can be disrupted in healthy volunteers by transcranial electromagnetic stimulation aimed at the superior cerebellar peduncle, and in disease of the lateral cerebellar lobe (see Clinical Panel 25.3).

CLINICAL DISORDERS OF THE CEREBELLUM

Diseases involving the cerebellum usually involve more than one lobe and/or more than one of the three sagittal strips. However, characteristic clinical pictures have been described in association with lesions of the vermis (Clinical Panel 25.1), of the anterior lobe (Clinical Panel 25.2), and of the neocerebellum (Clinical Panel 25.3).

THE CEREBELLUM AND HIGHER BRAIN FUNCTIONS

Positron emission tomography and functional MRI provide information about regional changes in blood flow and oxygen consumption. 'Movement maps' such as those in Figure 25.8 are derived from simple repetitive movements such as opening and closing a fist. A striking feature of movement maps is *how small* and *how medial* they are. Prior to PET, it was assumed that the lateral expansion of the human posterior lobe was necessary for manual dexterity. It now appears that the lateral expansion may be associated with cognitive functions (e.g. thinking), having an anatomic base in linkages with the lateral pre-frontal cortex of the cerebral hemisphere. Lateral cerebellar activity seems to be greatest during speech, with a one-sided predominance consistent with a possible linkage (via the thalamus) with the motor speech area of the dominant frontal cortex (Ch. 30). Something more than mere motor control may be involved, because lateral cerebellar activity is greater during functional naming (e.g. 'dig', 'fly') than during object identification (e.g. 'shovel', 'airplane').

Cerebellar cognitive affective syndrome is the summary term recently introduced to indicate cerebral functional deficits that follow sudden severe damage to the cerebellum, for example thrombosis of one of the three pairs of cerebellar arteries, or the unavoidable damage inflicted during removal of a cerebellar tumor. Such patients show *cognitive* defects in the form of diminished reasoning power, inattention, grammatical errors in speech, poor spatial sense, and patchy memory loss. If the vermis is included in the damage, *affective* (emotional) symptoms

Clinical Panel 25.1 Midline lesions: truncal ataxia

Lesions of the vermis occur most often in children, in the form of medulloblastomas in the roof of the fourth ventricle. These tumors expand rapidly and produce signs of raised intracranial pressure: headache, vomiting, drowsiness, papilledema. In the recumbent position, there may be no abnormality of motor coordination in the limbs. A dramatic feature is an inability to stand upright without support—a state of *truncal ataxia*. This tumor, which is highly sensitive to radiotherapy, attacks the pathway from the vermis to the nucleus of the vestibular nerve. The ataxia reflects malfunction of the lateral vestibular nucleus and consequently of the (lateral) vestibulospinal tract. Deficient antigravity function in this uncrossed pathway causes the child to fall to the more affected side on attempting to stand or walk.

Nystagmus can usually be elicited on visual tracking of the examiner's finger from side to side. Scanning movements of the eyes are also inaccurate owing to poor control of the gaze centers by the vermis.

Clinical Panel 25.2 Anterior lobe lesions: gait ataxia

Disease of the anterior lobe is most often observed in chronic alcoholics. Postmortem studies reveal pronounced shrinkage of the cortex of the anterior lobe, with up to 10% loss of granule cells, 20% loss of Purkinje cells, and 30% reduction in the thickness of the molecular layer. The lower limbs are most affected, and a staggering, drunken gait is evident even when the individual is sober. Some degree of correction may be exercised by voluntary control.

Instability of station with the feet together, and failure to 'toe the line' on walking, are present even when the eyes are open. A head tremor at 3 Hz is usually present. As the disease progresses, a peripheral sensory neuropathy may be added, giving rise to signs of sensory ataxia (Ch. 15) in addition. Tendon reflexes may be depressed in the lower limbs, owing to loss of tonic stimulation of fusimotor neurons via the pontine reticulospinal tract. Consequent reduction of monosynaptic reflex activity during walking may eventually result in stretching of soft tissues, with hyperextension of the knee joint during standing.

Clinical Panel 25.3 Neocerebellar lesions: incoordination of voluntary movements

Disease of the neocerebellar cortex, dentate nucleus, or superior cerebellar peduncle leads to incoordination of voluntary movements, particularly in the upper limb. When fine purposive movements are attempted (e.g. grasping a glass, using a key), an *intention tremor* (*action tremor*) develops: the hand and forearm quiver as the target is approached, owing to faulty agonist–antagonist muscle synergies around the elbow and wrist. The hand may travel past the target ('overshoot'). Because cerebellar guidance is lost, the normal smooth trajectory of reaching movements may be replaced by stepped flexions, abductions, etc. ('decomposition of movement').

Rapid alternating movements performed under command, such as pronation–supination, become quite irregular (*dysdiadochokinesia*). The 'finger-to-nose' and 'heel-to-knee' tests are performed with equal clumsiness whether the eyes are open or closed—in contrast to performance in posterior column disease, where performance is adequate when the eyes are open (Ch. 15).

Speech is impaired with regard to both phonation and articulation. Phonation (production of vowel sounds) is uneven and often tremulous owing to loss of smoothness of contraction of the diaphragm and the intercostal muscles. The terms 'explosive' and 'scanning' have been applied to this feature. Articulation is slurred because of faulty coordination of impulses in the nerves supplying the lips, mandible, tongue, palate, and infrahyoid muscles.

Signs of neocerebellar disorder sometimes originate in the midbrain or pons rather than in the cerebellum itself. The lesion responsible (usually vascular) interrupts one or other cerebellothalamic pathway (or both, if the lesion is at the decussation of the superior cerebellar peduncles).

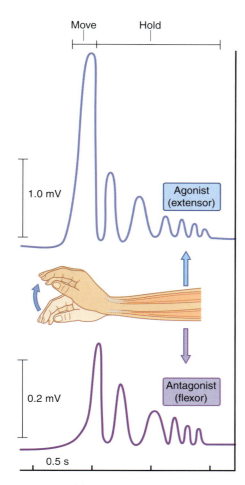

Figure 25.12 *Postural fixation*. The subject was instructed to perform sudden wrist extension and to briefly hold the extended posture. Electromyographic recordings show that wrist flexors come into action before completion of the movement. In the 'hold' position, note alternation of electrical activity between agonist and antagonist. Antagonist electromyographic activity is much weaker, as indicated by the scale bars on the left. (After Topke et al. 1999, with permission.)

appear, sometimes in the form of *flatness of affect* (dulling of emotional responses), other times in the form of aberrant emotional behavior. The cognitive affective syndrome is temporary, and it is of interest that it my be associated with reduction of blood flow (on PET) in one or more of the association areas linked to the cerebellum by corticopontocerebellar fibers. Recent studies in monkeys have shown that, in addition to its well-known thalamocortical projection to the motor cortex, the cerebellum also 'drives' thalamic neurons projecting to association areas serving cognitive and affective functions.

Posturography

Posturography is the instrumental recording of the erect posture. The subject stands on a platform, and spontaneous body sway is detected by strain gauges beneath the corners of the platform. Linkage of the strain gauge data to a computer can yield a graphic record of anteroposterior and side-to-side sway, first with the eyes open and then with the eyes closed. This is *static posturography*, and it helps to distinguish between different causes of ataxia.

Dynamic posturography provides information on the effects of an abrupt 4° backward tilt of the supporting platform. For this phase of the examination, surface electromyographic electrodes are applied over the calf muscles (ankle plantar flexors) and over the tibialis anterior (an ankle dorsiflexor). The normal response to the backward tilt is threefold: (a) a monosynaptic, spinal, stretch reflex contraction of the calf muscles after 45 ms; (b) a polysynaptic stretch reflex contraction of the calf muscles after 95 ms; and (c) a long-loop, reflex contraction of the ankle dorsiflexors after 120 ms. The ascending limb of the long loop is via the tibial–sciatic nerve and the posterior column–medial lemniscal pathway to the somatosensory cortex; the descending limb is via the corticospinal tract and the sciatic-peroneal nerve. Dynamic posturography helps to distinguish among a wide variety of disorders affecting different levels of the central nervous system and peripheral nervous system.

Core Information

The cerebellum is primarily concerned with coordination of movements on its own side of the body. Therefore disease in one cerebellar hemisphere leads to incoordination of limb movements on that side.

The cerebellar cortex contains a thick inner layer of tiny granule cells, a piriform layer of Purkinje cells, and a molecular layer containing granule cell axons and Purkinje dendrites. Granule cells are excitatory to Purkinje cells (via parallel fibers) but Purkinje cells—the only output cells of the cortex—are inhibitory to the central nuclei, which themselves are excitatory. Inhibitory purely cortical neurons are the stellate, basket, and Golgi cells.

The two types of afferents to the cortex are (a) mossy fibers from all sources except the olive—they excite granule cells; and (b) climbing fibers from the olive, which powerfully excite Purkinje cells.

The basic input–output circuit is mossy fibers → granule cells → Purkinje cells → deep nucleus → brainstem or thalamus. Olivocerebellar neurons are most active during novel learning; they elicit poststimulus depression of the Purkinje cell response to mossy fiber activity—a feature surely related to motor learning. The red nucleus is in a position to match the intended input to the cerebellum with the output achieved after passage through the basic circuit.

Functional parts

Vestibulocerebellum comprises vermis and fastigial nuclei, having two-way connections with the vestibular nucleus. It may be affected by midline tumors, yielding nystagmus and truncal ataxia.

Spinocerebellum, next to vermis and including much of the anterior lobe, includes the interposed nucleus. It receives spinocerebellar pathways, and it controls posture and gait. Lesions are characterized by ataxia of stance and gait.

Neocerebellum is largest and most lateral, receiving the corticopontocerebellar system. The dentate nucleus projects to the contralateral motor cortex via thalamus, and to the contralateral red nucleus. Lesions result in ipsilateral incoordination, notably of the upper limb, and to faulty phonation and articulation.

REFERENCES

Aumann TD. Cerebello-thalamic synapses and motor adaptation. Cerebellum 2002; 46:69–77.
Bastian AJ, Thach WT. Structure and function of the cerebellum. In: Manto U-B, Pandalfo M, eds. The cerebellum and its disorders. Cambridge: University Press; 2002:49–68.
Burke D, Gandevia SC. Muscle spindles, muscle tone and the fusimotor system. In: Gandevia SC, Burke D, Anthony M, eds. Science and practice in clinical neurology. Cambridge: Cambridge University Press; 1993:89–105.
Doyon J, Benali H. Reorganization and plasticity in the adult brain during learning of motor skills. Curr Opin Neurobiol 2005; 15:161–167.
Ebner TJ, Bloedel JR. Climbing fiber afferent system: intrinsic properties and role in cerebellar information processing. In: King JS, ed. New concepts in cerebellar neurobiology. New York: Alan R Lissl; 1987:371–386.
FitzGerald MJT. The cerebellum. In: Standring S, et al, eds. Gray's anatomy. 39th edn. Edinburgh: Churchill Livingstone; 2005:353–368.
Fox PT, Raichle ME, Thach WT. Functional mapping of the human cerebellum with positron emission tomography. Proc Natl Acad Sci USA 1985; 82:7462–7466.
Horne MK, Butler EG. The role of the cerebello-thalamocortical pathway in skilled movements. Prog Neurobiol 1995; 46:190–213.
Houk JC, Gibson AR. Sensorimotor processing through the cerebellum. In: King JS, ed. New concepts in cerebellar neurobiology. New York: Alan R. Liss; 1987:387–416.
Ito M. Cerebellar long-term depression: characterization, signal transduction, and functional roles. Phys Rev 2001; 81:1143–1195.
Kennedy PR. The rubro-olivo-cerebellar teaching circuit. Med Hypotheses 1979; 5:799–807.
Leiner HC, Leiner AL, Dow RS. The human cerebro-cerebellar system: its computing, cognitive, and language skills. Behav Brain Res 1991; 44:113–128.
Middleton FA, Strick PL. Cerebellar output channels. In: Schmahmann JD, ed. The cerebellum and cognition. International review of neurobiology, vol 41. San Diego: Academic Press; 1997:255–271.
Nitschke MF, Kleinschmidt A, Wessel K, et al. Somatotopic motor representation in the human anterior cerebellum. Brain 1996; 119:1023–1029.
Schmahmann JD, Sherman JC. The cerebellar cognitive affective syndrome. Brain 1998; 121:561–579.
Topke H, Mescheriakov S, Boose A, et al. A cerebellar-like terminal and postural tremor induced in normal man by transcranial magnetic stimulation. Brain 1999; 125:1551–1562.
Woogd J, Glickstein M. The anatomy of the cerebellum. Trends Neurosci 1998; 21:370–375.

Hypothalamus

STUDY GUIDELINES

1 Hypothalamic neuroendocrine cells fulfill the basic criteria both for neurons and for endocrine cells. Small neuroendocrine cells control release of hormones by the purely endocrine cells of the anterior pituitary gland into the cavernous sinus. Large ones have their terminals in the posterior pituitary, where they release hormones into the cavernous sinus directly.

2 Some neurons confined to the hypothalamus are involved in control of body temperature, food and fluid intake, and sleep. Others, involved in attack and defense responses and in memory, are controlled by the limbic system.

The hypothalamus develops as part of the limbic system, which is concerned with preservation of the individual and of the species. Therefore it is logical that the hypothalamus should have significant controls over basic survival strategies, including reproduction, growth and metabolism, food and fluid intake, attack and defense, temperature control, the sleep–wake cycle, and aspects of memory.

Most of its functions are expressed through its control of the pituitary gland and of both divisions of the autonomic nervous system.

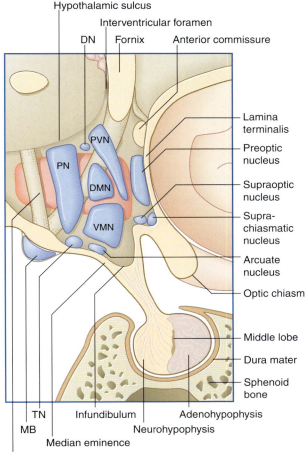

Figure 26.1 Hypothalamic nuclei and hypophysis, viewed from the right side. DMN, dorsomedial nucleus; DN, dorsal nucleus; MB, mammillary body; PN, posterior nucleus; PVN, paraventricular nucleus; TN, tuberomammillary nucleus; VMN, ventromedial nucleus. The lateral hypothalamic nucleus is shown in *pink*.

GROSS ANATOMY

The hypothalamus occupies the side walls and floor of the third ventricle. It is a bilateral, paired structure. Despite its small size—it weighs only 4 g—it has major functions in homeostasis and survival. Its homeostatic functions include control of the body temperature and the circulation of the blood. Its survival functions include regulation of food and water intake, the sleep–wake cycle, sexual behavior patterns, and defense mechanisms against attack.

Boundaries

The boundaries of the hypothalamus are as follow (see Figures 26.1 and 26.2).

* *Superior.* The **hypothalamic sulcus** separating it from the thalamus.

* *Inferior.* The **optic chiasm**, **tuber cinereum**, and **mammillary bodies**. The tuber cinereum shows a small swelling, the **median eminence**, immediately behind the **infundibulum** ('funnel') atop the pituitary stalk.
* *Anterior.* The lamina terminalis.
* *Posterior.* The tegmentum of the midbrain.
* *Medial.* The third ventricle.
* *Lateral.* The internal capsule.

Subdivisions and nuclei

In the sagittal plane, it is customary to divide the hypothalamus into three regions: *anterior* (supraoptic), *middle* (tuberal), and *posterior* (mammillary). The descriptive use of 'regions' has been convenient for animal experiments

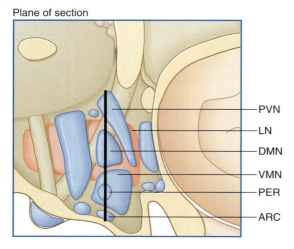

Plane of section

PVN
LN
DMN
VMN
PER
ARC

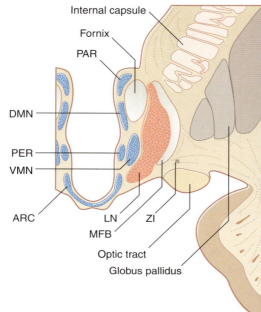

Internal capsule
Fornix
PAR
DMN
PER
VMN
ARC
LN   ZI
MFB
Optic tract
Globus pallidus

Figure 26.2 Hypothalamic nuclei, and related neural pathways, in a coronal section. ARC, arcuate nucleus; DMN, dorsomedial nucleus; LN, lateral nucleus; MFB, medial forebrain bundle; PAR, paraventricular nucleus; PER, periventricular nucleus; VMN, ventromedial nucleus; ZI, zona incerta.

involving placement of lesions. Named nuclei in the three regions are listed in Table 26.1.

In the coronal plane, the hypothalamus can be divided into *lateral*, *medial*, and *periventricular* regions. The full length of the lateral region is occupied by the **lateral hypo-**

Table 26.1 Hypothalamic nuclei

Posterior	Middle	Anterior
Posterior	Paraventricular	Preoptic
Mammillary	Dorsomedial	Supraoptic
Tuberomammillary	Lateral	Suprachiasmatic
Dorsal	Ventromedial	
	Arcuate	

thalamic nucleus. Merging with the lateral nucleus is the **medial forebrain bundle**, carrying aminergic fibers to the hypothalamus and to the cerebral cortex.

FUNCTIONS

Hypothalamic control of the pituitary gland

The arterial supply of the pituitary gland comes from hypophyseal branches of the internal carotid artery (Figure 26.3). One set of branches supplies a capillary bed in the wall of the infundibulum. These capillaries drain into **portal vessels** that pass into the adenohypophysis (anterior lobe). There they break up to form a second capillary bed that bathes the endocrine cells and drains into the cavernous sinus.

The neurohypophysis receives a direct supply from another set of hypophyseal arteries. The capillaries drain into the cavernous sinus, which delivers the secretions of the anterior and posterior lobes into the general circulation.

Secretions of the pituitary gland are controlled by two sets of **neuroendocrine cells**. Neuroendocrine cells are true neurons in having dendrites and axons and in conducting nerve impulses. They are also true endocrine cells, because they liberate their secretions into capillary beds (Figure 26.4). With one exception (mentioned below), the secretions are peptides, synthesized in clumps of granular endoplasmic reticulum and packaged in Golgi complexes.

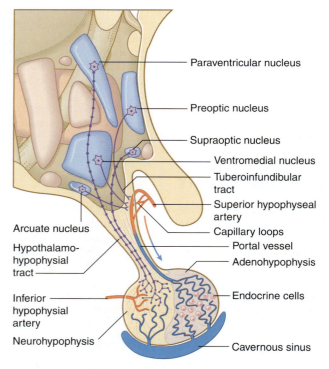

Paraventricular nucleus
Preoptic nucleus
Supraoptic nucleus
Ventromedial nucleus
Tuberoinfundibular tract
Superior hypophyseal artery
Capillary loops
Portal vessel
Adenohypophysis
Endocrine cells
Cavernous sinus
Arcuate nucleus
Hypothalamo-hypophysial tract
Inferior hypophysial artery
Neurohypophysis

Figure 26.3 Hypothalamic neuroendocrine cells. The blood supply to the hypophysis, including the endocrine cells of the adenohypophysis, is also shown (arrow indicates direction of blood flow in the portal system).

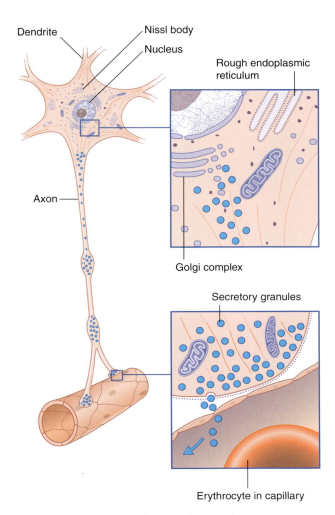

Dendrite
Nissl body
Nucleus
Rough endoplasmic reticulum
Axon
Golgi complex
Secretory granules
Erythrocyte in capillary

Figure 26.4 Morphology of a peptide-secreting neuroendocrine cell.

Table 26.2 Hypothalamic parvocellular releasing and inhibiting hormones	
Releasing or inhibiting hormone (RH, IH)	**Anterior lobe hormone**
Corticotropin RH	Adrenocorticotropic hormone
Thyrotropin RH	Thyrotropin
Growth hormone RH	Growth hormone
Growth hormone IH	Growth hormone
Prolactin RH	Prolactin
Prolactin IH	Prolactin
Gonadotropic hormone RH	Follicle-stimulating hormone, luteinizing hormone

dopamine, secreted from the arcuate (infundibular) nucleus.

The releasing and inhibiting hormones are not wholly specific: they have major effects on a single cell type, and minor effects on one or two others.

Multiple controls exist for parvocellular neurons of the hypophysiotropic area. The controls include depolarization by afferents entering from the limbic system and from the reticular formation; hyperpolarization by local circuit GABA neurons, some of which are sensitive to circulating hormones; and inhibition of transmitter release by opiate-releasing internuncials, which are numerous in the intermediate region of the hypothalamus. The picture is further complicated by the fact that opiates and other modulatory peptides may be released into the portal vessels and activate receptors on the endocrine cells of the adenohypophysis. *Stress* causes increased secretion of adrenocorticotropic hormone (ACTH), which in turn stimulates the adrenal cortex to raise the plasma concentration of glucocorticoids, including cortisol. Normally, cortisol exerts a negative feedback effect by exciting inhibitory hypothalamic neurons having glucocorticoid receptors. In patients suffering from major depression, this feedback system fails (Clinical Panel 26.1).

The magnocellular neuroendocrine system

Magnocellular neurons in the supraoptic and paraventricular nuclei give rise to the **hypothalamohypophyseal tract**, which descends to the neurohypophysis (posterior lobe) (Figure 26.3). Minor contributions to the tract are received from opiatergic and other peptidergic neurons in the periventricular region of the hypothalamus, and from aminergic neurons of the brainstem.

Two hormones are secreted by separate neurons located in both the supraoptic and paraventricular nuclei: *antidiuretic hormone* (ADH, vasopressin) and *oxytocin*. Axonal swellings containing the secretory granules for these hormones make up nearly half the volume of the neurohypophysis. The largest swellings, called Herring bodies, may be as large as erythrocytes. The Herring bodies provide a local depot of granules for release by smaller, terminal swellings into the capillary bed.

Antidiuretic hormone

Antidiuretic hormone continuously stimulates water uptake by the distal convoluted tubules and collecting

The peptides are attached to long-chain polypeptides called *neurophysins*. The capillaries concerned are outside the blood–brain barrier, and are fenestrated.

The somas of the neuroendocrine cells occupy the *hypophysiotropic area* in the lower half of the preoptic and tuberal regions. Contributory nuclei are the **preoptic**, **supraoptic**, **paraventricular**, **ventromedial**, and **arcuate** (infundibular). Two classes of neurons can be identified: **parvocellular** (small) **neurons** reaching the median eminence, and **magnocellular** (large) **neurons** reaching the posterior lobe of the pituitary gland.

The parvocellular neuroendocrine system

Parvocellular neurons of the hypophysiotropic area give rise to the **tuberoinfundibular** tract, which reaches the infundibular capillary bed. Action potentials traveling along these neurons result in calcium-dependent exocytosis of *releasing hormones* from some and *inhibiting hormones* from others, for transport to the adenohypophysis in the portal vessels. The cell types of the adenohypophysis are stimulated or inhibited in accordance with Table 26.2. In the left-hand column, the only non-peptide parvocellular hormone is the prolactin-inhibiting hormone, which is

Major depression is a state of depressed mood occurring without an adequate explanation in terms of external events. The condition affects about 4% of the adult population, and there is a genetic predisposition: about 20% of first-degree relatives have it too. Phases of depression may begin in childhood or adolescence.

Major depression is characterized by at least several of the following features.

- Depressed general mood, with loss of interest in normal activities and outside events.
- Diminished energy, easy fatigue, loss of appetite and of sex drive, constipation.
- Impairment of self-image, with a feeling of personal inadequacy.
- Disturbance of the sleep–wake cycle, typically shown by early morning wakefulness.
- Aches and pains. Recurrent abdominal pains may simulate organ disease.
- Periods of agitation, with restlessness and perhaps suicidal tendency.

Involvement of *monoamines* was first indicated by the chance observation that the use of reserpine in treatment of hypertension produced depression as a side effect. Reserpine depletes monoamine stores (serotonin, norepinephrine [noradrenaline], dopamine).

The symptoms listed above are also characteristic of *chronic stress*. It is therefore not surprising to find that the suprarenal cortex is hyperactive in depressed patients. Serum cortisol levels are elevated. As already mentioned, a rising serum cortisol level normally inhibits production of corticotropin-releasing hormone (CRH) by the hypothalamus. In depressed patients, the central glucocorticoid receptors are relatively insensitive. This change forms the basis of the *dexamethasone suppression test*. Dexamethasone is a potent synthetic glucocorticoid that reduces adrenocorticotropic hormone secretion in healthy individuals.

Some of the CRH neurons send branches into the brain itself. In the midbrain, CRH inhibits mesocortical dopaminergic neurons, which are normally associated with positive motivational drive. In the midbrain, they also inhibit raphe serotonergic neurons critically involved with diurnal rhythms, mainly through intense innervation of the suprachiasmatic nucleus.

The front line of therapy is dominated by drugs that enhance serotonergic transmission. The range of antidepressants is large, and their sites of action vary, for example some inhibit reuptake from the synaptic cleft, others inhibit degradation by monoamine oxidase (Ch. 13). They take several weeks to take effect; the latent interval is taken up with desensitizing (inhibitory) autoreceptors on serotonergic cell membranes.

Electroconvulsive therapy is at least as effective as the antidepressants. It seems to desensitize autoreceptors, to sensitize (excitatory) serotonin receptors on target neurons, and to depress noradrenergic transmission.

---

ducts of the kidneys. The chief regulator of electrical activity in the ADH-secreting neurons is the osmotic pressure of the blood. A rise of as little as 1% in the osmotic pressure causes the plasma to be diluted to normal levels by means of increased water uptake. The neurons are themselves sensitive to osmolar changes, but they are facilitated by inputs from osmolar and volume detectors elsewhere, notably from the **vascular** and **subfornical circumventricular organs** (Box 26.1).

Some ADH neurons also synthesize *corticotropin-releasing hormone*, the two hormones being released together from collateral branches into the capillary pool of the infundibulum. It is of interest that ADH neuronal activity is increased when the body is stressed, and that the output of ACTH is boosted by the presence of ADH in the adenohypophysis.

Withdrawal of ADH secretion results in *diabetes insipidus* (Clinical Panel 26.2).

Oxytocin

The principal function of oxytocin is to participate in a *neurohumoral reflex* when an infant is suckling at the breast. The afferent limb of this reflex is provided by impulses traveling from the nipple to the hypothalamus via the spinoreticular tract. Oxytocin is liberated by magnocellular neurons in response to suckling. Having entered the general circulation, it causes the expression of milk by stimulating myoepithelial cells surrounding the lactiferous ducts of the breast.

Oxytocin also has a mild stimulating action on uterine muscle during labor. The afferent stimulus in this case originates in the genital tract once labor gets under way.

Other hypothalamic connections and functions

Autonomic centers

In animals, stimulation of the anterior hypothalamic area produces *parasympathetic effects*: slowing of the heart, constriction of the pupil, salivary secretion, and intestinal peristalsis. On the other hand, stimulation of the posterior hypothalamic area produces sympathetic effects: increase in heart rate and blood pressure, pupillary dilatation, and intestinal stasis. Axons from both areas project to autonomic nuclei in the brainstem and spinal cord. In the midbrain and pons, this projection occupies the posterior longitudinal fasciculus, as seen in Chapter 17.

Box 26.1 Circumventricular organs

Six patches of brain tissue close to the ventricular system contain neurons and specialized glial cells abutting fenestrated capillaries. These are the **circumventricular organs (CVOs)** (Figure Box 26.1.1). The **median eminence** and **neurohypophysis** are described in the main text. The **vascular organ of the lamina terminalis** and the **subfornical organ** close to the interventricular foramen send axons into the supraoptic

and paraventricular nuclei of the hypothalamus and facilitate depolarization of neurons secreting antidiuretic hormone. In conditions of lowered blood volume, the kidney secretes renin, which, on conversion to angiotensin II, stimulates these two CVOs to complete a positive feedback loop.

The **pineal gland** synthesizes *melatonin*, an amine hormone implicated in the sleep–wake cycle. Melatonin is synthesized from serotonin, the requisite enzymes being unique to this gland. Melatonin is liberated into the pineal capillary bed at night and has a sleep-inducing effect; it may have other benefits, including clearance of harmful free radicals liberated from tissues during the aging process. Daytime secretion is suppressed by activity in sympathetic fibers reaching it from the superior cervical ganglia by way of the walls of the straight venous sinus. The relevant central pathway is from the paired suprachiasmatic nuclei via the posterior longitudinal fasciculus.

From the third decade onward, calcareous deposits ('pineal sand') accumulate within astrocytes in the pineal. Calcification is often detectable in plain radiographs of the head. A shift of the gland may denote a space-occupying lesion within the skull. However, a normal pineal may lie slightly to the left, because the right cerebral hemisphere is usually a little wider than the left at this level.

The **area postrema** is embedded in the roof of the fourth ventricle at the level of the obex. It is the *chemoreceptor trigger zone*, or *emetic* (vomiting) *center*. The emetic center contains neurons sensitive to a wide range of toxic substances, and it serves a protective function by reflexly eliciting emesis via connections with hypothalamus and reticular formation.

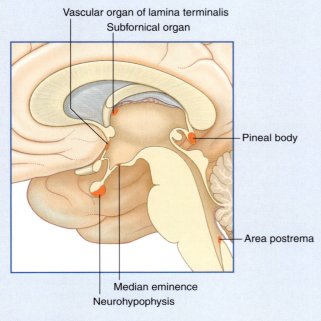

Vascular organ of lamina terminalis
Subfornical organ
Pineal body
Area postrema
Median eminence
Neurohypophysis

Figure Box 26.1.1 Circumventricular organs.

Clinical Panel 26.2 Hypothalamic disorders

The most dramatic disorder of hypothalamic function is *diabetes insipidus*, which is brought about by interruption of the hypothalamohypophyseal pathway—sometimes by tumors in the region, sometimes by head injury. The patient drinks upward of 10 L of water per day, and excretes a similar amount of urine. Historically, the term *insipidus* refers to the absence of taste sensation from the urine, in contrast to *diabetes mellitus*, in which the urine is sweet-tasting (mellitus) owing to its sugar content.

Hypophysectomy (surgical removal of the pituitary gland) can be performed in the treatment of other

diseases, without causing more than temporary diabetes insipidus, provided the pituitary stalk is sectioned at a low level. Within a short period, sufficient antidiuretic hormone is secreted into the capillary bed of the median eminence to ensure adequate water conservation.

A wide variety of hypothalamic dysfunctions have been reported in the clinical literature. Causes are also varied, and include tumors, congenital malformations, and head injury. Clinical manifestations include gross obesity, disturbances of autonomic control, excessive sleepiness, and memory loss.

Temperature regulation

The hypothalamus contains *thermosensitive neurons* that initiate appropriate responses to changes in the core temperature of the body. Activity of these neurons is reinforced by thermal information received (via the spinoreticular tract) from thermosensitive neurons supplying the skin (Ch. 11).

A slight change in the core temperature can usually be corrected by directing blood flow into or away from the skin, as appropriate. The requisite control of the sympathetic nervous system resides in the region of the posterior nucleus of the hypothalamus, which sends axons all the way to the lateral horn of the spinal cord.

Hypothalamic control of the sympathetic system diminishes with age. For this reason, the elderly are particularly prone to develop hypothermia in cold weather.

Hyperthermia is characteristic of *fevers*. Infectious agents (bacteria, viruses, parasites) cause tissue macrophages to liberate *endogenous pyrogen*, a protein that causes the hypothalamic 'thermostat' to be reset to a higher value. The chief mechanisms used to raise the body temperature to the new set point are cutaneous vasoconstriction and shivering.

Drinking

The chief center controlling the intake of water appears to be a ribbon of cells alongside the lateral nucleus known as the **zona incerta** (Figure 26.2). Stimulation of this region may produce excessive drinking; lesions may result in refusal to drink, with consequent severe dehydration.

Eating

Eating habits have obvious social and cultural components, causing dietary practice to vary widely among individuals and among communities. The hypothalamus provides a baseline for caloric and nutrient intake, in the form of interplay between the lateral and ventromedial nuclei. Together, they constitute the *appestat* (appetite set point). Stimulation of a lateral hypothalamic *feeding center* causes a cat or rat to eat excessively, whereas destruction of this center results in refusal to eat. Conversely, stimulation of a ventromedial *satiety center* inhibits the urge to eat, and bilateral ventromedial lesions result in persistent overeating and gross obesity. The satiety center is normally very sensitive to glucose levels in the blood. Of interest here is that serotonin is capable of altering the appetite set point by inhibiting the lateral nucleus. Anorexics tend to have a raised level of serotonin production, and bulimics a reduced level.

Rage and fear

The lateral and ventromedial nuclei are concerned with *mood* as well as food. Cats that are overweight in consequence of ventromedial lesions tend to be highly aggressive. Conversely, animals rendered underweight by ventromedial stimulation tend to be unduly docile (see also the amygdala in Ch. 34).

Sleeping and waking

The tiny (0.26 mm^3) **suprachiasmatic nucleus** embedded in the upper surface of the optic chiasm receives a direct input from the retina. It participates in setting the normal sleep–wake cycle through connections with the pineal gland. For reasons unknown, this nucleus contains peptidergic (vasopressin) neurons that are twice as numerous in homosexual men than in heterosexuals of either sex.

Lesions of the posterior hypothalamic area may cause hypersomnolence or even coma. This area contains the **tuberomammillary nucleus** (Figure 26.1), housing hundreds of *histaminergic neurons*, which project widely to the gray matter of the brain and spinal cord. Some of the fibers run rostrally within the medial forebrain bundle, in company with aminergic fibers of brainstem origin. Histaminergic fibers destined for the cerebral cortex fan out below the genu of the corpus callosum. They branch within the superficial layers of the frontal cortex, and run back to supply the cortex of the parietal, occipital, and temporal lobes.

In animals, there is abundant physiologic evidence in support of an *arousal function* for the histaminergic system. The tuberomammillary nucleus is normally activated during the awake state by the peptide *orexin* liberated by a small group of neurons in the lateral hypothalamus. Failure of orexin production appears to underlie the disabling sleep attacks characteristic of narcolepsy (Ch. 30).

Sexual arousal

A subset of neurons (known as INAH3) within the medial part of the preoptic nucleus is more than twice as large in males than in females. It is also rich in androgen receptors and activated by circulating testosterone. In females, estrogen-rich neurons are contained within the ventromedial nucleus. In laboratory animals, electrical stimulation of these nuclei elicits appropriate sexual responses.

Memory

The mammillary bodies belong to a limbic, *Papez circuit* involving the fornix, which sends fibers to it, and the mammillothalamic tract, which projects to the anterior nucleus of the thalamus. This circuit has a function in relation to memory (Ch. 34).

Core Information

The hypothalamus is a bilateral structure beside the third ventricle. In the sagittal plane, it can be divided into an anterior (supraoptic) region containing three nuclei, an intermediate (tuberal) region with five nuclei, and a posterior (mammillary) region with three. In the coronal plane, lateral, medial, and periventricular regions are described.

The pituitary gland is controlled by hypothalamic neuroendocrine cells, which are characterized by impulse transmission and hormonal secretion into capillary beds. Parvocellular neuroendocrine cells project to the median eminence. They secrete releasing and inhibiting hormones into the capillary bed there, to be taken to the adenohypophysis in a portal system of vessels. Large (magnocellular) neuroendocrine cells form the hypothalamohypophyseal tract, which liberates antidiuretic hormone and oxytocin into the capillary bed of the neurohypophysis.

Circumventricular organs comprise the median eminence and neurohypophysis; the vascular organ of lamina terminalis and subfornical organ (both of these involved in a feedback loop regulating plasma volume); the pineal gland, which secretes melatonin; the emetic area postrema; and the subfornical organ.

Anterior and posterior regions of the hypothalamus contain neurons that activate the parasympathetic and sympathetic system, respectively. Thermoregulatory neurons maintain the body temperature set point, mainly by manipulating the sympathetic system.

Stimulation of the lateral hypothalamic area provokes an increase in food and water consumption. Destruction of this area, or stimulation of a ventromedial satiety center, results in refusal to eat.

The suprachiasmatic nucleus participates in control of the sleep–wake cycle. The medial preoptic area contains androgen-sensitive neurons, and the ventromedial nucleus contains estrogen-sensitive neurons. The mammillary bodies receive inputs from the limbic system via the formix, having a function in relation to memory.

REFERENCES

Akil H, Watson SJ. Neuropeptides in brain and pituitary: overview. In: Meltzer HY, ed. Psychopharmacology: the third generation of progress. New York: Raven Press; 1987:367–371.

Gordon CJ. Integration and central processing in temperature control. Ann Rev Physiol 1986; 48:595–612.

Hatton GL. Emerging concepts of structure–function dynamics in adult brain: the hypothalamo-neurohypophysial system. Prog Neurobiol 1990; 34:337–504.

Lutten PGM, ter Horst TJ, Steffens AB. The hypothalamus: intrinsic connections and outflow pathways to the endocrine system in relation to the control of feeding and metabolism. Prog Neurobiol 1986; 28:1–54.

Rothwell NJ. CNS regulation of thermogenesis. Crit Rev Neurobiol 1994; 8:1–10.

Sawchenko PE. Toward a new neurobiology of energy balance, appetite, and obesity: the anatomists weigh in. J Comp Neurol 1998; 402:435–441.

Schwartz J-C, Arrang J-M, Garbarg M, et al. Histaminergic transmission in the mammalian brain. Physiol Rev 1991; 71:1–51.

Swaab DF, Hofman MA. Age, sex and light: variability in the human suprachiasmatic nucleus in relation to its functions. Prog Brain Res 1994; 100:261–265.

Weltzin K. Serotonin activity in anorexia and bulimia. J Clin Psychiatry 1991; 52(suppl):41–48.

Thalamus, epithalamus

STUDY GUIDELINES
1 The thalamus is an assembly of largely independent nuclear groups.
2 Five *specific (relay)* nuclei are reciprocally connected to motor or sensory areas of the cerebral cortex.
3 Six *association* nuclei are reciprocally connected to association areas of the cortex.
4 *Non-specific* nuclei (two described here) are not concerned with any specific motor or sensory function.

THALAMUS

The thalamus is the largest nuclear mass in the entire nervous system. It is a prominent feature in magnetic resonance imaging scans in each of the three planes in which slices are taken. The afferent and efferent connections of the main nuclear groups are listed in Table 27.1. The connections are so diverse that the thalamus cannot be said to have a unitary function.

As noted in Chapter 2, the two thalami lie at the center of the brain. Their medial surfaces are usually linked across the third ventricle, and their lateral surfaces are in contact with the posterior limb of the internal capsule. The upper surface of each occupies the floor of a lateral ventricle. The under-aspect receives sensory and cerebellar inputs as well as an upward continuum of the reticular formation (Table 27.1).

Thalamic nuclei

All thalamic nuclei except one (the reticular nucleus) have reciprocal excitatory connections with the cerebral cortex.

The Y-shaped **internal medullary lamina** of white matter divides the thalamus into three large cell groups: *medial dorsal*, *anterior*, and *lateral* (Figure 27.1A). The lateral group comprises *dorsal and ventral nuclear tiers*. At the back of the thalamus are the **medial** and **lateral geniculate bodies**. The **external medullary lamina** separates the thalamus from the shell-like **reticular nucleus**.

The thalamic nuclei are categorized into three functional groups: *specific* or *relay nuclei*, *association nuclei*, and *non-specific nuclei*.

Specific nuclei

The specific or relay nuclei are reciprocally connected to specific motor or sensory areas of the cerebral cortex. They comprise the nuclei of the ventral tier and the geniculate bodies (nuclei). Their afferent and efferent connections are indicated in Figure 27.1B.

The **anterior nucleus** receives the mammillothalamic tract and projects to the cingulate cortex. It is involved in a limbic circuit and has a function in relation to memory (Ch. 34).

The **ventral anterior nucleus** receives afferents from the globus pallidus, and it projects to the prefrontal cortex.

The anterior part of the **ventral lateral nucleus** (VL) receives afferents from the globus pallidus and projects to the supplementary motor area. The posterior part of VL is the principal target of the contralateral superior cerebellar peduncle, which originates in the dentate nucleus of the cerebellum; the posterior VL projects to the motor cortex.

The **ventral posterior nucleus** (VP) receives all the fibers of the medial, spinal, and trigeminal lemnisci (Figure 27.2). It projects to the somatic sensory cortex (SI). A smaller projection is sent to the second somatic sensory area (SII) at the foot of the postcentral gyrus (see Ch. 29).

The VP is somatotopically arranged, as indicated in Figure 27.3. The portion of the nucleus devoted to the face and head is called the **ventral posterior medial nucleus**, that for the trunk and limbs the **ventral posterior lateral nucleus**. Modality segregation is a feature of both nuclei, with proprioceptive neurons most anterior, tactile neurons

Table 27.1 Thalamic nuclei and their connections

Type	Nucleus	Afferents	Efferents
Specific	Anterior	Mammillary body	Cingulate gyrus
	Ventral anterior	Globus pallidus	Prefrontal cortex
	Ventral lateral		
	Anterior part	Globus pallidus	Supplementary motor area
	Posterior part	Cerebellum	Motor cortex
	Ventral posterior medial	Somatic afferents from head region	Somatic sensory cortex (SI)
	Ventral posterior lateral	Somatic afferents from trunk and limbs	Somatic sensory cortex
	Medial geniculate body	Inferior colliculus	Primary auditory cortex
	Lateral geniculate body	Superior colliculus, optic tract	Primary visual cortex
Association	Lateral dorsal	Parietal lobe	Cingulate cortex
	Posterior dorsal and pulvinar	Superior colliculus, parietal lobe	Visual association cortex
Non-specific	Intralaminar	Reticular formation	Cortex everywhere
	Reticular	Thalamus	Thalamus

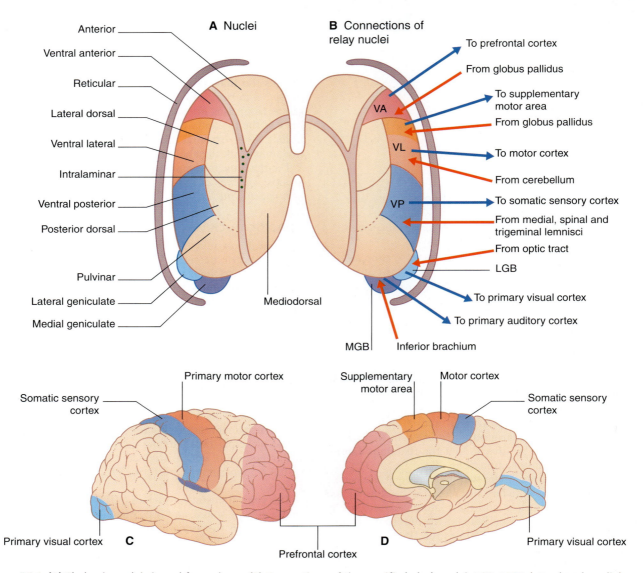

Figure 27.1 **(A)** Thalamic nuclei viewed from above. **(B)** Connections of the specific (relay) nuclei. LGB, MGB, lateral and medial geniculate bodies; VA, ventral anterior nucleus; VL, ventral lateral nucleus; VP, ventral posterior nucleus. **(C)** Lateral and **(D)** medial surface of hemisphere, showing cortical areas receiving projections from the relay nuclei.

in the midregion, and nociceptive neurons at the back. The nociceptive region is sometimes called the *posterior nucleus*.

There is no evidence in the VP of an antinociceptive mechanism comparable with that found in the substantia gelatinosa region of the spinal cord and spinal trigeminal nucleus. An unexplained disorder, the *thalamic syndrome*, may follow a vascular lesion that disconnects the posterior thalamic nucleus from the somatic sensory cortex. In this condition, a period of complete sensory loss may occur on the contralateral side of the body, to be replaced by bouts of severe pain occurring either spontaneously or in response to tactile stimuli.

The **medial geniculate body** (medial geniculate nucleus) is the thalamic nucleus of the auditory pathway. It receives the inferior brachium from the inferior colliculus (which carries auditory signals from both ears, Ch. 20), and it

projects to the primary auditory cortex in the superior temporal gyrus.

The **lateral geniculate body** (lateral geniculate nucleus) is the principal thalamic nucleus for vision. It receives retinal inputs from both eyes by way of the optic tract, and it projects to the primary visual cortex in the occipital lobe. The visual pathways are described in Chapter 28.

Association nuclei

The association nuclei are reciprocally connected to the association areas of the cerebral cortex.

The **lateral dorsal nucleus** has reciprocal connections with the posterior part of the cingulate cortex, which is involved in functions related to memory (Ch. 34).

The **mediodorsal nucleus** receives inputs from the olfactory and limbic systems, and is reciprocally connected with the entire prefrontal cortex. It has functions in relation to cognition (thinking), judgment, and mood.

The **lateral posterior nucleus** and the **pulvinar** belong to a single nuclear complex. They receive afferents from the superior colliculus, and project to the entire visual association cortex and to the entire parietal association cortex. An 'extrageniculate visual pathway' runs from the optic tract to the visual association cortex by way of the superior colliculus and the pulvinar. It has the function of drawing attention to objects of interest in the peripheral field of vision, but it is not itself a source of conscious visual perception.

Non-specific nuclei

The non-specific nuclei are so called because they are not specific to any one sensory modality. They include the intralaminar and reticular nuclei.

The **intralaminar nuclei** are contained within the internal medullary lamina of white matter. They can be regarded as a rostral continuation of the reticular formation of the midbrain (ascending reticular activating system in Ch. 24). They project widely to the cerebral cortex, as well as to the corpus striatum.

Afferents belonging to the ascending reticular activating system synapse in the intralaminar nuclei, also in the reticular nucleus and in the nucleus of Meynert in the basal forebrain (Ch. 34).

The **thalamic reticular nucleus** (TRN) is shaped like a shield around the front and lateral side of the thalamus. It is separated from the main thalamus by the external medullary lamina. All the thalamocortical projections from the specific thalamic nuclei pass through TRN and give collateral branches to it (Figure 27.4). *Fusiform neurons* within the innermost lamina (VI) of the cerebral cortex project to the thalamic nuclei and also give off collaterals to TRN. TRN is exclusively made up of inhibitory GABAergic neurons. Most of them project back into the corresponding nucleus and control (modulate) its rate of discharge to the cortex. There is general agreement, based on experimental recordings in rats and primates, that the primary function of TRN is that of *saliency*, which translates as helping to isolate any novel auditory, visual, or tactile experience from the normal background 'noise' of cortical activity in the awake state. The process is known as 'center-surround': corticothalamic feedback from lamina VI enhances activity of the stimulated patch of the sensory nucleus (the 'center') and simultaneously suppresses ongoing, random activity in the surrounding neurons not directly involved.

Tactile, visual, and auditory sensory modalities may be said to 'imprint' the areas of the TRN lattice that they collaterate as they pass through to reach the cerebral cortex. A minority of TRN cells project into *other* specific nuclei rather than the corresponding one mentioned above. This arrangement could enable TRN to participate in *combined processing*. This term refers to the simultaneous engagement of more than one sensory modality in a particular sensory task. As an example, an unexpected sound occurring within the lower right visual field, activating a topographically specific patch of the auditory TRN lattice overlying each medial geniculate body, could

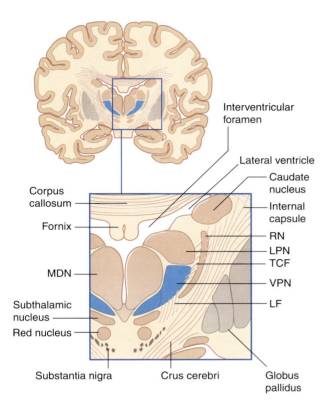

Figure 27.2 Coronal section through the thalamus and related structures. LF, lemniscal fibers; LPN, lateral posterior nucleus; MDN, mediodorsal nucleus; RN, reticular nucleus; TCF, thalamocortical fibers; VPN, ventral posterior nucleus.

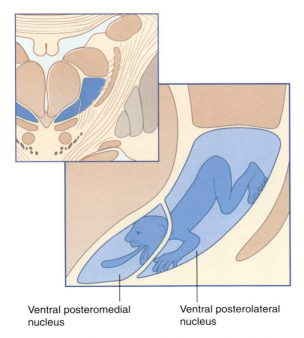

Figure 27.3 Somatic sensory map in the ventral posterior thalamic nucleus. (After Ohye 1990, with permission.)

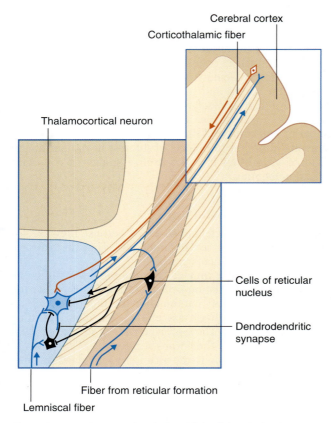

Figure 27.4 Basic synaptic relationships of the thalamic reticular nucleus. The 'sensory nucleus' includes somatic sensory, visual, and auditory thalamic nuclei. (After Pinault 2004, with permission.)

selectively disinhibit lateral geniculate neurons on the visual pathway from the upper left quadrants of both retinas. In this way, auditory inputs may facilitate selective visual attention to the area of interest.

Oscillation

A remarkable histologic feature of TRN neurons is the frequent occurrence of *dendritic sheaves*: these comprise bundles of dendrites belonging to different neurons, extending in the overall plane of TRN, and linked to one another by dendrodendritic synapses. This arrangement may form the anatomic basis of the phenomenon of *oscillation*. Oscillation is characterized by spontaneous burst-firing of large groups of TRNs at a rate of 5–15 Hz, usually for a few seconds at a time. The oscillations produce patches of surround inhibition in the underlying thalamocortical neurons, thereby evoking bursts of cortical activity known as *sleep spindles*, so called because they are

detectable by means of electroencephalography at the onset of sleep.

Sleep–wake cycles are described in Chapter 30.

Not represented in Table 27.1 are *aminergic afferents* passing to the ventral and intralaminar nuclei from the midbrain raphe (serotonergic) and cerulean nucleus (noradrenergic). The proven value of tricyclic anti-depressants in the therapy of chronic pain may be related to drug-induced prolongation of excitatory aminergic effects on thalamocortical neurons.

Thalamic peduncles

The reciprocal connections between the thalamus and the cerebral cortex travel in four thalamic peduncles, as shown in Figure 27.5. The **anterior thalamic peduncle** passes through the anterior limb of the internal capsule to reach the prefrontal cortex and cingulate gyrus. The **superior thalamic peduncle** passes through the posterior limb of the internal capsule to reach the premotor, motor, and somatic sensory cortex. The **posterior thalamic peduncle** passes through the retrolentiform part of the internal capsule to reach the occipital lobe and the posterior parts of the parietal and temporal lobes. The **inferior thalamic peduncle** passes below the lentiform nucleus to reach the anterior temporal and orbital cortex. Each of the four fans becomes incorporated into the corona radiata.

EPITHALAMUS

The epithalamus includes the pineal gland (considered in Ch. 26), and the habenula and stria medullaris, which are included with the limbic system in Chapter 33.

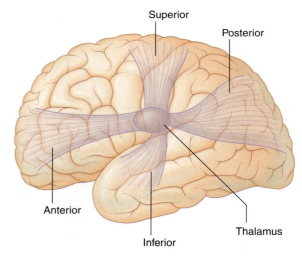

Figure 27.5 The thalamic peduncles (left hemisphere).

Core Information

Thalamus
The internal medullary lamina divides the thalamus anatomically into mediodorsal, anterior, and lateral nuclear groups, the lateral being separable into dorsal and ventral tiers. The thalamus may be divided functionally into specific, association, and non-specific nuclear groups.
  Of the specific nuclei:

- ventral anterior receives inputs from globus pallidus and projects to prefrontal cortex
- the anterior part of ventral lateral receives inputs from globus pallidus and projects to supplementary motor area, whereas the posterior part receives from contralateral cerebellum and projects to motor cortex
- ventral posterior receives the somatic sensory pathways and projects to the somatic sensory cortex
- the medial geniculate nucleus receives the inferior brachium and projects to the primary auditory cortex
- the lateral geniculate nucleus receives from the optic tract and projects to the primary visual cortex.

Of the association nuclei:

- the anterior receives the mammillothalamic tract and projects to the cingulate cortex
- the mediodorsal is reciprocally connected to all parts of the prefrontal cortex
- the lateral posterior–pulvinar complex receives from the superior colliculus and projects to the parietal association cortex.

Of the non-specific nuclei:

- the intralaminar nucleus receives inputs from the reticular formation and projects widely to the cerebral cortex, also to the corpus striatum
- the reticular nucleus (external to the thalamus proper) receives excitatory collaterals from all thalamocortical and corticothalamic neurons, and returns inhibitory fibers to all nuclei within the thalamus.

  Its best known function is generation of rhythmic electrical oscillations characteristic of early sleep. Reciprocal connections between thalamus and cortex travel in four thalamic peduncles that become incorporated into the corona radiata.

Epithalamus
The epithalamus includes the pineal gland, the habenula, and the stria terminalis.

REFERENCES

Alitto HJ, Usry WM. Corticothalamic feedback and sensory processing. Curr Opin Neurobiol 2003; 13:440–445.

Guillery RW. Anatomical evidence concerning the role of the thalamus in corticocortical communication: a brief review. J Anat 1995; 187:583–592.

Kultas-Ilinsky K, Ilinsky IA. Neuronal and synaptic organization of the motor nuclei of mammalian thalamus. In: Current topics in research on synapses, vol 3. New York: Alan R. Liss; 1986:77–145.

Lenz FA. Ascending modulation of thalamic function and pain. In: Sicuteri F, et al, eds. Advances in pain research and therapy. New York: Raven Press; 1992:177–196.

Mahe V, Chevalier F. Human circadian clock in disease. Presse Med 1995; 27:1041–1046.

Ohye C. Thalamus. In: Paxinos G, ed. The human nervous system. San Diego: Academic Press; 1990:439–468.

Pinault D. The thalamic reticular nucleus: structure, function and concept. Brain Res Rev 2004; 46:1–31.

Visual pathways

STUDY GUIDELINES
1 The great length of the visual pathways makes them vulnerable to disease at widely separate locations. The pattern of visual defect differs in accordance with the site of damage.
2 Visual defects can often be detected by merely wiggling a finger in different parts of the visual field of each eye in turn.
3 The most important item of anatomic information concerns the representations of the visual fields at successive steps along the visual pathways.

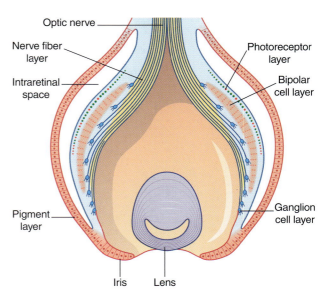

Figure 28.1 Embryonic retina. *Green* and *red* represent rods and cones, respectively.

The visual pathways are of outstanding importance in clinical neurology. They extend from the retinas of the eyes to the occipital lobes of the brain. Their great length makes them especially vulnerable to demyelinating diseases such as multiple sclerosis, to tumors of the brain or pituitary gland, to vascular lesions in the territory of the middle or posterior cerebral artery, and to head injuries.

The visual system comprises the retinas, the visual pathways from the retinas to the brainstem and visual cortex, and the cortical areas devoted to higher visual functions. The retinas and visual pathways are described in this chapter. Higher visual functions are described in Chapter 29.

RETINA

The retina and the optic nerves are part of the central nervous system. In the embryo, the retina is formed by an outgrowth from the diencephalon called the optic vesicle (Ch. 1). The optic vesicle is invaginated by the lens and becomes the two-layered optic cup.

The outer layer of the optic cup becomes the pigment layer of the mature retina. The inner, nervous layer of the cup gives rise to the retinal neurons.

Figure 28.1 shows the general relationships in the developing retina. The nervous layer contains three principal layers of neurons: **photoreceptors**, which become applied to the pigment layer when the intraretinal space is resorbed; **bipolar neurons**; and **ganglion cells**, which give rise to the optic nerve and project to the thalamus and midbrain.

Note that the retina is *inverted*: light must pass through the layers of optic nerve fibers, ganglion cells, and bipolar neurons to reach the photoreceptors. However, at the point of most acute vision, the **fovea centralis**, the bipolar and ganglion cell layers lean away all around a central pit

(fovea), and light strikes the photoreceptors directly (see *Foveal specialization*, later). In the mature eye, the fovea is about 1.5 mm in diameter and occupies the center of the 5 mm wide **macula lutea** ('yellow spot'), where many of the photoreceptor cells contain yellow pigment. The fovea is the point of most acute vision and lies in the *visual axis*—a line passing from the center of the visual field of the eye, through the center of the lens, to the fovea (Figure 28.2). To *fixate* or *foveate* an object is to gaze directly at it so that light reflected from its center registers on the fovea.

The axons of the ganglion cells enter the optic nerve at the **optic papilla** (*optic nerve head*), which is devoid of retinal neurons and constitutes the physiologic 'blind spot'.

The visual fields of the two eyes overlap across two-thirds of the total visual field. Outside this *binocular field* is a *monocular crescent* on each side (Figure 28.3). During passage through the lens, the image of the visual field is reversed, with the result that, for example, objects in the left part of the binocular visual field register on the right half of each retina, and objects in the upper part of the visual field register on the lower half. This arrangement is preserved all the way to the visual cortex in the occipital lobe.

From a clinical standpoint, it is essential to appreciate that *vision is a crossed sensation*. The visual field on one side of the visual axis registers on the visual cortex of the opposite side. In effect, the right visual cortex 'sees' the left visual field. Only half the visual information crosses in the optic chiasma, for the simple reason that the other half has already crossed the midline in space.

Visual defects caused by interruption of the visual pathway are always described *from the patient's point of*

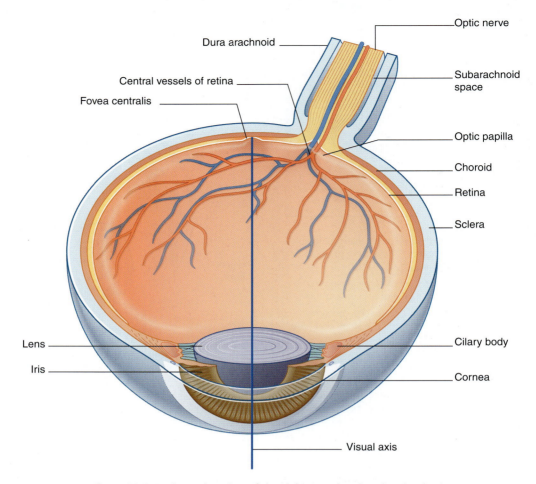

Figure 28.2 Horizontal section of the right eye, showing the visual axis.

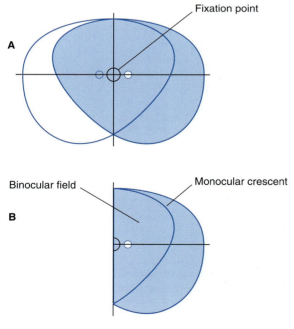

Figure 28.3 **(A)** Visual fields. Both eyes are targeted on the fixation point. The visual field of the right eye is shaded. **(B)** The right visual field. The *white spot* represents the blind spot of the right eye.

view, i.e. in terms of the visual fields, and not in terms of retinal topography.

Structure of the retina

In addition to the serially arranged photoreceptors, bipolar cells, and ganglion cells shown in *Figure 28.1*, the retina contains two sets of neurons arranged transversely: **horizontal cells** and **amacrine cells** (Figure 28.4). A total of eight layers are described for the retina as a whole.

Action potentials are generated by the ganglion cells, providing the requisite speed for conduction to the thalamus and midbrain. For the other cell types, distances are very short, and passive electrical change (electrotonus) is sufficient for intercellular communication, whether by gap junctional contact or transmitter release.

Photoreceptors

The photoreceptor neurons comprise **rods** and **cones**. Rods function only in dim light and are not sensitive to color. They are scarce in the outer part of the fovea and absent from its center. Cones respond to bright light, are sensitive to color (received in the form of electromagnetic wavelength energy) and to shape, and are most numerous in the fovea.

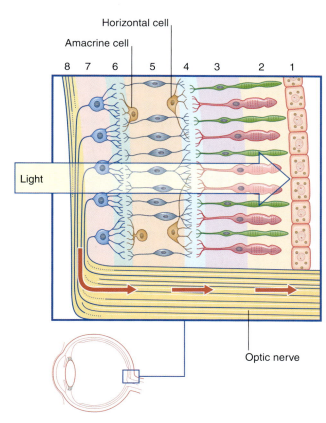

Figure 28.4 The layers of the retina. (1) Pigment layer, (2) photoreceptor layer, (3) outer nuclear layer, (4) outer plexiform layer, (5) inner nuclear layer, (6) inner plexiform layer, (7) ganglion cell layer, (8) nerve fiber layer.

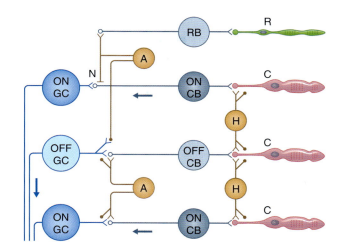

Figure 28.5 Retinal circuit diagram. A, amacrine cell; C, cone; CB, cone bipolar neuron; GC, ganglion cell; H, horizontal cell; N, nexus (gap junction); R, rod; RB, rod bipolar. (After Massey and Redburn 1987, with permission.)

Each photoreceptor has an outer and an inner segment and a synaptic end-foot. In the outer segment, the plasma membrane is folded to form hundreds of membranous disks that incorporate visual pigment (rhodopsin) formed in the inner segment. The synaptic end-foot makes contact with bipolar neurons and horizontal cell processes in the outer plexiform layer.

A surprising feature of the photoreceptors is that they are hyperpolarized by light. During darkness, Na$^+$ channels are opened, creating sufficient positive electrotonus to cause leakage of transmitter (glutamate) from the end-feet. Illumination causes the Na$^+$ channels to close.

Cone and rod bipolar neurons

Cone bipolar neurons
Cone bipolar neurons are of two types. ON bipolars are switched on (depolarized) by light, being inhibited by transmitter released in the dark. They converge on to ON ganglion cells. OFF bipolars have the reverse response and converge on to OFF ganglion cells (Figure 28.5).

Rod bipolar neurons
Rod bipolar neurons are all hyperpolarized by light. They activate ON and OFF ganglion cells indirectly, by way of amacrine cells (Figure 28.5).

Horizontal cells
The dendrites of horizontal cells are in contact with photo-receptors. The peripheral dendritic branches give rise to axon-like processes that make inhibitory contacts with bipolar neurons.

The function of horizontal cells is to inhibit bipolar neurons outside the immediate zone of excitation. The excited bipolars and ganglion cells are said to be on-line; the inhibited ones are off-line.

Amacrine cells
Amacrine cells have no axons. Their appearance is octopus-like, the dendrites all emerging from one side of the cell. Dendritic branches come into contact with bipolar neurons and ganglion cells.

More than a dozen different morphologic types of amacrine cells have been identified, as well as several different transmitters, including acetylcholine, dopamine, and serotonin. Possible functions include contrast enhancement and movement detection. For the rods, they convert large numbers of rods from OFF to ON with respect to ganglion cells.

Ganglion cells
The ganglion cells receive synaptic contacts from bipolar neurons in the inner plexiform layer. The typical response of ganglion cells to bipolar activity is 'center-surround'. An ON ganglion cell is excited by a spot of light, and inhibited by a surrounding annulus (ring) of light. The inhibition is caused by horizontal cells. OFF ganglion cells give the reverse response.

Coding for color
There are three types of cone with respect to spectral sensitivity. One is sensitive to red, one to green, and one to blue. Groups of each type are connected to ON or OFF ganglion cells.

The characteristic response of ganglion cells is one of *color opponency*.

- Ganglion cells that are on-line for green are off-line for red.
- Ganglion cells that are on-line for red are off-line for green.
- Ganglion cells that are on-line for blue are off-line for yellow, i.e. for green and red cones acting together.

Coding for black and white

White light is a mixture of green, red, and blue. In bright conditions, it is encoded by the three corresponding cones, all of them converging on to common ganglion cells. Both ON and OFF ganglion cells are involved in black-and-white vision, just as in color vision.

In very dim conditions, for example starlight, only rod photoreceptors are active, and objects appear in varying shades of gray. The rods are subject to the same rules as cones, showing center-surround antagonism between white and black, and being connected to ON or OFF ganglion cells.

Most rod and cone ganglion cells are small (parvocellular or 'P'), having small receptive fields and being responsive to color and shape. A minority are large (magnocellular or 'M'), having large receptive fields and being especially responsive to movements within the visual field.

Foveal specialization

The relative density of cones increases progressively, and their size diminishes progressively, from the edge of the fovea inward (Figure 28.6). The central one-third of the fovea, little more than 100 μm wide and known as the **foveola**, contains only *midget* cones. Two special anatomic features assist the foveal cones in general, and the midget cones in particular, in transducing the maximum amount of information concerning the form and color values of an object under direct scrutiny. First, the more superficial layers of the retina lean outward from the center, and their neurites are exceptionally long, with the result that the outer two-thirds of the fovea are little overlapped by bipolar cell bodies and the inner third is not overlapped at all; light reflected from the object strikes the cones of the foveola without any diffraction. Second, fidelity of central transmission is enhanced by one-to-one synaptic contact between the midget cones and *midget* bipolar neurons, and between these and *midget* ganglion cells. Outside the foveola, the amount of cone-to-bipolar-to-ganglion cell convergence increases progressively.

CENTRAL VISUAL PATHWAYS

Optic nerve, optic tract

The optic nerve is formed by the axons of the retinal ganglion cells. The axons acquire myelin sheaths as they leave the optic disk.

The number of ganglion cells varies remarkably between individuals, from 800 000 to 1.5 million. Because every ganglion cell contributes to the optic nerve, the number of axons in the optic nerve is correspondingly variable.

The retinal ganglion cells are homologous with the sensory projection neurons of the spinal cord. The optic nerve is homologous with spinal cord white matter, and is *not* a peripheral nerve. As explained in Chapter 9, true peripheral nerves, whether cranial or spinal, contain Schwann cells and collagenous sheaths, and are capable of regeneration. The optic nerve contains neuroglial cells of central type (astrocytes and oligodendrocytes) and is not capable of regeneration in mammals. In addition, the nerve is invested with meninges containing an extension of the subarachnoid space—a feature largely responsible for the changed appearance of the fundus oculi when the intracranial pressure is raised (*papilledema*, Ch. 4).

At the optic chiasm, fibers from the nasal hemiretina (medial half-retina) enter the contralateral optic tract, whereas those from the temporal (lateral) hemiretina remain uncrossed and enter the ipsilateral tract.

As already noted in Chapter 24, some optic nerve fibers enter the suprachiasmatic nucleus of the hypothalamus. This connection has been invoked to account for the beneficial effect of bright artificial light, for several hours per day, in the treatment of wintertime depression.

Each optic tract winds around the midbrain and divides into a medial and a lateral root.

Medial root of optic tract

The medial root contains 10% of the optic nerve fibers. It enters the side of the midbrain. It contains four distinct sets of fibers.

1 Some fibers, mainly from retinal M cells, enter the superior colliculus and provide for automatic scanning, for example reading this page.

2 Some fibers are relayed from the superior colliculus to the pulvinar of the thalamus; they belong to the extrageniculate pathway to the visual association cortex (Ch. 29).

3 Some fibers enter the pretectal nucleus and serve the pupillary light reflex (Ch. 23).

4 Some fibers enter the parvocellular reticular formation, where they have an arousal function (Ch. 24).

Lateral root of the optic tract and lateral geniculate body

The lateral root of the optic tract terminates in the lateral geniculate body (LGB) of the thalamus. The LGB shows six cellular laminae, three of which are devoted to crossed fibers and three to uncrossed fibers. The two deepest laminae (one for crossed and one for uncrossed fibers) are magnocellular and receive axons from retinal M ganglion cells concerned with detection of *movement*. The other four are parvocellular and receive the axons of P cells concerned with *particulars*, namely visual detail and color.

The circuitry of the LGB resembles that of other thalamic relay nuclei, and includes inhibitory (GABA) terminals derived from internuncial neurons and from the thalamic reticular nucleus. (The portion of the reticular nucleus serving the LGB is called the **perigeniculate nucleus**.) Corticogeniculate axons arise in the primary visual cortex and synapse on distal dendrites of relay cells

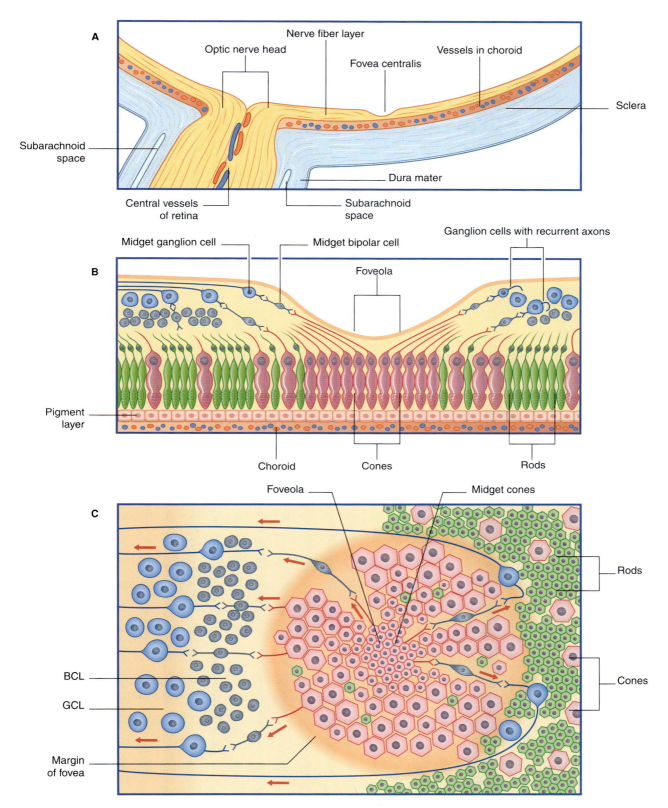

Figure 28.6 (A) Horizontal section of the right eyeball at the level of the optic disk and fovea centralis. **(B)** Enlargement from (A). Recurrent axons sweep around the fovea as shown in (C). **(C)** Surface view of fovea centralis and neighboring retina. Cones have been omitted at intervals to show the 'chain' sequence of neurons. BCL, bipolar cell layer; GCL, ganglionic cell layer.

as well as on inhibitory internuncials. Cortical synapses on relay cells are twice as numerous as those derived from retinal ganglion cells. Cortical stimulation usually enhances the response of relay cells to a given retinal input. A likely, but unproven, function could be that of selective enhancement of particular features of the visual scene, for example when searching for an object of known shape or color. Functional magnetic resonance imaging (MRI) (Ch. 29) is capable of detecting areas of increased neuronal activity in the brain. Functional MRI has shown that when volunteers expect to see an object of interest on-screen, metabolic activity in the LGB increases *before* the stimulus is presented.

Geniculocalcarine tract and primary visual cortex

The **geniculocalcarine tract**, or **optic radiation**, is of major clinical importance, because it is frequently compromised by vascular disorders or tumors in the posterior part of the cerebral hemisphere. It travels from the LGB to the primary visual cortex.

The anatomy of the optic radiation is shown in Figures 28.7–28.10. Fibers destined for the lower half

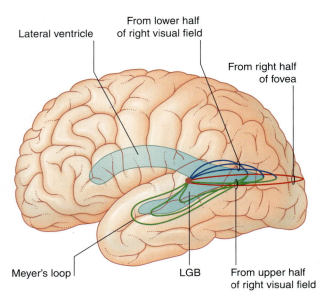

Figure 28.7 Left optic radiation. LGB, lateral geniculate body.

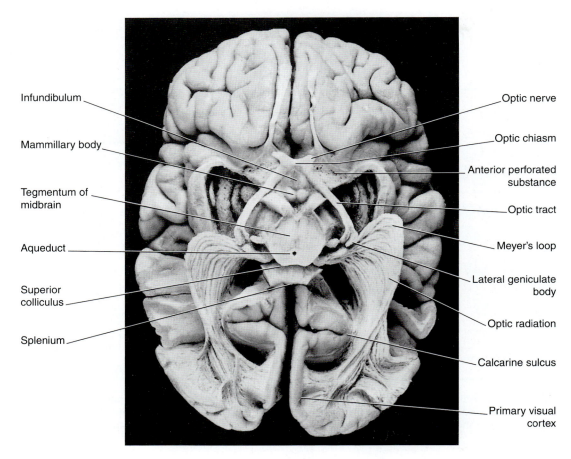

Figure 28.8 A dissection of the visual pathways, viewed from below. (From Gluhbegovic and Williams 1980, with permission of the authors and JB Lippincott.)

Visual fields

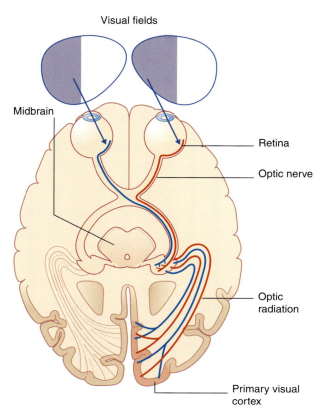

Midbrain

Retina

Optic nerve

Optic radiation

Primary visual cortex

Figure 28.9 Diagram of the visual pathways. The two visual fields are represented separately, without the normal overlap.

of the primary visual cortex sweep forward into the temporal lobe, as *Meyer's loop*, before turning back to accompany those traveling to the upper half. The tract enters the retrolentiform part of the internal capsule and continues in the white matter underlying the lateral temporal cortex. It runs alongside the posterior horn of the lateral ventricle before turning medially to enter the occipital cortex.

The **primary visual cortex** occupies the walls of the calcarine sulcus along its entire length (the sulcus is 10 mm deep). It emerges on to the medial surface of the hemisphere for 5 mm both above and below the sulcus, and on to the occipital pole of the brain for 10 mm. Its total area is about 28 cm^2. In the freshly cut brain, it is easily identified by a thin band of white matter (the *visual stria of Gennari*) within the gray matter—hence an alternative term, *striate* cortex. The left and right eyes are represented in the cortex in alternating stripes called *ocular dominance columns* (Figure 28.9).

Retinotopic map

The contralateral visual field is represented upside down. The plane of the calcarine sulcus represents the horizontal meridian. Retinal representation is posteroanterior, with a greatly magnified foveal representation in the posterior half of the calcarine cortex (Figure 28.10).

The clinical effects of various lesions of the visual pathway are described in Clinical Panel 28.1.

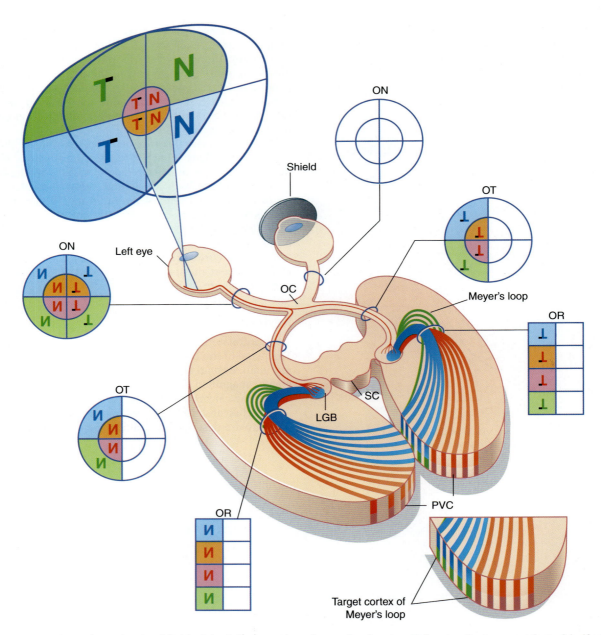

Figure 28.10 Pathway from the visual field of the left eye to the primary visual cortex. T denotes the temporal (outer) half of the left visual field; N denotes the nasal (inner) half of the left visual field.

In the left retina and optic nerve (ON), the neural representation of the image is reversed side to side. It is also inverted top to bottom. The right retina and optic nerve are inactive because this eye is shielded.

At the optic chiasm (OC), the axons forming the nasal half of the left optic nerve cross the midline and form the medial half of the right optic tract (OT). Those forming the lateral half of the nerve form the lateral half of the left optic tract. Each set synapses in the corresponding lateral geniculate body (LGB).

The optic radiations (OR) are fan-like (cf. Figure 28.7), with the axons carrying the foveal input initially in the middle of the fan.

As they approach the occipital pole, the foveal axons (red) in both hemispheres move to the back and enter the posterior part of the primary visual cortex (PVC). Note the striped pattern of delivery to the cortex on both sides. The blank intervals between are the same width and contain the axons and cortex responsible for the visual field of the *right* eye. SC, superior colliculus.

Clinical Panel 28.1   Lesions of the visual pathways

The following points arise in testing the visual pathways.

- The patient may be unaware of quite extensive blindness—sometimes even of a hemianopia.

- Large visual defects can often be detected by simple confrontation, as follows. The patient covers one eye at a time, and focuses on the examiner's nose. The examiner, seated opposite, looks the patient in the eye while bringing one or other hand into view from various directions, with the index finger wiggling.

- In a blind area, the patient does not see blackness; the patient does not see anything.

- Visual defects are described from the patient's viewpoint, in terms of the visual fields.

- Possible sites of injury to the visual pathways are shown in Figure CP 28.1.1. The effects produced correspond to the numbers in Table CP 28.1.1.

Notes on the numbered lesions

1 Eccentric lesions of the optic nerve produce scotomas in the nasal or temporal field of the affected eye. *When a young adult presents with a scotoma, multiple sclerosis must always be suspected.*

2 Total conduction blockage may follow head injury.

3 Compression of the middle of the chiasm is most often caused by an adenoma (benign tumor) of the pituitary gland.

4 Lesions of the optic tract are rare. Although homonymous (matching) visual fields are affected, the outer, exposed half of the tract tends to be more affected than the inner half, and the hemianopia is then described as incongruous.

5 Meyer's loop may be selectively caught by a tumor in the temporal lobe.

6 Lesions involving the optic radiation include tumors arising in the temporal, parietal, or occipital lobe. The visual fields of both eyes tend to be affected to an equal extent (*congruously*). Tumors impinging on the radiation from below produce an upper quadrantic defect at first, whereas tumors impinging from above produce a lower quadrantic defect. The stem of the radiation occupies the retrolentiform part of the internal capsule and is often compromised for some days by edema, following hemorrhage from a branch of the middle cerebral artery (classic stroke, Ch. 35).

7 Thrombosis of the posterior cerebral artery produces a homonymous hemianopia. The notches in field chart no. 7 represent macular sparing. Sparing of the macular hemifields is inconstant.

8 Bilateral central scotomas are most often caused by a backward fall with occipital contusion.

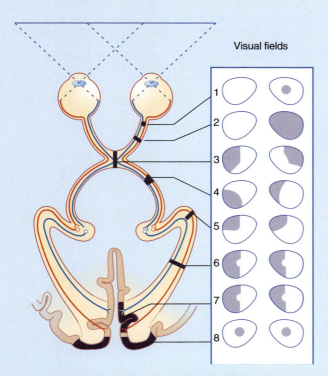

Visual fields

Figure CP 28.1.1 Visual field defects following various lesions of the visual pathways.

Table CP 28.1.1 Classification of dyphasia.		
Number	**Lesion**	**Field defect**
1	Partial optic nerve	Ipsilateral scotoma[a]
2	Complete optic nerve	Blindness in that eye
3	Optic chiasm	Bitemporal hemianopia
4	Optic tract	Homonymous[b] hemianopia
5	Meyer's loop	Homonymous upper quadrantanopia
6	Optic radiation	Homonymous hemianopia
7	Visual cortex	Homonymous hemianopia
8	Bilateral macular cortex	Bilateral central scotomas

[a]Patch of blindness.
[b]Matching.

Core Information

The embryonic retina is an outgrowth of the diencephalon. The embryonal optic cup is composed of an outer, pigment layer, an inner, nervous layer, with an intraretinal space between. The nervous layer contains three sets of radially disposed neurons, namely photoreceptors, bipolar cells, and ganglion cells, and two tangential sets, namely horizontal cells and amacrine cells. Except at the fovea centralis, light must pass through the other layers to reach the photoreceptors. The visual image is inverted and reversed by the lens. Two-thirds of the visual field are binocular, the outer one-sixth on each side being monocular. Visual defects are described in terms of visual fields.

Rod photoreceptors function in dim light and are absent from the fovea. Cones are most numerous in the fovea; they are responsive to shape and have three kinds of sensitivity to color. Ganglion cell responses are concentric, showing center-surround color opponency.

M ganglion cells are relatively large, are movement detectors, and project their axons to the two magnocellular layers of the lateral geniculate body (LGB). P ganglion cells signal particular features of the image as well as color, and project to the four parvocellular layers of LGB. LGB is binocular, receiving signals from the contralateral nasal hemiretina (via the optic chiasm) and from the ipsilateral temporal hemiretina. Both sets of axons arrive by the optic tract, which also gives offsets to the midbrain for lower-level visual reflexes.

The geniculocalcarine tract (optic radiation) arises from M and P cells of the LGB and swings around the side of the lateral ventricle to reach the primary visual cortex, in the walls of the calcarine sulcus.

Distinctive visual field defects occur following damage at any of the five major components of the visual pathway (optic nerve, optic chiasm, optic tract, optic radiation, visual cortex).

REFERENCES

Bynke H. The visual fields. In: Lessell S, van Dalen JTW, eds. Neuro-ophthalmology, vol 3. Amsterdam: Elsevier; 1984:348–357.

Celesia GG, DeMarco PJ. Anatomy and physiology of the visual system. J Clin Neurophysiol 1994; 11:482–492.

Curcio CA, Kimberly AA. Topography of ganglion cells in human retina. J Comp Neurol 1990; 300:5–28.

Frisen L. The neurology of visual acuity. Brain 1980; 103:639–670.

Gluhbegovic N, Williams, TW. The human brain. Philadelphia: JB Lippincott; 1980.

Karten HJ, Keyser KT, Brecha NC. Biochemical and morphological heterogeneity of retinal ganglion cells. In: Cohen B, Bodis-Wollner I, eds. Vision and the brain. New York: Raven Press; 1990:19–33.

Massey SC, Redburn DA. Transmitter circuits in the vertebral retina. Prog Neurobiol 1987; 28:55–96.

Sadun AA, Glaser JS, Bose S. Anatomy of the visual sensory system. In: Tasman W, ed. Duane's clinical ophthalmology, vol 2. Philadelphia: Lippincott, Williams and Wilkins; 2005:1–23.

Treue S. Visual attention: the where, what, how and why of saliency. Curr Opin Neurobiol 2003; 13:428–432.

Vaney DI. Patterns of neuronal coupling in the retina. Prog Retin Eye Res 1994; 13:301–355.

Wu SM. Synaptic transmission in the outer retina. Ann Rev Physiol 1994; 56:141–168.

Cerebral cortex

STUDY GUIDELINES

1 The cerebral cortex is the part of the body that makes us truly human. Its structure is enormously complex, and assignment of functions to different parts is made difficult, and often unrealistic, by the multiplicity of interconnections.

2 Sensory, motor, and cognitive areas of the cortex are taken in turn.

3 Although damage often leads to permanent disability, the *plasticity* of the cortex is of special interest to all concerned with neurorehabilitation. Examples are taken from sensory and motor areas.

STRUCTURE

The cerebral cortex, or *pallium* (*Gr.* 'shell'), varies in thickness from 2 to 4 mm, being thinnest in the primary sensory areas and thickest in the motor and association areas. More than half of the total cortical surface is hidden from view in the walls of the sulci. The cortex contains about 50 billion neurons, about 500 billion neuroglial cells, and a dense capillary bed.

Microscopy reveals the cortex to have both a laminar and a columnar structure. The general cytoarchitecture varies in detail from one region to another, permitting the cortex to be mapped into dozens of histologically different 'areas'. Although considerable progress has been achieved in relating these to specific functions, these areas are merely nodal points having widespread connections with other parts of the brain.

Laminar organization

A laminar (layered) arrangement of neurons is apparent in sections taken from any part of the cortex. Phylogenetically old elements, including the *paleocortex* of the uncus (concerned with olfaction), and the *archicortex* of the hippocampus in the medial temporal lobe (concerned with memory), are made up of three cellular laminae, whereas six laminae are seen in the **neocortex** (*neopallium*) covering the remaining 90% of the brain.

Cellular laminae of the neocortex
(Figure 29.1)

I The **molecular layer** contains the tips of the apical dendrites of pyramidal cells (see below), and the most distal branches of axons projecting to the cortex from the intralaminar nuclei of the thalamus.

II The **outer granular layer** contains small pyramidal and stellate cells.

III The **outer pyramidal layer** contains medium-sized pyramidal cells and stellate cells.

IV The **inner granular layer** contains stellate cells receiving afferents from the thalamic relay nuclei. Stellate cells are especially numerous in the primary somatic sensory cortex, primary visual cortex, and primary auditory cortex. The term *granular cortex* is applied to these areas. In contrast, the primary motor cortex contains relatively few stellate cells in lamina IV and is called *agranular cortex*.

V The **inner pyramidal layer** contains large pyramidal cells projecting to the corpus striatum, brainstem, and spinal cord.

VI The **fusiform layer** contains modified pyramidal cells projecting to the thalamus.

Columnar organization   (Figure 29.1)

In the somatic sensory cortex, the neurons were discovered (in monkeys) to be arranged functionally in terms of *columns* 50–100 µm in diameter extending radially through all laminae. Within each column, all the cells are modality-specific. For example, a given column may respond to movement of a particular joint but not to stimulation of the overlying skin. Subsequent research has shown that cell columns comprising several hundred neurons are the functional units or *modules* of the cortex. Some modules are activated by specific thalamocortical inputs, others by corticocortical inputs from the same hemisphere, others again by inputs from the opposite hemisphere. Aggregates of modules create a *cortical mosaic*.

Cell types

The three principal morphologic cell types are pyramidal cells, spiny stellate cells, and smooth stellate cells (Figure 29.2).

• **Pyramidal cells** have cell bodies ranging in height from 20 to 30 µm in laminae II and III to more than twice that height in lamina V. Tallest of all, at 80–100 µm, are the *giant cells of Betz* in the motor cortex. The single *apical* dendrite of each pyramidal cell reaches out to lamina I. The several *basal* dendrite branches arising from the basal 'corners' of the cell extend radially within their respective laminae. The apical and basal dendrites branch freely and are studded with dendritic spines. The axons of all pyramidal cells give off recurrent branches, capable of

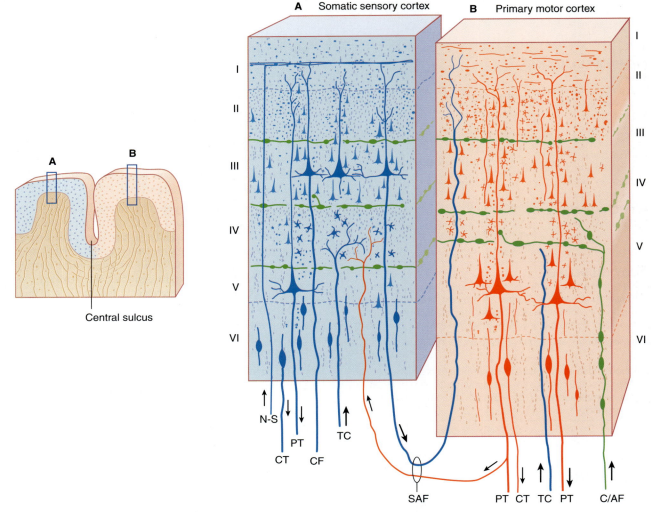

Figure 29.1 Cerebral isocortex. **(A)** Somatic sensory cortex. Cortical laminas I–VI are numbered on the left. Abbreviations, left to right: N-S, non-specific; CT, corticothalamic fiber passing from fusiform neuron to ventral posterior nucleus of thalamus; PT, pyramidal tract fiber running from large pyramidal neuron to a sensory relay nucleus; TC, thalamocortical afferent from ventral posterior nucleus of thalamus; SAF, short association fiber passing to the motor cortex. **(B)** Primary motor cortex. Cortical laminas are numbered on the right. Abbreviations, left to right: SAF, short association fiber collateral passing to somatic sensory cortex; PT, pyramidal tract fiber running to motor nucleus in brainstem or spinal cord; CT, corticothalamic fiber passing to ventral lateral nucleus of thalamus; TC, thalamocortical afferent from ventral lateral nucleus of thalamus; PT as above; C/AF, cholinergic or aminergic fiber.

exciting neighboring pyramidal cells, before leaving the gray matter. All pyramidal cells are excitatory, and use glutamate (or closely related aspartate) as transmitter.

- **Spiny stellate cells** have spiny dendrites and in general are excitatory. They receive most of the afferent input from the thalamus and from other areas of the cortex, and they form glutamatergic synapses on pyramidal cells.

- **Smooth stellate cells** have non-spiny dendrites and are inhibitory. They receive recurrent collateral branches from pyramidal cells, and they form GABAergic synapses on other pyramidal cells. Inhibitory, GABA-secreting neurons make up about 25% of all neurons in the cerebral cortex. Some synapse on the bases of

dendritic spines of pyramidal cells, some synapse on their somas, and some synapse on their initial axonal segments (Figure 29.3). The most powerful of the three types is the *chandelier cell*, so named because of the candle-shaped clusters of axoaxonic boutons. As is the case in the cerebellar cortex (Ch. 25), the GABA neurons exert a focusing action in the cerebral cortex by silencing weakly active cell columns.

- **Bipolar cells** are found mainly in the outer laminae. Most contain one or more peptides, such as vasoactive intestinal polypeptide, cholecystokinin, or somatostatin. Peptides are also coliberated with GABA from many smooth stellate cells.

Human cortical neurons can be captured in fragments of biopsies taken for other purposes. They can be kept

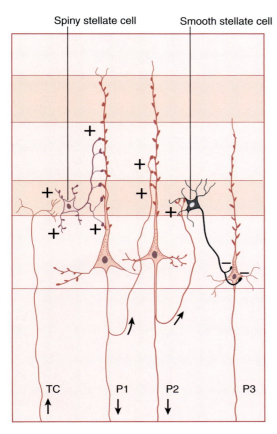

Spiny stellate cell   Smooth stellate cell

Figure 29.2 Input–output connections. Arrows indicate directions of impulse traffic. +,– signs denote excitation, inhibition. Pyramidal cell P1 is excited by the spiny stellate cell; it excites P2 within its own cell column. P3 within a neighboring column is inhibited by the smooth stellate cell. TC, thalamocortical fiber.

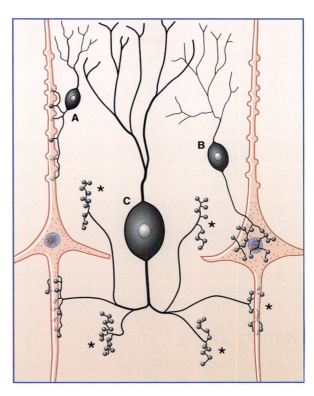

Figure 29.3 Three morphologic types of GABAergic inhibitory neuron. A, axodendritic cell synapsing on the shaft of the apical dendrite of a pyramidal cell; B, basket cell forming axosomatic synapses on a pyramidal cell; C, chandelier cell forming axoaxonic synapses (*) on the initial segments of the two pyramidal cell axons shown, and on four other initial segments not shown. (After DeFelipe 1999, with permission.)

alive for several hours and examined for responses to transmitters and transmitter analogs. It appears that a single pyramidal cell may have as many as 10 different kinds of receptor scattered over its surface. It also appears that the response of the neuron to a particular transmitter is not completely predictable, being modified by concurrent effects of other transmitters.

Afferents

Afferents to a given region of the cortex are derived from five sources.

1 Long and short *association fibers* from small and medium-sized pyramidal cells occupying other parts of the ipsilateral cortex.

2 *Commissural fibers* from medium-sized pyramidal cells projecting through the corpus callosum from matching areas in the opposite hemisphere.

3 *Thalamocortical fibers* from the appropriate specific or association nucleus, for example from the ventral posterior thalamic nucleus to the somatic sensory cortex, from the dorsomedial thalamic nucleus to the *prefrontal* cortex (defined below).

4 *Non-specific thalamocortical fibers* from the intralaminar nuclei.

5 *Cholinergic and aminergic fibers* from basal forebrain, hypothalamus, and brainstem. These fibers are represented in *green* in Figure 29.1.

The relevant nuclei of origin, and the transmitters or modulators involved, are as follow:

- basal forebrain nuclei, acetylcholine
- tuberoinfundibular (hypothalamus), histamine
- tegmentum (midbrain), dopamine
- raphe nucleus (midbrain), serotonin
- cerulean nucleus (pons), norepinephrine (noradrenaline).

These five sets of neurons are of particular relevance to psychiatry and are considered in Chapter 34.

Efferents

All efferents from the cerebral cortex are axons of pyramidal cells, and all are excitatory in nature.

Axons of some pyramidal cells contribute to short or long association fibers. Others form commissural or projection fibers.

- Examples of short association fiber projections are those entering the motor cortex from the sensory cortex and vice versa (Figure 29.1). Examples of long association fiber projections are the numerous backward projections from the *prefrontal cortex*—the cortex anterior to the motor areas (see below) to sensory association areas.

- The commissural fibers of the brain are *entirely* composed of pyramidal cell axons running across in the corpus callosum and anterior commissure (and in other, minor commissures) to matching areas in the opposite hemisphere.

- Projection fibers from the primary sensory and motor cortex form the largest input to the basal ganglia (Ch. 33). The thalamus receives projection fibers from all parts of the cortex. Other major projection systems are corticopontine (to the ipsilateral nuclei pontis), corticonuclear (to contralateral motor and somatic sensory cranial nerve nuclei in pons and medulla), and corticospinal (to anterior horn motor neurons).

CORTICAL AREAS

The most widely used reference map is that of Brodmann, who divided the cortex into 47 areas on the basis of cyto-architectural differences. Most of these areas are shown in Figure 29.4. Colored in that figure are the three principal primary sensory areas (somatic, visual, auditory) and the single primary motor area, together with the respective *unimodal association areas*. The rest of the neocortex comprises *multimodal (polymodal) association areas* receiving association fibers from more than one unimodal association area (e.g. receiving tactile and visual inputs, or visual and auditory).

Investigating functional anatomy

Two dominant methods are in use for localization of functions in the human brain. Both techniques depend on the local increases in blood flow that meet the additional oxygen demand imposed by localized neural activity.

Positron emission tomography

Positron emission tomography (*PET*) measures oxygen consumption following injection of water labeled with oxygen-15 into a forearm vein. ^{15}O is a positron-emitting isotope of oxygen; the positrons react with nearby electrons in the blood to create gamma rays that are counted by gamma ray detectors. Alternatively, fluorine-18–labeled deoxyglucose may be used to measure glucose consumption. ^{18}F-deoxyglucose is taken up by neurons as readily as glucose is.

Image subtraction and *image averaging* are required for meaningful interpretation of PET studies, as explained in the caption to Figure 29.5.

For specialized investigations, radiolabeled drugs are used to quantify receptor function, for example radiolabeled dopamine in the corpus striatum in relation to Parkinson's disease (Ch. 33), radiolabeled serotonin in brainstem and cortex in relation to depression (Ch. 26),

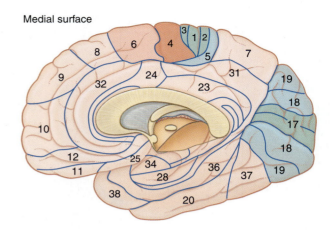

Medial surface

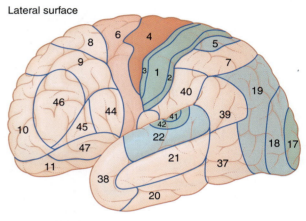

Lateral surface

Figure 29.4 Cytoarchitectural areas of Brodmann.

Colored areas
Motor:
4, primary motor cortex
6 on medial surface, supplementary motor area
6 on lateral surface, premotor cortex.

Sensory:
3/1/2 primary somatic sensory cortex, SI
40, secondary somatic sensory cortex, SII (stippled) on deep surface above insula
17, primary visual cortex
18, 19, visual association cortex
41, 42, primary auditory cortex
22, auditory association cortex.
(*Note*: The primary auditory cortex is not in fact visible from the side, being entirely on the *upper* surface of the superior temporal gyrus.)

and radiolabeled acetylcholinesterase in relation to Alzheimer's disease (Ch. 34).

Functional magnetic resonance imaging

Functional magnetic resonance imaging (Figure 29.6) does not require introduction of any extraneous material. It depends on the different magnetic susceptibility of oxygenated versus deoxygenated blood. As it happens, the local increases in blood flow are more than sufficient to meet oxygen demands, and it is the increase in the ratio of oxyhemoglobin to deoxyhemoglobin that is exploited to generate the magnetic resonance signal.

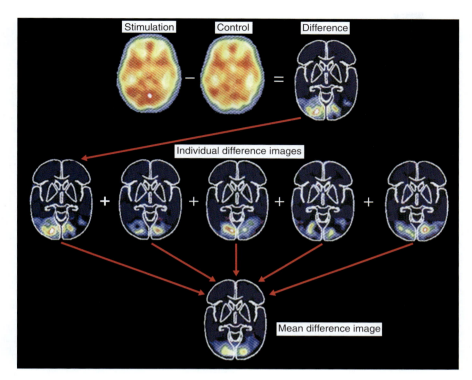

Figure 29.5 Image subtraction and image averaging in positron emission tomography scans.
Top: The middle image is from a control mode, where the subject lies at rest. Uptake of ^{15}O is active throughout the cortex and subcortical gray matter. The left image is from the same subject staring at dots moving on a screen. The high level of background activity obscures the effect. The right image is produced by subtracting the control value to reveal the additional activity in the visual cortex produced by the staring task.
Middle: Four other subjects have performed the same task. Subtraction of background 'noise' reveals varying differences among the five. Because brains vary in size between individuals, activities in all five brains have been projected on to a common, 'average' brain (hence the identical brain profiles in this row).
Bottom: A mean value for the five brains produces a 'mean difference image' representative of the five as a group. (After Posner and Raichle 1994, with permission.)

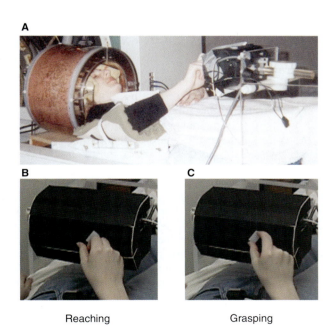

Reaching   Grasping

Figure 29.6 Patient with visual impairment undergoing fMRI investigation during object recognition and object grasping tasks. (Kindly provided by Dr. T.W. James, Vanderbilt University Visual Research Center, Nashville, Tennessee.)

Sensory areas

Somatic sensory cortex (areas 3, 1, 2)

Components

The somatic sensory or *somesthetic cortex* occupies the entire postcentral gyrus (Figure 29.7). Representation of contralateral body parts is inverted except for the face, and the hand, lips, and tongue have disproportionately large representations. The original of the homunculus diagram shown in Figure 29.7A was intended to be only schematic, and ignored the extensive overlap of body part representations.

In vertical sections, represented by Figure 29.8, the somesthetic cortex is divisible into areas 3, 2, and 1. Area 3 is divided into a smaller area 3a, in receipt of information relayed from muscle spindles, and a larger area 3b, receiving information relayed from cutaneous receptors. Area 3b is highly granular and is regarded as the true primary somatosensory cortex (S1).

Modules in area 1 have peripheral receptive fields confined to a single digit (Figure 29.8D). In monkeys, needle electrodes recording from area 1 reveal *feature extraction*, for example some modules are rapidly adapting, some are

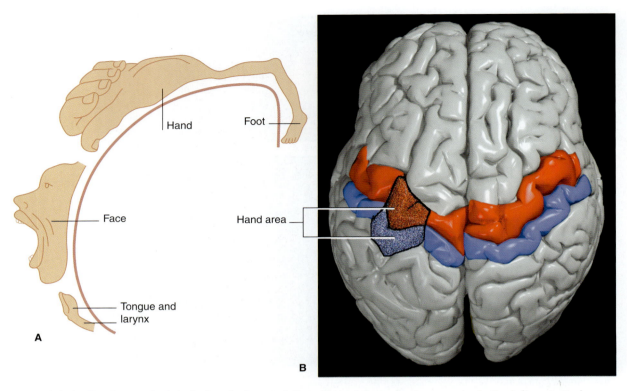

Figure 29.7 **(A)** The figurine on the left depicts the inverted disposition of the motor homunculus in the left precentral gyrus excepting the face. Overlap among the various body parts is not represented in the figurine. (After Penfield and Rasmussen 1960, with permission of Hafner). **(B)** Three-dimensional computerized reconstruction of postmortem brain: the primary motor cortex (*red*) and primary somatosensory cortex (*blue*) viewed from above. The relatively larger representation of the motor and sensory areas in the left hemisphere is typical of right-handed individuals. On the right hemisphere is shown the extensive overlap of the (left) face and tongue representations. (After Kretschmann and Weinrich 1998, with permission of Thieme.)

slowly adapting, some respond only to skin stroking in a specific direction, and some respond only to noxious stimulation of the skin. Modules in area 2 (Figure 29.8E) have multidigit receptive fields and receive from muscles and joint capsules in addition to skin.

Afferents

In addition to thalamic afferents from the ventral posterior nucleus (Figure 29.8B), the somesthetic cortex receives commissural fibers from the opposite somatic sensory cortex through the corpus callosum, and short association fibers from the adjacent primary motor cortex. Many of the fibers from the motor cortex are collaterals of corticospinal fibers traveling to the anterior horn of the spinal cord, and they may contribute to the *sense of weight* when an object is lifted.

It is not unusual for the somesthetic cortex to be compromised by bleeding from a striate branch of the middle cerebral artery supplying the sensory thalamocortical projection, within the upper part of the internal capsule. *Cortical-type sensory loss* in such cases is shown by a reduction in sensory acuity on the opposite side of the body, especially in forearm and hand, evidenced by a raised sensory threshold, poor two-point discrimination, and impaired vibration sense and position sense. The term *stereoanesthesia* is sometimes used with reference to

inability to identify an unseen object held in the hand, as a consequence of cortical-type sensory loss.

Efferents

Efferents from the somesthetic cortex comprise association, commissural, and projection fibers. *Association fibers* pass to the ipsilateral motor cortex, to area 5 and to area 40 (the supramarginal gyrus). *Commissural fibers* pass to the contralateral somesthetic cortex. *Projection fibers* descend within the posterior part of the pyramidal tract (PT) and terminate on internuncial neurons in sensory relay nuclei, namely the ventral posterior nucleus of the thalamus of the same side, and the posterior column and spinal posterior gray horn of the opposite side. As explained in Chapter 16, sensory transmission in the spinothalamic pathway may be suppressed (via inhibitory internuncials) during vigorous activities such as running, whereas in the posterior column–medial lemniscal pathway, transmission may be enhanced (via excitatory internuncials) during exploratory activities such as palpation of textured surfaces.

Somatic sensory association area

This term is used with respect to area 5, directly behind the somatic sensory cortex. Most area 5 modules are

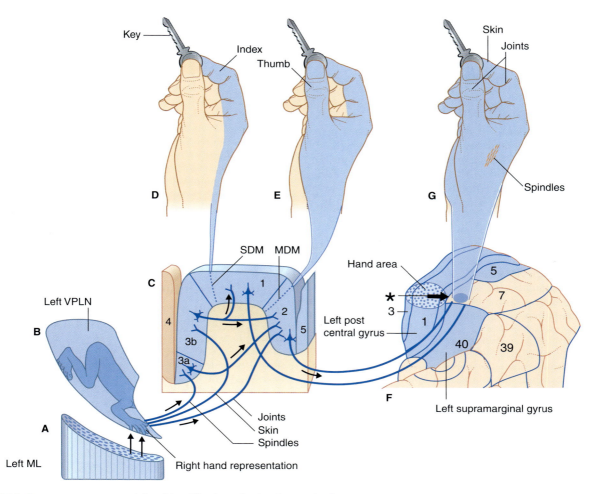

Figure 29.8 Sensory sequence enabling identification of a key by touch alone.
(A) Coded information from the *right* hand is traveling along second-order sensory (crossed) neurons in the medial part of the *left* posterior column–medial lemniscal pathway in the upper brainstem. ML, medial lemniscus.
(B) The hand area of the ventral posterior lateral nucleus (VPLN) of the thalamus contains somas of third-order sensory neurons.
(C) The third-order neurons project to areas 3, 1 (indirectly), and 2 of the somatic sensory cortex.
(D) SDM, single digit module.
(E) MDM, multidigit module.
(F) Surface view of the left parietal lobe. The asterisk indicates the level of section C through the hand area. Area 7 receives short association fibers from areas 1, 2, and 5, and integrates information from skin, muscle spindles, and joint capsules.
(G) Connections with tactile memory stores here and in area 5 enable an image of the key to be perceived without the aid of vision.

active during reaching movements of the contralateral arm taking place under visual guidance (the *dorsal visual pathway* is discussed later).

Superior parietal lobule

Clinically, the term *superior parietal lobule* is equated with area 7. The *lower* part of area 7 receives inputs from areas 1, 2, and 5. Receipt of tactile and proprioceptive information from skin, muscles, and joints causes area 7 to tap into its own memory stores concerning the shapes of objects held in the (opposite) hand, whereby an unseen object can be identified (Figure 29.7). This ability is known as *stereognosis*, a term signifying ability to perceive objects 'in the round'.

The *upper* part of area 7 contains a cell station in the 'Where?' visual pathway (see later).

Inferior parietal lobule

The inferior parietal lobule comprises areas 39 (angular gyrus) and 40 (supramarginal gyrus). Both are concerned with language, a mainly left hemisphere function described in Chapter 32.

Intraparietal cortex

The cortex in the walls of the intraparietal sulcus is especially active during tasks involving visuomotor coordination, for example reaching for and grasping objects identified in the contralateral visual field and subjecting them to simultaneous visual and tactile three-dimensional analysis. This area includes the parietal eye field (PEF) (see later).

Secondary somatic sensory area

On the medial surface of the parietal operculum of the insula is a small *secondary somatic sensory area* (SII). It receives a nociceptive projection from the thalamus, and it is highlighted during PET scans of the brain during peripheral painful stimulation (Ch. 34). SII also appears to collaborate with SI in aspects of tactile discrimination.

Plasticity of the somatic sensory cortex

In monkeys, cortical sensory representations of the individual digits of the hand can be defined very exactly by recording the electrical response of cortical cell columns to tactile stimulation of each digit in turn. These digital maps can be altered by peripheral sensory experience, as the following experiments indicate.

- The median nerve supplies the ventral surface of the outer three digits of the hand, whereas the radial nerve supplies their dorsal surfaces. If the median nerve is crushed, the representation of the dorsal surface on the digital map increases at the expense of the ventral representation. The increase begins within hours and progresses slowly over a period of weeks. With regeneration of the median nerve, the cortical map reverts to normal.

- If the middle digit is denervated, the corresponding cortical area is unresponsive for a few hours, then becomes progressively (over weeks) taken over by expansion of the representations of the second and fourth digits.

- If the pad skin of a digit is chronically stimulated, for example by having to press a rotating sanded disk in order to release pellets of food, representation of the pad may increase to twice its original size over a period of weeks, reverting to normal after the experiment is discontinued.

These experiments show that somatic sensory maps are *plastic*, being modified by peripheral events. A purely anatomic explanation (e.g. sprouting of nerve branches within the central nervous system, or peripherally) is not appropriate for the earliest changes, which begin within hours. Instead, they can be accounted for on the basis of sensory competition.

Sensory competition

Sensory maps made at the level of the posterior gray horn, posterior column nuclei, thalamus, and somesthetic cortex all show evidence of anatomic overlap. For example, the thalamocortical somesthetic projection for the third digit overlaps the projections for the second and fourth. Within the zone of overlap, cortical columns are shared by afferents from two adjacent digits. As already explained, smooth stellate cells exert lateral inhibition on weakly stimulated columns. Under experimental conditions (in cats), the number of columns responding to a particular

thalamocortical input can be increased by local infusion of a GABA antagonist drug (bicuculline), which suppresses lateral inhibition. The effect of removal of a peripheral sensory field may be comparable: if one set of thalamocortical neurons falls silent owing to loss of sensory input, it no longer exerts lateral inhibition, and cortical columns within its territory are taken over by neighboring, active sets.

In the human somatosensory body map, the digits are represented next to the face. In several well-documented cases of upper limb amputation, patients had later experiences of 'phantom finger' sensations on touching their face on that side with an implement such as a comb held in the other hand. This illusion may occur within 2 weeks of amputation. It can be explained on the basis of the unmasking of preexisting overlap of thalamocortical neurons.

Visual cortex (areas 17, 18, 19)

The visual cortex comprises the *primary visual cortex* (area 17) and the *visual association cortex* (areas 18 and 19).

Primary visual cortex   (Figure 29.4)

As noted in Chapter 28, the primary visual cortex is the target of the geniculocalcarine tract, which relays information from the ipsilateral halves of both retinas, and therefore from the contralateral visual field. This myelinated tract creates a pale **visual stria** within the primary visual cortex before synapsing on spiny stellate cells of the highly granular lamina IV. The visual stria (first noted by medical student Francesco Gennari circa 1775) has provided the alternative name, *striate cortex*, for area 17.

The spiny stellate cells belong to *ocular dominance columns*, so named because alternating columns are dominated by inputs from the left and right eyes (Ch. 28). In a surface view of the visual cortex, the columnar arrangement takes the form of whorls, resembling fingerprints. The geniculocalcarine projection is so ordered that matching points from the two retinas are registered side by side in contiguous columns. This arrangement is ideal for binocular vision, because modules at the edge of a column respond to inputs from both eyes.

Under experimental conditions (in monkeys), spiny stellate cells of the primary visual cortex give 'simple' responses to slits of light of a particular orientation. Some of the pyramidal cells give 'complex' responses to bars (broad slits) of a particular orientation; for many cells, the bar must be moving broadside in a specific direction. Other pyramidal cells are 'hypercomplex', responding to L shapes. This hierarchy of responses can be explained on the basis of convergence of several simple cell axons on to complex cells and convergence of complex cell axons on to hypercomplex cells.

Plasticity of the primary visual cortex

The basic pattern and balance of ocular dominance columns is now known to be established before birth, and it is preserved in animals reared in complete darkness. If one eye is deprived of sensory experience in childhood, the corresponding cortical columns remain small and those from the visually experienced eye become larger than normal.

Visual association cortex (Figure 29.4)

The visual association cortex comprises areas 18 and 19, which are also conjointly called the *peristriate* or *extrastriate* cortex. Afferents are received mainly from area 17, but they include some direct thalamic projections from the pulvinar. The cell columns are concerned with *feature extraction*. Some columns respond to geometric shapes, some respond to color, and some are involved in stereopsis (depth perception).

Many of the peristriate columns have large receptive fields. Some of these straddle the physiologic 'blind spot' (optic nerve head) and may be responsible for 'covering up' the blind spot during monocular vision.

The projection from the pulvinar to the visual association cortex is considered to be part of the pathway involved in 'blind sight'. This remarkable condition has been observed in patients following thrombosis of the calcarine branch of the posterior cerebral artery. Although blindness in the contralateral field appears complete, these patients are nonetheless able to point to a moving spot of light—without any perception of it, merely a 'feeling' that it is there. The likely pathway concerned is via the medial root of the optic tract, the superior colliculus, and the pulvinar.

The most functionally advanced modules occupy the lateral and medial parts of area 19. The lateral set of modules is colloquially described as belonging to a dorsal, 'Where?' visual pathway. The medial set belongs to a ventrally placed 'What?' pathway.

The 'Where?' visual pathway (Figure 29.9)

Consistent with electrical recordings taken from alert monkeys, PET scans of human volunteers reveal the lateral part of area 19 to be especially responsive to *movement*

taking place in the contralateral visual hemifield. The main projection from this area is to area 7, known to clinicians as the *posterior parietal cortex*. In addition to movement perception, area 7 is involved in *stereopsis* (three-dimensional vision) and with *spatial sense*, defined as perception of the position of objects in relation to one another.

Area 7 receives 'blind sight' fibers from the pulvinar, and it projects via the superior longitudinal fasciculus to the ipsilateral frontal eye field (FEF) and premotor cortex (PMC).

In monkeys, cell columns in area 7 are activated when a significant object (e.g. fruit) appears in the contralateral visual hemifield. Through association fibers, the active cell columns increase the resting firing rate of columns in the FEF and PMC, but without producing movement. The effect is called *covert attention*, or *covert orientation*. It becomes *overt* when the animal responds with a saccade with or without a reaching movement directed toward the object. Following a lesion to area 7, the motor responses to significant targets occur late, and reaching movements of the contralateral arm are inaccurate.

In human volunteers, PET scans show increased cortical metabolism in area 7 in response to object movement in the contralateral visual hemifield. During reaching of the opposite arm toward an object, areas 5 and 7 are both active. In humans (as in monkeys), a lesion that includes area 7 is associated with clumsy, inaccurate reaching into the contralateral visual hemifield.

In volunteers, two additional areas of cortex become active when items of special interest appear. Shown in Figure 29.9, and mentioned again later, is the *dorsolateral prefrontal cortex* (DLPFC), a significant decision-making area, notably in relation to an *approach* or *withdraw* decision. Shown in Figure 29.10 is a patch in the cortex of the anterior cingulate gyrus. This area is considered in Chapter 34, but it may be mentioned here that it is activated by the dorsolateral cortex when subjects are *paying attention* to a visual task.

The 'What?' visual pathway (Figure 29.10)

The ventral visual pathway converges on to the anteromedial part of area 19, mainly within the fusiform gyrus part of the occipitotemporal gyrus. This region is concerned with three kinds of visual identification, indicated in Figure 29.10B.

1 Relatively lateral are modules activated by the *forms* (shapes) of objects of all kinds, including the shapes of letters. It is regarded as a center for generic (categoric or canonic) object identification (e.g. a dog *as such*, without connotation).

2 In the midregion are modules specifically devoted to the generic identification of human faces.

3 Relatively medial is the *color recognition area*, essential for recognition of all colors except black and white. A state of *achromatopsia* may occur following a sustained fall in blood pressure within both posterior cerebral arteries, caused, for example, by an embolus blocking the top of the parent basilar artery. Such patients see everything only in black and white (gray scale).

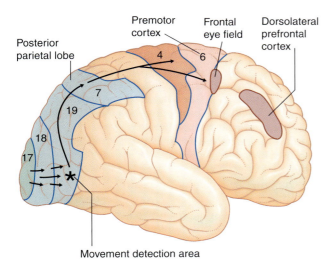

Posterior parietal lobe

Premotor cortex

Frontal eye field

Dorsolateral prefrontal cortex

Movement detection area

Figure 29.9 Lateral surface of right hemisphere, showing the 'Where?' visual pathway from the visual cortex to the parietal and frontal lobes. The asterisk marks the area for detection of movement in the left visual field. Activity of the right frontal eye field facilitates a saccade toward the left visual field.

- V4 includes the three sets of identification modules in the fusiform gyrus (anteromedial Brodmann area 19).
- V5 equates with the movement detection modules in the lateral occipital cortex (anterolateral Brodmann area 19).

Auditory cortex (areas 41, 42, 22)

The *primary auditory cortex* occupies the anterior transverse temporal gyrus of Heschl, described in Chapter 20. Heschl's gyrus corresponds to areas 41 and 42 on the upper surface of the superior temporal gyrus. Columnar organization in the primary auditory cortex takes the form of *isofrequency stripes*, each stripe responding to a particular tonal frequency. Higher frequencies activate lateral stripes in Heschl's gyrus, lower frequencies activate medial stripes. Because of incomplete crossover of the central auditory pathway in the brainstem (Ch. 20), *each ear is represented bilaterally*. In experimental recordings, the primary cortex responds equally well from both ears in response to monaural stimulation, but the contralateral cortex is more responsive during simultaneous binaural stimulation.

The auditory association cortex corresponds to area 22, for speech perception (considered in Ch. 32). Visual and auditory data are brought together in the polymodal cortex bordering the superior temporal sulcus (junction of areas 21 and 22).

Excision of the entire auditory cortex (in the course of removal of a tumor) has no obvious effect on auditory perception. The only significant defect is loss of *stereoacusis*: on testing, the patient has difficulty in appreciating the direction and the distance of a source of sound.

Motor areas
Primary motor cortex

The primary motor cortex (area 4) is a strip of agranular cortex within the precentral gyrus. It gives rise to 60–80% (estimates vary) of the PT. The remaining PT fibers originate in the premotor and supplementary motor areas and in the parietal cortex, as illustrated in Chapter 16.

There is an inverted somatotopic representation of contralateral body parts except the face, with relatively large areas devoted to the hand, circumoral region, and tongue (Figure 29.7). The hand area can usually be identified as a backward projecting knob 6–7 cm from the upper margin of the hemisphere.

Ipsilateral body parts are also represented in the somatotopic map, ipsilateral motor neurons being supplied by the 10% of PT fibers that remain uncrossed.

A computerized graphic reconstruction of the motor cortex and corticospinal tracts from a postmortem brain is shown in Figure 29.11.

Direct stimulation of the human motor cortex indicates that the cell columns control *movement direction*. Individual PT fibers are known to branch extensively as they approach the anterior gray horn, and to terminate on motor neuronal dendrites in nuclei serving several different muscles. The pattern of distribution of PT fibers is directed toward *movement synergy*, which in this context means the simultaneous contraction of all the muscles

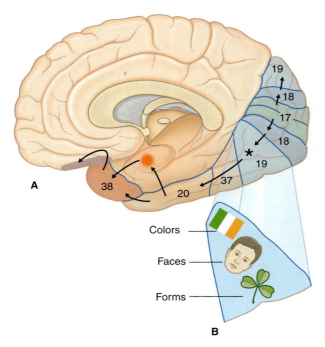

Figure 29.10 (A) Medial view of right hemisphere, showing the 'What?' pathway. The asterisk marks the visual identification area within the fusiform gyrus on the inferior surface. Ventral area 19 is enlarged in **(B)**.

Recognition of individual objects and faces is a function of the anterior part of the 'What?' pathway in the *inferotemporal cortex* (area 20) and in the cortex of the temporal pole (area 38). These two areas are engaged during identification of, for example, *Mary's* face or *my* dog. Failure of facial recognition (*prosopagnosia*) is a frequent and distressing feature of Alzheimer disease (Ch. 34), where the patient may cease to recognize family members despite retaining the sense of familiarity of common objects.

Threatening sights or faces cause areas 20 and 38 to activate the *amygdala*, especially in the right (emotional) hemisphere; the right amygdala in turn activates the fear-associated *right orbitofrontal cortex* (see Ch. 30).

How are visual association areas activated, for example in execution of a decision to look for an apple in a bowl of mixed fruit, or for a particular word in a page of text? In PET studies, the frontal lobe is active whenever attention is being paid to a task at hand. The DLPFC is particularly active during visual tasks involving form and color. During visual searching, the role of the frontal lobe seems to be to activate memory stores within the visual association areas, so that the relevant memories are held online during the search. The anterior part of the cingulate cortex is also active.

The V1–V5 nomenclature
Specialists in vision research use the following designations in relation to cortical visual processing.

- V1 equates with Brodmann area 17.
- V2 and V3 equate with Brodmann areas 18 and 19, respectively.

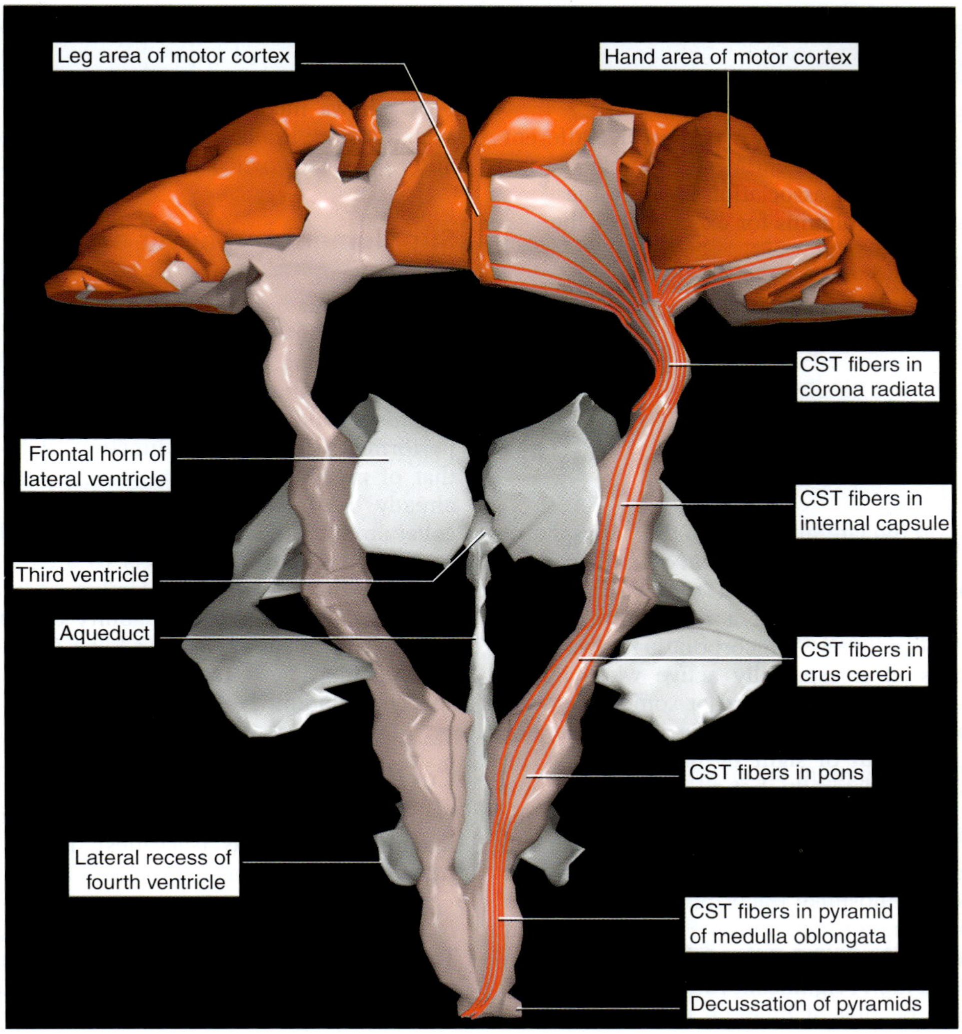

Figure 29.11 Computerized graphic reconstruction of postmortem brain; anterior view showing the precentral gyri and the relationships of the corticospinal tracts (CST) to the ventricular system. Some corticospinal fibers have been added to the original. (After Kretschmann and Weinrich 1998, with permission of Thieme.)

concerned, with a bias among them suited to the task at hand. The act of picking up a pen, for example, requires a moderate contraction of opponens pollicis as prime mover, a matching level of contraction of the portion of flexor digitorum profundus providing the tendon to the terminal phalanx of the index finger, and lesser levels of contraction of adductor and flexor brevis pollicis. Steadying the upper limb as a whole during any kind of manipulative activity is a function of the PMC (see later).

Plasticity in the motor cortex
In monkeys and in lower mammals, small lesions of the motor cortex produce an initial paralysis of the corresponding body part, followed within a few days (sometimes within hours) by progressive recovery. The recovery is attributable to a change of allegiance of cell columns close to the lesion, which take on the missing motor function. Instead of inflicting a lesion, it is possible to enlarge the motor territory of a patch of cortex merely by injecting a GABA antagonist drug locally into the cortex.

Expansion of motor territories at spinal cord level is already provided for by extensive overlap of projections from area 4 to the motor cell columns in the ventral gray horn.

Sources of afferents to the primary motor cortex

1 The *opposite motor cortex*, through the corpus callosum. The strongest commissural linkages are between matching cell columns that control the vertebral and abdominal musculature. This is to be expected, because these muscle groups routinely act bilaterally in maintaining the upright position of the trunk and head. The weakest commissural linkages are between cell columns controlling the distal limb muscles, where the two sides tend to act independently.

2 *Somatosensory cortex*. *Cutaneous* cell columns in areas 1, 2, and 3 feed forward via short association

fibers. Linkages for the hand are especially numerous; the distance is short, because the hand areas of the motor and somatic sensory cortex mainly occupy the

contribute to recovery of function in cases of pure motor hemiplegia (Ch. 35) following a vascular lesion confined to the corticospinal tract within the corona radiata. The

Core Information

The cerebral cortex has both a laminar and a columnar organization. The two basic cell types are pyramidal and stellate. Pyramidal cells occupy laminae II, III, V, and (as fusiform cells) lamina VI. Lamina IV is rich in spiny stellate cells. Small pyramidal cells link the gyri within the hemisphere; medium-sized pyramidal cells link matching areas of the two hemispheres; the largest ones project to thalamus, brainstem, and spinal cord. Spiny stellate cells are excitatory to pyramidal cells, smooth ones are inhibitory. Columnar organization takes the form of cell columns 50–100 µm wide.

The somatic sensory cortex contains an inverted representation of body parts (except the head). Important inputs come from the ventral posterior nucleus of thalamus; important outputs go to the primary motor and inferior parietal cortex. The primary visual cortex receives the geniculocalcarine tract. Cellular responses of differing complexity depend on convergence of simpler on to more complex cell types.

The visual association areas are characterized by feature extraction, for example motion, color, shape. Form and color extraction continues into the cortex on the underside of the temporal lobe, motion into the posterior parietal lobe. The primary auditory cortex occupies the upper surface of the superior temporal gyrus, and the auditory association cortex is lateral to it.

The primary motor cortex occupies the precentral gyrus. It gives rise to most of the pyramidal tract, the body parts (except the head) being represented upside down. Its main inputs are from somatosensory cortex, cerebellum (via the ventral posterior nucleus of thalamus), and premotor and supplementary motor areas. The premotor area operates mainly in response to external cues, the supplementary motor area in response to internally generated cues. Under control of the dorsolateral prefrontal cortex, four distinct cortical areas are involved, in different contexts, in producing contraversive saccades.

REFERENCES

Ashe J, Ugurbil K. Functional imaging of the motor system. Curr Opin Neurobiol 1994; 4:832–839.

Behrmann M, Geng JJ, Shomstein S. Parietal cortex and attention. Curr Opin Neurobiol 2004; 14:210–217.

Celesia CG. Anatomy and physiology of the visual system. J Clin Neurophysiol 1994; 11:482–492.

Crowley JC, Katz LC. Ocular dominance development revisited. Curr Opin Neurobiol 2002; 12:104–109.

Damasio AR. The somatic marker hypothesis and the possible functions of the prefrontal cortex. In: Roberts AC, Robbins TW, Weiskrantz L, eds. The prefrontal cortex: executive and cognitive functions. Oxford: Oxford University Press; 1998:36–50.

DeFelipe J. Chandelier cells and epilepsy. Brain 1999; 122:1807–1822.

Desomone R, Duncan J. Neural mechanisms of selective attention. Ann Rev Neurosci 1995; 18:193–222.

Donoghue JP, Saines JN. Motor areas of the cerebral cortex. J Clin Neurophysiol 1994; 11:382–396.

Edeline J-M. Learning-induced physiological plasticity in the thalamocortical sensory systems. Prog Neurobiol 1999; 57:165–224.

Elliott LL. Functional brain imaging and hearing. J Acoust Soc Am 1994; 96:1397–1408.

Frith CD, Dolan RJ. Higher cognitive processes. In: Frackowiak RSJ, Friston KJ, Frith CD, et al, eds. Human brain function. London: Academic Press; 1997:329–366.

Garoff RJ, Slotnick SD, Shacter DL. The neural origins of specific and general memory: the role of the fusiform cortex. Neuropsychologia 2005; 43(6):847–859.

Ghose GM. Learning in mammalian sensory cortex. Curr Opin Neurobiol 2004; 14:513–518.

Goel V, Gold B, Kapur S, et al. Neuroanatomical correlates of human reasoning. J Cognit Neurosci 1998; 10:293–302.

Grefkes C, Fink GR. The functional organization of the intraparietal sulcus in humans and monkeys. J Anat 2005; 207:3–17.

Halsband U, Freund H-J. Motor learning. Curr Opin Neurobiol 1993; 3:940–949.

Innocenti G. Some new trends in the study of the corpus callosum. Behav Brain Res 1994; 64:1–8.

Kaas JH. Plasticity of sensory and motor maps in adult mammals. Ann Rev Neurosci 1991; 14:137–167.

Kleinschmidt A, Nitschke MF, Frahm J. Somatotopy in the human motor cortex hand area. Eur J Neurosci 1997; 9:2178–2186.

Klintsova AY, Greenough WT. Synaptic plasticity in cortical systems. Curr Opin Neurobiol 1999; 9:203–208.

Kretschmann H-J, Weinrich W. Neurofunctional systems: 3D reconstructions with correlated neuroimaging: text and CD-ROM. New York: Thieme; 1998.

Krimer LS, Muly EC, Williams GV, et al. Dopaminergic regulation of cerebral cortical microcirculation. Nature Neurosci 1998; 1:296–299.

Mesulam M-M. From sensation to cognition. Brain 1998; 121:1013–1052.

Munoz DP. Commentary: saccadic movements: overview of neural circuitry. NeuroReport 2002; 13:2325–2330.

Passingham D, Sakai K. The prefrontal cortex and working memory: physiology and brain imaging. Curr Opin Neurobiol 2004; 14:163–168.

Paulesu P, Frackowiak RSJ, Bottini G. Maps of somatosensory systems. In: Frackowiak RSJ, Friston KJ, Frith CD, et al, eds. Human brain function. London: Academic Press; 1997:367–404.

Penfield W, Rasmussen T. The cerebral cortex of man. New York: Hafner; 1960.

Posner MI, Raichle ME. Images of the brain. In: Posner MI, Raichle ME, eds. Images of the mind. New York: Scientific American Library; 1994:57–82.

Posner MI. Attention: the mechanisms of consciousness. Proc Natl Acad Sci USA 1994; 91:7398–7403.

Salin P-A, Bullier J. Corticocortical connections in the visual system: structure and function. Physiol Rev 1995; 75:107–154.

Schwartz AB. Distributed motor processing in cerebral cortex. Curr Opin Neurobiol 1994; 4:840–846.

Somogyi P, Tamas G, Lujan R, et al. Salient features of synaptic organisation in the cerebral cortex. Brain Res Rev 1998; 29:111–135.

Tanji J. The supplementary motor area in the cerebral cortex. Neurosci Res 1994; 19:251–298.

Ungerlieder LG. Functional brain imaging studies of cortical mechanisms for memory. Science 1995; 270:769–775.

Weinberger NM. Dynamic regulation of receptive fields and maps in the adult sensory cortex. Ann Rev Neurosci 1995; 18:129–158.

Yousry TA, Schmid UD, Alkadhi H, et al. Localization of the motor hand area to a knob on the precentral gyrus. Brain 1997; 120:141–157.

Zilles K, et al. Mapping of human and macaque sensorimotor areas by integrating architectonic, transmitter receptor, MRI and PET data. J Anat 1995; 187:515–538.

Zilles K, Palomero-Gallagher N, Schleicher A. Transmitter receptors and functional anatomy of the cerebral cortex. J Anat 2004; 205:417–432.

Electroencephalography

STUDY GUIDELINES
The electroencephalogram is an important neurologic tool for three main reasons.

1 It is central to the study of the physiology of sleep and the pathophysiology of sleep disorders.

2 It is essential for analysis of seizures and for monitoring progress during their treatment.

3 It is a valuable aid to localization of space-occupying lesions, including brain tumors.

We suggest you review the amino acid transmitters in Chapter 8 before reading about drug therapy in the Clinical Panel on seizures.

NEUROPHYSIOLOGIC BASIS OF THE EEG

Since its initial development, electroencephalography has remained a unique tool for the study of cortical function and a valuable supplement to history, physical examination, and information gained by radiologic studies.

When small metallic disk electrodes are placed on the surface of the scalp, oscillating currents of 20–100 μV can be detected and are referred to as an *electroencephalogram (EEG)*. Their origin is a direct consequence of the additive effect of groups of cortical pyramidal neurons being arranged in radial (outward-directed) columns. The columns relevant here are those beneath the surface of the cortical gyri. As the membrane potentials of these columns fluctuate, an electrical dipole (adjacent areas of opposite charge) develops. The dipole results in an electrical *field potential* as current flows through the adjacent extracellular space as well as intracellularly through the neurons (Figure 30.1). It is the extracellular component of this current that is recorded in the EEG, and variations in both the strength and density of the current loops result in its characteristic sinusoidal waveform.

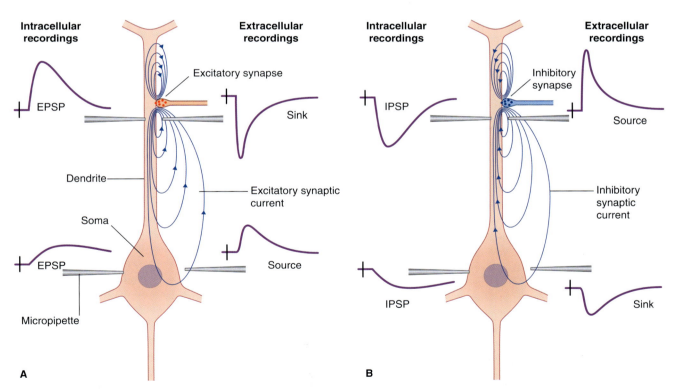

Figure 30.1 The contribution of individual excitatory and inhibitory synaptic currents to the extracellular field potentials. Micropipettes are being used to sample intracellular and extracellular events. **(A)** *Intracellular recordings* show that the excitatory synapse generates a rapid excitatory postsynaptic potential (EPSP) at the synaptic site on the dendrite, and a slower and smaller EPSP at the soma. *Extracellular recordings* show that the *source* (positive) of excitatory synaptic current flows outward through the membrane of proximal dendrite and soma, and inward (the *sink*) at the synaptic site. **(B)** An inhibitory synapse is seen to have the opposite effect. The inhibitory postsynaptic potential (IPSP) is associated with a current source at the synaptic site and a sink along the proximal dendrite and soma.

The oscillations of the EEG (measured in mV) are thought to be generated by reciprocal excitatory and inhibitory interactions of neighboring cortical cell columns.

TECHNIQUE

After careful preparation of the skin of the scalp to ensure good contact, electrodes are affixed in a placement that is in conformity with the *10–20 International System of Electrode Placement*, when the scalp is divided into a grid in accordance with Figure 30.2.

By defining a consistent placement of electrodes, direct comparison with follow-up studies is feasible, as is a method to compensate for differences in head size. Each electrode placement allows it to preferentially record over a cortical surface area of approximately 6 cm². The nomenclature employed to define each electrode position combines a letter with a number, as shown in Figure 30.2.

Actual EEG recordings are made from all sites simultaneously. The potential difference between electrode pairs is recorded (as a rule), and this is displayed as a separate individual graph or channel. Often, other physiologic recordings are performed at the same time (e.g. an electrocardiograph and/or a surface electromyograph).

If varying pairs of electrodes are used, the *montage* (output) is termed *bipolar* (Figure 30.3A). If they have one recording site in common (auricle, or mastoid area) it is called *referential* (Figure 30.3B).

TYPES OF PATTERN

Normal EEG rhythms
Awake-state EEG

The EEG demonstrates prominent changes both with the level of alertness and during the various stages of sleep.

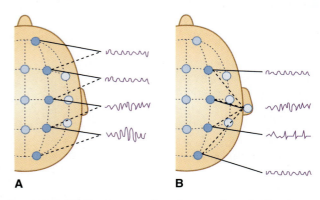

Figure 30.3 (A) Bipolar recording. A succession of adjacent pairs of electrodes is used. Only four sample tracings are shown. **(B)** Referential recording. The reference electrode is attached to the ear in this example. Again, only four sample tracings are shown.

Each of these patterns is specific and is taken into account during the EEG interpretation. A routine EEG study will usually take 30–45 min and will include recordings made during wakefulness and during early stages of sleep, because specific abnormalities (especially epileptiform ones) may be detected only during the sleep portion of the recording.

In the alert awake state (Figure 30.5A), the montage is described as *desynchronized*, because the wave forms are quite irregular. The *background frequency* is usually around 9.5 Hz. A *beta frequency* of more than 14 Hz may be superimposed over anterior head regions.

In a relaxed state with the eyes closed, rhythmic waveforms appear in the *alpha frequency* (8–14 Hz), notably over the parietooccipital area (Figure 30.5B).

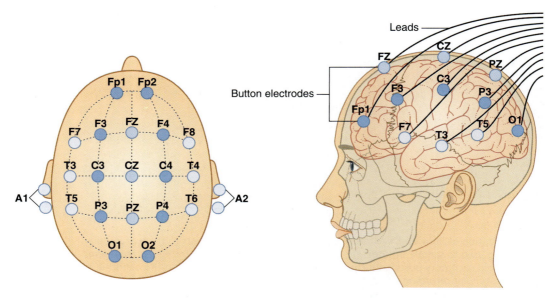

Figure 30.2 Deployment of surface electrodes on the scalp. *Letters*: C, coronal; F, frontal; Fp, frontopolar; O, occipital; P, parietal; T, temporal; Z, midline. *Numbers*: Odd numbers, left side; even numbers, right side. A1, A2, reference electrode positions (see text).

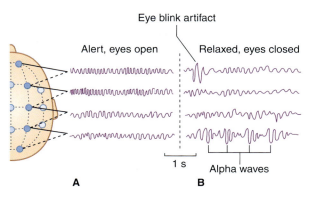

Figure 30.5 Electroencephalogram in the awake state. **(A)** Subject is alert with eyes open. Beta waves are seen in Fp2–F4 and F4–C4. **(B)** Subject is relaxed with eyes closed. An eye blink artifact is seen in the Fp2–F4 tracing. Alpha waves are seen in C4–P6. The beta waves are characterized by low amplitude and high frequency, the alpha waves by rhythmic waxing and waning.

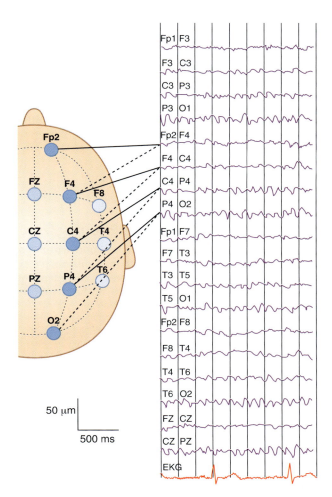

Figure 30.4 A complete set of normal tracings is shown, tagged in accordance with the nomenclature in Figure 30.2. An electrocardiogram has been taken simultaneously. Note the low amplitude of the waves (20 mV or less) and their high frequency in this 2-s sample.

Normal sleep EEG

Glossary

- **REM sleep**. Dreamy light sleep accompanied by *rapid eye movements*; also called *paradoxical sleep* because of EEG resemblance to the awake state.

- **NREM sleep**. *Non–rapid eye movement* sleep (stages 1–4); also called *slow-wave sleep*.

People normally pass through three to five sleep cycles per night. The sequence of events is summarized in Figure 30.6. Alpha rhythm becomes apparent (on occipital leads) during drowsiness with eyes closed.

By general agreement, sleep proper is associated with slow-wave patterns in the EEG. There is a rapid descent through stage 1, characterized by a steady theta rhythm, into stage 2, characterized by theta waves interrupted by sinusoidal waveforms called *sleep spindles*, and by occasional *K-complex* spikes. Stages 3 and 4 are characterized by slow, *delta waves*—hence the term *slow-wave sleep* for these two stages.

It is generally agreed that the waxing and waning of cortical activity during slow-wave sleep has its origin in

the thalamus, where the relay nuclei projecting to the cortex also enter a rhythmic discharge mode during slow-wave sleep. This rhythm is characterized by a succession of hyperpolarized states succeeded by depolarized states exhibiting bursts of firing. The vigorous firing is triggered by momentary opening of voltage-gated calcium channels. The *transient* (momentary) opening accounts for the term *T channels* applied to these.

As described in Chapter 27, thalamocortical projections pass through an inhibitory shell in the form of the thalamic reticular nucleus, with reciprocal connections to parent relay cells as shown in Figure 27.4. Burst firing excites the reticular nucleus, which in turn causes the relay neurons to become hyperpolarized by opening the GIRK potassium channels (Figure 8.11).

The rhythmic waxing and waning of thalamic neurons is attributed to a pulsatile discharge pattern inherent to the cells of the reticular nucleus. Exaggeration of the normal spike and wave pattern is found in a common form of epilepsy known as *absence* seizures (see later).

After about an hour's sleep, the stage 2 wave pattern is repeated and is succeeded by a longer period of slow-wave sleep. Up to a minute may then be spent in REM sleep, a dream state accompanied by:

- visual imagery
- rapid eye movements generated by the extraocular muscle contractions
- electromyographic silence in the musculature of trunk and limbs
- an EEG beta rhythm characteristic of the waking state—hence the term *paradoxical sleep*.

Rapid eye movement sleep is the dominant state during the final two of the 8 h spent in bed. Although the significance of dreams is a matter of endless debate, activation of the visual cortex is brought about by the *PGO pathway*, from pontine reticular formation to lateral geniculate body to occipital cortex. In individuals blind from birth, dreams have a purely auditory content, perhaps associated with activation of the lateral geniculate body.

In the clinic, because of the brief time that an EEG is recorded, a patient does not often cycle through REM sleep patterns. It should be noted that, in normal circumstances, REM sleep almost never occurs during the *first* descent into sleep. Should it do so, the strange sleep disorder called *narcolepsy* should come to mind (Clinical Panel 30.1).

EEG activation procedures

In order to check for susceptibility to seizures, a routine EEG includes a brief period of hyperventilation and another of photic stimulation using strobe (flickering) light at varying frequencies (Figure 30.6).

Maturation of wave format

Accurate interpretation of printouts necessitates familiarity on the part of the electroencephalographer with those patterns that are age-specific and hence represent normal maturational stages of development. Earliest recordings, from premature babies, show only imtermittent electrical activity that is asynchronous between the hemispheres. Continuous symmetric activity emerges during childhood, with increasingly distinctive wakeful and sleeping patterns. During early teenage years, the adult EEG pattern is established.

Abnormal EEG rhythms
Focal abnormalities without seizures

Focal slowing
Focal slowing, in the form of delta waves (Figure 30.8), indicates presence of a mass of some kind.

Phase reversal
Focal spike or sharp wave discharges are occasionally seen over localized areas of the cortex. Both appear as abrupt events, individual spikes lasting 80–100 msec and sharp waves 100–200 msec. Such discharges project to the surface with a negative polarity and the point of *phase reversal* between adjacent electrodes defines their EEG localization (Figure 30.9). In anterior temporal and frontal areas, they may be indicative of an *ictal focus*, a term denoting a locus of origin of seizures. In the occipital region any correlation is usually with visual impairment.

Generalized abnormalities without seizures

Disorders that cause generalized dysfunction within cortical or subcortical structures result in a diffuse pattern of abnormalities on EEG. Such disorders include hypoglycemia, hypoxia, and dementia. This can be manifested by replacement of normal background frequency by diffuse slowing. Disorders involving the white matter of the brain are more often associated with delta waves and are often polymorphic (of variable appearance and less sinusoidal).

Seizures

See Clinical Panel 30.2.

Clinical Panel 30.1   Narcolepsy

Narcolepsy (*Gr.* 'sleep seizure') is a sleep disorder characterized by daytime sleep attacks. These attacks form part of the *narcoleptic syndrome*. The complete syndrome has characteristic features:

- An irresistible desire to sleep, for periods of up to an hour, several times during the day. This is accompanied by *sleep paralysis*, at either the beginning or the end of an attack, where the patient is in a state of complete awareness, but is unable even to open the eyes, for 1–2 min.

- *Hypnogogic* (*Gr.* 'accompanying sleep') *hallucinations* at sleep onset, characterized by striking visual imagery. The hallucinations are a distorted version of rapid eye movement (REM) sleep.

- During the awake state, there occur brief episodes of muscle paralysis, known as *cataplexy*, a term used to signify sudden paralysis, triggered by emotion, notably by surprise of any kind. Cataplexy occurs in three out of four patients, and may range from mild (e.g. dropping the head or jaw, dropping something held in the hand) to severe, with collapse due to flaccid paralysis of the trunk and limbs with consciousness fully preserved. Occasionally, cataplexy may be the *only* presenting symptom.

The 'distorted REM' nature of narcolepsy appears obvious from the occurrence of the two kinds of paralysis during the awake state rather than during the vivid dream period. Narcolepsy has a familial incidence and may be an immune disorder. It can be very distressing: the patient may be accused of laziness or incompetence at work, and is at risk of accidents when driving a car or merely crossing the street.

Narcolepsy occasionally occurs in dogs, and it can be induced in rodents. The key problem is failure of production of the excitatory peptide *orexin* by a group of neurons in the lateral hypothalamus. Orexin receptors are normally present on the histaminergic neurons of the tuberomammillary nucleus (TMN). As mentioned in Chapter 26, the TMN projects widely to the cerebral cortex and maintains the awake state by activating H_1 receptors on cortical neurons. The drug *modafinil* is under clinical trial as a replacement; in animal experiments, it activates the TMN. The current conventional approach is to reduce sleep attacks by means of noradrenergic drugs such as amphetamine in low dosage, and/or monoamine oxidase inhibitors; both of these prolong the action of norepinephrine (noradrenaline) released by the cerulean nucleus, and reduce REM sleep.

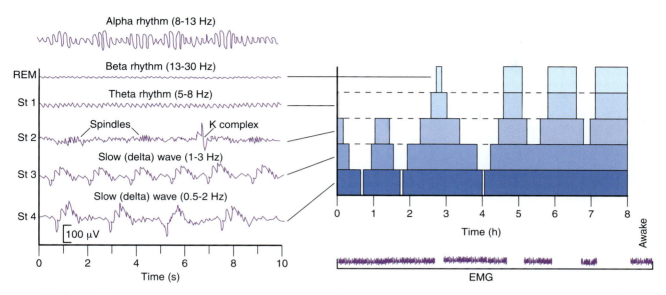

Figure 30.6 Typical 8-h sleep and electroencephalogram patterns. Note that the rapid eye movement (REM) period (in pink) is not one of the official four sleep stages, despite being routinely referred to as REM sleep. The electromyography (EMG) trace shows that skeletal muscles are 'paralyzed' during REM sleep.

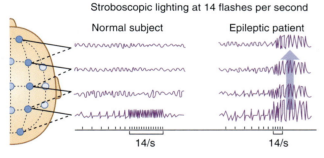

Figure 30.7 In normal subjects, stroboscopic lighting induces matching spikes in occipital recordings. Many epileptic patients are at risk of generalized tonic–clonic seizures caused by forward spread of strobe effects. (After Binnie 1988, with permission.)

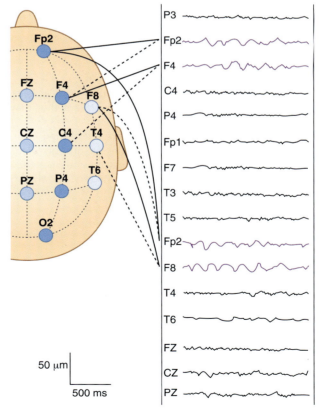

Figure 30.8 Focal slowing. Printouts from right lateral frontal areas display delta waves in this 60-year-old woman suffering from headache and drowsiness for several months without overt physical signs. Coronal magnetic resonance imaging slices showed compression of the right lateral ventricle. Surgery revealed the cause to have been an astrocytoma.

Clinical Panel 30.2 Seizures

Next to cerebrovascular disease, seizures (epileptic attacks) are the most common group of problems encountered in clinical neurology. Some 3% of the population suffer two or more attacks during their lifetime.

The term *seizure* or *ictus* refers to a transient alteration of behavior brought about by abnormal burst firing of neurons in the cerebral cortex. The *interictal period* is the time interval between seizures.

Seizures are categorized as shown in Figure CP 30.2.1.

Generalized seizures

Generalized, tonic–clonic seizures (formerly known as *grand mal* seizures) are characterized by sudden onset of unconsciousness. The individual is 'struck down'. The body stiffens, for up to a minute (tonic stage), and then exhibits jerky movements of all four limbs, and chewing movements of the mouth for about another minute (clonic stage). Usually, a third minute is spent in more relaxed unconsciousness. Electroencephalogram (EEG) recordings taken at the onset of this kind of *ictus* (attack) show simultaneous bilateral burst firing all over the cortex (Figure CP 30.2.2).

In those at risk, hyperventilation, or photic stimulation by strobe lighting (Figure 30.6), can precipitate tonic–clonic attacks.

Absence seizures (formerly *petit mal*) are characterized by a generalized, 3-Hz spike-and-wave activity (Figure CP 30.2.3). These seizures usually occur between the ages of 4 and 14. In later years, they may be replaced by tonic–clonic attacks.

The typical patient is a child who, on relaxing after some physical or mental activity, passes through 'blank' periods (absences) of 10–30 s, usually with detectable twitching of muscles of the face or fingers. Dozens of such episodes may occur over a period of several hours, often so mild that low-level activities such as walking

are not interrupted. The child is unaware of individual episodes, which may occur a hundred or more times in 1 day. The blank periods are brought about by prolonged inhibitory postsynaptic potentials on sensory thalamic relay neurons, generated by thalamic reticular neurons made hyperactive by corticothalamic excitatory discharges.

Focal (partial) seizures

Note: The term *partial* refers to seizure origin in a particular lobe (focus) of the brain, and not to seizures confined to one part of the body. Complex partial seizures become generalized.

Simple focal seizures are almost always motor or sensory. Loci of origin are shown in Table CP 30.2.1.

A *jacksonian seizure* (named after neurologist Hughlings Jackson) involves sequential activation of adjacent areas of the motor cortex, for example ankle, knee, hip, shoulder, elbow, hand, lips, tongue, larynx. A jacksonian seizure may be followed by weakness or paralysis of the affected limb(s) for a period of hours or days; it is known as *Todd paralysis*.

Benign rolandic epilepsy, a relatively common disorder in childhood, has an ictal focus of origin in front of or behind the fissure of Rolando (central sulcus). *Motor attacks* originate in front of the rolandic fissure and usually involve only the contralateral arm or leg or face, although some become jacksonian. *Somatosensory attacks* (Figure CP 30.2.4) originate behind the fissure and are described in Table CP 30.2.1.

A diagnosis of benign rolandic epilepsy requires EEG confirmation while the child is quietly asleep during an interictal period. Here, a normal background montage is interrupted by high-voltage spikes that occur at short intervals in the area of the

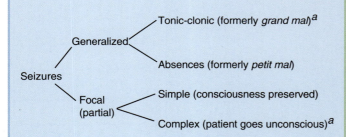

Figure CP 30.2.1 Classification of seizures. [a]*Notes on conventional terminology*: A complex seizure has an ictal focus of origin, usually in the temporal lobe. It spreads to induce a *secondarily generalized tonic–clonic seizure*. In this context, seizures that are generalized from the start are called *primary generalized tonic–clonic seizures*.

Table CP 30.2.1 Loci of origin of simple focal seizures	
Motor	Movement of any part of the motor homunculus, sometimes with aphasia
Somatosensory	Contralateral numbness or tingling of face, fingers, or toes
Primary visual cortex	Flashes of light or patches of darkness in contralateral visual field
Visual association cortex	Twinkling light images in contralateral visual field
Basal occipitotemporal junction	Formed visual images of people or places, sometimes accompanied by sounds
Superior temporal gyrus (unusual)	Tinnitus, sometimes garbled word sounds

Clinical Panel 30.2 *Continued*

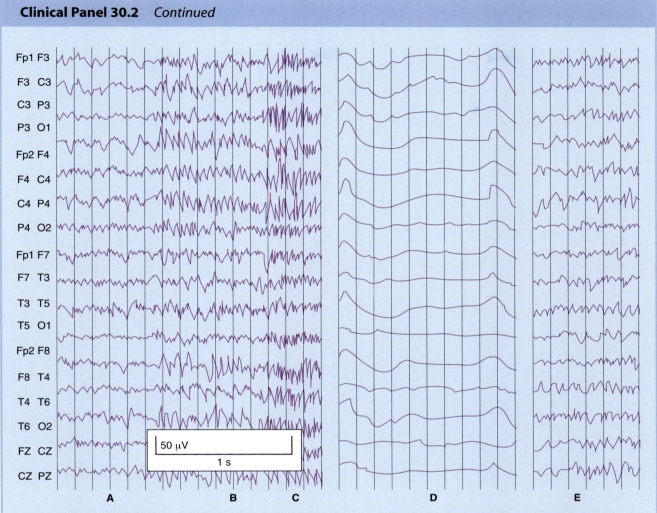

Figure CP 30.2.2 Characteristic 'frenzied' pattern of a tonic–clonic seizure. **(A)** End of interictal period. **(B)** Generalized seizure pattern involving all electrode positions. **(C)** In less than 1 s, the seizure pattern is 'scribbled' by generalized tonic muscle spasm. **(D)** Immediate postictal period, with slow waveform pattern throughout. **(E)** Resumption of normal waveforms.

ictal focus. Frequency of attacks dwindles during maturation, and the EEG becomes normal by the 16th year.

Complex focal seizures are synonymous with *temporal lobe epilepsy*. They will be taken up in Chapter 34 following a description there of the relevant areas of the temporal lobe.

Drug therapy

Anticonvulsant drugs, administered in appropriate amounts, control about half of all seizures completely and another quarter almost completely. The drugs are of a broadly predictable nature. Most of them either reduce glutamatergic activity or enhance GABA-mediated inhibitory processes. For tonic–clonic and complex partial seizures, some current drugs of choice are as follow.

- *Glutamate antagonists. Phenytoin* and *carbamazepine* (Tegretol) reduce high-frequency repetitive firing by blocking glutamate ionotropic receptors (Ch. 8), making their ion channels less permeable to sodium and/or calcium.

- *GABA agonists. Benzodiazepines* and *barbiturates* enhance the hyperpolarizing effect of GABA on glutamate neurons, as illustrated in Figure 8.7. *Sodium valproate* blocks the transaminase enzyme that converts GABA to glutamate within adjacent astrocytes (Ch. 8), thereby extending GABA's time in the synaptic cleft.

The most widely used and most effective drug for absence seizures is *ethosuximide*. Ethosuximide is a specific T-channel blocker. At appropriate dosage, excitability of thalamic relay neurons is sufficiently reduced to prevent them entering burst firing mode.

Clinical Panel 30.2   *Continued*

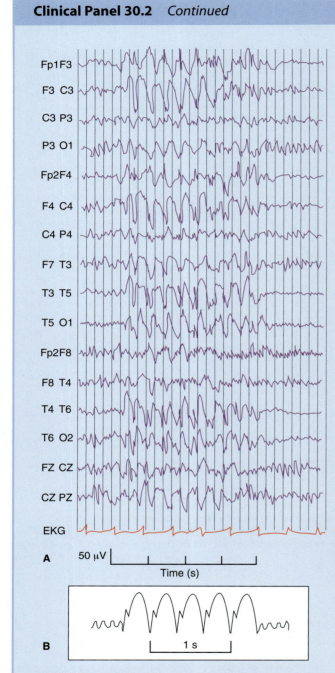

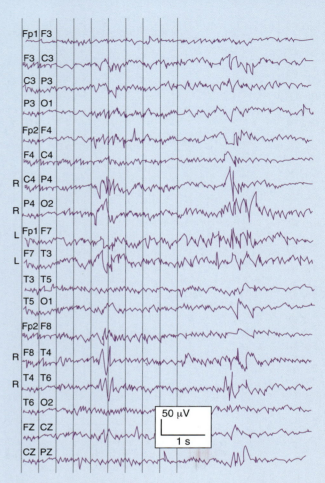

Figure CP 30.2.4 Simple partial seizure originating in the right somatic sensory cortex (R) and expressed by 'pins and needles' along the left arm. There is some spread to the left sensory cortex (L). (Refer to Figure 30.2 for R and L electrode positions.)

Figure CP 30.2.3 (A) Absence seizure. This episode was recorded over a period of 4 s. The spike-and-slow wave pattern is bilateral and generalized. In this patient, the pattern is rather faster than usual. **(B)** The spike-and-slow-wave prototype.

REFERENCE

Oliviero A, Della Marca G, Tonali PA, et al. Functional involvement of the cerebral cortex in human narcolepsy. J Neurol 2005; 252:56–61.

Core Information

The oscillations recorded in the electroencephalogram (EEG) are produced by excitatory and inhibitory postsynaptic potentials collectively produced by cortical cell columns. The grid-like standard arrangement of the recording electrodes permits overall sampling of electrical activity in a manner that is applicable to both children and adults.

Sleep
The awake-state EEG displays a desynchronized (irregular) background frequency, sometimes with a beta frequency of 9.5 Hz superimposed in frontal montages. During drowsiness with eyes closed, alpha rhythms (8–15 Hz) appear in parietooccipital tracings. Stage 1 of sleep exhibits theta rhythm at 5–8 Hz. During stage 2, theta rhythm is interrupted by sleep spindles and K-complex spikes. Stages 3 and 4 are characterized by 'slow-wave sleep' in the delta range (3 Hz or less). The final four sleep cycles are capped by 'paradoxical sleep' associated with wake-type desynchronization, rapid eye movements (REM), and dreams with high visual content.

Narcolepsy is a sleep disorder characterized by an irresistible desire to sleep for up to an hour several times during the day, hypnogogic hallucinations at sleep onset, cataplexy (momentary muscle paralysis), and sleep paralysis at the beginning or end of an attack.

Abnormal EEG rhythms
Focal abnormalities without seizures include focal slowing and phase reversal. Generalized abnormalities without seizures include hypoglycemia, hypoxia and dementias, any of which may replace normal background activity by diffuse slowing expressed in theta frequency. Disorders involving cerebral white matter may be expressed by delta activity with polymorphic interference.

Seizures
Seizures are of two broad kinds: generalized, manifested by either tonic–clonic or absence attacks;

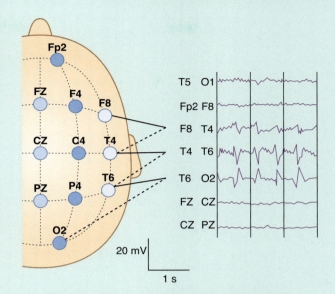

Figure 30.9 Phase reversal. This patient suffered from primary focal seizures. Phase reversal between the T4 and T6 electrodes suggests an ictal focus located in the right posterior temporal lobe.

and focal (partial), where consciousness is preserved during simple motor or sensory attacks but lost during complex attacks. Complex focal seizures refer to temporal lobe epilepsy.

Anticonvulsant drugs have one or more effects. Most enhance GABA-mediated inhibitory activity; some inhibit glutamate activity by either inhibiting glutamate synthesis or by blocking sodium channels at glutamatergic nerve terminals.

REFERENCES

Binnie CD. Electroencephalography. In: Laidlaw J, Richins A, Oxley J, eds. A textbook of epilepsy. Edinburgh: Churchill Livingstone; 1988:236–306.

Leonard BE. Drug treatment of the epilepsies. In: Fundamentals of psychopharmacology. 3rd edn. Chichester: Wiley; 2003:295–318.

Nestler EJ, Hyman SE, Malenka RC. Sleep, arousal and attention. In: Nestler EJ, Hyman SE, Malenka RC. Molecular neuropharmacology: a foundation for clinical neuroscience. New York: McGraw-Hill; 2001:409–432.

Pedley T, Traub R. Physiologic basis for EEG. In: Daly DD, Pedley TA, eds. Current practice of clinical electroencephalography. New York: Raven Press; 1990:100.

Smith SJM. EEG in the diagnosis, classification, and management of patients with epilepsy. J Neurol Neurosurg Psychiatry 2005; 76(suppl II):ii2–ii7.

Steriade M. Sleep, epilepsy and thalamic reticular nucleus. Trends Neurosci 2005; 28:317–324.

STUDY GUIDELINES

1 *Sensory evoked potentials* are the waveforms detected by specific electroencephalogram recording electrodes in response to different kinds of peripheral sensory stimulation. The objective is to assess the integrity of central sensory pathways.

2 *Motor evoked potentials* are elicited in skeletal muscles by magnetic stimulation of the motor cortex. The main clinical objective is to ascertain the conduction velocity of the corticospinal tract.

A physiologic observation of general interest revealed by this technique is that acquisition of a particular motor skill can be accelerated by mental rehearsal.

SENSORY EVOKED POTENTIALS

The term *sensory evoked potentials* is used to define the response of the central nervous system (CNS) to specific sensory stimulation. In clinical neurophysiology, the specific stimuli relate to vision, hearing, and touch.

A difficulty with these evoked potentials is that their low amplitudes, of 20 µV or even less, render them undetectable in routine electroencephalography (EEG) recordings because of the background wave pattern. Advantage is taken of the regularity of the response to repeated stimuli of the same type. With repetitious stimulation followed by computer averaging, irregular background rhythms cancel each other out and the evoked potentials can be clearly seen.

The three basic kinds of sensory evoked potentials are described as *visual*, *auditory*, and *somatosensory*.

Visual evoked potentials

The speed and amplitude of impulse conduction in the visual pathway are tested by the technique known as *pattern reversal* or *pattern shift*. With one eye covered at a time, the patient stares at a spot in the center of a screen illuminated in a black-and-white checkerboard pattern. Once or twice per second, the pattern is reversed (to white and black), over a period of 100 repetitions. Averaging is performed on the first 500 ms of data from a bipolar recording at the occipital and parietal midline EEG sites (OZ and PZ).

The wave peak of interest is called P100. In healthy subjects, it is a positive deflection 100 ms poststimulus (Figure 31.1). In the clinical example shown, taken from a patient with a presumptive diagnosis of multiple sclerosis, the normal P100 wave from the right-eye test indicated that both optic tracts and both optic radiations were clear. The P100 wave from the left eye was both delayed and of reduced amplitude, suggesting presence of one or more plaques of myelin degeneration in the left optic nerve. (*Note*: On screen and in printouts, it is now customary for the waveforms to be 'flipped', with positive responses registering as upward deflections.)

Conduction defects caused by demyelination are more often expressed in the form of *latency delays* of the kind shown, than in the form of amplitude abnormalities.

In the absence of any evidence for multiple sclerosis elsewhere, an abnormal P100 from one eye may be caused by an ocular disease such as glaucoma or by compression or ischemia of the optic nerve.

Bilateral abnormal P100 recordings usually indicate pathology in one or both optic radiations. In such a situation, it is usual to take recordings from electrode pairs placed a few centimeters to one side of the midline, and then a few centimeters to the other side. Should the (say) right optic radiation be at fault, any P100 abnormality is likely to be more pronounced in recordings from that side of the midline.

Brainstem auditory evoked potentials

Remarkably, it is possible to follow the sequence of electrical events in the auditory pathway, step by step, from cochlea to primary auditory cortex. Following placement of temporal scalp recording electrodes, 0.1-ms click sounds are presented at 10 Hz to each ear in turn through conventional audiometric earphones. Click intensity is adjusted to 65–70 dB above click hearing threshold for the ear being tested. The contralateral ear is 'masked' by white noise.

A sequence of seven averaged-out waves (I–VII) constitutes the brainstem auditory evoked response (BAER). They are accounted for in the caption to Figure 31.2.

Pathology anywhere along the auditory pathway results in reduction or abolition of the wave above that level. *The technique is the most sensitive screening test available for acoustic neuroma.* A diagnostic feature here is *I–III latency separation. Latency* refers to the time interval between stimulus and response; *separation* refers to extension of the interval between waves I and III, caused by delay during passage along the affected cochlear nerve during a characteristically reduced amplitude wave II.

In about 30% of patients who have multiple sclerosis with no clinical evidence of brainstem lesions, the BAER is abnormal. Most frequent abnormalities are reduced amplitude of wave V and overall slowing of conduction indicated by increased interwave intervals.

Another clinical application of the BAER technique is the assessment of cochlear function in infants under suspicion of congenital deafness.

Assessment of brainstem auditory evoked potentials is also important in the medicolegal domain, to assess veracity of claims of deafness induced by environmental noise in industry.

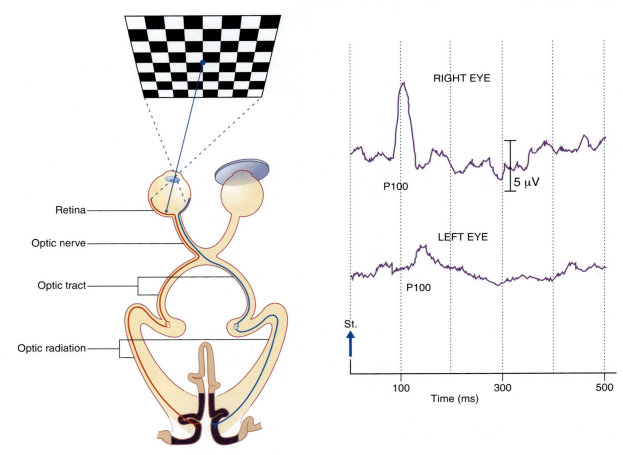

Figure 31.1 Visual evoked potentials. The patient's right eye has been tested and is now shielded. The left eye is fixated on the spot in the center of the checkerboard during pattern reversal episodes. The pattern from the right eye is normal, showing a positive (downward) deflection at 100 ms poststimulus. In the recording from the left eye, the P100 is both delayed and reduced in amplitude. The combined results indicate presence of a lesion in the left eye or left optic nerve.

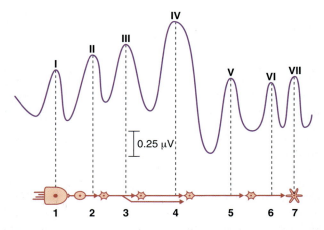

Figure 31.2 Brainstem auditory evoked potentials. Sources of evoked potentials:
1 cochlear hair cells
2 cochlear nerve
3 cochlear nucleus
4 lateral lemniscus
5 inferior colliculus (inferior brachium)
6 medial geniculate body (auditory radiation)
7 primary auditory cortex.

Evidence for a 'Where?' auditory pathway

When recording electrodes are specifically deployed over the temporoparietal region, and brief sounds are emitted from loudspeakers placed in the left and right visual fields, a cortical response can be detected over the posterior part of the temporal plane, close to the temporoparietal junction. The right posterior temporal plane gave a stronger response, suggesting a right-sided dominance for auditory as well as visual space analysis.

Somatosensory evoked potentials

Somatosensory evoked potentials are the waveforms recorded at surface landmarks en route from the point of stimulation of a peripheral nerve to the contralateral somatic sensory cortex. The rate and amplitude of impulse conduction provide valuable information about the status of myelinated nerve fibers in both peripheral nerves and central pathways.

The nerve of choice for stimulation in the upper limb is the median at the wrist, in the lower limb the common peroneal at the knee. Repetitive electrical pulses are delivered to the nerve through a surface or needle electrode. The larger myelinated fibers are stimulated. Computer averaging is required to distinguish the stimulated responses from background noise, notably within the CNS. In the example shown in Figure 31.3, impulse traffic along the median nerve is detected by a sequence of active electrodes attached to the skin for the purpose of recording speed and amplitude of nerve conduction in sequential segments, as follows:

1 over the brachial plexus, to assess the median nerve segment extending from wrist to anterior triangle of neck

2 over the spine of vertebra C2, for the segment 1 plus transit through nerve roots and ipsilateral posterior column (fasciculus cuneatus)

3 over the contralateral scalp at some distance away from the somatic sensory cortex, to 'pick up' stimulus traffic ascending the medial lemniscus

4 directly over the hand area of the somatic sensory cortex, to detect activity in the thalamocortical projection.

In the various peripheral neuropathies mentioned in Chapter 9, the first segment (wrist to brachial plexus) reveals slowing, usually with a reduction of amplitude. The second segment (brachial plexus to nucleus gracilis) may be affected in the first few milliseconds of its time course as a result of posterior nerve root compression by osteo-phytes in patients with cervical spondylosis. A little later, the curve may be affected by posterior column disease (Ch. 15). Abnormality in the third segment (contralateral medial lemniscus) is found in 9 out of 10 patients suffering from multiple sclerosis in the presence of sensory symptoms, and in 6 out of 10 in the absence of sensory symptoms.

MOTOR EVOKED POTENTIALS

Motor evoked potentials are motor unit action potentials detected in surface electromyographic recordings following controlled excitation of the corticospinal tract. The technique was mentioned in Chapter 18, because it revealed that the pyramidal tract pathway to sternomastoid spinal motor neurons is essentially crossed, rather than being ipsilateral as previously thought. The most frequent objective is to determine *central motor conduction time* along the corticospinal tract. The procedure is both safe and painless. It uses a subtraction approach comparable in principle with that used to determine peripheral nerve conduction times (Figure 31.3).

The procedure is known as *transcranial magnetic stimulation (TMS)*. Figure 31.4 illustrates the concept in action. Stimulation is by means of a magnet in the form of a circular coil about 10 cm in diameter. To stimulate the

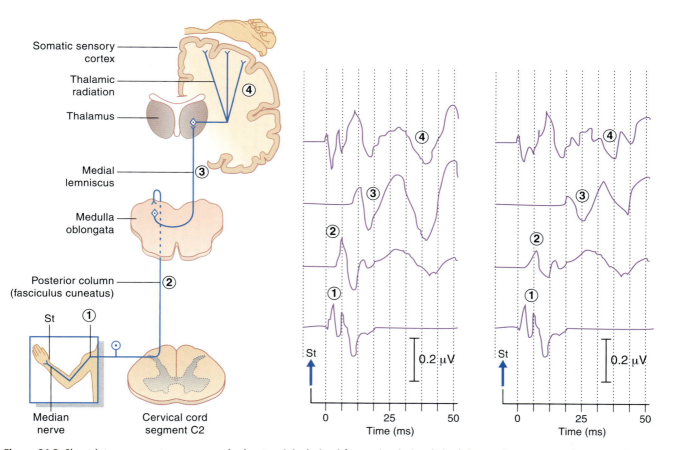

Figure 31.3 Short-latency somatosensory evoked potentials derived from stimulation (St) of the median nerve at the wrist. The pathway has four segments: **1** is purely peripheral nervous system (PNS), **2** is PNS from brachial plexus to spinal cord and central nervous system (CNS) within the cord, **3** and **4** are purely CNS. In the recordings from a patient with multiple sclerosis, trace 1 is normal, trace 2 shows reduced amplitude of the negative (upward) and positive peaks within the posterior column, traces 3 and 4 show slowing as well as reduced amplitude. (In part after Adams and Victor 1993, with permission.)

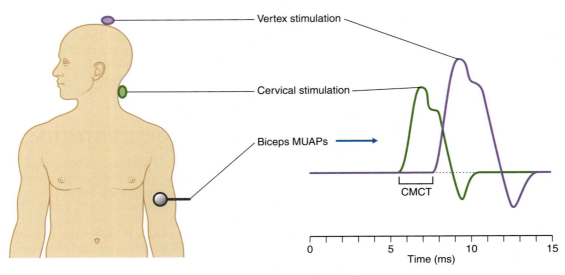

Figure 31.4 Laboratory estimation of central motor conduction time. The (*red*) coil over the vertex has delivered a 200-ms current to pyramidal tract neurons serving upper limb spinal motor neurons. A large compound motor action potential (CMAP) is elicited in the biceps. The (*green*) coil over the cervical spine has delivered a weaker pulse to cervical anterior nerve roots, eliciting a smaller CMAP. The difference between the two latencies (stimulus–response intervals) represents the central motor conduction time (CMCT). MUAP, motor unit action potential.

pyramidal cells of the corticospinal tract supplying left anterior horn motor neurons, the magnet is handheld a little to the right side of the vertex, and the patient maintains the selected limb muscle (biceps brachii in this example) in a state of slight contraction. A few very brief (200 ms) currents are pulsed at an intensity comfortably above the threshold required to elicit a twitch. The patient feels only a small 'tap' sensation on the scalp. The procedure is then repeated with the magnet touching the skin of the neck overlying the spine of vertebra C5, again eliciting a 'tap' sensation. It is generally agreed that the second pulse depolarizes the axons of anterior nerve roots exiting the vertebral canal.

The latencies and amplitudes of the compound motor action potentials are measured. The right biceps can then be activated in the same manner. The same procedure can be performed for the lower limb, the spinal stimulus being delivered in the lumbar region.

In neurophysiology units, central motor conduction time is estimated where there is reason to suspect the presence of plaques of multiple sclerosis in the white matter of brain or spinal cord; and where muscle wasting in the arms and/or legs leads to suspicion that upper as well as lower motor neurons may be degenerating; and in patients where moderate muscle weakness on one side, associated with brisk tendon reflexes, raises suspicion of a stroke.

Motor training

A remarkable degree of plasticity in the healthy motor cortex has been demonstrated by TMS studies. Figure 31.5 represents outcomes of five-finger piano-playing exercises. A small magnetic coil was used over the scalp to locate the modules primarily involved in flexion and extension of the fingers of the right hand. This small scalp area was marked in three sets of volunteers, and the baseline size

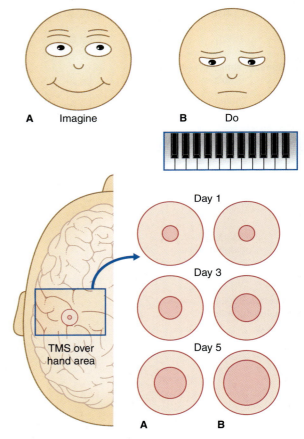

Figure 31.5 Five-finger exercise. As explained in the main text, group A volunteers imagined performing a daily five-finger piano exercise for the right hand, whereas group B actually did perform it. Transcranial magnetic stimulation (TMS) showed remarkable enlargement of the area of motor cortex activating the finger flexors and extensors in both groups. (Based on data from Pascual-Leone et al. 1995.)

was measured for each subject. Group A *imagined* doing the five-finger exercise for 2 h per day for 5 days; group B *did* the exercises for the same periods; group C did not participate in any way, prior to attempting the task once on day 5. As indicated in the figure, merely thinking about performance led to a major increase in the number of modules that activated the fingers when stimulated on days 3 and 5. Group B—the actual performers—showed the greatest increase of participating motor modules. The performance skills on day 5 were substantially better in group B than in group A, and group A's performance was better than that of group C.

There is general agreement that dramatic alterations such as those shown in this group experiment are best explained in terms of *unmasking of preexisting connec-*

tions, as in the case of rapid expansion of the cortical sensory territory of one thalamocortical projection following experimental inactivation of a neighboring projection. The most likely mechanism of additional pyramidal cell recruitment appears to be one of *disinhibition*, probably by the premotor cortex, involving activation of sequential pairs of GABAergic neurons in the manner illustrated in Figure 6.2.

In this general context, it has also been shown that performance improvement in weightlifting is optimal when subjects *mentally rehearse* weightlifting during the days between performing the exercises.

Finally, Box 31.1 includes an experiment in which TMS has been used to assess the supposed usefulness of acupuncture in improving motor performance.

Box 31.1   Acupuncture

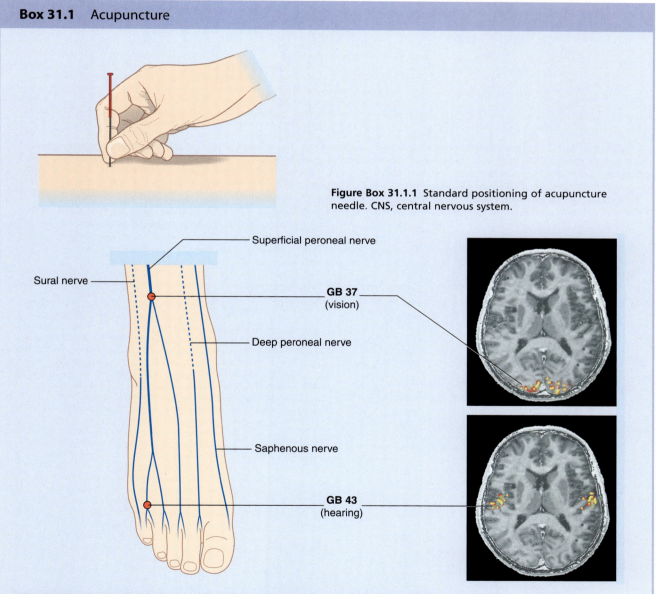

Figure Box 31.1.1 Standard positioning of acupuncture needle. CNS, central nervous system.

Superficial peroneal nerve

Sural nerve

GB 37 (vision)

Deep peroneal nerve

Saphenous nerve

GB 43 (hearing)

Figure Box 31.1.2 Bilateral needling of acupoint GB 37 generates increased blood flow in the visual cortex. Bilateral needling of acupoint GB 43 generates increased blood flow in the auditory cortex. (Adapted from Cho et al. 2000, with permission of Pabst.)

Box 31.1 *Continued*

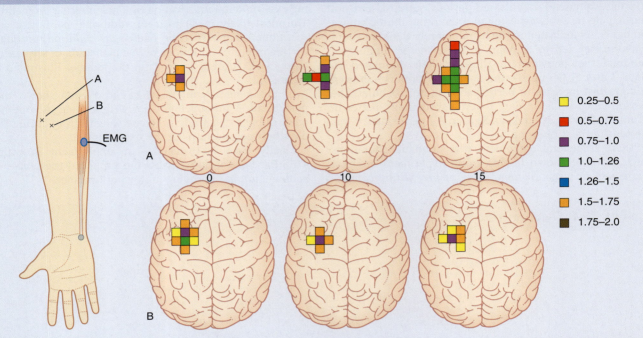

Figure Box 31.1.3 Surface electromyographic activity recorded over flexor carpi ulnaris muscle in response to transcranial magnetic stimulation (TMS) over and beyond the arm area of the left motor cortex. **A** marks an acupoint traditionally used to improve motor function, **B** marks the sham needling point (needling in the same manner 2 cm medial to point A). Both points are well clear of ulnar nerve motor and sensory territories. Numbers 0, 10, and 15, respectively, refer to baseline (no needling), 10 min with needle in place, and 15 min after needle removal. The table records voltage of motor unit action potentials in response to TMS over equal areas of the frontal lobe at each time interval. (After Lo et al. 2005, with permission.) (The assistance of Dr. Y.L. Lo, Department of Neurology, National Neuroscience Institute, Singapore, is gratefully acknowledged.)

REFERENCES

Cho Z-H, Na C-S, Wang EK, et al. Functional magnetic resonance imaging of the brain in the investigation of acupuncture. In: Litscher G, Cho Z-H, eds. Computer-controlled acupuncture. Lengerich: Pabst; 2000:45–64.

Lin YC. Acupuncture and traditional Chinese medicine. In: Oken PS, ed. Complementary therapies in neurology. Boca Raton: Parthenon; 2004:113–126.

Lo YL, Cui SL, Fook-Chong S. The effect of acupuncture on motor cortex excitability and plasticity. Neurosci Lett 2005; 384:385–389.

Pomeranz P, Berman B. Scientific basis of acupuncture. In: Stux G, Berman B, Pomeranz B, eds. Basics of acupuncture. 5th edn. Berlin: Springer; 2003:52–86.

Tageshige C. Mechanisms of acupuncture analgesia produced by low frequency electrical stimulation of acupuncture points. In: Stux G, Hammerschlag R, eds. Clinical acupuncture: scientific basis. Berlin: Springer; 2003:29–50.

Yan B, Li K, Xu J, et al. Acupoint-specific fMRI patterns in human brain. Neurosci Lett 2005; 383:236–240.

Core Information

Sensory evoked potentials

Recordings of sensory evoked potentials are used to assess conduction rates and amplitudes in central sensory pathways that may be under suspicion on clinical grounds.

Potentials in the visual cortex are evoked by means of checkerboard pattern reversal, one eye being tested at a time. Conduction deficits caused by demyelination are usually expressed in the form of latency delays.

Potentials in the auditory cortex are evoked by click sounds. The montage normally shows seven successive waveforms generated by the seven cell groups involved in the pathway from cochlea to cortex. The auditory evoked potential technique is the most sensitive test available for detection of an acoustic neuroma.

Potentials in the somatosensory cortex are elicited by electrical pulses delivered to a peripheral nerve, for example median at wrist, peroneal at knee, with recording electrodes in place to detect waveforms in the brachial or lumbar plexus, posterior column of spinal cord, medial lemniscus, and thalamocortical projection.

Different disorders impair conduction in different segments of the pathway from skin to cortex.

Motor evoked potentials

Transcranial magnetic stimulation is used clinically to estimate conduction time in the pyramidal tract in patients with motor weakness originating in the central nervous system. Surface electromyographic recordings of compound action potentials are taken from selected muscles, while a magnetic coil delivers very brief currents over the scalp to excite the motor cortex and repeats the technique over the cervical and/or lumbar spine to excite anterior nerve roots innervating the selected muscle(s). The central motor conduction time is provided by subtraction of the peripheral nerve time segment from the total cortex-to-muscle time.

Transcranial magnetic stimulation is also used in neurophysiology laboratories to study activity changes in the motor cortex occurring in the course of training for motor skill or strength tasks.

REFERENCES

Adams RD, Victor M. Principles of neurology. Philadelphia: Saunders; 1993.

Chiappa KH, Hill RA, Jayaka P. Evoked potentials in clinical medicine. In: Joynt RJ, Griggs RC, eds. Clinical neurology, vol 1. New York: Lippincott Williams & Wilkins; 1998:1–45.

Cho Z-H, Na C-S, Wang EK, et al. Functional magnetic resonance imaging of the brain in the investigation of acupuncture. In: Litscher G, Cho Z-H, eds. Computer-controlled acupuncture. Lengerich: Pabst; 2000:45–64.

Hallett M, Chokroverty S. Magnetic stimulation in clinical neurophysiology. 2nd edn. London: Butterworths-Heinemann; 2005.

Maccabee PJ, Amassian VE, Ziemann U, et al. Emerging applications in neuromagnetic stimulation. In: Levin KH, Lüders HO, eds. Comprehensive clinical neurology. Philadelphia: Saunders; 2000:325–347.

Pascual-Leone A, Dang N, Cohen LG. Modulation of muscle responses evoked by transcranial magnetic stimulation during the acquisition of new motor skills. J Neurophysiol 1995; 74:1037–1045.

Tata MS, Ward LM. Spatial attention modulates activity in a posterior 'where' pathway. Neuropsychologia 2005; 43:509–516.

Walsh P, Kane N, Butler S. The clinical role of evoked potentials. J Neurol Neurosurg Psychiatry 2005; 76(suppl II):ii16–ii22.

Hemispheric asymmetries

STUDY GUIDELINES
1 Language functions, with their strikingly asymmetric distribution in the majority of the population, are of great significance in the contexts of clinical diagnosis and speech rehabilitation.
2 Also asymmetric as a rule are the functional emphases of the parietal lobes.
3 A first reading of relevant parts of Chapter 35 should follow, in order to integrate the clinical information with the regional distribution of the cerebral arteries.

The two cerebral hemispheres are *asymmetric* in certain respects. Some of the asymmetries have to do with handedness, language, and complex motor activities; other, more subtle differences come under the general rubric of *cognitive style*. (Limbic asymmetries are described in Ch. 34.)

HANDEDNESS AND LANGUAGE

Handedness indicates the hemisphere that is dominant for motor control. Left hemisphere, right-hand dominance is the rule. Advances in ultrasound technology have made it possible to observe motor behavior in the fetus to be observed, and it has been noted that handedness is already established before birth, on the basis of the preferred hand used for thumb sucking during fetal life.

The best indicator available for population estimates of handedness is the preferred hand for writing: this criterion indicates a left-hemisphere dominance for motor control in about 90%—at any rate for literate communities!

In 90% of subjects, the left hemisphere is dominant for language. In a further 7.5%, the right hemisphere is dominant in both sexes, and in the remaining 2.5%, the two hemispheres have an equal share. Although the left hemisphere is dominant in respect of both motor control and language, the two features are statistically independent: many left-handers have their language areas in the left hemisphere.

Language areas

Although several areas of the cortex, notably in the frontal lobe, are active during speech, two areas are specifically devoted to this function.

Broca's area   (Figure 32.1)

The French pathologist Pierre Broca assigned a motor speech function to the inferior frontal gyrus of the left side in 1861. The principal area concerned occupies the opercular and triangular parts of the inferior frontal gyrus, corresponding to areas 44 and 45 of Brodmann. Both areas are larger on the left side in right-handers. The main output of Broca's area is to cell columns in the face and tongue areas of the adjacent motor cortex. Lesions involving Broca's area are associated with expressive aphasia (see Clinical Panel 32.1). Some workers believe that expressive aphasia requires that the lesion should also include the lower end of the precentral gyrus.

Wernicke's area   (Figure 32.1)

The German neurologist Karl Wernicke made extensive contributions to language-processing in the late nineteenth century. He designated the posterior part of area 22 in the superior temporal gyrus of the left hemisphere as a sensory area concerned with understanding the spoken word. The upper surface of Wernicke's area is called the **temporal plane** (Figure 32.2). The volume of cerebral cortex in the temporal plane is larger on the left side in 60% of subjects. The horizontal part of the lateral fissure is longer in consequence—a feature readily identified on magnetic resonance imaging (MRI) scans. Lesions involving Wernicke's area in adults are associated with receptive aphasia (see Clinical Panel 32.1).

Wernicke's area is linked to Broca's area by association fibers of the arcuate fasciculus curving around the posterior

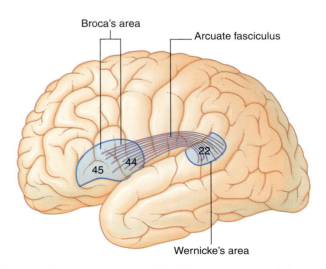

Figure 32.1 Broca's and Wernicke's language areas and the arcuate fasciculus.

Clinical Panel 32.1 The aphasias

Aphasia is a disturbance of language function caused by a lesion of the brain. The usual cause is a stroke produced by vascular occlusion in the anterior cortical territory of the left middle cerebral artery.

Motor (anterior) aphasia

Patients having a lesion that includes Broca's area suffer from motor aphasia. These patients have difficulty in expressing what they want to say. Speech is slow, labored, and characteristically 'telegraphic' in style. The important nouns and verbs are spoken, but prepositions and conjunctions are omitted. The patient comprehends what other people are saying and is well aware of being unable to speak fluently. There is usually an associated agraphia (inability to express thoughts in writing).

If the lesion involves a substantial amount of the cortical territory of the middle cerebral artery, there will be a motor weakness of the right lower face and right arm. Because the lips and tongue are affected on the right side, the patient will also have *dysarthria* (difficulty in speech articulation) in the form of slurring of certain syllables. In the example shown in Figure CP32.1.1A, Broca's aphasia would be associated with right-sided weakness of the lower face but not of the arm, together with some dysarthria.

Sensory (posterior) aphasia

A lesion in Wernicke's area is accompanied by a deficit of auditory comprehension. (If the lesion includes the angular gyrus, the ability to read will also be compromised.) In addition to their difficulty in understanding the speech of others, these patients lose the ability to monitor their own conversation, and usually have difficulty in retrieving correct descriptive names. Speech fluency is normal, but two kinds of abnormality occur in the use of nouns.

1 Verbal paraphrasia (use of words usually of allied meaning): instead of 'use a knife', 'use a fork'.

2 Phonemic paraphrasia (use of made-up but similar-sounding syllables): instead of 'knife and fork', 'bife and dork'.

The most striking feature of Wernicke's aphasia is that, despite garbling to the point of being unintelligible (*jargon aphasia*), the patient may be quite unaware of making mistakes.

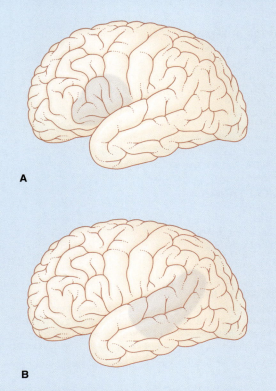

Figure CP 32.1.1 (A) Vascular lesion involving Broca's area. **(B)** Vascular lesion involving Wernicke's area.

In the example shown in Figure CP 32.1.1B, Wernicke's aphasia would be associated with alexia (angular gyrus), with ideomotor apraxia (supramarginal gyrus), and probably with a right upper quadrant visual deficit (lower fibers of left optic radiation in the temporal white matter).

Aprosodia

Lesions of the right hemisphere may affect speech in subtle ways. Lesions that include area 44 (corresponding to Broca's area on the left) tend to change the patient's speech to a dull monotone. On the other hand, lesions that involve area 22 (corresponding to Wernicke's area) may lead to listening errors, for example being unable to detect inflections of speech; the patient may not know whether a particular remark is intended as a statement or as a question.

end of the lateral fissure within the underlying white matter (Figure 32.1). The two areas are also linked through the insula.

It is difficult to assess the significance of the asymmetry of the temporal plane. The 60% incidence of left-sided relative enlargement does not match the 95% left-hemisphere dominance for speech. Moreover, the *overall* length of the lateral sulcus is much the same on both sides. The *parietal plane* of that sulcus is longer on the right side, because the right supramarginal gyrus is larger than the left one. This feature has been advanced as an explanation for the shorter temporal plane on the right.

Maldevelopment of the left temporal plane is a significant feature in cases of schizophrenia (Clinical Panel 32.2).

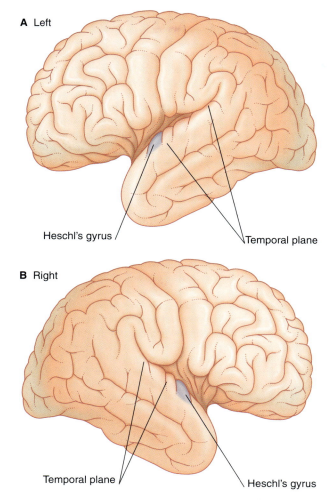

A Left

Heschl's gyrus

Temporal plane

B Right

Temporal plane

Heschl's gyrus

Figure 32.2 (A and B) Views of the opened lateral sulcus, showing the upper surface of the temporal lobes.

Right hemisphere contribution

During normal conversation, there is some increase in blood flow in areas of the right hemisphere matching those of the left. These areas are believed to be concerned with melodic aspects of speech—the cadences, emphases, and nuances collectively called *prosody*. Disturbances of the melodic function are called *aprosodias* (Clinical Panel 32.1).

Recovery of speech function—when it occurs—depends on the age of the subject, and in adults on the extent of the lesion. Occasional cases have been reported of recovery of near-normal speech in right-handed patients, 7 years of age or less, following complete removal of the left hemisphere as a treatment for intractable epilepsy. This can be explained only by language-processing, including speech, being not fully lateralized at the time of operation. In adults, positron emission tomography (PET) studies have shown increased activity in Broca's and Wernicke's equivalents on the right side following cerebrovascular accidents on the left. However, significant improvement is possible only if the left temporal plane is sufficiently viable to be able to process signals passed to it from the right side through the corpus callosum.

Angular gyrus

The angular gyrus (area 39) belongs descriptively to the inferior parietal lobule. The *left* angular gyrus receives a projection from the inferior part of area 19 (the lingual gyrus, shown in Figure 2.6), and itself projects to the temporal plane. It is commonly included as a part of Wernicke's area.

The angular gyrus seems to contain a neural lexicon (dictionary) of words, syllables, and numeric or other symbols that can be retrieved by visual inputs—or even by visual imagery—and forwarded in the form of impulse trains to Wernicke's area within area 22. During reading, it is engaged in the conversion of written syllables (*graphemes*) into the corresponding sound equivalents (*phonemes*). The angular gyrus is also active during listening to spoken words.

Listening to spoken words

Figure 32.3 contrasts regional increases in blood flow during PET scanning when a volunteer listens to words (*active listening*) versus random tone sequences (*passive listening*). As expected, tone sequences activate the primary auditory cortex (bilaterally). Wernicke's area (left side) also becomes active, probably in screening out this non-verbal material from further processing. Area 9 in the

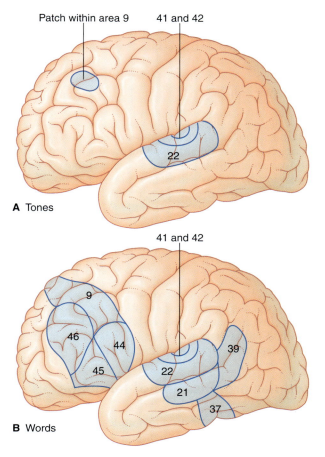

Patch within area 9   41 and 42

22

A Tones

41 and 42

9

46

44

39

45

22

21

37

B Words

Figure 32.3 Regions of increased blood flow during listening **(A)** to tones, and **(B)** to words.

Clinical Panel 32.2 Frontal lobe dysfunction

Symptoms of *early* frontal lobe disease typically involve subtle changes in personality and social function rather than diminution of cognitive performance on objective tests. Lack of foresight (failure to anticipate the consequences of a course of action), distractibility (poor concentration), loss of willpower (*abulia*), and difficulty in 'switching cognitive sets' (e.g. inability to switch easily from one subject of conversation to another) are characteristic. These general symptoms are more often associated with bilateral disease with impending dementia than with a brain tumor. With increasing disease, especially if bilateral, the *gait* is affected. *Marche á petit pas* ('Walk with small steps') refers to a characteristic short, shuffling gait often associated with disequilibrium (tendency to fall) and 'freezing' (especially when turning). This syndrome may give rise to a mistaken suspicion of Parkinson's disease.

Large *dorsolateral lesions* are associated with slowing of mental processes of all kinds, leading to hypokinesia, apathy, and indifference to surrounding events. The picture resembles that of the 'withdrawn' type of schizophrenia, and it is of interest that, in 'withdrawn' schizophrenic patients, cortical blood flow may not show the anticipated increase in the dorsolateral region in response to appropriate psychologic tests.

Large *orbitofrontal* lesions are associated with hyperkinesia, with increased instinctual drives in relation to food and sexual behavior. With disease more pronounced (or only) in the right orbitofrontal cortex, the 'fearful' side of the patient's nature may be lost, leading to puerile jocularity and compulsive laughter. Compulsive crying may be a clue to left-sided disease. A well-known cause of orbitofrontal disturbance is a meningioma arising in the groove occupied by the olfactory nerve; *anosmia* (loss of the sense of smell) may be discovered on testing, and optic atrophy may follow pressure on the optic nerve where it emerges from the optic canal. Hyperkinetic frontal lobe disorders have been treated in the past by means of *lobotomy*—a surgical procedure in which the white matter above the orbital cortex was severed through a supraorbital incision.

Gliomas within the frontal lobe may become large before any cognitive or physical defects appear. Eventually, a left-sided tumor may invade or compress Broca's area and cause motor aphasia. On either side, a progressive hemiparesis may supervene.

REFERENCE

Kertesz A, Ferro JM. Lesion size and location in ideomotor apraxia. Brain 1984; 107:921–933.

---

frontal lobe is thought to be part of a supervisory, vigilance system.

During active listening to words, areas 21 (middle temporal lobe), 37 (posteroinferior temporal lobe), and 39 (angular gyrus) all participate in auditory word processing. Area 39 identifies phonemes. Areas 21 and 37 identify words in the sound sequence and tap into lexicons (dictionaries) stored in memory in a search for meaning—a process called *semantic retrieval*.

Activity in the left dorsolateral prefrontal cortex (DLPFC) expands to include area 46. Engagement of Broca's area is thought to signify 'subvocal articulation' of words heard (see *Neuroanatomy of reading*, later).

When listening to one's own voice, the areas of the temporal lobe identified above become active. An important function being served here is *metanalysis* (post hoc analysis) of speech, whereby 'slips of the tongue' can be identified. Speech metanalysis is singularly lacking in cases of receptive aphasia (Clinical Panel 32.1).

Modular organization of language

In alert subjects, electrical studies of the cortex exposed during neurosurgical procedures indicate the presence of a vast cortical mosaic for language. The mosaic of modules extends along the entire length of the frontoparietal operculum above the lateral sulcus, and of the temporal operculum below the sulcus. The frontoparietal operculum is predominantly concerned with the motor functions of speaking and writing, and the temporal operculum with the sensory functions of hearing and reading.

It is well known that children have greater facility than adults in acquiring a second language. Functional MRI and other approaches have shown that the loci of second-language acquisition before the age of 7 years overlap extensively with those processing the native language. A second language learned in later years is non-overlapping with the first. One possibility is that, in children, the syntactic systems for processing nouns, verbs, etc. are able to cope with two languages simultaneously.

It is of interest that, following a small vascular lesion in an adult, *either* a late-acquired language *or* the native language may be lost, leaving the other relatively intact.

COGNITIVE STYLE

Hemispheric specializations in relation to information-processing have been revealed by various forms of visual, auditory, and tactile tests. Results show that the left hemisphere is superior in processing information that is susceptible to *sequential analysis* of its parts, whereas the right is superior in respect of *shapes* and *spatial relationships*. Accordingly, the left hemisphere is described as being *analytic* and the right as *holistic*. The right is also 'musical': there is a relative increase in blood flow in the right auditory association area when listening to music, versus a left-sided increase for words.

The analytic character of the left hemisphere may be due to its unique capacity to perform the 'inner speech' that usually accompanies problem-solving.

Neuroanatomy of reading (Figure 32.4)
Glossary

- **Graphemes**. Written syllables.
- **Orthography** (*Gr.* 'correct writing'). Word and sentence construction.
- **Phonemes** (*Gr.* 'sounds'). The sounds of syllables. 'Cat' is a single syllable containing three phonemes: [k], [a], and [t].
- **Phonology**. The rules governing the sounds of words. Testing could include: 'How many of these words have two syllables?' or 'How many of these words rhyme with one another?'

- **Retrieval**. Matching words, phrases, and sentences with those previously entered into memory.
- **Semantics** (*Gr.* 'meaning'). Meaning of words and sentences.

Reading sequence

A *Carry out visual processing*
B *Perform orthographic processing*
C *Perform phonologic assembly*
D *Perform semantic retrieval*
E *Execute motor plans*

A *Visual processing* is performed bilaterally in areas 17, 18, and 19. It includes analysis of letter shapes for their identification; distinguishing between letters in upper versus lower case, and between real letters and meaningless shapes ('false fonts'). Processed information in the right extrastriate cortex (areas 18 and 19) is transferred to the left side through the forceps major traversing the splenium of the corpus callosum—a point of clinical significance (see later).

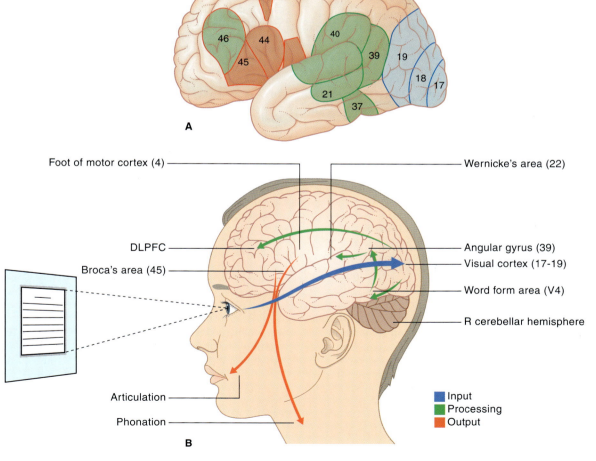

Figure 32.4 (A) Areas of increased cortical blood flow in the left hemisphere observed in positron emission tomography scans during reading aloud. **(B)** Input (*blue*), processing (*green*), and output (*red*) pathways active during reading aloud. DLPFC, dorsolateral prefrontal cortex.

hand. They have *astereognosis*. Patients with a comparable

B *Orthographic processing* means discerning whether or not each letter string in a sentence represents a real word or a pseudoword, for example 'word' versus 'wurd'. Medial area 19 (V4) is especially involved.

C *Phonologic processing* means the conversion of graphemes to phonemes. The angular gyrus (area 39) and middle temporal gyrus (area 21) participate.

D *Semantic retrieval* means performance of a memory search, using both orthographic and phonologic cues from the text to extract the meaning of words and sentences. The anterior part of Broca's area (area 45) becomes active at this advanced stage, together with area 37 in the posterior temporal lobe and area 40 (supramarginal gyrus) in the inferior parietal lobule.

E *Execution of motor plans (phonologic execution)* is the performance of inner speech (subvocal articulation). Both parts of Broca's area become active, as do the adjacent parts of premotor and motor cortex, the supplementary motor area (medial area 6), and the contralateral cerebellar hemisphere. The same four areas become much more active during reading aloud.

The left lateral prefrontal cortex, in and around area 46, is 'switched on' throughout A–E. Also active is part of area 32 in the left anterior cingulate cortex, which is involved in all cognitive activities requiring attention.

Developmental dyslexia is considered in Clinical Panel 32.3.

Schizophrenia, a psychiatric disorder involving the left hemisphere more than the right, is described in Chapter 34.

Figure 32.5 Brodmann's areas in the parietal lobe: **(A)** lateral view, **(B)** medial view. 3/1/2, somesthetic cortex; 5, somesthetic association area; 7, posterior parietal cortex; 39, angular gyrus; 40, supramarginal gyrus.

PARIETAL LOBE   (Figure 32.5)

The parietal lobe—especially the *right* one—is of prime importance for appreciation of spatial relationships. There is also evidence that the parietal lobe—especially the *left* one—is concerned with initiation of movement.

Parietal lobe and the body schema

The term *body schema* refers to an awareness of the existence and spatial relationships of body parts, based on previous (stored) and current sensory experience. The reality of body schema has been established by the condition known as *hemineglect*, in which a patient with a lesion involving the superior parietal lobule ignores the contralateral side of the body. The lesion may be in the superior or inferior lobule or (usually) both, implying collaboration of both lobules in the healthy state.

Hemineglect is much more common following a right parietal lobe lesion than a left one. Under normal conditions, however, each parietal lobe exchanges information freely with its partner through the corpus callosum, and the left and right hand are equally adept at distinguishing a key from a coin in a coat pocket without the aid of vision (*stereognosis*, Ch. 29).

Patients with a right hemisphere lesion involving the superior parietal lobule have difficulty in distinguishing

lesion in the left hemisphere are able to make this distinction using the right hand, but they have difficulty in announcing the *function* of a selected object. The left supramarginal gyrus participates in phonologic retrieval, as already noted, and the deficit, although a semantic one, may be related to interference with the inner speech that usually accompanies problem-solving.

Both deficits are forms of *tactile agnosia*. The right hemisphere deficit has become known as *apperceptive* tactile agnosia ('apperception', awareness of perception), and the left one as *associative* tactile agnosia (failure to identify functional associations).

Handedness and balance

In Chapter 18, we noted that the pathway from the vestibular nucleus to the parietoinsular vestibular cortex (PIVC) is mainly ipsilateral. Formal testing of static and dynamic vestibular functions, under functional MRI monitoring, has revealed that maximal activation of the PIVC is produced in the *minor* hemisphere, as illustrated in Figure 32.6. The limited data available indicate a tight linkage between handedness and vestibular cortical activation, to the extent that the 'preferred' vestibular hemisphere, in view of its phylogenetic antiquity and early ontogenesis (e.g. presence of postural vestibular reflexes in the newborn) may *determine* handedness.

Clinical Panel 32.3 Developmental dyslexia

It is generally agreed that reading is a more skilled activity than speech, because it requires an exquisite level of integration of visual scanning and auditory ('inner speech') comprehension. Reading is thought to activate two pathways in parallel: one passes via the angular gyrus to Wernicke's area and accesses a phonologic representation of every syllable in a temporal lobe memory store; the other passes to the left dorsolateral prefrontal cortex and accesses a semantic (meaning) memory store for every word.

Developmental dyslexia is a specific and pronounced reading difficulty in children who are the match of their peers in other respects. The anomaly is widespread, affecting 10–15% of children and about half that number of adults. There is a 30% incidence in siblings of affected children, and a similar incidence in one or other parent. There is a slightly higher incidence in boys, and in left-handers of either gender.

A consistent finding in positron emission tomography (PET) and functional magnetic resonance imaging studies during reading is diminished activity (compared with peers) in the left temporoparietal region (areas 22, 39, and 40).

Two commonly used classroom tests are *rhyming*, for example, in the alphabet, to identify the eight letters that rhyme with the letter B; and to pronounce *non-words* (*pseudowords*) within a word string, for example 'door', 'melse', 'farm', 'duve', miss'. Both tests are used to detect *phonologic impairment*, characterized by slow and inaccurate processing of the sound structure of language. The diagnostic label *phonologic dyslexia* is usually used, although other language-processing difficulties may also be present.

The performance of dyslexic children can often be improved by special training. Nevertheless, severe dyslexia tends to be associated with developmental deficiencies in one or more relevant parts of the brain, implicating as many as four different chromosomes.

- Magnetic resonance imaging scans tend to show a smaller than average left temporal plane. The cortex of the temporal plane (in dyslexic children dying from unrelated causes) may reveal evidence of incomplete migration of neurons from the ventricular zone to the temporal cortex during fetal life, with abnormally few neurons in laminae I–III, an excess in laminae IV–VI, and occasional clusters stranded in the subjacent white matter.
- The magnocells (M cells) of the ganglionic layer of the retina and lateral geniculate body are often smaller than normal, and their smaller axons conduct more slowly. One consequence is that the scanning movements used in reading, controlled by M-cell inputs to the superior colliculus (Ch. 28), are so inefficient that written syllables tend to run together and individual letters may appear to be transposed.

- Magnocellular neurons of the medial geniculate body, projecting to the primary auditory cortex (Ch. 20), may also be smaller. The presumed effect is one of reduced detection of the frequency and amplitude of sounds, leading to below-normal perception of words that are read aloud.
- The right cerebellum tends to be smaller than average, and during PET scans to be less active during performance of simple rapid motor tasks. Given that the cerebellovestibular system is responsible for keeping the eyes trained on visual targets, a cerebellar deficit could account for wobbly eye movements seen in about 1 in 10 dyslexics during reading. These children often try to stabilize their gaze by running an index finger below the line they are reading.
- Some dyslexic children are rather clumsy in executing fine movements such as buttoning a coat or tying a shoelace. The action tremor (intention tremor) characteristic of cerebellar disease is absent, and the label *dyspraxia* is used instead, to signify difficulty in controlling the rate and accuracy of fine learned movements in the absence of detectable loss of neuromuscular function. Here the fault appears to be at the interface between the processing and output pathways.
- A surprising finding is that the *insula* is inactive on certain language tasks on PET scans in dyslexics compared with in controls. It has been postulated that the insula may be a station linking the supramarginal and angular gyri to Broca's area, and that a defect in this linkage may be significant in relation to the inner speech that normally takes place during leisurely reading.

Commonly used classroom tests or indications on 5- to 7-year-olds include the following.

Speech
- Slow, laborious reading with little or no expressive intonation.
- *Vowel sounds*, for example, in the alphabet, to identify the eight letters that rhyme with the letter B.
- *Phoneme sounds*, i.e. the sounds of normal word syllables; also ability to pronounce *non-words* (*pseudowords*) within a word string, for example 'door', 'melse', 'farm', 'duve', miss'.
- *Syllable sequencing*, for example a dyslexic child may say 'emeny' for 'enemy'.

Writing
Dyslexia tends to be accompanied by slow and poor-quality handwriting, and spelling errors, for example letter transposition (e.g. 'pelcin' for 'pencil'), reversal (e.g. 'pad' for 'bad'), or inversion (e.g. 'wad' for 'mad').

Clinical Panel 32.3 *Continued*

REFERENCES

Habib M. The neurological basis of developmental dyslexia.
   Brain 2000; 123:2373–2399.
Leonard CM, Eckert MA, et al. Anatomical risk factors for
   phonological dyslexia. Cereb cortex 2001; 11:148–157.
Nicholson N, Fawcett A. Dyslexia early screening test (DEST).
   Online. Available: http://www.psychocorp.com.au/dest.htm
   2004.

Temple E. Brain mechanisms in normal and dyslexic readers.
   Curr Opin Neurobiol 2002; 12:178–183.

The assistance of Rena Lyons, Department of Speech and
Language Therapy, National University of Ireland, Galway, is
gratefully acknowledged.

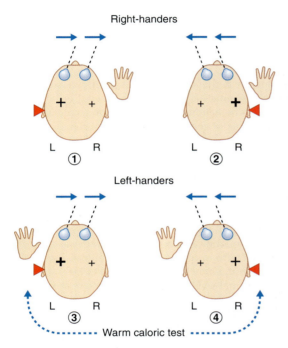

Figure 32.6 Hemispheric dominance in response to warm-water irrigation of the outer ear canals of pronounced right-handers and left-handers. (The head was tilted back 20° to put the lateral semicircular canals in the horizontal position.) Arrows indicate the slow phase of caloric nystagmus. Note pronounced minor hemisphere response to ipsilateral caloric irrigation (2 and 3), and less marked major hemisphere response (1 and 4). (After Dieterich et al. 2003, with permission.)

Parietal lobe and movement initiation

There are several sites for movement initiation in different behavioral contexts. The present context is the performance of learned movements of some complexity; examples would include turning a door knob, combing one's hair, blowing out a match, and clapping. It is logical to anticipate a starting point within the dominant hemisphere, because they can all be performed in response to verbal command (oral or written). This notion receives support from the observation that, if the corpus callosum has been severed surgically, the patient can perform a learned movement on command using the right hand, but not on attempting it with the left hand.

Failure to perform a learned movement on request is called *ideomotor apraxia* or *limb apraxia*. It has been repeatedly observed immediately following vascular lesions at the sites listed and described in Figure 32.7. In that figure, a vascular stroke at area no. 2, injuring corticospinal fibers descending from the hand area of the motor cortex, would cause clumsiness of movement of both hands, whereas a similar lesion on the right side would compromise only movements of the left hand.

Ideomotor apraxia can be accounted for if the dominant parietal lobe is considered to contain a repertoire of learned movement programs that, on retrieval, elicit appropriate responses by the premotor cortex on one or both sides under directives from the prefrontal cortex.

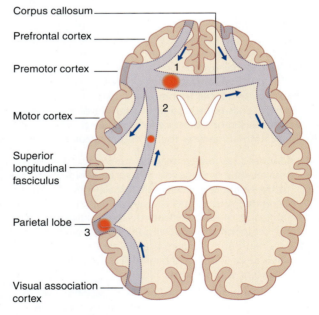

Figure 32.7 Association and commissural pathways serving motor responses to sensory cues. The premotor cortex is under higher control by the prefrontal cortex. Lesions at site 1 effectively sever the anterior part of the corpus callosum and produce ipsilateral limb apraxia (left lesion, left limb). Lesions at either site 2 (superior longitudinal fasciculus) or site 3 (angular gyrus) may produce *bilateral* limb apraxia. In practice, a large lesion may render right-limb apraxia impossible to assess because of associated right hemiplegia or receptive aphasia. (After Kertesz and Ferro 1984, with permission.)

(The basal ganglia would also be involved, as described in Ch. 33.)

Ideomotor apraxia is a transient phenomenon. Because parietal blood flow increases almost equally on both sides during reaching movements, the right hemisphere seems to be able to assume a full role for the left arm when no longer overshadowed.

Clinical Panel 32.4 provides a brief account of parietal lobe dysfunction.

PREFRONTAL CORTEX

The prefrontal cortex has two-way connections with all parts of the neocortex except the primary motor and sensory areas, with its fellow through the genu of the corpus callosum, and with the mediodorsal nucleus of the thalamus. It is uniquely large in the human brain and is concerned with the highest brain functions, including abstract thinking, decision-making, anticipating the effects of particular courses of action, and social behavior.

The DLPFC, centered in and around area 9, is strongly active in both hemispheres during waking hours. It has been called the *supervisory attentional system*. It participates in all cognitive activities and is essential for conscious learning of all kinds. During conscious learning, it operates *working memory*, whereby memories appropriate to the task (work) in hand are retrieved and 'held in the mind'.

The medial prefrontal cortex has auditory and verbal associations. The orbitofrontal cortex has been described as the *neocortical representative of the limbic system*, being richly connected to the amygdala, septal area, and cortex of the temporal pole—three limbic structures described in Chapter 34.

Clinical Panel 32.4   Parietal lobe dysfunction

Anterior parietal cortex
Lesions of the somatic sensory cortex and area 5 tend to occur together, causing cortical-type sensory loss and inaccurate reaching movements into contralateral visual hemispace (e.g. at mealtimes, the patient tends to knock things over).

Supramarginal gyrus
Lesions affecting the supramarginal gyrus (area 40) are usually vascular (middle cerebral artery) and are usually concomitant with contralateral hemiplegia with or without hemianopia. However, the blood supply to the gyrus is sometimes selectively occluded.

The characteristic result of damage to the supramarginal gyrus is *personal hemineglect*. The patient ignores the opposite side of the body unless attention is specifically drawn to it. A male patient will shave only the ipsilateral side of the face; a female patient will comb her hair only on the ipsilateral side. The patient will acknowledge a tactile stimulus to the contralateral side when tested alone; simultaneous testing of both sides will be acknowledged only ipsilaterally (*sensory extinction*).

Angular gyrus
Lesions of the anterior part of the angular gyrus (area 39, Figure 32.4.1) are notably associated with *extrapersonal hemineglect*. The patient tends to ignore the contralateral visual hemispace, even if the visual pathways remain intact, and there is *visual extinction* (contralaterally) to simultaneous bilateral stimuli, for example when the clinician wiggles his or her index fingers in both visual fields simultaneously.

Hemineglect is at least five times more frequent following lesions on the *right* side, irrespective of handedness.

An isolated vascular lesion of the posterior part of the left angular gyrus (very rare) produces *alexia*

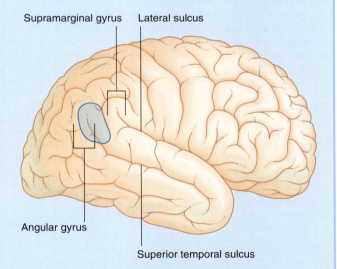

Figure CP 32.3.1 Angular gyrus.

(complete inability to read) and *agraphia* (inability to write); letters on the page are suddenly without any meaning. If the temporal plane has survived, patients can still name words spelt aloud to them.

For *ideomotor apraxia*, see main text.

REFERENCES

Chokron S, Colliot P, Bartolomeo P, et al. Visual, proprioceptive and tactile performance in left neglect patients. Neuropsychologia 2002; 40:1965–1976.
Hanna-Pladdy B, Heilman KM, Foundas AL. Cortical and subcortical contributions to ideomotor apraxia. Brain 2001; 124:2513–2527.
Mort DJ, Malhotra P, Mannan SK, et al. The anatomy of visual neglect. Brain 2003; 126:1986–1987.

In general terms, the left prefrontal cortex has an 'approach' bias, being engaged in all language-related activities, including the inner speech that accompanies investigative activities. The right prefrontal cortex has a 'withdraw' bias, being particularly activated by fearful contexts, whether real or imagined.

Aspects of frontal lobe dysfunction are described in Clinical Panel 32.2.

Core Information

Hemispheric asymmetries mainly concern handedness, language, and cognitive style. Some 10% of people are left-handers. Language areas are left-sided in 90%, right-sided in 2.5%, and bilateral in 2.5%. Broca's motor speech area occupies the inferior frontal gyrus; lesions here give rise to motor aphasia with difficulty in writing. Wernicke's sensory speech area in the temporal plane is required for understanding the spoken word; lesions here result in receptive aphasia, plus difficulty in reading if the angular gyrus is involved. The left hemisphere is usually superior in processing information susceptible to sequential analysis; the right hemisphere is superior for analysis of shapes and spatial relationships. The inferior parietal lobule is concerned with the body schema; lesions here may result in neglect of personal and extrapersonal space on the opposite side. Finally, the left parietal lobe may initiate complex motor programs; lesions here may be associated with ideomotor apraxia.

The prefrontal cortex is involved in highest brain functions. The dorsolateral prefrontal cortex (DLPFC) contains a supervisory attentional system especially involved in conscious learning, where it operates working memory appropriate to the task at hand. The orbitofrontal cortex is a neocortical representative of the limbic system. The left prefrontal cortex has investigative, 'approach' characteristics; the right has 'withdraw' characteristics. General signs of frontal lobe disease include lack of foresight, distractibility, and difficulty in switching cognitive sets. The gait may take the form of short shuffling steps with instability and 'freezing'. DLPFC lesions lead to slowing of mental process, apathy, and indifference. Orbitofrontal lesions tend to produce a hyperkinetic state with increased instinctual drives and puerile behavior.

REFERENCES

Behrmann M, Geng JJ, Shomstein S. Parietal cortex and attention. Curr Opin Neurobiol 2004; 14:212–217.

Binder JR, Frost JA, Hammeke TA, et al. Function of the left planum temporale in auditory and linguistic processing. Brain 1996; 119:1239–1247.

Bottini G, Cappa SF, Sterzi R, et al. Intramodal somaesthetic recognition disorders following right and left hemisphere damage. Brain 1995; 118:395–399.

Buchsbaum BR, Olsen RK, Koch PF, et al. Reading, hearing, and the planum temporale. Neuroimage 2005; 24:444–454.

Dieterich M, Bense S, Lutz S, et al. Dominance for vestibular cortical function in the non-dominant hemisphere. Cereb Cortex 2003; 13:994–1007.

Eden GF, Zeffiro TA. Neural systems affected in developmental dyslexia revealed by functional neuroimaging. Neuron 1998; 21:329–282.

Farne A, Roy AC, Paulignan Y, et al. Visuo-motor control of the ipsilateral hand: evidence from right brain-damaged patients. Neuropsychologia 2003; 41:739–757.

Heiss WD, Kessler J, Thiel A, et al. Differential capacity of left and right hemispheric areas for compensation of poststroke aphasia. Ann Neurol 1999; 45:430–438.

Hepper PG, Wells DL, Lynch C. Prenatal thumb sucking is related to postnatal handedness. Neuropsychologia 2005; 43:313–315.

Hynd GW, Marshall R, Hall J, et al. Learning disabilities: neuroanatomic asymmetries. In: Davidson RJ, Hugdahl K, eds. Brain asymmetry. Cambridge: MIT Press; 1995:617–636.

Jäncke L, Schlaug G, Huang Y, et al. Asymmetry of the planum parietale. NeuroReport 1994; 5:1161–1163.

Jäncke L, Steinmetz H. Anatomical brain asymmetries and their relevance for functional asymmetries. In: Hugdahl K, Davison RJ, eds. The asymmetrical brain. Cambridge: MIT Press; 2003:187–230.

Kertesz A, Polk M, Black SE, et al. Anatomical asymmetries and functional laterality. Brain 1992; 115:589–605.

Liotti M, Gay CT, Fox PT. Functional imaging and language. J Clin Neurophysiol 1994; 11:175–190.

Passingham RE. Attention to action. In: Roberts AC, Robbins TW, Weiskranz L, eds. The prefrontal cortex: executive and cognitive functions. Oxford: Oxford University Press; 1998:131–143.

Perelle IB, Ehrman ND. An international study of human handedness: the data. Behav Genet 1994; 24:217–225.

Price CJ, Mechelli A. Reading and reading disturbance. Curr Opin Neurobiol 2005; 15:232–238.

Price CJ. The anatomy of language: contributions from functional imaging. J Anat 2000; 197:335–339.

Pugh KR, Shaywitz BA, Shaywitz SE, et al. Cerebral organization of the functional processes in reading. Brain 1996; 119:1221–1238.

Reed CL, Caselli RJ, Farah MJ. Tactile agnosia. Brain 1996; 119:875–888.

Rumsey JM, Donohue BC, Brady DR, et al. An MRI study of planum temporale asymmetry in males with developmental dyslexia. Arch Neurol 1997; 54:1481–1489.

Russell SM, Elliott R, Forshaw D, Kelly PJ, Golfinos JG. Resection of parietal lobe gliomas: incidence and evolution of neurological deficits in 28 consecutive patients correlated to the location and morphological characteristics of the tumor. J Neurosurg 2005; 103:1010–1017.

Sabaté M, González B, Rodríguez M. Brain lateralization of motor imagery: motor planning asymmetry as a cause of movement lateralization. Neuropsychologia 2004; 42(8):1041–1049.

Sakata H, Taira M. Parietal control of hand action. Curr Opin Neurobiol 1994; 4:847–856.

Silbersweig DA, et al. A functional neuroanatomy of hallucinations in schizophrenia. Nature 1995; 378:176–179.

Stein JF. Developmental dyslexia, neural timing, and hemispheric lateralization. Int J Neurophysiol 1994; 18:241–249.

Thompson-Schill SL, Bedny M, Goldberg RF. The frontal lobes and the regulation of mental activity. Curr Opin Neurobiol 2005; 15:219–224.

Tzourio N, Crivello F, Mellet E, et al. Functional anatomy of dominance for speech comprehension in left handers vs right handers. Neuroimage 1998; 8:1–16.

Basal ganglia

STUDY GUIDELINES
The contributions of the basal ganglia to motor control systems is the subject of intensive research at basic and clinical levels.

In this account, the open-loop pathways linking different areas of the cerebral cortex through the basal ganglia are highlighted. Particular emphasis is put on the breaking and release mechanisms inherent in the arrangement of sequential sets of inhibitory neurons.

Pathologic changes within the basal ganglia are associated with several kinds of motor disorders, both in children and in adults.

The term *basal ganglia* is used to designate the areas of basal forebrain and midbrain known to be involved in the control of movement (Figure 33.1). It includes the following.

- The **striatum** (caudate nucleus, putamen of lentiform nucleus, nucleus accumbens).
- The **pallidum** (globus pallidus of lentiform nucleus), which comprises a **lateral segment** and a **medial**

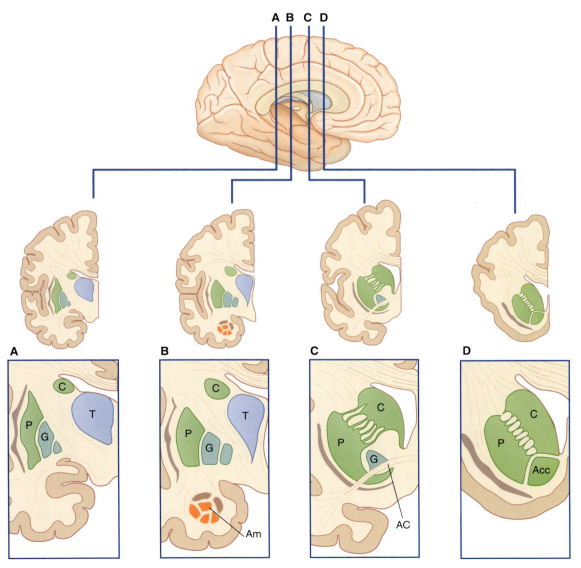

Figure 33.1 (A–D) Four coronal sections of the brain, viewed from behind. Ventral parts are enlarged below. AC, anterior commissure; Acc, nucleus accumbens; Am, amygdala; C, caudate nucleus; G, globus pallidus; P, putamen; T, thalamus.

segment. The medial segment has a midbrain extension known as the **reticular part (pars reticulata)** of the **substantia nigra.**

- The **subthalamic nucleus (STN).**
- The pigmented, **compact part (pars compacta)** of **substantia nigra.**

BASIC CIRCUITS

It is possible to demonstrate at least four circuits that commence in the cerebral cortex, traverse the basal ganglia, and return to the cortex. The four comprise:

1 a *motor loop*, concerned with learned movements
2 a *cognitive loop*, concerned with motor intentions
3 a *limbic loop*, concerned with emotional aspects of movement
4 an *oculomotor loop*, concerned with voluntary saccades.

Motor loop

The motor loop commences in the sensorimotor cortex and returns there via striatum, thalamus, and supplementary motor area (SMA).

Figure 33.2 is derived from Figure 33.1A. It is a schematic coronal section including the posterior part of the striatum, depicting component parts of the motor loop. Two pathways are known. The 'direct' pathway traverses the corpus striatum and thalamus and involves five consecutive sets of neurons (Figure 33.2A). The 'indirect' pathway engages the STN in addition and involves seven sets of neurons (Figure 33.2B). The anatomy of the two thalamic projections from the medial pallidum (**ansa lenticularis** and **lenticular fasciculus**) is shown in Figure 33.3.

All projections from the cerebral cortex arise from pyramidal cells and are excitatory (glutaminergic). So, too, is the projection from thalamus to SMA. Those from striatum and from both segments of pallidum arise from medium-sized spiny neurons and are inhibitory. They are GABAergic, and also contain neuropeptides of uncertain function.

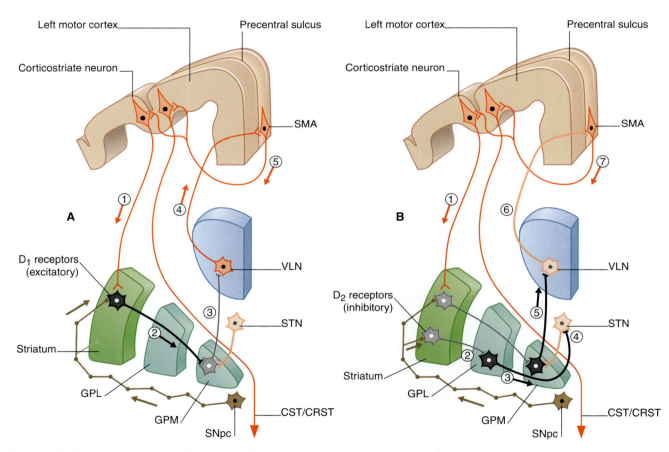

Figure 33.2 Coronal section through the motor loop, based on Figure 33.1A. **(A)** The sequence of five sets of neurons involved in the 'direct' pathway from sensorimotor cortex to thalamus with final return to sensorimotor cortex via supplementary motor area. **(B)** The sequence of seven sets of neurons involved in the 'indirect' pathway.
The *red–pink* neurons are excitatory, utilizing glutamate. The *black–gray* neurons are inhibitory, utilizing γ-aminobutyric acid. The *brown*, nigrostriatal neuron utilizes dopamine, which is excitatory via D₁ receptors on target striatal neurons, and inhibitory via D₂ receptors on the same and other striatal neurons. CST/CRST, corticospinal, corticoreticular fibers; GPL, GPM, lateral and medial segments of globus pallidus; SMA, supplementary motor area; SNpc, compact part of substantia nigra; STN, subthalamic nucleus; VLN, ventral lateral nucleus of thalamus.

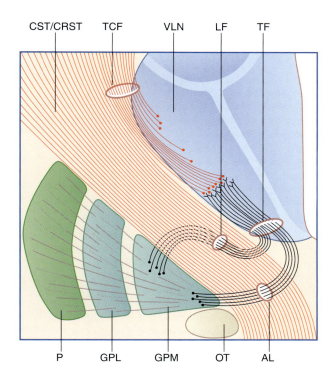

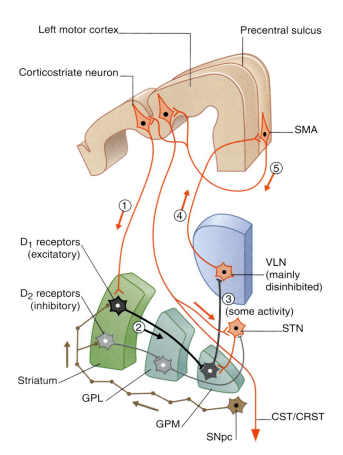

Figure 33.3 Part of the projection from the medial segment of globus pallidus (GPM) to the ventral lateral nucleus (VLN) of the thalamus sweeps around the base of the internal capsule as the **ansa lenticularis (AL)**; the remainder traverses this region as the **lenticular fasciculus (LF)**. The two parts come together as the **thalamic fasciculus (TF)** before entering the thalamus. CST/CRST, corticospinal and corticoreticular fibers; GPL, lateral segment of globus pallidus; OT, optic tract; P, putamen; TCF, thalamocortical fibers.

Figure 33.4 Activities in the striatal motor loops, prior to movement.

The supplementary motor area (SMA) is activated through the 'direct' pathway as follows. **(1)** Corticostriate fibers from the sensorimotor cortex activate those GABAergic spiny neurons in striatum having D_1 receptors tonically facilitated by nigrostriatal inputs. **(2)** The activated striatal neurons inhibit medial pallidal (GPM) neurons **(3)** with consequent *disinhibition* of ventral lateral nucleus (VLN) thalamocortical neurons **(4)** and activation of SMA **(5)**, which both modifies ongoing corticostriate activity and initiates impulse trains along corticospinal (CST) and corticoreticular (CRST) fibers. Activity along the 'indirect' pathway is relatively slight, because of tonic dopaminergic inhibition of the relevant striatal neurons via D_2 receptors. However, the subthalamic nucleus (STN) is tonically activated by corticosubthalamic fibers, curtailing the inhibition of GPM. GPL, lateral segment of globus pallidus; SNpc, compact part of substantia nigra. (Cerebellothalamocortical projection is not shown.)

The *nigrostrial pathway* projects from the compact part of the substantia nigra to the striatum, where it makes two kinds of synapses on the projection neurons there (Figure 33.4). Those on direct pathway neurons are facilitatory, by way of dopaminergic type 1 (D_1) receptors on the dendritic spines; those on indirect pathway neurons are inhibitory, by way of type 2 (D_2) receptors. Cholinergic internuncial neurons within the striatum are excitatory to projection neurons, and they are inhibited by dopamine.

A healthy substantia nigra is tonically active, favoring activity in the direct pathway. Facilitation of this pathway is necessary for the SMA to become active before and during movement. SMA activity immediately prior to movement can be detected by means of recording electrodes attached to the scalp. This activity is known as the (electrical) *readiness potential*, and its manner of production is described in the caption to Figure 33.4. Impulses pass from SMA to the motor cortex, where a cerebellothalamocortical projection selectively enhances pyramidal and corticoreticular neurons within milliseconds prior to discharge.

The putamen and globus pallidus are somatotopic, permitting selective facilitation of neurons relevant to (say) arm movements via the direct route, with simultaneous disfacilitation of unwanted (say) leg movements via the indirect route. For suppression of unwanted movements, the STN, acting on the body map in the medial pallidal segment, is especially important, because we know that destruction of STN results in uncontrollable flailing movements of one or more body parts on the opposite side (see later).

Progressive failure of dopamine production by the compact part of substantia nigra is the precipitating cause of *Parkinson's disease* (*PD*) (Clinical Panel 33.1).

Clinical Panel 33.1   Hypokinesia: Parkinson's disease

Parkinson's disease (PD) affects about 1% of people over 65 years of age in all countries. The primary underlying pathology is degeneration of nigrostriatal neurons, resulting in diminished dopamine content within the striatum. *[¹⁸F]fluorodopa* is a mildly radioactive compound that, when injected intravenously, binds with dopamine receptors in the striatum. In symptomatic PD, a significant reduction of [¹⁸F]fluorodopa binding (and therefore of receptors) is revealed by means of positron emission tomography scanning (Figure CP 33.1.1). One consequence is *increased* striatal activity, with a shift from the direct to the indirect motor pathway (Figure CP 33.1.2).

Nigrostriatal degeneration seems to take the form of a *dying-back neuropathy*, because dopamine is lost from the striatum earlier than in the midbrain. The spiny striatal neurons also deteriorate, with reduction in the length of dendrites and the numbers of spines. It may be that the spiny neurons depend on dopaminergic inputs for protection against potentially toxic effects of ongoing glutamate activity.

Some 60% of nigral neurons have been lost before the first symptoms appear. This delay is accounted for by (a) increased dopamine production by surviving neurons, and (b) increased production (up-regulation) of dopamine receptors in the target striatal neurons.

The following symptoms and signs are characteristic: tremor, bradykinesia, rigidity, and impairment of postural reflexes. Not all are expressed in every patient.

Tremor
Tremor, at 3–6 Hz (times per second) in one limb is the initial feature in two-thirds of cases. The commonest

sequence of limb involvement is from one upper limb to ipsilateral lower limb within 1 year, followed by contralateral limb involvement within 3 years. Rhythmic tremor of lips and tongue, pronation–supination of the forearm, and flexion–extension of the fingers may be obvious. A pill-rolling movement of index and middle fingers against the thumb pad is characteristic. Typically, the tremor involves only muscle groups that are 'at rest', and vanishes during voluntary movement. A patient with an exclusively resting tremor has no difficulty in raising and draining a tumblerful of water. The term *resting tremor* used to distinguish it from the *intention tremor* of cerebellar disease. Intention tremor is absent at rest (unless cerebellar dysfunction is severe) and is brought on by voluntary movement.

Tremor is associated with rhythmic bursting activity within all five cell groups of the direct motor loop (Figure 28.2A) and in anterior horn cells of the spinal cord. The contribution of disordered autogenetic inhibition to both resting tremor and rigidity is described below.

A fine *action tremor* is often detectable in patients having pronounced resting tremor, and it is more pronounced on the side more affected by resting tremor. The action tremor is best seen in the fingers when the arms are fully outstretched, and it may be manifested by tremulous handwriting. Note particularly that, in the *absence* of resting tremor, a fine action tremor is indicative of *benign essential tremor* (see later).

Rigidity
Rigidity affects all the somatic musculature simultaneously, but a predilection for flexors imposes a

Head of caudate nucleus   Putamen

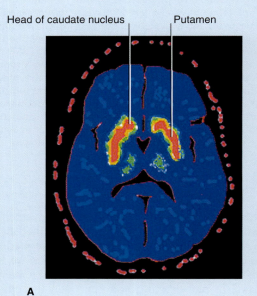 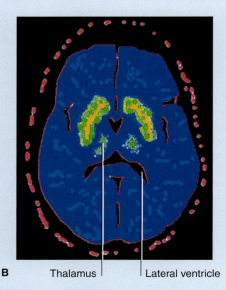

A

B   Thalamus | Lateral ventricle

Figure CP 33.1.1 Typical results of brain scans following intravenous injection of [¹⁸F]fluorodopa. Intensity of uptake is indicated as *red* (greatest), *yellow, green, blue* (least). **(A)** Control; **(B)** Parkinson's disease.

Clinical Panel 33.1   *Continued*

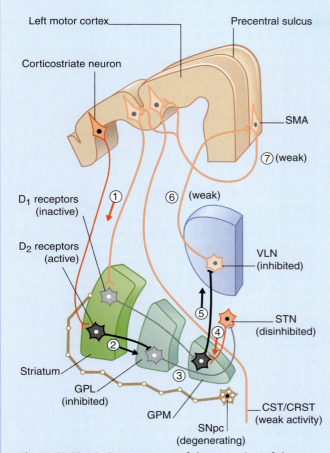

Left motor cortex

Precentral sulcus

Corticostriate neuron

SMA

⑦ (weak)

D₁ receptors
(inactive)

①

⑥ (weak)

D₂ receptors
(active)

VLN
(inhibited)

⑤

STN
(disinhibited)

④

②

Striatum

③

GPL
(inhibited)

GPM

CST/CRST
(weak activity)

SNpc
(degenerating)

Figure CP 33.1.2 Consequences of degeneration of the pathway from the compact part of the substantia nigra (SNpc) to the striatum in Parkinson's disease. The effects arise from loss of tonic facilitation of spiny striatal neurons bearing D₁ receptors, together with loss of tonic inhibition of those bearing D₂ receptors. The 'direct' pathway is disengaged; the 'indirect' pathway is activated by default. **(1)** Corticostriate neurons from the sensorimotor cortex now strongly activate those GABAergic neurons **(2)** in the striatum that synapse on others **(3)** in the lateral pallidal segment (GPL). The double effect is *disinhibition* of the subthalamic nucleus (STN). STN discharges strongly **(4)** on to the GABAergic neurons of the medial pallidal segment (GPM); these in turn discharge strongly **(5)** into the ventral lateral nucleus (VLN) of thalamus, resulting in reduced output along thalamocortical fibers **(6)** traveling to the supplementary motor area (SMA). Inputs **(7)** from SMA to corticospinal and corticoreticular fibers (CST, CRST) become progressively weaker, with pathetic consequences for initiation and execution of movements.

stooped posture. Passive flexion and extension of the major joints evince resistance through the full range of movement. The term *lead pipe rigidity* is used to distinguish this type of resistance from the 'clasp knife rigidity' of the spastic state that accompanies upper

motor neuron lesions. The clinician may detect a subtle underlying tremor in the form of ratchety, 'cogwheel' sensation.

Historically, rigidity has been abolished by section of dorsal nerve roots, thus proving its peripheral sensory origin. It can also be alleviated by a surgical lesion of the pallidum or of the ventral lateral nucleus of the thalamus. Because muscle spindle stretch reflexes are not exaggerated in PD, attention has focused on the Golgi tendon organ afferents responsible for autogenetic inhibition. As illustrated in Chapter 10, these afferents synapse on inhibitory, 1b internuncials that, when activated by muscle contraction, dampen activity of motor neurons supplying the same muscle and any homonymous contributors to the same movement (e.g. impulses generated in biceps brachii tendon organs will depress both brachialis and biceps motor neurons). In PD patients, autogenetic inhibition is reduced, and it is also delayed to the extent that it becomes entrained with the pulses descending from the brain, with the effect of contributing to the tremor. It may also contribute to the rigidity, because in PD there is some degree of cocontraction of prime movers and antagonists.

Given that muscular contraction is required to activate tendon organs, why do patients display resting tremor with supposedly inactive muscles? It transpires that, when the forearms (say) are resting on the lap or on the arms of a chair, the forearm and hand muscles are *not* fully at rest. If the limb is properly supported at elbow and wrist, the tremor disappears. The tremor also disappears during sleep.

Normally, both corticospinal and reticulospinal fibers are tonically facilitatory to 1b inhibitory internuncials. In PD, activation of the primary motor cortex by the supplementary motor area (SMA) is known to be both reduced and oscillatory, thus accounting for the pronounced effects in the forearm and hand. Impaired reticulospinal activity is more likely to be significant with respect to the lower limbs.

In addition to its massive projections into the pallidum, the putamen projects to another group of GABAergic neurons, namely the reticular part of the substantia nigra. The compact part of the substantia nigra also projects to the reticular part. The reticular part projects in turn to the brainstem locomotor center (Ch. 24). In PD, the overactive putamen would be expected to have the knock-on effect of inhibiting impulse traffic in the projections from the locomotor area to the pontine and medullary reticular reticulospinal tracts.

Difficulty in writing is a common early feature. The individual written letters become small and irregular. Loss of writing skill is attributable to cocontraction of wrist flexors and extensors, owing to marked reduction of supraspinal activation of 1a internuncials synapsing on antagonist motor neurons.

Clinical Panel 33.1 *Continued*

Bradykinesia

Bradykinesia means slowness of movement. Patients report that routine activities, such as opening a door, require deliberate planning and consciously guided execution. Electromyographic studies of the limb musculature show a reduction of the 'initial agonist burst' of electrical charge accompanying the first contraction of relevant prime movers. Normally, the basal ganglionic contribution to movement comes on stream some milliseconds after the premotor cortex and cerebellum have raised the firing rate of motor–cortex neurons to threshold at spinal lower motor neuron level. In PD, the boost to lower motor neuron activation is weak because of the weakened contribution from SMA.

Impairment of postural reflexes

Patients go off balance easily, and tend to fall stiffly ('like a telegraph pole') in response to a mild accidental push. The underlying fault is an impairment of anticipatory postural adjustments; normally, a push to the upper part of the body elicits immediate contraction of lower limb muscles appropriate for the maintenance of equilibrium.

Two other symptoms, *oculomotor hypokinesia* and *dementia*, are mentioned in the main text.

Misdiagnosis

Parkinson's disease has two principal kinds of presentation. In one, tremor is the predominant feature. In the other, akinesia and rigidity predominate. It is now known that more than one in five people initially diagnosed and treated as suffering from PD either do not have PD at all, or have a 'Parkinson plus' syndrome.

Benign essential tremor is more than twice as prevalent as PD and is often mistaken for it. It is characterized initially by a faint trembling, most noticeable when the arms are fully outstretched. Later, head-bobbing—not a feature of PD—and orthostatic (when upright) trunk tremor may appear, and a tremulous diaphragm may impart a vocal tremor. Benign essential tremor is sometimes called *familial tremor* because of autosomal dominant inheritance; it commonly becomes manifest during the fifth decade. When observed in the elderly, it may be called *senile tremor*.

The cause is unknown. Levodopa (L-dopa) (see below) is ineffective, whereas it relieves both kinds of tremor in PD.

Multisystem atrophy is a Parkinson plus degenerative disorder of brainstem, basal ganglia, and central autonomic neurons. Patients present with one or more of the following.

- Akinesia or rigidity with little or no tremor.
- One or more signs of autonomic failure: postural hypotension, bladder or bowel dysfunction, impotence, dry eyes and mouth, pupillary abnormalities, impaired sweating.
- Bilateral pyramidal tract degeneration leading to pseudobulbar palsy (Ch. 18) and 'upper motor neuron signs' (Ch. 16) in the limbs.
- Poor ocular convergence.

L-dopa is of little value.

Clinical neurology texts describe other relevant disorders, for example *progressive supranuclear palsy* and *corticobasal degeneration*.

Treatment of Parkinson's disease

Drugs

The first line of treatment of PD is administration of L-dopa, which can cross the blood–brain barrier and is metabolized to dopamine by surviving nigral neurons. Some 75% of patients benefit, with a reduction of symptoms by 50% or more. After several years of L-dopa therapy, many patients develop spontaneous choreiform movements (described in Clinical Panel 33.2) owing to excessive striatal response. After a year or more, the effectiveness of L-dopa declines with the progressive loss of nigral neurons, and dopamine agonist drugs are often used instead to stimulate striatal postsynaptic dopamine receptors.

Anticholinergic drugs reduce activity of the cholinergic internuncials in the striatum. They ameliorate tremor (of both kinds) in particular, but the required dosage is liable to produce one or more of the autonomic side effects listed in Clinical Panel 12.3.

Surgery

The optimal approach at present is to *paralyze* subthalamic nucleus (STN) neurons by high-frequency (133 Hz) stimulation through implanted electrodes. The paralysis is caused by specific inactivation of Ca^{2+} and Na^+ voltage-dependent channels. This approach is being used in the STN to dramatic effect in patients with full tremor, bradykinesia, and rigidity. All three are greatly ameliorated in most cases. Unilateral stimulation may be sufficient for bilateral relief, of rigidity in particular. The bilateral effect of *deep brain stimulation* has been accounted for (in monkey experiments) by paralysis of an excitatory projection from STN to the GABAergic reticular part of the substantia nigra, which, as already mentioned, gives an inhibitory supply to the cross-connected right and left locomotor center. According to this interpretation, STN blockage *disinhibits* the locomotor center. It is also possible that deep brain stimulation generates a new type of discharge endowed with beneficial effects.

Other approaches are under intense investigation, including grafts of fetal substantia nigra, striatal infusion of growth factors, and gene therapy.

Clinical Panel 33.1   *Continued*

REFERENCES

Benabid AL. Deep brain stimulation for Parkinson's disease. Curr Opin Neurobiol 2003; 13:696–706.

Buhmann C, Glauche V, Stuhrenburg HJ, et al. Pharmacologically modulated fMRI: cortical responsiveness to levodopa in drug-naive hemiparkinsonian patients. Brain 2003; 126:154–161.

Colnat-Coulbois S, Gauchard GC, Maillard L, et al. Bilateral subthalamic nucleus stimulation improves balance control in Parkinson's disease. J Neurol Neurosurg Psychiatry 2005; 76: 780–787.

Garcia L, D'Alessandro G, Bioulac B, et al. High frequency stimulation in Parkinson's disease: more or less? Trends Neurosci 2005; 28:209–216.

Grafton ST. Contributions of functional imaging to understanding parkinsonian symptoms. Curr Opin Neurobiol 2004; 14:715–719.

Grillner S, Hellgren J, Menard A, et al. Mechanisms for selection of basic motor programs—roles for the striatum and pallidum. Trends Neurosci 2005; 28:364–370.

Louis ED, Levy G, Cote LJ, et al. Clinical correlates of action tremor in Parkinson's disease. Arch Neurol 2001; 58:1633–1634.

Mendez I, Sanchez-Pernaute R, Cooper O, et al. Cell type analysis of functional fetal dopamine cell suspension transplants in the striatum and substantia nigra of patients with Parkinson's disease. Brain 2005; 128:1498–1510.

Romanelli P, Esposito V, Schaal DW, et al. Somatotopy in the basal ganglia: experimental and clinical evidence for segregated sensorimotor channels. Brain Res Rev 2005; 48:112–128.

Stephens B, Mueller AJ, Shering AF, et al. Evidence of a breakdown of corticostriatal connections in Parkinson's disease. Neuroscience 2005; 132:741–754.

Timmermann L, Gross J, Dirks M, et al. The cerebral oscillatory network of parkinsonian resting tremor. Brain 2003; 126:199–212.

The assistance of Dr. Tim Counihan, Department of Medicine, University College Hospital, Galway, is gratefully acknowledged.

What are the normal functions of the motor loop?

Although movements can be produced on the opposite side of the body by direct electrical stimulation of the healthy putamen, the basal ganglia do not normally initiate movements. Nevertheless, they are active during movements of all kinds, whether fast or slow. They seem to be involved in *scaling the strength* of muscle contractions and, in collaboration with SMA, in *organizing the requisite sequences* of excitation of cell columns in the motor cortex. They come into action after the corticospinal tract has already been activated by 'premotor' areas including the cerebellum. Because patients with PD have so much difficulty in performing internally generated movement sequences, it is believed that the putamen provides a reservoir of learned motor programs that it is able to assemble in appropriate sequence for the movements decided on, and to transmit the coded information to SMA.

Cognitive loop

The head of the caudate nucleus receives a large projection from the prefrontal cortex, and it participates in *motor learning*. Positron emission tomography scan studies have demonstrated increased contralateral blood flow through the head of the caudate when novel motor actions are performed with one hand. There is also increased activity in the anterior part of the contralateral putamen, globus pallidus, and ventral anterior (VA) nucleus of the thalamus. The VA nucleus completes an 'open' cognitive loop through its projection to the premotor cortex, and a 'closed' loop through a return projection to the prefrontal cortex. The cortical connections of the caudate suggest that it participates in *planning ahead*, particularly with respect to

complex motor intentions. When the novel motor task has been practiced to the level of automatic execution, the motor loop becomes active instead.

Limbic loop

Figure 33.5 depicts the *limbic* basal ganglia loop. This loop passes from inferior prefrontal cortex through **nucleus accumbens** (anterior end of the striatum, Figure 33.1D) and ventral pallidum, with return via mediodorsal nucleus of thalamus to inferior prefrontal cortex.

The limbic loop is likely to be involved in giving motor expression to emotions, for example through smiling or

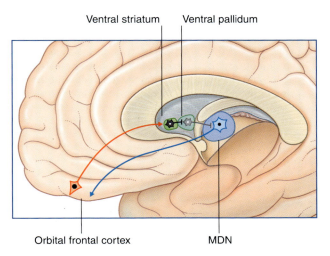

Ventral striatum   Ventral pallidum

Orbital frontal cortex   MDN

Figure 33.5 The limbic basal ganglia loop, right hemisphere. The medial dorsal nucleus of thalamus (MDN) is being released by means of disinhibition.

gesturing, or adoption of aggressive or submissive postures. The loop is rich in dopaminergic nerve endings, and their decline may account for the mask-like facies and absence of spontaneous gesturing characteristic of PD, and for the *dementia* that may set in after several years.

Oculomotor loop

The oculomotor loop commences in the *frontal eye field* and *posterior parietal cortex* (area 7). It passes through the caudate nucleus and through the reticular part of the substantia nigra (SNpr). It returns via the VA nucleus of the thalamus to the frontal eye field and prefrontal cortex. SNpr sends an inhibitory GABAergic projection to the superior colliculus, where it synapses on cells controlling automatic saccades (Ch. 23). These cells are also supplied directly from the frontal eye field.

While the eyes are fixated, SNpr is tonically active. Whenever a deliberate saccade is about to be made toward another object, the oculomotor loop is activated and the superior colliculus is *disinhibited*. The superior colliculus then discharges to reinforce the activity of the direct pathway. Maximum speed (80 km/h) is achieved instantly, the eyeballs are flicked to the target, and SNpr resumes its vigilance.

In PD, *oculomotor hypokinesia* can be revealed by special tests. Saccades toward targets in the peripheral visual field tend to be slow and sometimes inadequate. This hypokinesia can be explained on the basis of faulty disinhibition of the superior colliculus following associated neuronal degeneration within SNpr.

Other disorders involving the basic ganglia include several *hyperkinetic states* briefly described in Clinical Panel 33.2.

Core Information

The basal ganglia are nuclear groups involved in movement control. They comprise the striatum (including nucleus accumbens); pallidum; subthalamic nucleus (STN); substantia nigra; and thalamic motor nuclei ventral lateral (VL), ventral anterior (VA), and medial dorsal. The pallidum has a lateral and a medial segment (GPL, GPM), the latter tapering into the midbrain as the reticular part of substantia nigra (SNpr). Four circuits commence in the cerebral cortex, pass through the basal ganglia, and return to the cortex. The compact part of substantia nigra (SNpc) stands aside of the circuits but influences them by way of the nigrostriatal pathway.

Cortical inputs to striatum and STN are excitatory. Striatal outputs are inhibitory to the pallidum; so, too are the pallidal outputs to STN and thalamus. STN is excitatory to GPM.

The 'direct' pathway, striatum → GPM, is facilitated by the normal tonic activity of nigrostriatal dopaminergic neurons. The 'indirect' pathway, striatum → GPL → GPM, is inhibited. In the *motor loop*, facilitation of the direct pathway is necessary for the supplementary motor area (SMA) to become active before and during movement. SMA activity immediately prior to movement is detectable as the readiness potential, and is produced by silencing of GPM neurons with consequent liberation (disinhibition) of thalamocortical neurons to SMA, with follow-through to the motor cortex for initiation of movement.

Striatum and pallidum are somatotopically organized, permitting selective activation of body parts; STN is especially important for inhibition of unwanted movements.

The main function of the motor loop seems to be the appropriate sequencing of serial order actions for the execution of learned motor programs. In Parkinson's disease (PD), the loss of nigrostriatal dopaminergic neurons causes the indirect pathway to become dominant, with follow-through suppression of VL and reduced SMA activity, thus accounting for the characteristic bradykinesia. PD symptomatology also includes rigidity, tremor, and impairment of postural reflexes. Benign essential tremor and multisystem atrophy are too often misdiagnosed as PD.

The *cognitive loop* begins in the association cortex, and returns via VA nucleus of thalamus to the premotor and prefrontal cortex. It is actively engaged during motor learning, and also seems concerned with planning ahead for later movements.

The *limbic loop* begins in cingulate cortex and amygdala, passes through nucleus accumbens, and returns to SMA; it is probably involved in giving physical expression to the current emotional state.

The *oculomotor loop* disinhibits SNpr, thereby liberating the superior colliculus to execute a saccade.

Hyperkinetic states include many cases of cerebral palsy; also Huntington chorea and hemiballism.

Clinical Panel 33.2 Other extrapyramidal disorders

Cerebral palsy

Cerebral palsy is an umbrella term covering a variety of motor disorders arising from damage to the brain during fetal life or in the perinatal period. The incidence is about 2 per 1000 live births in all countries.

The most frequent type of congenital motor disorder is *spastic diplegia*. During the early postnatal months, affected children are usually 'floppy' (atonic), changing to a spastic state (of the lower limbs in particular) by the end of the first year. Remarkably, many children who are spastic at the age of 2 will be completely normal by the age of 5. Most of the remainder 'grow into their disability' and become more severely affected.

The ventricular system in spastic diplegia is dilated, owing to maldevelopment of periventricular oligodendrocytes in the sixth to eighth month of gestation, notably those myelinating corticospinal fibers destined for lumbosacral segments of the spinal cord. Intrauterine infection (Figure CP 33.2.1), ischemia, and metabolic disorders are etiologic suspects.

Extrapyramidal or *dyskinetic* cerebral palsy is statistically correlated with perinatal asphyxia. In this condition, the striatum is particularly affected, perhaps because it is normally highly active metabolically in establishing synaptic connections with the pallidum.

Choreoathetosis is characteristic. *Chorea* refers to momentary spontaneous twitching of muscle groups in a more or less random manner, interfering with voluntary movements. *Athetosis* refers to writhing movements that are continuous except during sleep and may be so severe as to prevent sitting or standing. Waxing and waning of muscle tone commonly cause the head to roll about. Both movements are regarded as escape phenomena resulting from damage to the striatum.

Huntington chorea

Huntington chorea is an autosomal (chromosome 4) dominant, inherited disease that occurs in 50% of the offspring of affected families. Onset of symptoms is usually delayed until the forties. The clinical history is one of chronic, progressive chorea, often with athetoid movements superimposed. Sooner or later, a progressive dementia sets in.

Hemiballism

Hemiballism (or hemiballismus) tends to occur in the elderly. It is known to result from thrombosis of a small branch of the posterior cerebral artery supplying the subthalamic nucleus (STN). The condition is perhaps the most remarkable one in the whole of clinical neurology. It is marked by the abrupt onset of wild, flailing movements of the contralateral arm, sometimes of the leg as well. The appearances suggest that the thalamocortical pathway from ventral lateral nucleus (VLN) to supplementary motor area has become intensely overactive.

This question may well be asked: If vascular destruction of STN results in hemiballism, why does STN paralysis by high-frequency stimulation not have the same effect? Occasionally it does, requiring immediate withdrawal of the electrode. But, in general, the VLN–SMA pathway remains sufficiently inhibited by underactivity in the 'direct' pathway.

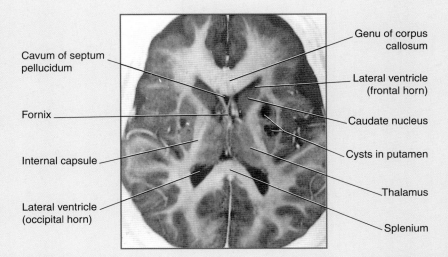

Cavum of septum pellucidum

Fornix

Internal capsule

Lateral ventricle (occipital horn)

Genu of corpus callosum

Lateral ventricle (frontal horn)

Caudate nucleus

Cysts in putamen

Thalamus

Splenium

Figure CP 33.2.1 Horizontal magnetic resonance imaging slice at the level of the corpus striatum from a 2-year-old girl suffering from severe choreoathetosis as a result of intrauterine damage to her basal ganglia by toxoplasmosis. The putamen on both sides has been partly replaced by cysts. (Magnetic resonance image kindly provided by Professor J. Paul Finn, Director, MRI Facility, Northwestern University School of Medicine, Chicago.)

REFERENCES

Berger W, Discher M, Trippel M, et al. Developmental aspects of stance regulation, compensation and adaptation. Exp Brain Res 1992; 90:610–619.

Brooks DJ. The role of the basal ganglia in motor control: contributions from PET. J Neurol Sci 1995; 133:1–13.

Brown P, Steiger MJ. Basal ganglia disorders. In: Bronstein AM, Brandt T, Woollacott M, eds. Clinical disorders of balance, posture and gait. London: Arnold; 2004; 156–166.

Burne JA, Lippold OCJ. Loss of tendon organ inhibition in Parkinson's disease. Brain 1996; 119:1115–1121.

Ceballos-Baumann AO, Boecker H, Bartenstein P, et al. A PET study of subthalamic nucleus stimulation in Parkinson's disease. Arch Neurol 1999; 56:997–1003.

Chase TN, Oh JD, Blanchet PJ. Neostriatal mechanisms in Parkinson's disease. Neurology 1998; 51(suppl 2):S33–S35.

Crossman AR. Functional anatomy of movement disorders. J Anat 2000; 196:519–525.

Henderson JM, Dunnett SB. Targeting the subthalamic nucleus in the treatment of Parkinson's disease. Brain Res Bull 1998; 46:467–474.

Hornykiewicz O. Biochemical aspects of Parkinson's disease. Neurology 1998; 51(suppl 2):S2–S9.

Kuban KCK, Leviton A. Cerebral palsy. New Engl J Med 1994; 333:188–195.

Le W-D, Rowe DB, Jankovic J, et al. Effects of cerebrospinal fluid from patients with Parkinson's disease on dopaminergic cells. Arch Neurol 1999; 56:194–202.

Lozano AM, Lang AE, Hutchison WD, et al. New developments in understanding the etiology of Parkinson's disease and its treatment. Curr Opin Neurobiol 1998; 8:783–790.

Meunier S, Pol S, Houeto JL, et al. Abnormal reciprocal inhibition between antagonist muscles in Parkinson's disease. Brain 2000; 123:1017–1026.

Pfann KD, Penn RD, Shannon KM, et al. Pallidotomy and bradykinesia. Neurology 1998; 51:796–803.

Stanley FJ. The aetiology of cerebral palsy. Early Hum Dev 1994; 36:81–88.

Summers JJ. The pathogenesis of gait hypokinesia in Parkinson's disease. Brain 1994; 117:1169–1181.

Weiss P, Stelmach GE, Hefter H. Programming a movement sequence in Parkinson's disease. Brain 1997; 120:91–102.

Wichmann T, DeLong MR. Functional and pathophysiological models of the basal ganglia. Curr Opin Neurobiol 1996; 6:751–758.

Olfactory and limbic system

STUDY GUIDELINES

Olfactory system
In most vertebrates, the olfactory system is altogether more important than it is in humans. Clinically, its main interest is that damage to the olfactory pathway on one side is associated with anosmia on that side.

Limbic system
Cortical and subcortical limbic areas are prominent features of the brain in primitive mammals, where they are intimately concerned with mechanisms of attack and defense, procreation, and feeding. The principal effector elements of the limbic system are the hypothalamus and the reticular formation.

The elements, pathways, and transmitters of the human limbic system provide the bedrock on which most of psychiatry and clinical psychology are built.

OLFACTORY SYSTEM

The olfactory system is remarkable in four respects.

1 The somas of the primary afferent neurons occupy a surface epithelium.
2 The axons of the primary afferents enter the cerebral cortex directly; second-order afferents are not interposed.
3 The primary afferent neurons undergo continuous turnover, being replaced from basal stem cells.
4 The pathway to the cortical centers in the frontal lobe is entirely ipsilateral.

The olfactory system comprises the olfactory epithelium and olfactory nerves, the olfactory bulb and tract, and several areas of olfactory cortex.

Olfactory epithelium

The olfactory epithelium occupies the upper one-fifth of the lateral and septal walls of the nasal cavity. The epithelium contains three cell types (Figure 34.1).

1 **Olfactory neurons.** These are bipolar neurons, each with a dendrite extending to the epithelial surface and an unmyelinated axon contributing to the olfactory nerve. The dendrites are capped by immotile cilia containing molecular receptor sites. The axons run upward through the cribriform ('sieve-like') plate of the ethmoid bone and enter the olfactory bulb. The axons (some 3 million on each side) are grouped into fila (bundles) by investing Schwann cells. The collective fila constitute the **olfactory nerve**.
2 **Sustentacular cells** are interspersed among the bipolar neurons.
3 **Basal stem cells** lie between the other two cell types. Olfactory bipolar neurons are unique in that they undergo a continuous cycle of growth, degeneration, and replacement. The basal cells transform into fresh bipolar neurons, which survive for about a month. Replacement declines over time, accounting for the general reduction in olfactory sensitivity with age.

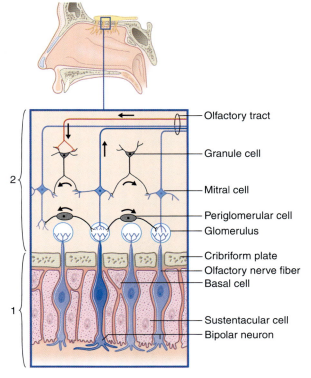

Figure 34.1 Connections of olfactory epithelium and olfactory bulb. The second glomerulus from the left is 'online' (see text).

Olfactory bulb (Figure 34.1)

The olfactory bulb consists of a three-layered allocortex surrounding the commencement of the olfactory tract. The chief cortical neurons are some 50 000 **mitral cells**, which receive the olfactory nerve fibers and give rise to the olfactory tract.

Contact between olfactory fibers and mitral cell dendrites takes place in some 2000 **glomeruli**, which are sites of innumerable synapses and have a glial investment. Glomeruli that are 'online' (active) inhibit neighboring, 'off-line' glomeruli through the mediation of GABAergic **periglomerular cells** (cf. the horizontal cells of the retina). Mitral cell activity is also sharpened at a deeper level by **granule cells**, which are devoid of axons (cf. the amacrine cells of the retina). The granule cells receive excitatory dendrodendritic contacts from active mitral cells, and they suppress neighboring mitral cells through inhibitory (GABA) dendrodendritic contacts.

Central connections

Mitral cell axons run centrally in the **olfactory tract** (Figure 34.2). The tract divides in front of the anterior perforated substance into **medial** and **lateral olfactory striae**.

The medial stria contains axons from the **anterior olfactory nucleus**, which consists of multipolar neurons scattered within the olfactory tract. Some of these axons travel to the septal area via the diagonal band (see later, under *Limbic system*). Others cross the midline in the anterior commissure and inhibit mitral cell activity in the contralateral bulb (by exciting granule cells there). The result is a relative enhancement of the more active bulb, providing a directional cue to the source of olfactory stimulation.

The lateral olfactory stria terminates in the **piriform lobe** of the anterior temporal cortex. The human piriform lobe includes the cortical part of the amygdala, the uncus,

and the anterior end of the parahippocampal gyrus. The highest center for olfactory discrimination is the posterior part of the orbitofrontal cortex, which receives connections from the piriform lobe via the mediodorsal nucleus of the thalamus.

The medial forebrain bundle links the olfactory cortical areas with the hypothalamus and brainstem. These linkages trigger autonomic responses such as salivation and gastric contraction, and arousal responses through the reticular formation.

Points of clinical interest are mentioned in Clinical Panel 34.1.

LIMBIC SYSTEM

The limbic system comprises the limbic cortex (so-called *limbic lobe*) and related subcortical nuclei. The term *limbic* (Broca, 1878) originally referred to a *limbus* or rim of cortex immediately adjacent to the corpus callosum and diencephalon. The limbic cortex is now taken to include the three-layered *allocortex* of the hippocampal formation and septal area, together with transitional *mesocortex* in the parahippocampal gyrus, cingulate gyrus, and insula. The principal subcortical component of the limbic system is the amygdala, which merges with the cortex on the medial side of the temporal pole. Closely related subcortical areas are the hypothalamus and reticular formation, and the nucleus accumbens. Cortical areas closely related to the limbic system are the orbitofrontal cortex and the temporal pole (Figure 34.3).

Figure 34.4 is a graphic reconstruction of mainly subcortical limbic areas.

Parahippocampal gyrus

The parahippocampal gyrus is a major junctional region between the cerebral neocortex and the allocortex of the hippocampal formation. Its anterior part is the **entorhinal**

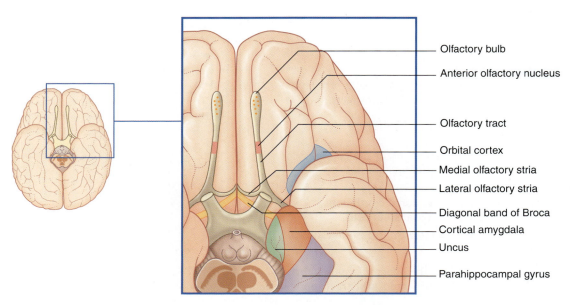

Figure 34.2 Brain viewed from below, showing cortical olfactory areas.

Labels (top to bottom):
- Olfactory bulb
- Anterior olfactory nucleus
- Olfactory tract
- Orbital cortex
- Medial olfactory stria
- Lateral olfactory stria
- Diagonal band of Broca
- Cortical amygdala
- Uncus
- Parahippocampal gyrus

Clinical Panel 34.1   Olfactory disturbance

A routine test of olfactory function is to ask the patient to identify strong-smelling substances such as coffee and chocolate through each nostril in turn. Loss of smell, or **anosmia**, may not be detected by the patient without testing if it is unilateral. If it is bilateral, the complaint may be one of loss of taste, because the flavor of foodstuffs depends on the olfactory qualities of volatile elements; in such cases, the four primary taste sensations (sweet, sour, salty, bitter) are preserved.

Unilateral anosmia may be caused by a *meningioma* compressing the olfactory bulb or tract, or by a head injury with fracture of the anterior cranial fossa. Anosmia may be a clue to a fracture, and should prompt tests for leakage of cerebrospinal fluid into the nasal cavity.

Olfactory auras are a typical prodromal feature of *uncinate epilepsy* (see Clinical Panel 34.3).

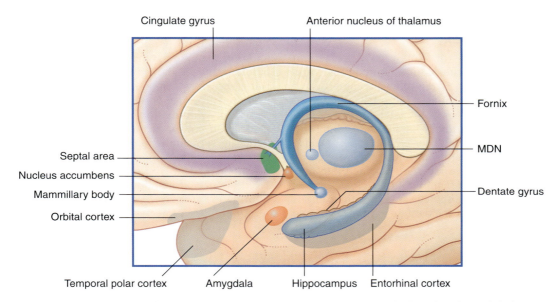

Figure 34.3 Medial view of cortical and subcortical limbic areas. MDN, mediodorsal nucleus of thalamus.

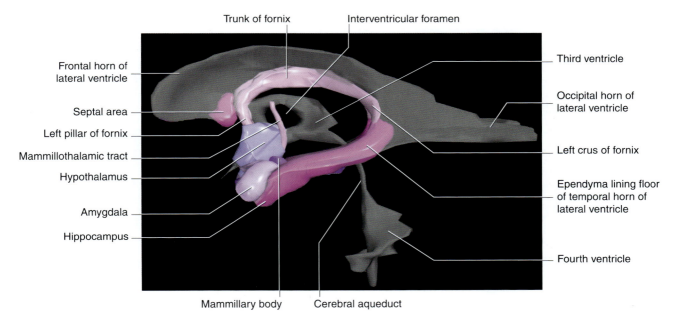

Figure 34.4 Three-dimensional computerized reconstruction of postmortem brain, showing components of the limbic system in relation to the ventricular system. (After Kretschmann and Weinrich 1998, with kind permission of Thieme and the authors.)

cortex (area 28 of Brodmann), which is six-layered but has certain peculiar features. The entorhinal cortex can be said to face in two directions. Its *neocortical face* exchanges massive numbers of afferent and efferent connections with all four association areas of the neocortex. Its *allocortical face* exchanges abundant connections with the hippocampal formation. In the broadest terms, the entorhinal cortex receives a constant stream of cognitive and sensory information from the association areas, transmits it to the hippocampal formation for consolidation (see later), retrieves it in consolidated form, and returns it to the association areas where it is encoded in the form of memory traces. The fornix and its connections form a second, circuitous pathway from hippocampus to neocortex.

Hippocampal complex

The hippocampal complex (or hippocampal formation) comprises the **subiculum**, the **hippocampus proper**, and the **dentate gyrus** (Figure 34.5). All three are composed of temporal lobe allocortex that has tucked itself into an S-shaped scroll along the floor of the lateral ventricle. The band-like origin of the fornix from the subiculum and hippocampus is the **fimbria**. The hippocampus is also known as *Ammon's horn* (after an Egyptian deity with a ram's head). For research purposes, it is divided into four *cornu ammonis* (*CA*) zones (Figure 34.6A).

The principal cells of the subiculum and hippocampus are **pyramidal cells**; those of the dentate gyrus are **granule**

cells. The dendrites of both granule and pyramidal cells are studded with dendritic spines. The hippocampal complex is also rich in inhibitory (GABA) internuncial neurons.

It should be mentioned that, in general discussions related to memory, it is customary to use the term *hippocampus* as synonymous with *hippocampal complex*.

Connections

Afferents

The largest afferent connection of the hippocampal complex is the **perforant path**, which projects from the entorhinal cortex on to the dendrites of dentate granule cells (Figure 34.6B). The subiculum gives rise to a second, *alvear* path, which contributes to a sheet of fibers on the ventricular surface of the hippocampus, the **alveus**.

The axons of the granule cells are called **mossy fibers**; they synapse on pyramidal cells in the CA3 sector. The axons of the CA3 pyramidal cells project into the fimbria; before doing so, they give off *Schaffer collaterals*, which run a recurrent course from CA3 to CA1. CA1 projects into the entorhinal cortex.

Auditory information enters the hippocampus from the association cortex of the superior and middle temporal gyri. The supramarginal gyrus (area 40) transmits coded information about personal space (the *body schema* described in Ch. 32) and extrapersonal (visual) space. From the occipitotemporal region on the inferior surface, information concerning object shape and color, and facial recognition, is projected to cortex called *perirhinal*, or

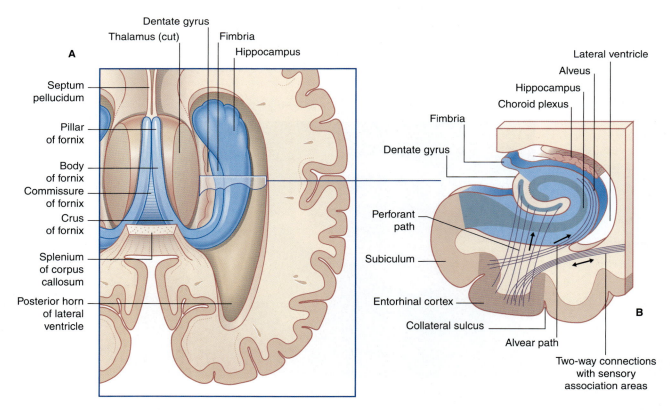

Figure 34.5 Hippocampal formation. **(A)** View from above. **(B)** Enlargement from (A) showing the entorhinal cortex and the three component parts of the hippocampal formation.

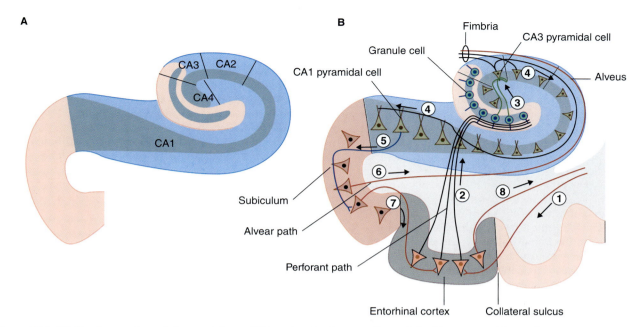

Figure 34.6 **(A)** The four sectors of Ammon's horn. **(B)** Input–output connections of the hippocampal formation.

1 Afferent from sensory association cortex.
2 Entorhinal cortex projecting perforant path fiber to dentate gyrus.
3 Dentate granule cell projecting to CA3.
4 CA3 principal neuron projecting into fimbria and CA1.
5 CA1 principal cell projecting to subiculum.
6 Subicular principal cell projecting into fimbria.
7 Subicular principal cell projecting into entorhinal cortex.
8 Entorhinal pyramidal cell projecting to sensory association cortex.

transrhinal, immediately lateral to the entorhinal cortex. From here, it enters the hippocampus. A return projection from entorhinal to perirhinal cortex is linked to the temporal polar and prefrontal cortex.

In addition to the discrete afferent connections mentioned above, the hippocampus is diffusely innervated from several sources, mainly by way of the fornix.

- A dense *cholinergic* innervation, of particular significance in relation to memory, is received from the septal nucleus.
- A *noradrenergic* innervation is received from the cerulean nucleus.
- A *serotonergic* innervation enters from the raphe nuclei of the midbrain. The linkage between serotonin depletion and major depression is mentioned in Chapter 26.
- A *dopaminergic* innervation enters from the ventral tegmental area of the midbrain. The linkage between dopamine and schizophrenia is discussed in Clinical Panel 34.2.

Efferents

The largest efferent connection is a massive projection via the entorhinal cortex to the association areas of the neocortex. A second, forward projection is the **fornix** (Figure 34.5A). The fornix is a direct continuation of the **fimbria**, which receives axons from the subiculum and hippocampus proper. The **crus** of the fornix arches up beneath the corpus callosum, where it joins its fellow to form the **trunk**, and links with its opposite number through a small **hippocampal commissure**. Anteriorly, the trunk divides into two **pillars**. Each pillar splits around the anterior commissure, sending *precommissural* fibers to the septal area, and *postcommissural* fibers to the anterior hypothalamus, mammillary body, and medial forebrain bundle. The mammillary body projects into the anterior nucleus of thalamus, which projects in turn to the cingulate cortex, completing the *Papez circuit* from cingulate cortex to hippocampus, with return to cingulate cortex via fornix, mammillary body, and anterior thalamic nucleus (Figure 34.7).

The term *medial temporal lobe* is clinically inclusive of the hippocampal complex, parahippocampal gyrus, and amygdala. The term is most often used in relation to seizures (Clinical Panel 34.3).

Memory function of the hippocampal complex

The evidence for a *mnemonic* (memory-related) function in the hippocampal formation is discussed at considerable length in psychology texts. Some insights are given below.

Glossary

- **Short-term memory**. Holding one or more items of new information briefly in mind (e.g. a new telephone number while pressing the buttons).

Clinical Panel 34.2 Schizophrenia

Schizophrenia occurs in about 1% of the population in all countries where the incidence has been studied. In about 10% of cases, there is some evidence of a schizophrenic personality in one or more close relatives. Magnetic resonance imaging brain-imaging studies reveal some degree of atrophy of frontal and temporal parts of the cortex, especially on the left side. There is a reduction or even a reversal of the usual left–right difference in the size of the temporal plane on the upper surface of the temporal lobe. In postmortem studies, a substantial failure of development of the neurons that project from the medial geniculate nucleus to the primary auditory cortex on the left side has also been detected; this may be attributed to failure of the lost cells to establish proper connections with target neurons in the primary auditory cortex. The various anatomic changes can be accounted for, theoretically, on a basis of disordered cell migrations into the developing cortex during the middle trimester of gestation, with consequent failure to establish a full range of connectivities during postnatal growth. The underlying pathology remains relatively stable in adult life, and schizophrenia is described as a psychosis rather than a dementia.

   The mode of presentation is quite variable, but the behavioral changes permit most patients to be categorized into two classes: those in whom positive symptoms predominate, and those in whom negative symptoms predominate.

- *Positive psychotic symptoms* include hallucinations, delusions, and bizarre behavior. Hallucinations are typically auditory (the patient hears voices, and commonly converses with them aloud). Delusions often take a paranoid form, with a belief that one's thoughts and actions are being controlled by some outside agency. Bizarre behavior may include physical aggression in response to the hallucinations or delusions. The positive symptoms are believed to originate in the temporal lobe. Although positive symptoms may cause great alarm, they are much more responsive to treatment than the negative ones.

- *Negative (deficit) symptoms* are those of withdrawal from society into a private world. The patient has

little to say, and in conversation rambles from one inconsequential theme to another. There is a loss of emotional responsiveness (*flattening of affect*), including inability to experience pleasure (*anhedonia*). Personal hygiene is a matter of indifference. The negative symptoms are attributed to 'hypofrontality', i.e. to diminished frontal lobe function. Positron emission tomography scans support this idea by demonstrating failure of the normal response of the left dorsolateral prefrontal cortex to standard tests of cognitive function.

   Drugs used to treat psychotic disorders such as schizophrenia are called *antipsychotics, neuroleptics*, or *major tranquillizers*. Treatment of schizophrenia is by means of one of the antipsychotic drugs that block dopamine D_2 receptors (e.g. chlorpromazine or haloperidol). In the normal brain, the D_2 receptors are on spiny (excitatory) stellate cells in the territory of the mesocortical territory of the ventral tegmental nucleus (Figure 34.15). D_2 receptors are inhibitory, for one or more of three possible reasons noted in Chapter 8. Another reason is the fact that symptoms closely resembling the positive psychotic ones of schizophrenia may be induced by a 'binge' of amphetamine ('speed'); amphetamine is known to increase the amount of dopamine in the forebrain extracellular space (Clinical Panel 34.5). In schizophrenia, dopaminergic overactivity seems not to be a matter of overproduction but of greater effectiveness through an increased number of postsynaptic dopamine receptors on the spiny stellate neurons.

REFERENCES

Fallon JH, Opole IO, Potkin SG. The neuroanatomy of schizophrenia: circuitry and neurotransmitter systems. Clin Neurosci Res 2003; 3:77–107.
Grace A. Gating within the limbic–cortical circuits and its alteration in a developmental disruption model of schizophrenia. Clin Neurosci Res 2003; 3:343–348.
Leonard BE. Drug treatment of schizophrenia and the psychoses. In: Leonard BE. Fundamentals of psychopharmacology. Chichester: Wiley; 2003:255–294.

- **Long-term (remote) memory**. Stored information capable of retrieval at appropriate moments. Two kinds of long-term memory are recognized: explicit and implicit.

   Explicit memory has to do with recollections of facts and events of all kinds that can be explicitly stated—or declared, hence the term *declarative memory*. The term *episodic memory* is also used, in the autobiographic sense of recollection of episodes involving personal experience. Yet another term,

semantic memory, was devised in the context of memory for the meaning of written and spoken words, but it is now also used to include knowledge of facts and concepts.

Implicit memory, having to do with performance of learned motor procedures, for example riding a bicycle, assembling a jigsaw puzzle. The term *procedural memory* is commonly used.

- **Working memory**. Effortless, brief simultaneous retrieval of several items from long-term memory

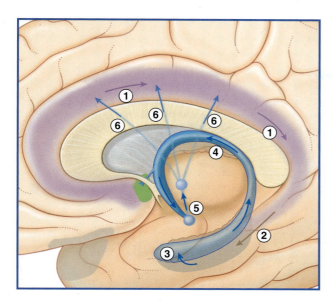

Figure 34.7 The Papez circuit.

1 Backward-projecting neurons in the cingulate gyrus.
2 Projection into the entorhinal cortex.
3 Projection into hippocampus.
4 Fornix.
5 Mammillothalamic tract.
6 Projections from anterior nucleus of thalamus to cingulate cortex.

stores for a task in hand, for example driving a car along a familiar route while making appropriate decisions based on previous experience.

• **Consolidation.** The process of storing new information in long-term memory. Novel factual information is relayed from the relevant sensory association areas to the hippocampal complex for encoding. Following a prolonged period of processing, the encoded information is relayed back to the same association areas, and (with the exception of strongly autobiographic episodes) no longer depends on the hippocampal complex for retrieval.

Clinical and experimental observations

Bilateral damage or removal of the anterior part of the hippocampal formation is followed by *anterograde amnesia*, a term used to denote absence of conscious recall of newly acquired information for more than a few minutes. When asked to name a commonplace object, the subject will have no difficulty, because access to long-term memories does not require the anterior hippocampus. However, when the same object is shown a few minutes later, the subject will not remember having seen it. There is loss of explicit or declarative memory.

Procedural ('How to do') memory is preserved. If asked to assemble a jigsaw puzzle, the subject will do it in the normal way. When asked to repeat the exercise the next day, the subject will do it faster, although there will be no recollection of having seen the puzzle previously. *The hippocampus is not required for procedural memory*. We

have previously noted that the basal ganglia are the storehouse of routine motor programs, and the cerebellum the storehouse of motor adaptations to novel conditions.

Long-term potentiation (*LTP*) is uniquely powerful in dentate gyrus and hippocampus. It is regarded as vital for preservation (consolidation) of memory traces. Under experimental conditions, LTP is most easily demonstrated in the perforant path–dentate granule cell connections and in the Schaffer collateral–CA1 connections. A strong, brief (milliseconds) stimulus to the perforant path or Schaffer collaterals induces the target cells to show long-lasting (hours) sensitivity to a fresh stimulus. LTP is associated with a cascade of biochemical events in the target neurons, following activation of appropriate glutamate receptors, as described in Chapter 8 in the context of pain sensitization (Figure 34.8). Repetitive stimuli may cause cyclic AMP to increase its normal rate of activation of protein kinases involved in phosphorylation of proteins that regulate gene transcription. The outcome is increased production of proteins (including enzymes) required for transmitter synthesis, and of other proteins for construction of additional channels and synaptic cytoskeletons.

Long-term potentiation is described as an *associative* phenomenon, because the required expulsion of the magnesium plug from the NMDA receptor (*Figure 8.8*) is facilitated when the powerful depolarizing stimulus is coupled with a weaker stimulus to the depolarized neuron from another source. *Norepinephrine* (*noradrenaline*) and *dopamine* are suitable associative candidates, one or both being released during elevation of the attentional or motivational state at the appropriate time.

Cholinergic activity in the hippocampus is also significant for learning. In human volunteers, central acetylcholine blockade (by administration of scopolamine) severely impairs memory for lists of names or numbers, whereas a cholinesterase inhibitor (physostigmine) gives above-normal results. Clinically, hippocampal cholinergic

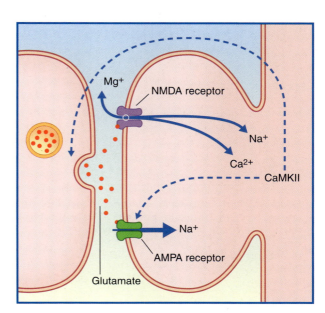

Figure 34.8 Long-term potentiation.

Clinical Panel 34.3   Temporal lobe epilepsy

Complex focal (partial) seizures are synonymous with *temporal lobe epilepsy*. The initial event, or *aura*, may be a simple partial seizure whose electrical activity escapes into the temporal lobe. Many originate in a focus of runaway neural activity within the temporal lobe, and spread over the general cortex within seconds to trigger a *secondarily generalized* tonic–clonic seizure (Figure CP 34.3.1), as mentioned in Chapter 30. Types of temporal lobe auras include well-formed visual or auditory hallucinations (scenes, sound sequences), a sense of familiarity with the surrounding scene (*déjà vu*), a sense of strangeness (*jamais vu*), or a sense of fear. Attacks originating in the uncus are ushered in by unpleasant olfactory or gustatory auras. Bizarre *psychic auras* can occur, where patients have an 'out-of-body experience' in the form of a sensation of floating in the air and looking down at themselves and any others present.

Following accurate localization of the ictal (seizure) focus by means of recording electrodes inserted into the exposed temporal lobe, a tissue block including the focus may be removed, with abolition of seizures in four out of five cases. Histologic examination of the surgical biopsy typically reveals *hippocampal sclerosis*: the picture is one of glial scarring, with extensive neuronal loss in CA2 and CA3 sectors. The granule cells of the dentate gyrus are relatively well preserved. Loss of inhibitory, GABA internuncials has been blamed in the past, but these cells have recently been shown to persist. Instead, the granule cells appear to be *disinhibited*, because of loss of minute, inhibitory basket cells from among their dendrites.

Because 30% of sufferers from temporal lobe epilepsy have first-degree relatives similarly afflicted, often from childhood, a genetic influence must be significant. One possibility could be 'faulty wiring' of the hippocampus during mid-fetal life. Histologic preparations show areas of congenital misplacement of hippocampal pyramidal cells, some lying on their sides or even in the subjacent white matter.

The sclerosis is regarded as a typical central nervous system healing process following extensive loss of neurons. The neuronal loss in turn seems to be inflicted by *glutamate toxicity*—a known effect of excessively high rates of discharge of pyramidal cells in any part of the cerebral cortex. Dentate granule cells are the main source of burst firing, which is no surprise in view of their natural role in long-term potentiation and kindling (see main text).

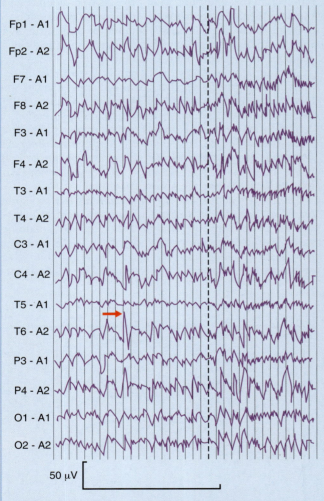

50 μV

Figure CP 34.3.1 Complex focal seizure. The ictal focus (*arrow*) occupies the middle–posterior junctional zone of the right temporal lobe (cf. electrode positions in Figure 30.2). Within a second, the entire cortex exhibits a secondarily generalized seizure (to the right of the dashed line).

REFERENCES

Bernasconi N, Bernasconi A, Caramanos Z, et al. Mesial temporal damage in temporal lobe epilepsy: a volumetric MRI study of the hippocampus, amygdala and parahippocampal region. Brain 2003; 126:462–469.

Emerson RG, Pedley TA. EEG and evoked potentials. In: Bradley WG, Daroff RB, Finichel GM, et al, eds. Neurology in clinical practice: principles of diagnosis and management. Philadelphia: Elsevier; 2004:465–490.

Jay V, Becker LE. Surgical pathology of epilepsy. Pediatr Pathol 1994; 14:731–750.

Meldrum BS. The role of glutamate in epilepsy and other CNS disorders. Neurology 1994; 44(suppl 8):S14–S22.

Shin C. Mechanism of epilepsy. Ann Rev Med 1994; 45:379–389.

Willis WD, Westland KR. Pain system. In: Paxinos G, Mai JK, eds. The human nervous system. 2nd edn. Amsterdam: Elsevier; 2004:1125–1170.

activity is severely reduced in Alzheimer's disease (AD), which is particularly associated with amnesia (see Clinical Panel 34.4).

Kindling ('lighting a fire') is a property unique to the hippocampal formation and amygdala, although its relationship to learning is not obvious. Kindling is the progressively increasing group response of neurons to a repetitive stimulus of uniform strength. In both humans and experimental animals, it can spread from mesocortex to neocortex and cause generalized convulsive seizures.

The contribution of the fornix projection to memory is uncertain. Indirect evidence has been adduced from *diencephalic amnesia*, a state of anterograde amnesia that may follow bilateral damage to the diencephalon. Such damage may interrupt the *Papez circuit* linking the fornix to the cingulate gyrus by way of the mammillary body and the anterior nucleus of the thalamus. Particularly impaired is relational memory (e.g. recollection of the sight and sound of a particular waterfall with the wind blowing spray over some fleeing viewers; recollection of the structure and the function of the (say) left vestibular labyrinth, along with inevitable symptoms and signs that follow sudden occlusion of the left labyrinthine artery.)

Left versus right hippocampal functions

In keeping with known hemispheric asymmetries, the left anterior hippocampus and dorsolateral prefrontal cortex (DLPFC) are engaged in encoding novel material involving language function. Also consistent is the finding that the right hippocampus and right inferior parietal lobe are

---

Clinical Panel 34.4   Alzheimer's disease

Dementia is defined as a severe loss of cognitive function without impairment of consciousness. Alzheimer's disease (AD) is the commonest cause of dementia, afflicting 5% of people in their seventh decade and 20% of people in their ninth. AD patients fill 20% of all beds in psychiatric institutions.

Magnetic resonance imaging brain scans usually reveal severe atrophy of the cerebral cortex, with widening of the sulci and enlargement of the ventricular system. The primary sensory and motor areas, and the upper regions of the prefrontal cortex, are relatively well preserved.

Postmortem histologic studies of the cerebral cortex reveal the following.

- Extensive loss of pyramidal neurons throughout the brain.
- *Amyloid plaques* and *neurofibrillary tangles*, notably in the hippocampus and amygdala. The plaques begin in the walls of small blood vessels, and have been explained in terms of an enzyme defect resulting in abnormal, *beta-amyloid* protein production. The tangles are made up of clumps of microtubules associated with an abnormal variant of a microtubule-associated *tau* protein. The tangles are progressively replaced by amyloid.
- Loss of up to 50% of the cholinergic neurons from the basal nucleus of Meynert and from the septal area, together with their extensive projections through the cerebral isocortex and mesocortex. Indeed, degenerating acetylcholine (ACh) terminals seem to contribute to the neurofibrillary tangles in the temporal lobe.

Hypometabolism can be shown on positron emission tomography scans arranged to detect glucose utilization. This is attributable in part to loss of pyramidal cells, and in part to loss of cholinergic innervation of the pyramidal cells remaining. Healthy pyramidal neurons have excitatory ACh receptors in their cell membranes.

Although the pattern of degeneration varies from case to case, its general trend is to commence in the medial temporal lobe and to travel upward and forward. The following clinical features are explained in that sequence.

- *Dwindling hippocampal function*. Anterograde amnesia leads to *forgetfulness*, for example recounting a personal event within minutes of telling it (loss of present-time episodic memory); difficulty in finding one's way around familiar streets, or alarming misjudgments while driving an automobile (hippocampal activity is required to sustain parietal lobe *spatial sense*); *attentional deficit*, whose earliest manifestation is an inability to switch attention from one thing to another.
- *Dwindling occipitotemporal function*. Damage to area 37 leads to an inability to read and write. Damage to the temporal polar region leads to distressing failure to recognize the faces of family and friends. Involvement of the supramarginal and angular gyri leads to an inability to write.
- *Dwindling frontal lobe function*. Usually within 3 years of onset, the patient is 'spaced out', staring at walls and seemingly unaware of what is going on in the room. This 'vacant' state lasts for up to 5 or 6 years antemortem.

An unusual variant, known as *early-onset AD*, shows clear evidence of an autosomal dominant trait. The illness appears during the fourth or fifth decade. Chromosomal analyses have revealed a specific mutation in the gene coding for amyloid precursor protein on the long arm of chromosome 21. This mutation is also found in Down syndrome, where sufferers surviving into middle age usually develop AD.

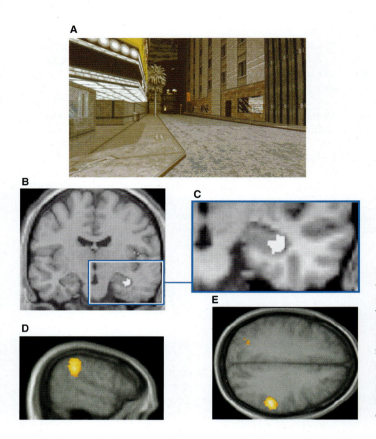

Figure 34.9 Navigation in a virtual environment. **(A)** Scene from a virtual town. Subjects navigated through the town using a keypad to stay clear of obstacles. Positron emission tomography scans taken during the virtual journey showed increased activation, **(B)** and **(C)**, within the right hippocampus and, **(D)** and **(E)**, within the right supramarginal gyrus. Orientation (L, R) of the magnetic resonance imaging scans is for a more general readership. (Kindly provided by Dr. Eleanor Maguire, Wellcome Department of Cognitive Neurology, Institute of Neurology, University College, London, UK, and with permission of the editor of *Current Opinion in Neurobiology*.)

engaged in spatial tasks such as driving a car (Figure 34.9). Blood flow in the DLPFC increases more on the left side during driving, presumably because of the 'inner speech' that occurs when exploring novel territory.

Anterior versus posterior hippocampal functions

The hippocampus is about 8 cm in length, and there is evidence for anteroposterior functional specialization with respect to novelty versus familiarity; for example, when novel material is being read on a screen, the left anterior hippocampus is especially active, but with development of familiarity with repeated exposure, activity shifts to the posterior part, suggesting that this region is involved in encoding material into long-term memory.

Long-term medial temporal lobe dependency

Autobiographic recollections typically are mainly visual, whereby we revisit scenes from the past, sometimes from childhood. Clinical studies indicate that medial temporal lobe damage impairs or even deletes such ego-centered (personal) memories, whereas allocentered (impersonal), map-like recall of past scenes is preserved. On the other hand, damage to the visual association cortex has the opposite effect.

Prefrontal cortex and working memory

Volunteers have been examined under functional magnetic resonance imaging while preparing to give a motor response to one or more sensory cues. The midregion of the prefrontal cortex (area 46) tends to be especially active. It is in a position to tap into memory stores in

relevant sensory association areas, and to organize motor responses including speech.

Insula

The anterior insula is a cortical center for pain (Box 34.1). The central region is continuous with the frontoparietal and temporal opercular cortex, and it seems to have a *language* rather than a limbic function. During language tasks, positron emission tomography (PET) scans show activity there as well as in the opercular speech receptive and motor areas—but not in people with congenital dyslexia, where it remains silent (Ch. 32). The posterior insula is interconnected with the entorhinal cortex and the amygdala, and is therefore presumed to participate in emotional responses—perhaps in the context of pain evaluation.

Cingulate cortex and posterior parahippocampal gyrus

The cingulate cortex is part of the Papez circuit, receiving a projection from the anterior nucleus of the thalamus and becoming continuous with the parahippocampal gyrus behind the splenium of the corpus callosum.

The *anterior* cingulate cortex belongs to the *rostral limbic system*, which includes the amygdala, ventral striatum, orbitofrontal cortex, and anterior insular cortex.

Six functional areas can be discerned in the anterior cingulate cortex (Figure 34.10).

Box 34.1   Pain and the brain

The International Association for the Study of Pain has given the following definition: *Pain is an unpleasant sensory and emotional experience associated with actual or potential tissue damage or described in terms of such damage.*

This definition emphasizes the *affective* (emotional) component of pain. Its other component is *sensory-discriminative* ('Where and how much?').

Table Box 34.1.1 is a glossary of conventional terms used in relation to pain.

Peripheral pain pathways

As already noted in Chapter 9, pain is served by finely myelinated (A) and unmyelinated (C) fibers belonging to unipolar spinal ganglion cells. These fibers are loosely known as 'pain fibers', although others of similar diameters are purely mechanoreceptors and others again elicit pain only when discharging at high frequency, notably mechanical nociceptors and thermoreceptors. The latter are referred to as *polymodal nociceptors* in the general context of pain.

From somatic tissues including skin, parietal pleura and parietal perineum, muscle, joint capsules, and bone, the distal processes of the ganglion cells travel in all the spinal nerves. The proximal processes branch within the dorsal root entry zone and span five or more segments of the spinal cord within the posterolateral tract of Lissauer before terminating in laminae I, II, and IV of the posterior gray horn. The corresponding fibers of the trigeminal nerve terminate in the spinal nucleus of that nerve.

From the viscera, the distal processes share perineural sheaths with postganglionic fibers of the sympathetic system. The proximal processes mingle with the somatic fibers within Lissauer's tract and terminate in the same region. As noted in Chapter 13, overlap of somatic and visceral afferent terminals on the dendrites of central pain-projecting neurons (CPPNs) is thought to account for *referred pain* in visceral disorders such as myocardial infarction and acute appendicitis.

Sensitization of nociceptors

Injured tissue liberates molecules, including bradykinin, prostaglandin, and leukotrienes, which lower the activation threshold of nociceptors. Injured C fibers also initiate axon reflexes (Ch. 11), whereby substance P ± calcitonin gene-related peptide is liberated into the adjacent tissue, causing histamine release from mast cells. Histamine receptors may develop on the nerve terminals and (as already noted in Ch. 8) produce

Table Box 34.1.1 Conventional terms used in relation to pain

Term	Meaning
Allodynia	Pain produced by normally innocuous stimuli (e.g. stroking sunburned skin, moving an inflamed joint).
Central pain-projecting neurons (CPPNs)	An inclusive conventional term denoting all posterior horn neurons projecting pain-encoded information to contralateral brainstem and thalamic nuclei. Pathways included are *spinothalamic*, the lateral pain pathway to posterior nucleus of thalamus; *spinoreticulothalamic*, medial pain pathway to the medial and intralaminar nuclei of thalamus via brainstem reticular formation; *spinoamygdaloid*, to the amygdala via the reticular formation; and *spinotectal*, to the superior colliculus.
Central pain state	A state of chronic pain, resistant to therapy, sustained by hypersensitivity of peripheral and/or central neural pathways.
Wind-up phenomenon	Sustained state of excitation of CPPNs induced by glutamate activation of NMDA receptors.
Fast pain	Stabbing pain perceived following activation of Aδ nociceptors.
Hyperalgesia	Hypersensitivity to stimulation of injured tissue and of surrounding uninjured tissue. Causes include mechanical or thermal damage, bacterial or viral inflammation, small-fiber peripheral axonal neuropathy, radiculopathy (posterior nerve root injury).
Neurogenic inflammation	Inflammation caused by liberation of substance P (in particular) following antidromic depolarization of fine peripheral nerve fibers.
Neuropathic pain	Chronic stabbing or burning pain resulting from injury to peripheral nerves (e.g. postherpetic neuralgia, amputation neuroma).
Nociceptors	Peripheral receptors whose activation generates a sense of pain. These receptors occupy the plasma membrane of fine nerve endings and contain transduction channels that convert the requisite physical or chemical stimulus into trains of impulses decoded by the brain as a sense of pain.
Polymodal nociceptors	Peripheral nociceptors (notably in skin) responsive to noxious thermal, mechanical, or chemical stimulation.
Sensitization	Lowering the threshold of peripheral nociceptors by histamine (in particular) following peripheral release of peptides via the axon reflex.
Slow pain	Aching pain perceived following activation of C-fiber nociceptors.

Box 34.1 *Continued*

arachidonic acid by hydrolysis of membrane phospholipids. The enzyme *cyclooxidase* converts arachidonic acid into a prostaglandin. (The main action of aspirin and other non-steroidal antiinflammatory analgesics is to inactivate that enzyme, thereby reducing synthesis of prostaglandins.)

The net result is sustained activation of large numbers of C-fiber neurons and sensitization of mechanical nociceptors, manifested by *allodynia*, where even gentle stroking of the area may elicit pain; and by *hyperalgesia*, where moderately noxious stimuli are perceived as very painful.

As already noted in Chapter 13, *irritable bowel syndrome* is characterized by sensitization of nociceptive *interoceptors* in the bowel wall. That event also underlies the painful urinary bladder condition known as *interstitial cystitis*.

Sensitization of C-fiber neurons may include *gene transcription* (Ch. 8), whereby abnormal sodium channels are inserted into the cell membrane of the parent neurons in the posterior root ganglion. Spontaneous trains of impulses generated here are thought to account for occasional failure of quite high-level nerve blocks to abolish the pain.

Neuropathic pain
When a peripheral nerve is severed, and the proximal and distal stumps separated by developing scar tissue, trapped regenerating axons form thread-like balls, called *neuromas*, that are exquisitely sensitive to pressure. Repetitive activation may prolong the victim's suffering by engendering a central pain state (see below). *Postherpetic neuralgia* is a neuropathic pain that may be a sequel to *herpes zoster* ('creeping girdle'), manifested by clusters of watery blisters, usually along the cutaneous territory of an intercostal nerve. The virus concerned may perpetuate the pain by precipitating the gene transcription mentioned above.

Central pain pathways
Central pain-projecting neurons are of two kinds, as described in Chapter 15: nociceptive-specific, with small peripheral sensory fields (about 1 cm²), and wide dynamic range, with fields of 2 cm² or more; these are mechanical nociceptors encoding tactile stimuli by low-frequency impulses and noxious stimuli by high-frequency impulses.

The current consensus is that there exists a *lateral, sensory-discriminative pathway* and a *medial, affective pathway* in relation to pain (Figure Box 34.1.2).

Lateral pain pathway
For the trunk and limbs, the lateral pathway arises in the posterior gray horn of the spinal cord, and projects as the lateral spinothalamic tract to the posterior part of the contralateral ventral posterior lateral nucleus of thalamus. For the head and neck, it commences in the

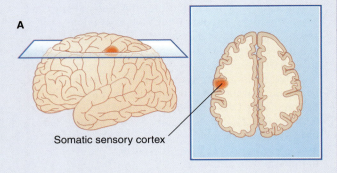
Somatic sensory cortex

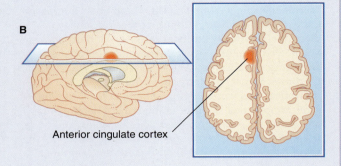
Anterior cingulate cortex

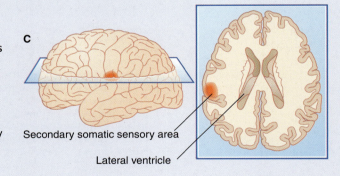
Secondary somatic sensory area

Lateral ventricle

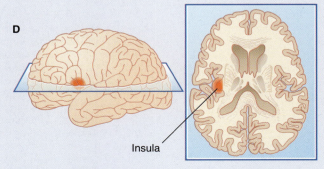
Insula

Figure Box 34.1.1 Areas showing increased metabolic activity following application of noxious heat to the right forearm.

Box 34.1   *Continued*

spinal nucleus of the trigeminal nerve and occupies the trigeminothalamic projection to the contralateral posterior medial thalamic nucleus. The onward projection is mainly to the primary somatic sensory cortex (SI), partly to the upper bank of the lateral sulcus (SII). The arrangement is somatotopic, as can be seen on positron emission tomography scanning when a noxious heat stimulus is applied to different parts of the body. Animal investigations demonstrate intensity-responsive, nociceptive-specific neurons in SI, having appropriately small peripheral receptive fields—ideal candidates for encoding the 'Where and how much?' aspects of pain.

Onward projections to the posterior parietal cortex and SII are indicated in Figures Box 34.1.1 and 34.1.2.

Not surprisingly, the spinothalamic early warning system stimulates orientation of head and eyes toward the source of pain. As mentioned in Chapter 15, the

spinotectal tract ascends alongside the spinothalamic and terminates in the superior colliculus. Its imprint is somatotopic, and it elicits a *spinovisual reflex* to orient the eyes, head, and trunk toward the area stimulated. In addition to activation of this phylogenetically ancient (reptilian) reflex, the 'Where?' visual channel (Ch. 29) is engaged by association fibers passing to the posterior parietal cortex from SI.

Nociceptive neurons in SII are less numerous, and many also receive visual inputs. They are linked to the insula, which also receives direct inputs from the thalamus. Insular stimulation may elicit autonomic responses such as a rapid pulse rate, vasoconstriction, and sweating. Surprisingly, preexisting lesions of the insula may abolish the aversive quality of painful stimuli while preserving the location and intensity aspects. The condition is known as *asymbolia for pain*.

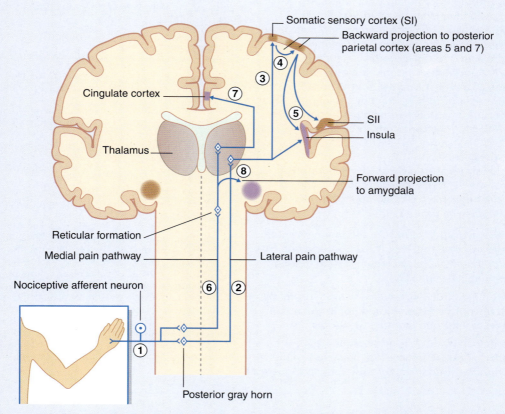

Figure Box 34.1.2 Pain pathways. *Note*: Violet signifies having emotional significance. The amygdala is in fact anterior to the plane of section in the diagram, and areas 5 and 7 are posterior to it. **(1)** Peripheral nociceptive neurons project to posterior gray horn. **(2)** 'Fast ' central pain-projecting neurons (CPPNs) project direct to contralateral posterolateral thalamus and **(3)** relay to somatic sensory cortex (SI). **(4)** Association fibers connect SI to posterior parietal cortex both for 'Where?' and 'How much?' tactile analysis, and to provide a 'Where?' visual alert. **(5)** Posterior parietal cortex projects to SII for tactile–visual integration. Onward relay to the insular cortex, supplemented by some direct thalamic inputs there, may elicit autonomic and emotional responses. **(6)** 'Slow' CPPNs relay via the reticular formation to the medial thalamus, with forward projection to the prefrontal cortex (not shown here) for overall evaluation. **(7)** Upward projection to the cingulate cortex normally generates an aversive (L. 'turn away') emotional evaluation. **(8)** Some CPPNs excite reticular neurons projecting to the amygdala, where they are likely to generate a sense of fear.

Box 34.1   *Continued*

Medial pain pathway

The medial pathway is polysynaptic, via spinoreticular and trigeminoreticular tracts to the contralateral medial dorsal thalamic nucleus (among others), with onward projection to the anterior cingulate cortex. That this area is concerned with the affective component of pain experience is strongly supported by the effect of surgical undercutting (cingulotomy) or removal (cingulectomy) as a treatment for chronic pain. Patients report that the intensity of their pain is unchanged, but that it has lost its aggressive nature. Precisely the same result follows morphine injection—presumably because the anterior cingulate has the greatest number of opiate receptors in the cerebral cortex.

Following cingulotomy, edema of the bladder control area frequently causes temporary urinary incontinence. More importantly, more than half of all patients show permanent 'flatness of affect', i.e. low experience of either elation or depression.

An unexpected stab of pain from any source is likely to generate an immediate sense of fear. This is attributable to activation of *spinomesencephalic* fibers projecting to the midbrain reticular formation, with onward projection to the amygdala, a nucleus particularly associated with the sense of fear (see main text). Some of the fibers are believed to ascend in or alongside the posterolateral tract of Lissauer; they may account for the persistence of pain perception in some patients following the cordotomy procedure.

Central pain states

Central pain states are almost always generated by *wind-up* of the CPPNs of the spinothalamic and spinoreticular pathways. One or more of three mechanisms may be responsible.

1 Repetitive activation of NMDA glutamate receptors by posterior nerve root inputs, over a period of weeks or months, tends to induce a state of long-term potentiation of CPPNs.

2 The threshold of CPNNs may be lowered further by gene transcription, whereby additional glutamate receptors are inserted into their dendrites.

3 The term 'paradoxical' seems appropriate for the third mechanism. Reference was made in Chapter 24 to *supraspinal antinociception*, whereby serotonergic neurons projecting from the medullary magnus raphe nucleus (NRM) to the posterior gray horn may *inhibit* CPNNs by activating internuncial enkephalinergic neurons. Evidence from animal experiments now indicates that, while either of the first two mechanisms may initiate a central pain state, its maintenance requires that non-serotonergic neurons in or near NRM *facilitate* CPNNs by a direct excitatory transmitter of uncertain nature. Following limb amputation, an ultimate expression of wind-up is *phantom limb pain*, where, following limb amputation, severe pain may be experienced in the distal part of the missing limb.

As mentioned in Chapter 27, the central pain state known as *thalamic syndrome* may develop following a vascular lesion in the white matter close to the ventroposterior nucleus of thalamus. Explanation of the bouts of severe contralateral pain sensation may lie in elimination of the normal inhibitory feedback to the posterior thalamus from the surrounding thalamic reticular nucleus.

REFERENCES

Brooks J, Tracey I. From nociception to pain perception: imaging the spinal and supraspinal pathways. J Anat 2005; 207:19–33.

Froth M, Mauguiere M. Dual representation of pain in the operculo-insular cortex in humans. Brain 2003; 126:438–450.

Gatchel RJ. Clinical essentials of pain management. Washington: American Psychological Association; 2005.

Lewin GR, Lu Y, Park T. A plethora of painful molecules. Curr Opin Neurobiol 2004; 14:443–449.

Merskey H. Classification of chronic pain. Descriptions of chronic pain syndromes and definitions of pain terms. Pain 1986; 3(suppl 1):S1–S225.

Richardson JD, Vasco MR. Cellular mechanisms of neurogenic inflammation. J Pharmacol Exp Ther 2002; 302:839–845.

Treede RD, Kenshalo DR, Gracely RH, et al. The cortical representation of pain. Pain 1999; 79:105–111.

Vanegas H, Schaible H-G. Descending control of persistent pain: inhibitory or facilitatory? Brain Res Rev 2004; 46:295–309.

Villanueva L, Nathan PW. Multiple pain pathways. In: Progress in pain research and management, vol 16. Seattle: IASP Press; 2000:371–386.

Willis WD, Westlund KR. Pain system. In: Paxinos G, Mai JK, eds. The human nervous system. 2nd edn. Amsterdam: Elsevier; 2004:1125–1170.

1 An *executive area* is connected directly with the DLPFC and with the supplementary motor area (SMA). The executive area becomes active prior to execution of willed movements, including voluntary saccades (Ch. 29)—and even prior to the SMA itself. The executive area is thought to have special significance, together with the DLPFC, in generating *appropriate motor plan selection* by the SMA.

2 A *pain perception area* receives afferents from the medial dorsal nucleus of the thalamus (Box 34.1).

3 An *emotional area* lies close to the pain perception area. When volunteers 'think happy' while undergoing PET scans, the anterior cingulate cortex 'lights up' and the amygdala 'switches off'. A reverse result occurs when volunteers 'think sad'. Anterior cingulectomy has often been performed in the past, yielding successful control of aggressive psychiatric disorders.

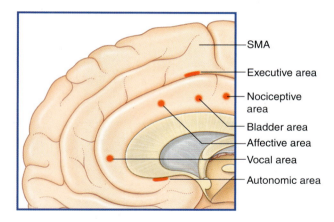

Figure 34.10 Functional areas in the anterior cingulate cortex. SMA, supplementary motor area.

4 A *bladder control area* becomes increasingly active during bladder filling (Ch. 24).

5 A *vocalization area* becomes active, together with the DLPFC, during *decision-making* about appropriate sentence construction for speech activity. Electrical stimulation of this area causes jumbling of speech. *Stammering* in children is associated with *reduced* blood flow in the left anterior cingulate gyrus during speech. Blood flow there is also reduced in people suffering from *Tourette syndrome*, which is characterized by brief, loud utterances of a single syllable or phrase, often offensive to the ear.

6 An *autonomic area*, below the rostrum of the corpus callosum, elicits autonomic and respiratory responses when stimulated electrically. This area is thought to participate in eliciting the visceral responses typical of emotional states.

The posterior cingulate gyrus (area 23 of Brodmann) merges with the posterior parahippocampal gyrus (area 36). This cortical complex is richly interconnected with visual, auditory, and tactile–spatial association areas. The complex evidently contains memory stores related to these functions, because PET studies reveal increased activity there when scenes or experiences are conjured up in the mind. The complex is also engaged during reading (Ch. 32).

Amygdala

The **amygdala** (*Gr.* 'almond'; also called the *amygdaloid body* or *amygdaloid complex*) is a large group of nuclei above and in front of the temporal horn of the lateral ventricle, and anterior to the tail of the caudate nucleus. *The amygdala is primarily associated with the emotion of fear*, as illustrated by the effect of looking at an angry or fearful face (Figure 34.11). Current clinical and basic science ambition is to gain diagnostic and therapeutic insights into the role of the amygdala with regard to various *phobias* and *anxiety states* prevalent in both the young and the adult population. The connections of the amygdala (inasmuch as these are understood) are consistent with the present perception of a 'bottleneck' position in the perception and expression of fear.

Afferent pathways

Within the amygdala, nuclear groups receiving afferents are predominantly laterally placed and are usually referred to collectively as the **lateral nucleus**. In Table 34.1 and related figures, the afferents are segregated into subcortical and cortical.

Subcortical access, depicted in Figure 34.12, is thought to be especially important in infancy and childhood, at a time when the amygdala is developing faster than the hippocampus and is capable of acquiring fearful memory traces without hippocampal participation. Such memories cannot be *consciously* recalled at any later time, despite generating physical responses of an 'escape' nature. The general sense and special sense pathways listed and depicted are sufficiently comprehensive to account for the acquisition of almost any specific 'unexplained' phobia (e.g. enclosed spaces, smoke, heights, dogs, faces).

As indicated in Table 34.2 and Figure 34.13, all sensory association areas of the cortex have direct access to the lateral nucleus of the amygdala. These areas are also linked to the prefrontal cortex through long association fiber bundles, rendering all conscious sensations subject to cognitive evaluation.

Activity of the visual association cortex is especially important in connection with phobias and anxiety states. Area V4 at the inferior surface of anterior area 19 is a link in the object or face recognition pathway. V5, at the lateral surface of anterior area 19, is a link in the movement detection pathway. Both are connected to the amygdala

Table 34.1 Afferents to lateral nucleus of amygdala

Nature	Subcortical source	Cortical source
Tactile	Ventral posterior nucleus of thalamus	Parietal lobe
Auditory	Medial geniculate body	Superior temporal gyrus
Visual	Lateral geniculate body[a]	Occipital cortex
Olfactory	–	Piriform lobe
Mnemonic	–	Hippocampus or entorhinal cortex
Cardiac	Hypothalamus	Insula
Nociceptive	Midbrain reticular formation	–
Cognitive	–	Orbital cortex
Attention-related	Cerulean nucleus	Basal nucleus of Meynert

[a]Afferents from lateral geniculate body to amygdala have yet to be clearly identified.

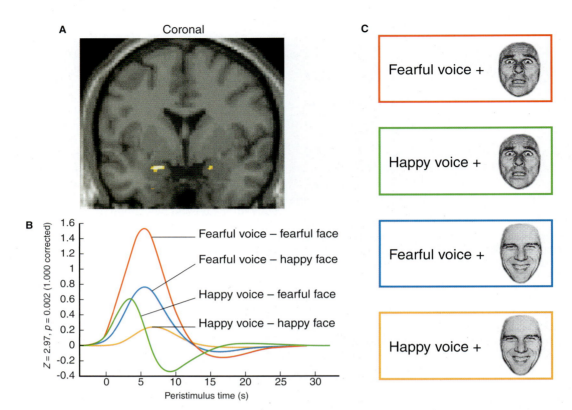

A Coronal

C

Fearful voice +

Happy voice +

Fearful voice +

Happy voice +

B Fearful voice – fearful face
Fearful voice – happy face
Happy voice – fearful face
Happy voice – happy face

Peristimulus time (s)

Figure 34.11 Cross-modal emotional responses. **(A)** Coronal structural magnetic resonance imaging (MRI) image showing a superimposed functional MRI map of bilateral activation of the amygdala in a volunteer observing a fearful face accompanied by a fearful voice (see (C)). At the opposite end of the spectrum, amygdalar activity was below normal level in the presence of a happy face and voice. **(B)** The associated graph shows experimental condition-specific functional MRI responses in the amygdala. **(C)** Stimuli. (Kindly provided by Professor R.J. Dolan, Wellcome Department of Cognitive Neurology, Institute of Neurology, University College, London, UK.)

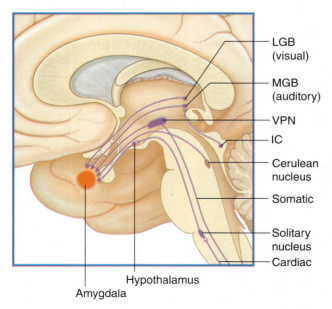

Figure 34.12 Subcortical afferents to lateral nucleus of amygdala. IC, inferior colliculus; LGB, MGB, lateral, medial geniculate body; VPN, ventral posterior nucleus of thalamus.

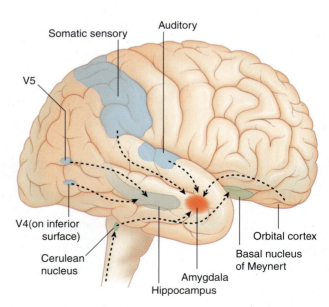

Figure 34.13 Cortical afferents to lateral nucleus of amygdala. V4, object or face recognition area; V5, motion detection area.

Table 34.2 Efferents from central nucleus of amygdala

Target nucleus or pathway	Function or effect
Periaqueductal gray matter (to medulla or raphespinal tract)	Antinociception
Periaqueductal gray matter (to medullary reticulospinal tract)	Freezing
Cerulean nucleus	Arousal
Norepinephrine (noradrenaline) medullary neurons (projection to lateral gray horn)	Tachycardia or hypertension
Hypothalamus or dorsal nucleus of vagus (to heart)	Bradycardia or fainting
Hypothalamus (liberation of corticotropin-releasing hormone)	Stress hormone secretion
Parabrachial nucleus (to medullary respiratory nuclei)	Hyperventilation

via the hippocampus, where fearful visual memories may be recalled by the current visual scene. The visual association cortex is also important in that fearful visual images conjured in the mind independently of current sensation may activate the amygdala. This capability has resonance in relation to *posttraumatic stress disorder*, where a seemingly innocent scene may cause the afflicted individual to 'relive' a horrific visual experience up to 20 years or more after the event. In the multimodal anterior region of the superior temporal gyrus, where sound and vision coalesce, a door banged shut may induce a 'virtual reality' reenactment of a horrific encounter (e.g. of a haunting war experience).

The orbital prefrontal cortex of the right side, with its bias toward 'withdrawal' rather than 'approach' (Ch. 32), is commonly active (in PET scans) along with the right amygdala in fearful situations, for example when a specific phobia is presented to a susceptible subject. On the one hand, this offers the 'downside' potential to 'feed on one's fear'. On the other hand, expert social or psychologic conditioning may eventually suffice to reduce the 'negative drive' of the orbital cortex. When conditioning is combined with use of anxiolytic drugs, specific phobias may be abolished completely.

The insula is omitted from Figure 34.13 but, as noted earlier, its posterior part also has direct access to the amygdala, probably related to the emotional evaluation of pain.

Finally, the basal nucleus of Meynert is listed. The cholinergic projection from this nucleus is thought to be of significance in facilitating cortical cell columns in the context of situations having negative emotional valence. Meynert activity appears to be heightened in association with *anxiety*, generating a raised level of autonomic activity involving the amygdala (and/or the adjacent bed nucleus of the stria terminalis, mentioned below).

Efferent pathways   (Table 34.2)

Easily identified in the postmortem brain is the **stria terminalis** (Figure 34.14), which, on emerging from the central nucleus of the amygdala, follows the curve of the caudate nucleus and accompanies the thalamostriate vein along the upper surface of the thalamus. The stria sends

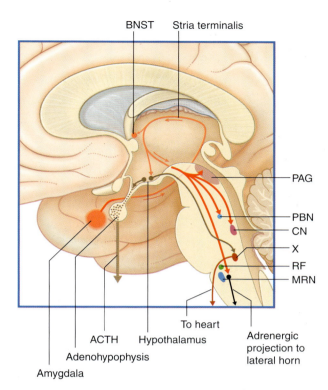

Figure 34.14 Efferents from central nucleus of amygdala via stria terminalis. The only postsynaptic pathways shown are autonomic. The periaqueductal gray matter (PAG) projects to the magnocellular reticular formation, giving rise to the medullary reticulospinal tract; PAG also projects to magnus raphe neurons, giving rise to the raphespinal tract. ACTH, adrenocorticotropic hormone; BNST, bed nucleus of stria terminalis; CN, nucleus ceruleus; MRN, magnus raphe nucleus; PBN, parabrachial nucleus; RF, reticular formation; X, dorsal nucleus of vagus.

fibers to the septal area and hypothalamus before entering the medial forebrain bundle and (downstream) the central tegmental tract. Some fibers of the stria terminate in a *bed nucleus* above the anterior commissure. The bed nucleus is regarded by some workers as part of the 'extended amygdala'; it may be more active than the amygdala proper, on PET scans, in anxiety states.

A second efferent projection, the **ventral amygdalofugal pathway**, passes medially to synapse within the nucleus accumbens (Figure 34.15). This connection is considered in the context of schizophrenia (Clinical Panel 34.2).

Notes on the efferent target connections
Periaqueductal gray matter (PAG). A source of *supraspinal antinociception* was described in Chapter 24, namely the opioid-containing axons from the hypothalamus, which disinhibit the excitatory projection from the PAG to the serotonergic cells of origin of the raphespinal tract. The excitatory cells of the *dorsal* PAG are directly stimulated by axon terminals entering from the amygdala via the medial forebrain bundle.

In laboratory animals, stimulation of the *ventral* PAG causes *freezing*, where a fixed, flexed posture is adopted. The ventral PAG contains neurons projecting to the cells

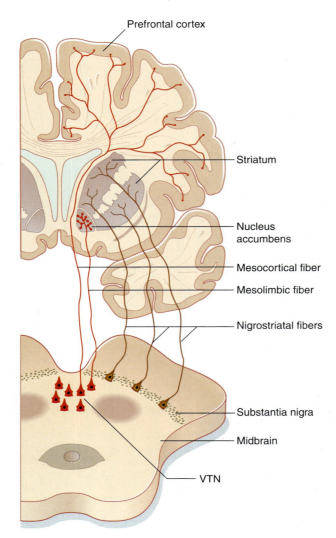

Figure 34.15 Coronal section at the level of the nucleus accumbens, highlighting distribution of dopaminergic fibers arising in the ventral tegmental nuclei (VTN) of the midbrain.

of origin of the medullary reticulospinal tract. This tract activates flexor motor neurons during the walking cycle, and intense activation may cause a frightened person to 'go weak at the knees' and perhaps fall down.

Cerulean nucleus. Facilitation of excitatory cortical neurons by the noradrenergic projection from this pontine nucleus is to be expected.

Medullary adrenergic neurons. As noted in Chapter 24, these neurons are a component of the baroreflex pathway sustaining the blood pressure against gravitational force. Sudden stimulation by the direct projection from the amygdala may send the heart dullthudding and cause a major elevation of systemic blood pressure.

Hypothalamus. Fibers of the stria terminalis synapse on two sets of hypothalamic neurons. The first, located in the anterolateral region, sends axons into the dorsal longitudinal fasciculus to synapse in cells of origin of the vagal supply to the heart. The well-known condition, referred to by psychiatrists as *blood trauma phobia* (fainting at the sight of blood at the scene of an accident), is characterized

by initial sympathetic excitation followed by vagus-induced bradycardia causing the individual to collapse (faint).

The second set of neurons secrete corticotropin-releasing hormone (CRH) into the adenohypophysis via the hypophysial portal system, with consequent release of adrenocorticotropin. Curiously, these CRH neurons send collateral branches into the central nucleus of the amygdala, with positive feedback enhancement of its activity.

Parabrachial nucleus. In individuals subject to *panic attacks*, hyperventilation, together with a sense of fear, may be triggered by what may appear to be relatively trivial environmental challenges. Normally, the respiratory alkalosis produced by washout of carbon dioxide reduces the respiratory rate causing the blood pH to return to normal, whereas susceptible individuals continue to hyperventilate. Because specific serotonin reuptake inhibitors are highly successful in treatment, the prevailing view is that the normal inhibitory role of serotonergic terminals within the nucleus accumbens (see below) has become deficient. However, overactivity of the cerulean nucleus has also been implicated, because the drug yohimbine can induce a panic attack, apparently through norepinephrine release.

Limbic striatal loop. This circuit is depicted in Chapter 33, passing from the prefrontal cortex through the nucleus accumbens and medial dorsal nucleus of thalamus, with return to the prefrontal cortex. However, the central nucleus of the amygdala participates in this circuit through an excitatory projection to the nucleus accumbens. In the right hemisphere, this projection is likely to facilitate a withdrawal response; in the left, it may facilitate an approach response.

Bilateral ablation of the amygdala has been carried out in humans for treatment of *rage attacks*, characterized by irritability, building up over several hours or days to a state of dangerous aggressiveness. This controversial operation has been successful in eliminating such attacks. In monkeys, bilateral ablation leads to placidity, together with a tendency to explore objects orally and to exhibit hypersexuality (*Kluver–Bucy syndrome*). A comparable syndrome has occasionally been observed in humans.

At the other end of the spectrum, PET studies of incarcerated murderers have revealed that the amygdala of the majority remains 'silent' even when gruesome scenes are presented on screen.

Nucleus accumbens

The full name is *nucleus accumbens septi pellucidi*, 'the nucleus leaning against the septum pellucidum'. More accurately, the nucleus abuts septal nuclei located in the base of the septum. Figures 34.15 and 34.18C show this relationship. The accumbens is one of many deep-seated brain areas where electrodes have been inserted on a therapeutic trial basis, notably in the hope of providing pain relief. Stimulation of the accumbens induces an intense sense of well-being (*hedonia*), comparable with that experienced by intake of drugs of addiction such as heroin (see Clinical Panel 34.5). This 'high' feeling is attributed to flooding of the nucleus, and of the medial prefrontal cortex, by synaptic and volume release of dopamine from the neurons projecting from the ventral

Clinical Panel 34.5   Drugs of dependency

Experimental evidence from the injection of drugs of abuse has yielded the following results (Figures CP 34.5.1 and 34.5.2).

- Cocaine binds with the dopamine reuptake transporter, blocking reuptake of the normal secretion, with consequent dopamine accumulation in the extracellular space.

- Amphetamine and methamphetamine are potent dopamine-releasing agents and also tend to block the reincorporation of dopamine into synaptic vesicles. These two drugs are also significantly active within the terminal dopaminergic network in the prefrontal cortex.

- Cannabinoids activate specific, excitatory, cannabinoid receptors on dopamine nerve endings.

- Nicotine attaches to specific excitatory receptors in the plasma membrane of parent somas in the midbrain.

- Opioids such as morphine and dihydromorphine (heroin) activate specific *inhibitory* receptors located

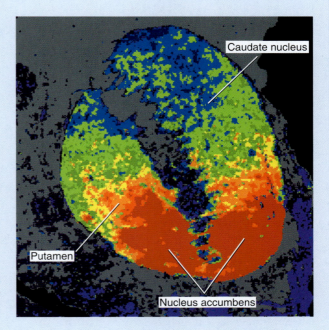

Figure CP 34.5.2 Intense activation (*red*) of D_3 receptors (D_2 variants) in the nucleus accumbens of a cocaine addict. (From Staley and Mash 1996, with permission.)

in the plasma membrane of GABAergic internuncial neurons within the nucleus. These neurons normally exert a tonic braking action on the projection cells of the ventral tegmental nuclei. Opioid-induced hyperpolarization of the internuncials leads to functional disinhibition of the projection cells, with consequent increased activity of both mesolimbic and mesocortical neurons.

- Ethanol also interferes with normal GABAergic activity. It binds to postsynaptic GABA membrane receptors throughout the brain without activating them; again, the target neurons become more excitable.

Serotonergic and noradrenergic neurons projecting to limbic system and hypothalamus have also been implicated in connection with drug dependency, notably in expressing some of the effects of abrupt drug withdrawal.

REFERENCE

Staley JK, Mash DC. Adaptive increase in D_3 dopamine receptors in the brain reward circuits of human cocaine fatalities. J Neurosci 1996; 16:610–616.

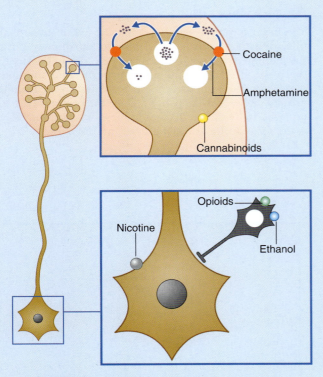

Figure CP 34.5.1 Mesolimbic neuron supplying nucleus accumbens, showing sites of action of some drugs of dependency.

tegmental area. Normally, dopamine is released in small amounts and quickly retrieved from the extracellular space by a specific dopamine reuptake transporter.

Septal area

The septal area comprises the **septal nuclei**, merging with the cortex directly in front of the anterior commissure, together with a small extension into the septum pellucidum (Figure 34.16).

Afferents to the septal nuclei are received from:

- the amygdala, via the *diagonal band* (of Broca), a slender connection passing alongside the anterior perforated substance
- the olfactory tract, via the medial olfactory stria
- the hippocampus, via the fornix
- brainstem monoaminergic neurons, via the medial forebrain bundle.

The two chief *efferent* projections are as follow.

- The **stria medullaris**, a glutamatergic strand running along the junction of side wall and roof of the third ventricle to synapse on cholinergic neurons in the **habenular nucleus**. The habenular nuclei of the two sides are connected through the **habenular commissure** located close to the root of the pineal gland, as shown earlier, in Figure 17.20. The habenular nucleus sends the cholinergic **habenulointerpeduncular tract** (**fasciculus retroflexus**) to synapse in the **interpeduncular nucleus** of the reticular formation in the midbrain (Figure 17.19). The interpeduncular nucleus is believed to participate in the sleep–wake cycle together with the cholinergic neurons beside the cerulean nucleus, identified earlier (Figure 24.3).

- The **septohippocampal pathway** running to the hippocampus by way of the fornix (Figure 34.17). It is responsible for generating the slow-wave *hippocampal theta rhythm* detectable in electroencephalography (EEG) recordings from the temporal lobe. Glutamatergic neurons in this pathway are pacemakers determining the *rate* of theta rhythm; cholinergic neurons determine the *size* of the theta waves. Theta rhythm is produced by synchronous discharge of groups of hippocampal pyramidal cells, and is significant in the development of biochemical alterations within pyramidal glutamate receptors during the LTP involved in laying down episodic memory traces. The strength of theta rhythm is greatly reduced in AD, reflecting the substantial loss of both cholinergic neurons and episodic memory formation and retrieval in this disease.

Electrical stimulation of the human septal area produces sexual sensations akin to orgasm. In animals, an electrolytic lesion may evince signs of extreme displeasure (so-called 'septal rage'). This surprising response may be due to destruction of a possible inhibitory projection from septal area to amygdala.

Basal forebrain

The basal forebrain extends from the bifurcation of the olfactory tract as far back as the infundibulum, and from the midline to the amygdala (Figure 34.18). In the floor of the basal forebrain is the **anterior perforated substance**, pierced by anteromedial central branches arising from the arterial circle of Willis (Ch. 5). Here the cerebral cortex is replaced by scattered nuclear groups, of which the largest is the **magnocellular basal nucleus** of Meynert.

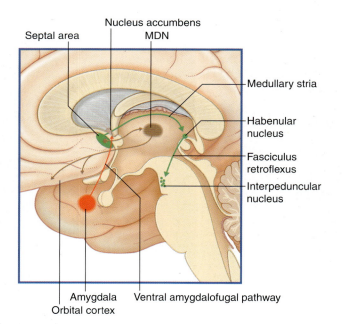

Figure 34.16 Connections of the septal area. MDN, mediodorsal nucleus of thalamus.

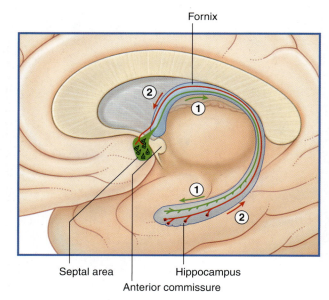

Figure 34.17 Septohippocampal pathway **(1)** with return projection from hippocampus **(2)**.

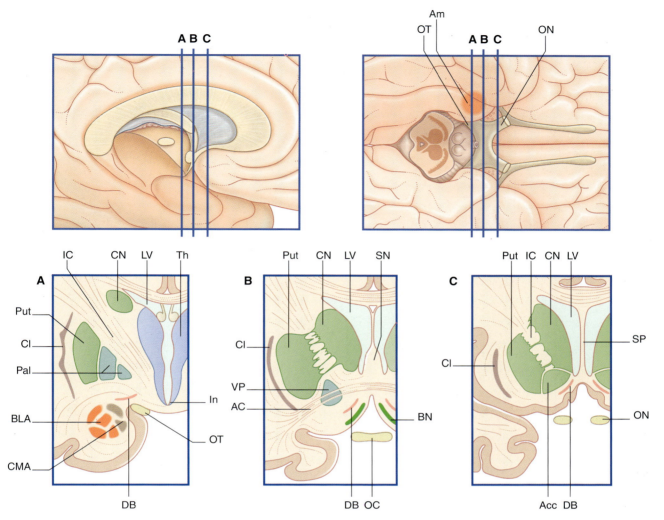

Figure 34.18 Coronal sections of the basal forebrain in the planes indicated. AC, anterior commissure; Acc, nucleus accumbens; Am, amygdala; BLA, basolateral amygdala; BN, basal nucleus of Meynert; Cl, claustrum; CMA, corticomedial amygdala; CN, caudate nucleus; DB, diagonal band of Broca; In, infundibulum; IC, internal capsule; LV, lateral ventricle; OC, optic chiasm; ON, optic nerve; OT, optic tract; Pal, pallidum; Put, putamen; SN, septal nucleus; SP, septum pellucidum; Th, thalamus; VP, ventral pallidum.

The *cholinergic neurons of the basal forebrain* have their somas mainly in the septal nuclei and basal nucleus of Meynert (Figure 34.19). The basal nucleus projects to all parts of the cerebral neocortex, which also contains scattered intrinsic cholinergic neurons.

The septal and basal nuclei, and small numbers contained in the diagonal band of Broca, are often referred to as the *basal forebrain nuclei*.

In the neocortex, the cholinergic supply from Meynert's nucleus is tonically active in the waking state, contributing to the 'awake' pattern on EEG recordings. All areas of the neocortex are richly supplied. Tonic liberation of acetylcholine tonically activates muscarinic receptors on cortical neurons, causing a reduction of potassium conductance, making them more responsive to other excitatory inputs. The cholinergic supply promotes LTP and training-induced synaptic strengthening of neocortical pyramidal cells.

The general psychic slow-down often observed in patients following a stroke may be accounted for by interruption of cholinergic fiber bundles in the subcortical white matter, caused by arterial occlusion within the territory of the anterior or middle cerebral artery. The result may be virtual cholinergic denervation of the cortex both at and posterior to the site of the lesion.

Neurogenesis in the adult brain

The term *neurogenesis* signifies the development of neurons from stem cell precursors. It is now well established that neurogenesis within the brain continues into adult life and, at a much lower rate, into old age. In the brains of laboratory animals including monkeys, and in biopsies taken from human brains during neurosurgery, mitotic neuronal stem cells have been detected in two regions.

1 In the subventricular zone, i.e. the zone immediately deep to the ependymal lining of the lateral ventricles. This is the original source of the stem cells of the olfactory bulb referred to earlier. In the adult, the stem cells of the subventricular zone generate cells that are incorporated into the gray matter of the

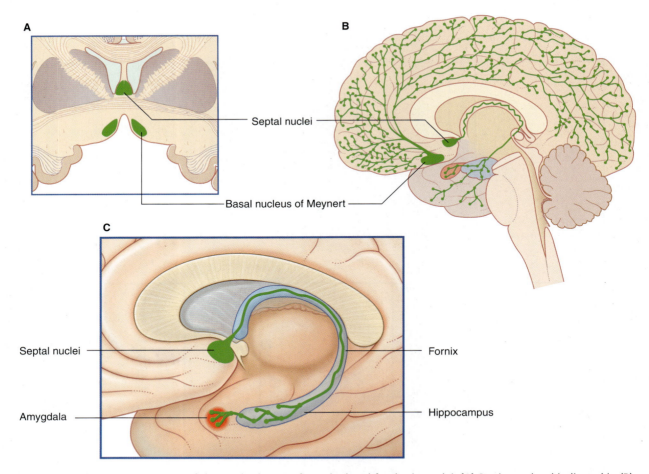

Figure 34.19 Cholinergic innervation of the cerebral cortex from the basal forebrain nuclei. **(A)** Section at level indicated in **(B)**. **(B)** Cortical innervation. **(C)** Septohippocampal pathway via the fornix. The amygdala is also supplied via this route.

frontal, parietal, and temporal lobes; however, whether they are destined to become neurons or neuroglia is uncertain.

2 Within the hippocampal formation, in the zone immediately deep to the granule cell layer of the dentate gyrus. In all species examined, including humans, these stem cells, when followed in cell cultures, exhibit branching and acquire electrical activity. Serial histologic studies in rats prove that they become mature, integrated granule cells.

In adult rats, the numbers of mitotic stem cells may increase dramatically in response to appropriate sensory stimulation. For example, the numbers in the olfactory bulb increase fivefold in the presence of an odor-rich environment; in the dentate subgranular zone, a sharp increase is observed in the presence of learning opportunities provided by tread-wheels and mazes. These observations lend credence to the belief that continuing to exercise body and mind is beneficial to humans entering their retirement years.

There is evidence from animal experiments that existing pharmacologic therapies for neurodegenerative and neuropsychiatric disorders exert beneficial neurotropic effects. A high level of serotonin in the extracellular fluid of the dentate gyrus stimulates the proliferation of neurons there, notably following administration of serotonin reuptake or monoamine oxidase inhibitors.

Core Information

Olfactory system

The olfactory system comprises the olfactory epithelium in the nose, the olfactory nerves, olfactory bulb and olfactory tract, and several patches of olfactory cortex. The epithelium comprises bipolar olfactory neurons, supporting cells, and basal cells that renew the bipolar neurons at a diminishing rate throughout life. Central processes of the bipolar neurons form the olfactory nerves, which penetrate the cribriform plate of the ethmoid bone and synapse on mitral cells in the bulb. Mitral cell axons form the olfactory tract, which has several low-level terminations in the anterior temporal lobe. Olfactory discrimination is a function of the orbitofrontal cortex, which is reached by way of the mediodorsal nucleus of the thalamus.

Limbic system

The limbic system comprises the limbic cortex and related subcortical nuclei. The limbic cortex includes the hippocampal formation, septal area, parahippocampal gyrus, and cingulate gyrus. The principal subcortical nucleus is the amygdala. Closely related are the orbitofrontal cortex, temporal pole, hypothalamus and reticular formation, and nucleus accumbens.

The anterior part of the parahippocampal gyrus is the *entorhinal cortex*, which receives cognitive and sensory information from the cortical association areas, transmits it to the hippocampal formation for consolidation, and returns it to the association areas where it is encoded in the form of memory traces.

The hippocampal formation comprises the subiculum, hippocampus proper, and dentate gyrus. Sectors of the hippocampus are called cornu ammonis (CA) 1–4.

The perforant path projects from the entorhinal cortex on to the dendrites of dentate granule cells. Granule cell axons synapse on CA3 pyramidal cells, which give Schaffer collaterals to CA1. CA1 back-projects to the entorhinal cortex, which is heavily linked to the association areas.

The fornix is a direct continuation of the fimbria, which receives axons from the subiculum and hippocampus. The crus of the fornix joins its fellow to form the trunk. Anteriorly, the pillar of the fornix divides into precommissural fibers entering the septal area, and postcommissural fibers entering anterior hypothalamus, mammillary bodies, and medial forebrain bundle.

Bilateral damage to or removal of the hippocampal formation is followed by anterograde amnesia, with loss of declarative memory. Procedural memory is preserved. Long-term potentiation of granule and pyramidal cells is regarded as a key factor in the consolidation of memories.

The insula has functions in relation to pain and to language. The anterior cingulate cortex has functions in relation to motor response selection, emotional tone, bladder control, vocalization, and autonomic control. The posterior cingulate responds to the emotional tone of what is seen or felt.

The amygdala, above and in front of the temporal horn of the lateral ventricle, is the principal brain nucleus associated with the perception of fear. Its afferent, lateral nucleus receives inputs from olfactory, visual, auditory, tactile, visceral, cognitive, and mnemonic sources. The central, efferent nucleus sends fibers via the stria terminalis to the hypothalamus, activating corticotropin release and vagus-mediated bradycardia, and to the brainstem, activating dorsal and ventral periaqueductal gray matter, and influencing respiratory rate and autonomic activity. The amygdalofugal pathway from the central nucleus facilitates defensive or evasive activity via the limbic striatal loop.

The nucleus accumbens is a clinically important component of the mesolimbic system in the context of drug dependency, based on its abundance of dopaminergic nerve terminals derived from ventral tegmental nuclei. Dopamine levels in the extracellular space in nucleus accumbens and medial prefrontal cortex are raised by cocaine and amphetamines, which interfere with local dopamine recycling, and by cannabinoids, which activate specific terminal receptors. Nicotine activates specific receptors in the parent tegmental neurons. Opioids and ethanol interfere with the normal braking action of GABA tegmental internuncials.

The septal area comprises two main nuclear groups. One sends a set of glutamatergic fibers in the stria medullaris thalami to the habenular nucleus, which in turn sends the cholinergic fasciculus retroflexus to the interpeduncular nucleus, which participates in the sleep–wake cycle. The other forms the septohippocampal pathway to synapse on hippocampal pyramidal cells. Glutamatergic and cholinergic elements govern the rate and strength, respectively, of hippocampal theta rhythm that facilitates formation of episodic memories.

The basal forebrain is the gray matter in and around the anterior perforated substance. It includes the cholinergic, nucleus basalis of Meynert, which projects to all parts of the neocortex, and the cholinergic, septal nucleus projecting to the hippocampus. Both lose about half of their neurons in Alzheimer disease, and the neocortical distribution is vulnerable to stroke.

REFERENCES

Bayley PJ, Gold JJ, Hopkins RO, et al. The neuroanatomy of remote memory. Neuron 2005; 46:799–810.

Braak H, Braak E. Evolution of neuronal changes in the course of Alzheimer's disease. J Neural Transm 1998; 53(suppl):127–140.

Campo P, Maistu F, Ortiz T, et al. Is medial temporal lobe activation specific for encoding long-term memories? Neuroimage 2005; 25:34–42.

Coplan JD, Lydiard RB. Brain circuits in panic disorder. Biol Psychiatry 1998; 44:1264–1266.

Cummings JL, Vinters HV, Cole GM, et al. Alzheimer's disease: etiologies, pathophysiology, cognitive reserve, and treatment opportunities. Neurology 1998; 51(suppl):S2–S17.

Dekker AJAM, Connor DJ, Thal LJ. The role of cholinergic projections from the nucleus basalis in memory. Neurosci Behav Rev 1991; 15:349–347.

Doetch F, Hen R. Young and excitable: the function of new neurons in the adult mammalian brain. Curr Opin Neurobiol 2005; 15:121–128.

Dolan RJ, Paulesu E, Fletcher P. Human memory systems. In: Frackowiak RSJ, Friston KJ, Frith CD, et al, eds. Human brain function. London: Academic Press; 1997:367–404.

Duvernoy HM. The human hippocampus. 2nd edn. Berlin: Springer; 1998.

Francis PT, Palmer AM, Snape M, et al. The cholinergic hypothesis of Alzheimer's disease: a review of progress. J Neurol Neurosurg Psychiatry 1999; 66:137–147.

Hyman SE. Brain neurocircuitry of anxiety and fear. Biol Psychiatry 1998; 44:1401–1403.

Kemperrmann G, Wiskott L, Gage FH. Functional significance of adult neurogenesis. Curr Opin Neurobiol 2004; 14:186–191.

Kim E, Shirvalkar P, Herrera DG. Regulation of neurogenesis in the aging vertebrate brain: role of oxidative stress and neuroactive trophic factors. Clin Neurosci Res 2003; 2:285–343.

Kretschmann H-J, Weinrich W. Neurofunctional systems: 3D reconstructions with correlated neuroimaging: text and CD-ROM. New York: Thieme; 1998

Kuhl DE, Koeppe RA, Minoshima S, et al. In vivo mapping of cerebral acetylcholinesterase activity in aging and Alzheimer's disease. Neurology 1999; 52:691–699.

Leonard BE. Fundamentals of psychopharmacology. 3rd edn. Chichester: Wiley; 2003.

Liddell BL, Brown KJ, Kemp AH, et al. A direct brainstem–amygdala–cortical 'alarm' system for subliminal signals of fear. Neuroimage 2005; 24:235–243.

Maguire EA, Burgess N, O'Keefe J. Human spatial navigation: cognitive maps, sexual dimorphism, and neural substrates. Curr Opin Neurobiol 1999; 9:171–177.

Moscovitch M, Rosenbaum RS, Gilboa A, et al. Functional neuroanatomy of remote episodic, semantic and spatial memory: a unified account based on multiple trace theory. J Anat 2005; 207:35–66.

Passingham D, Sakai K. The prefrontal cortex and working memory: physiology and brain imaging. Curr Opin Neurobiol 2004; 14:163–168.

Perry RJ, Hodges JR. Attention and executive deficits in Alzheimer's disease. Brain 1999; 122: 383–404.

Phelps EA. Human emotion and memory: interactions of the amygdala and hippocampal complex. Curr Opin Neurobiol 2004; 14:198–202.

Richardson MP, Strange B, Dolan RJ. Encoding of emotional memories depends on the amygdala and hippocampus and their interactions. Nat Neurosci 2000; 7:278–285.

Selden NR, Gitleman DR, Salammon-Murayama N, et al. Trajectories of cholinergic pathways within the cerebral hemispheres of the human brain. Brain 1998; 121:2249–2257.

Staley JK, Mash DC. Adaptive increase in D_3 dopamine receptors in the brain reward circuits of human cocaine fatalities. J Neurosci 1996; 16:610–616.

Van der Werf YD, Scheltens P, Lindeboom J, et al. Disorders of memory, executive functioning and attention following infarction in the thalamus: a study of 22 cases with localised lesions. Neuropsychologia 2003; 41:1330–1344.

Cerebrovascular disease

STUDY GUIDELINES

1 This final chapter touches on a large range of neurologic symptoms and signs, because it deals with vascular damage to every part of the brain. The great majority of the symptoms and signs have already been mentioned, individually, in other contexts.

2 The Clinical Panels may convey the impression that clinical diagnosis is rather straightforward. However, following vascular insults, patients are seldom alert and cooperative.

3 An objective of this chapter is to demonstrate the value of understanding the regional as well as the systems anatomy of the brain, because cerebrovascular accidents cause injury to regions, with consequent effects on multiple systems.

4 The increasing range of diagnostic aids does not shrink the need for clinical acumen. The more accurate the tentative diagnosis, the more likely the most appropriate technology will be selected for further elucidation.

5 Perhaps refresh the blood supply (Ch. 5).

Cerebrovascular disease is the third leading cause of death in adults, being superseded only by heart disease and cancer. The most frequent expression of cerebrovascular disease is that of a *stroke*, which is defined as a focal neurologic deficit of vascular origin that lasts for more than 24 h if the patient survives. The most frequent example is a hemiplegia caused by a vascular lesion of the internal capsule. However, it will be seen that many varieties of stroke symptomatology are recognized, based on place and size.

The chief underlying disorders are atherosclerosis within the large arteries supplying the brain, heart disease, hypertension, and 'leaky' perforating arteries.

- Atherosclerosis signifies fatty deposits in the intimal lining of the internal carotid and vertebrobasilar system—most notably in the internal carotid trunk or in one of the vertebral arteries. The deposits pose a dual threat: in situ enlargement may cause progressive occlusion of a main artery, and breakaway deposits may form emboli (plugs) blocking distal branches within the brain. However, gradual occlusion is often redeemed by routing of blood through alternative channels. For example, an internal carotid artery (ICA) may be progressively occluded over a period of 10 years or more without apparent brain damage; the contralateral ICA utilizes the circle of Willis to perfuse both pairs of anterior and middle cerebral arteries, and it is not unusual in such cases for external carotid blood to assist, by retrograde flow from the facial artery through the ophthalmic artery on the affected side. Similarly, occlusion of the stem of one of the three cerebral arteries may be compensated for through small (< 0.5 mm) anastomotic arteries in the depths of cortical sulci, perfused by the other two cerebrals. The number of such small arteries varies greatly between individuals. The crescent-shaped anastomotic region is known as the *border zone* (Figure 35.1). On the other hand, all the arteries penetrating the brain substance are end arteries, i.e. their communications with neighboring penetrating arteries are too fine to save brain tissue in the event of blockage.

- Many cerebral emboli originate as blood clots in the left side of the heart, in association with coronary or valvular disease.

- Hypertension is obviously associated with cerebral hemorrhage, which may be so massive as to rupture into the ventricular system and cause death within minutes or hours.

- Less obvious are lacunae ('small pools'), up to 2 cm in diameter, in the white matter adjacent to one or more perforating end arteries. The source is now believed to be a 'leakiness' of the vascular endothelium (blood–brain barrier), permitting extravasation of plasma into the perivascular space. The fundamental cause is not understood; there is only a marginal association with hypertension and none with atherosclerosis or stenosis. Lacunar strokes are the most likely to recur, and such recurrence has a high association with a state of *multiinfarct dementia*. An *infarct* is an area of brain destruction produced by vascular occlusion, hemorrhage, or extravasation.

Cerebral infarcts become swollen after a few days, because of osmotic activity. Some become large enough to produce distance effects by causing subfalcal or tentorial herniation of the brain, in the manner of a tumor (Ch. 4).

It is usually easy to distinguish the symptoms and signs of vascular disease from those of a tumor. A vascular stroke takes up to 24 h to evolve, whereas the time frame for tumors is usually several months or more. However, hemorrhage into a tumor may cause it to expand suddenly and to mimic the effects of a stroke. Very often, the hemorrhage is into a metastatic tumor, notably from lung, breast, or prostate; in fact, a stroke may be the first manifestation of a cancer in one of those organs.

Some 10% of vascular strokes are caused by rupture of a 'berry' aneurysm into the brain. As explained later, berry aneurysms usually bleed directly into the subarachnoid space, because they originate in or near the circle of Willis,

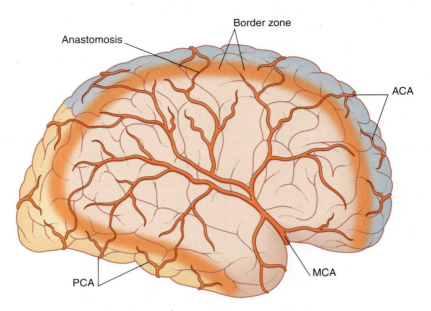

Figure 35.1 Border zone of anastomotic overlap between the middle cerebral artery (MCA) and the anterior and posterior cerebral arteries (ACA, PCA).

but some arise at an arterial bifurcation point within the brain. A ruptured aneurysm is always a prime suspect when a stroke comes 'out of the blue' in someone less than 40 years old.

ANTERIOR CIRCULATION OF THE BRAIN

Clinicians refer to the ICA and its branches as the *anterior circulation* of the brain, and the vertebrobasilar system (including the posterior cerebral arteries) as the *posterior circulation*. The anterior and posterior circulations are connected by the posterior communicating arteries (Figure 35.2).

About 75% of cerebrovascular accidents originate in the anterior circulation.

Internal capsule

The following details supplement the account of the arterial supply of the internal capsule in Chapter 5.

The blood supply is shown in Figure 35.3. The three sources of supply are the **anterior choroidal**, a direct branch of the internal carotid; the **medial striate**, a branch of the anterior cerebral; and **lateral striate** (**lenticulostriate**) branches of the middle cerebral artery.

The contents of the internal capsule are shown in Figures 35.4 and 35.5. The anterior choroidal branch of the ICA supplies the lower part of the posterior limb and the retrolentiform part of the internal capsule, and the inferolateral part of the lateral geniculate body. Some of its branches (not shown) supply a variable amount of the temporal lobe of the brain and the choroid plexus of the inferior horn of the lateral ventricle.

The medial striate branch of the anterior cerebral artery (recurrent artery of Heubner) supplies the lower part of the anterior limb and genu of the internal capsule.

The lateral striate arteries penetrate the lentiform nucleus and give multiple branches to the anterior limb, genu, and posterior limb of the internal capsule.

POSTERIOR CIRCULATION OF THE BRAIN

Additional information is confined to the stem branches of the posterior cerebral artery shown in Figure 35.6.

TRANSIENT ISCHEMIC ATTACKS

Transient ischemic attacks (TIAs) are episodes of vascular insufficiency that cause temporary loss of brain function, with total recovery within 24 h. Most TIAs last for less than half an hour, with no residual signs at the time of clinical examination. Diagnosis is therefore usually based on reported symptoms alone.

Most attacks follow lodgment of fibrin clots or detached atheromatous tissue at an arterial branch point, with subsequent dissolution.

- Transient symptoms originating in the anterior circulation include motor weakness (a 'heavy feeling') in an arm or leg, hemisensory deficit (a 'numb feeling'), dysphasia, and monocular blindness from occlusion of the central artery of the retina.
- Transient symptoms originating in the posterior circulation include vertigo, diplopia, ataxia, and amnesia.

Recognition of TIAs involving the anterior circulation is important, because they serve notice of impending major illness. Without treatment, one patient in four will die from a heart attack within 5 years, and one in six will suffer a stroke.

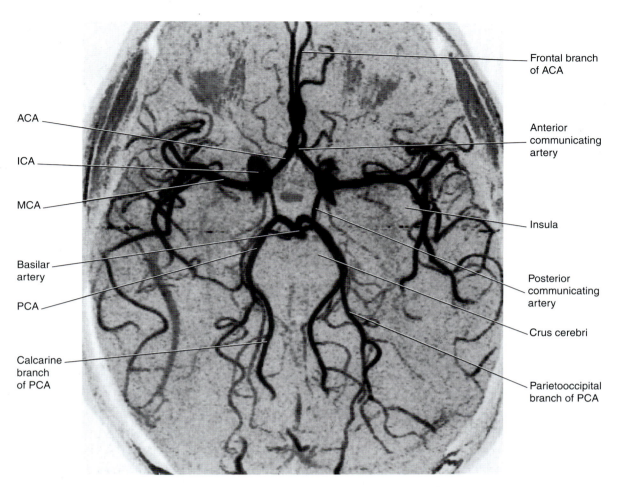

ACA

ICA

MCA

Basilar
artery

PCA

Calcarine
branch
of PCA

Frontal branch
of ACA

Anterior
communicating
artery

Insula

Posterior
communicating
artery

Crus cerebri

Parietooccipital
branch of PCA

Figure 35.2 Circle of Willis and its branches. This is a magnetic resonance angiogram based on the principle that flowing blood generates a different signal to that of stationary tissue, without injection of a contrast agent. Conventional angiograms, for example those in Chapter 5, require arterial perfusion with a contrast agent. The vessels shown here are contained within a single thick magnetic resonance 'slice'. Some, for example the calcarine branch of the posterior cerebral artery, could be followed further in adjacent slices. ACA, anterior cerebral artery; ICA, internal carotid artery; MCA, middle cerebral artery; PCA, posterior cerebral artery. (From a series kindly provided by Professor J. Paul Finn, Director, Magnetic Resonance Research, Department of Radiology, David Geffen School of Medicine at UCLA, California.)

CLINICAL ANATOMY OF VASCULAR OCCLUSIONS

In the Clinical Panels, the term *occlusion* encompasses all causes of regional arterial failure other than aneurysms. Symptoms of occlusions within the anterior circulation are summarized in Clinical Panels 35.1–35.4, within the posterior circulation in Clinical Panel 35.5, specifically within the territory of the posterior cerebral artery in Clinical Panel 35.6. Subarachnoid hemorrhage is considered in Clinical Panel 35.7.

It should be emphasized that the majority of strokes originate in the territory of the middle cerebral artery.

Finally, recovery of motor function after a stroke is discussed in Clinical Panel 35.8.

Clinical Panel 35.1   Anterior choroidal artery occlusion

A complete anterior choroidal artery syndrome is produced by occlusion of the proximal part of the artery, compromising the lower part of the posterior limb and retrolentiform part of the internal capsule. The clinical picture is one of contralateral hemiparesis, hemisensory loss of cortical type (Ch. 29), and hemianopia. Damage to the (crossed) cerebellothalamocortical pathway may add evidence of intention tremor in the contralateral upper limb, yielding so-called *ataxic hemiparesis.*

Isolated occlusion of the branch to the lateral geniculate body results in a contralateral upper quadrant hemianopia.

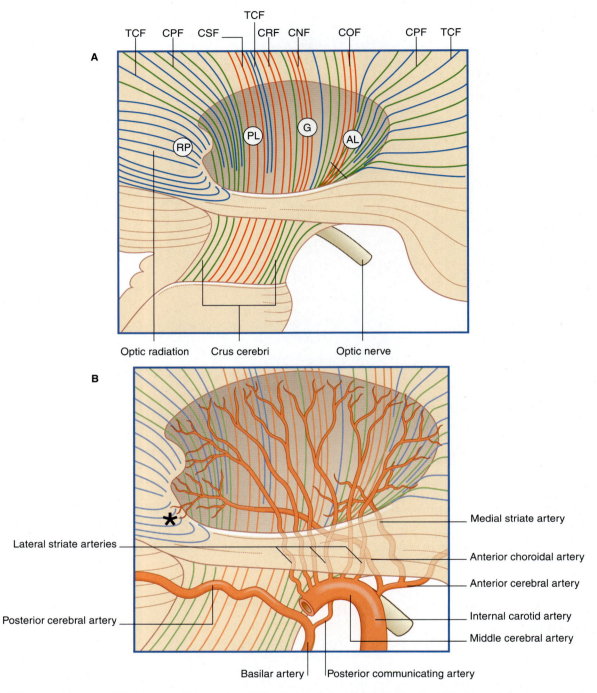

Figure 35.3 Internal capsule. **(A)** Pathways. Lateral view of the right cerebral hemisphere, showing the oval depression in the white matter following removal of the lentiform nucleus. The internal capsule occupies the floor of the depression. CNF, corticonuclear fibers; COF, corticooculomotor fibers; CPF, corticopontine fibers; CRF, corticoreticular fibers; CSF, corticospinal fibers; TCF, thalamocortical fibers. Other abbreviations as in Figure 35.4. **(B)** Blood supply. The medial striate branch of the anterior cerebral artery is the recurrent artery of Heubner. Only two of the six lateral striate branches of the middle cerebral artery shown are labeled. The asterisk indicates arterial supply from the anterior choroidal artery to the inferolateral part of the lateral geniculate body.

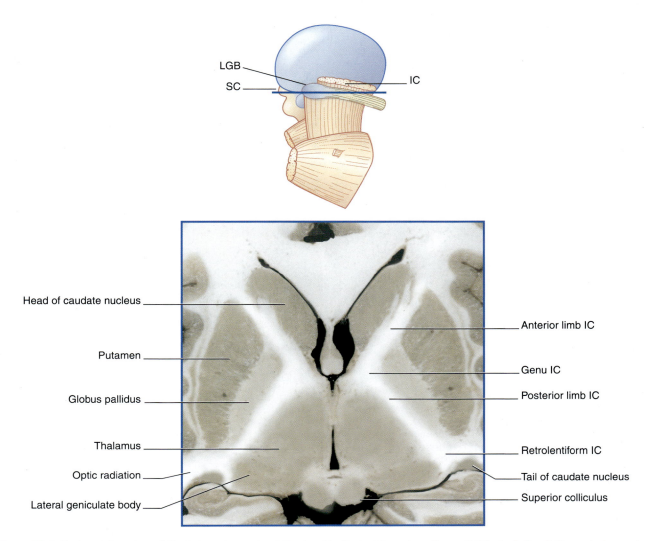

Head of caudate nucleus

Putamen

Globus pallidus

Thalamus

Optic radiation

Lateral geniculate body

Anterior limb IC

Genu IC

Posterior limb IC

Retrolentiform IC

Tail of caudate nucleus

Superior colliculus

LGB

SC

IC

Figure 35.4 Horizontal section of the internal capsule at the level indicated (based on Figure 2.11), depicting its boundaries and parts (*left*) and stroke-relevant motor contents (*right*). IC internal capsule; LGB, lateral geniculate body; SC, superior colliculus.

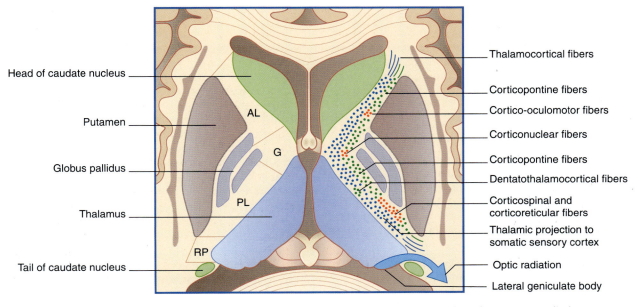

Head of caudate nucleus

Putamen

Globus pallidus

Thalamus

Tail of caudate nucleus

Thalamocortical fibers

Corticopontine fibers

Cortico-oculomotor fibers

Corticonuclear fibers

Corticopontine fibers

Dentatothalamocortical fibers

Corticospinal and corticoreticular fibers

Thalamic projection to somatic sensory cortex

Optic radiation

Lateral geniculate body

AL

G

PL

RP

Figure 35.5 Horizontal section of the internal capsule, depicting its parts (*left*) and contents (*right*). AL, anterior limb; G, genu: PL, posterior limb; RP, retrolentiform part.

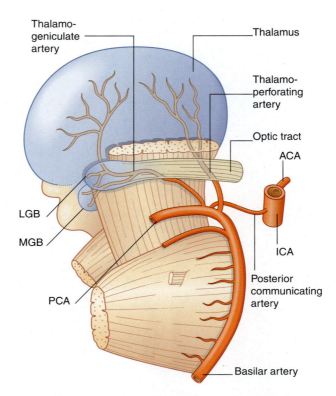

Figure 35.6 Central branches of the posterior cerebral artery (PCA). Although only two arteries are shown, each in fact comprises several branches from the PCA. The thalamoperforating artery shown pierces the posterior perforated substance and supplies the anterior one-third of the thalamus. The thalamogeniculate artery shown supplies the geniculate bodies and the posterior two-thirds of the thalamus. ACA, anterior cerebral artery; ICA, internal carotid artery; LGB, MGB, lateral, medial geniculate bodies.

Clinical Panel 35.2   Anterior cerebral artery occlusion

Complete interruption of flow in the proximal anterior cerebral artery is rare, because the opposite artery has direct access to its distal territory through the anterior communicating artery. However, branch occlusions are well recognized, with corresponding variations in the clinical picture.

- *Orbital or frontopolar branch.* The usual result is an apathetic state with some memory loss.

- *Medial striate artery* (recurrent artery of Heubner) occlusion may result in dysarthria owing to compromise of the motor supply to the contralateral nuclei supplying the muscles of the mandible (V), lips (VII), and tongue (XII). Hoarseness and dysphagia are also present if the supranuclear supply to the nucleus ambiguus is interrupted.

- *Callosomarginal.* This branch supplies the dorsomedial prefrontal cortex, the supplementary motor area (SMA), and the lower limb and perineal areas of the sensorimotor cortex and the supplementary sensory

area (SSA). The commonest manifestation of occlusion is motor weakness and some cortical-type sensory loss in the contralateral lower limb, as a result of infarction within the paracentral lobule. Urinary incontinence may occur for some days, owing to contralateral weakness of the pelvic floor. Damage to the prefrontal cortex results in abulia (lack of initiative). A left-sided infarct of the SMA may produce mutism, because SMA normally collaborates with Broca's area in the initiation of speech. Finally, damage to the SSA may result in inability to reach with the contralateral arm toward the side of the lesion.

- *Pericallosal.* Infarction of the anterior part of the corpus callosum may result in ideomotor apraxia. (The lesion would be comparable with lesion 1 in Figure 32.7). Infarction of the midregion may cause tactile anomia owing to blocked transfer of tactile information from right to left parietal lobe.

Clinical Panel 35.3   Middle cerebral artery occlusion

Embolic and lacunar infarcts are frequent in late middle life and in the elderly. Hemorrhage from one of the striate branches is also a frequent event.

Embolism

An embolus may lodge in the stem of the artery, in the upper division, in the lower division, or in a cortical branch of either division.

Stem

Occlusion of the stem affects the central as well as the cortical branches. The complete picture includes contralateral hemiplegia, severe contralateral sensory loss, contralateral homonymous hemianopia, and drifting of the eyes toward the side of the infarct. Left-sided lesions are usually accompanied by global aphasia, right-sided ones with contralateral sensory neglect. Many patients die in coma following midbrain compression by a swollen infarct.

The condition of some patients with stem occlusion improves markedly within days, as explained in Clinical Panel 35.8.

Upper division

An embolus occluding the upper division gives rise to contralateral paresis (weakness) and cortical-type sensory loss in the face and arm, together with dysarthria arising from damage to supranuclear pathways involved in speech articulation. Left-sided lesions are usually accompanied by Broca's aphasia, right-sided lesions by contralateral neglect.

Lower division

Embolism of the lower division produces contralateral homonymous hemianopia, and sometimes a confused, agitated state attributed to involvement of limbic pathways in the temporal lobe. Left-sided lesions are also accompanied by Wernicke's aphasia, alexia, and sometimes by ideomotor apraxia (corresponding to lesion 3 in Figure 32.7).

Branch embolism

The following isolated deficits are attributable to an embolus lodged in one of the cortical branches.

- Orbitofrontal: elements of a prefrontal syndrome (Ch. 32) may be present.
- Precentral (prerolandic): Broca's aphasia (left lesion), monotone speech (right lesion).
- Central (rolandic): contralateral loss of motor and/or sensory function in the face and arm.
- Inferior parietal: contralateral hemineglect (especially with right lesion), tactile anomia.
- Angular: contralateral homonymous hemianopia; alexia with left lesion.
- Posterior or middle temporal: Wernicke's aphasia (left lesion), sensory aprosodia (right lesion).

Lacunar infarcts

Lacunar infarction is suspected where the clinical evidence suggests a small lesion. Well-recognized are the following.

- *Pure motor hemiparesis*, caused by a lacuna in the corona radiata or internal capsule (see also pons, later). The weakness is mainly in the lower face and arm, and there are no sensory or higher cortical disturbances.
- *Pure sensory syndrome*, produced by a lacuna in the ventral posterior nucleus of the thalamus. There is severe impairment of tactile discrimination (Ch. 15) in the contralateral limbs, together with sensory ataxia.
- *Dysarthria–clumsy hand syndrome*, produced by a lacuna among fibers descending to, or within, the genu of the internal capsule, containing (a) corticonuclear fibers descending to contralateral motor nuclei of pons and medulla oblongata, and (b) fibers from the premotor cortex involved in contralateral manual control. The most apparent results are dysarthria owing to paresis of lip, tongue, and jaw musculature, and clumsiness of hand movement.

Hemorrhage

The commonest source of a cerebral hemorrhage is one of the lateral striate branches of the middle cerebral artery. The commonest location is the putamen, with spread into the anterior and posterior limbs of the internal capsule. The usual cause is a preexisting systemic hypertension. The hematoma may be as small as a pea or as big as a golf ball. Large hemorrhages rupture into the lateral ventricle and are usually fatal within 24 h.

A typical clinical case is one in which a sudden, severe headache is followed by unconsciousness within a few minutes. The eyes tend to drift toward the side of the lesion, as noted in Chapter 29. With recovery of consciousness, there is a complete, flaccid hemiplegia (apart from the upper part of the face). Tendon reflexes are absent on the hemiplegic side, and a Babinski sign is present.

Following any kind of stroke involving the left internal capsule, right-handers often notice some initial clumsiness in the left hand. Functional magnetic resonance imaging studies indicate that, in healthy right-handers, the left motor cortex is more active during movements of the left hand than the right motor cortex is during movements of the right hand. In other words, the left motor cortex has a greater degree of bilateral control.

The end result of capsular stroke is often one of ambulatory spastic hemiparesis with hemihypesthesia (reduced sensation). Figure CP 35.3.1 shows the typical posture during walking: the elbow and fingers are flexed, and the leg has to be circumducted during the

Clinical Panel 35.3 *Continued*

Figure CP 35.3.1 Hemiplegic gait. The patient's right side is affected.

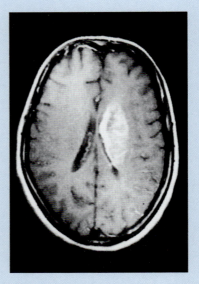

Figure CP 35.3.2 Contrast-enhanced magnetic resonance image taken from a patient 11 days after an embolic stroke (see text). (From Sato 1991, with kind permission of Dr. S. Takahashi, Department of Radiology, Tohoku University School of Medicine, Sendai, Japan, and the editors of *Radiology*.)

swing phase (unless an ankle brace is worn) because of the antigravity tone of the musculature. During the early rehabilitation period, an arm sling is required in order to protect the shoulder joint from downward subluxation (partial dislocation). This is because the supraspinatus muscle is normally in continuous contraction when the body is upright, preventing slippage of the humeral head.

Figure CP 35.3.2 is from a magnetic resonance study of a patient who had suffered a right hemiplegia with sensory loss 11 days previously. The picture shows extensive infarction of the white matter on the left side, at the junctional region between the corona radiata and internal capsule, with compression of the lateral ventricle.

Internal carotid artery

In addition to being a source of cerebral emboli, atheromatous plaques may cause partial or complete occlusion of the internal carotid artery itself (see Clinical Panel 35.4).

REFERENCE

Mohr JP, Lazar RM. Middle cerebral artery disease. In: Mohr JP, Choi DW, Grotta JC, et al, eds. Stroke pathophysiology, diagnosis and management. 4th edn. Philadelphia: Churchill Livingstone; 2004:123–166.

Clinical Panel 35.4   Internal carotid artery occlusion

The lumen of the internal carotid artery may become progressively obstructed by atheromatous deposits. Common sites of obstruction are the point of commencement in the neck, and the cavernous sinus. A slowly progressive obstruction may be compensated for by the opposite internal carotid artery, through the circle of Willis. Additional blood may also be provided through the orbit from the facial artery. At the other extreme, sudden occlusion may cause death from infarction of the entire anterior and middle cerebral territories, and sometimes the posterior cerebral also.

Warning signs of carotid occlusion take the form of transient ischemic attacks (TIAs) lasting for up to a few hours. As with TIAs elsewhere, the physician is unlikely to be present during an attack and must interpret the account given by the patient or relative. The territory of the middle cerebral artery is most often affected. Individual symptoms tend to occur in isolation, and include any of the following: a feeling of heaviness, weakness, numbness, or tingling in one arm or leg; halting or slurring of speech. Disturbance of flow in the ophthalmic artery may cause transient monocular blindness (one eye may be perceived as filled with fog or white steam).

Clinical Panel 35.5   Occlusions within the posterior circulation

The clinical phrase *long tract signs* is most often used in the context of brainstem lesions. It refers to evidence of a lesion in one or more of the three long tracts, namely the pyramidal tract, the posterior column–medial lemniscal pathway, and the spinothalamic pathway. All the long tract signs occur in the limbs on the side opposite to the lesion.

Small brainstem infarcts may yield the following features.

- *Midbrain*. Ipsilateral third nerve paralysis and/or bilateral cerebellar ataxia caused by damage to the decussation of the superior cerebellar peduncles; 'crossed' third nerve paralysis featuring ipsilateral paralysis combined with contralateral hemiplegia.

- *Pons*. A tegmental infarct may cause ipsilateral facial and/or abducens and/or mandibular nerve paralysis and/or anesthesia of the face. A basilar infarct may produce a contralateral pure motor hemiplegia whose brainstem origin may be indicated by transient ipsilateral functional impairment of the abducens, facial, or mandibular nerve passing through the tegmentum.

- *Medulla oblongata*. Most characteristic is the lateral medullary syndrome, described in Chapter 19, caused by occlusion of the posterior inferior cerebellar artery. Occlusion of the labyrinthine branch of the anterior inferior cerebellar artery causes immediate destruction of the inner ear; sudden deafness in that ear is accompanied by vertigo, with a tendency to fall to that side.

Large brainstem infarcts in pons or medulla oblongata are usually fatal, because of damage to the vital centers of the reticular formation. In the midbrain, they may produce a permanent state of coma.

Cerebellar ataxia of the limbs on one side, without brainstem damage, is more often due to occlusion of the top end of the vertebral artery on that side than to occlusion of one of the three cerebellar arteries.

The posterior cerebral arteries are usually perfused through the basilar bifurcation. Occlusion is more common in branches than in either main stem (Clinical Panel 35.6).

Clinical Panel 35.6   Posterior cerebral artery occlusion

A variety of effects may follow occlusion of branches of the posterior cerebral artery. Usually, the occlusion is limited to a branch to the midbrain, or to the thalamus, or to the subthalamic nucleus, or to the cerebral cortex.

Midbrain
The classic picture of a unilateral infarct of the midbrain is that of a crossed third nerve palsy, i.e. a complete oculomotor paralysis (Ch. 23) on one side with a hemiplegia on the other side (Weber syndrome). The hemiplegia is due to infarction of the crus cerebri, which contains corticospinal and corticonuclear fibers in its midportion. The hemiplegia usually shows early improvement, but ataxia may appear on that side because of damage to dentatothalamic fibers bypassing the red nucleus.

Thalamus
Occlusion of a thalamogeniculate branch may cause infarction of the posterior lateral nucleus of the thalamus (which receives the spinothalamic tract and medial lemniscus), and sometimes of the lateral geniculate nucleus also. The usual result is a contralateral sense of numbness, perhaps with hemianopia. The rare and unpleasant thalamic syndrome (Ch. 27) may supervene.

Subthalamic nucleus
Occlusion of a thalamoperforating branch may destroy the small subthalamic nucleus and give rise to ballism on the contralateral side, usually affecting the arm (Ch. 33).

Corpus callosum
Infarction of the splenium of the corpus callosum blocks transfer of written information from the right visual association cortex to the left. The result of infarction is alexia for written material presented to the left visual field.

Cortex
Occlusion of the stem of the posterior cerebral artery behind the midbrain gives rise to a homonymous hemianopia in the contralateral field. Macular vision may be spared. One view of macular sparing is that it signifies bilateral representation of the fovea in the primary visual cortex. Another view is that the occipital pole is supplied by a long branch from the middle cerebral artery supplying the angular gyrus.

Occlusion of the left artery also produces alexia, the left visual field being the only area detectable by the patient.

A pure alexia, without agraphia, may follow a lesion of the left lingual gyrus.

Bilateral occlusion
Partial or complete cortical blindness may result from a thrombus arrested where the lumen of the basilar artery normally narrows below the basilar bifurcation, with consequent blockage of both posterior cerebral arteries. It has also been recorded following cardiac arrest with resuscitation.

Temporary cessation of flow in both posterior cerebral arteries sometimes affects only the anterior parts of their territories. If damage is confined to the occipitotemporal junctions, prosopagnosia (inability to identify faces) may occur alone. (Prosopagnosia has been recorded with purely right-sided perfusion failure.) If the entorhinal cortex–hippocampus is compromised on both sides, anterograde and/or retrograde amnesia may follow.

Clinical Panel 35.7   Subarachnoid hemorrhage

Blister-like 'berry' aneurysms 5–10 mm in diameter are a routine autopsy finding in about 5% of people. Most are in the anterior half of the circle of Willis. Spontaneous rupture of an aneurysm into the interpeduncular cistern usually occurs in early or late middle age. The characteristic presentation is a sudden blinding headache, with collapse into semiconsciousness or coma within a few seconds. On physical examination, a diagnostic feature (absent in one-third of cases) is nuchal (neck) rigidity. This is caused by movement of blood into the posterior cranial fossa, where the dura mater is supplied by cervical nerves 2 and 3 (Ch. 4). The term *meningismus* is sometimes used for this sign.

The massive rise in intracranial pressure may be fatal within a few hours or days. Recovery may be impeded by a secondary elevation of intracranial pressure caused by blood clot obstruction of cerebrospinal fluid circulation through the tentorial notch, or even within the arachnoid granulations.

About a quarter of all patients develop a neurologic deficit 4–12 days after the initial attack. The deficit is fatal in a quarter of those who get it. The immediate cause is spasm of the main, conducting segments of the cerebral arteries. The amount of spasm is proportionate to the size of the surrounding blood clot in the interpeduncular cistern.

It is usual practice to define the aneurysm by means of carotid angiography, and to ligate it surgically. Without operation, most aneurysms will leak again at some future date.

Clinical Panel 35.8   Motor recovery after stroke

Very early recovery—up to 24 h

Within hours of a stroke associated with severe hemiplegia, caused by embolic occlusion of a major artery of supply to the corona radiata or internal capsule, some patients show remarkable recovery of motor function, to a level where only a moderate weakness of an arm or a leg may persist. One or both of two explanations are possible.

1 The embolus has undergone fragmentation, freeing up some or all of the primary branches of the artery.

2 Collapse of the blood pressure within the territory of the blocked artery has permitted retrograde filling of the peripheral branches through the small-artery anastomoses along the border zone illustrated in Figure 35.1.

Early recovery—the first few days

A more limited improvement, during a period of a week or more, is attributable to resolution of the surrounding edema, permitting resumption of oxygen and glucose supply to viable neurons.

Later recovery

During the ensuing months, slow but progressive recovery of motor function is the rule, especially with the assistance of remedial exercises supervised by a physical therapist. Because the majority of strokes result from damage to white matter rather than cortex, attention has been directed bilaterally to all areas of the cortex known to influence corticospinal output. (The parietal lobe contribution is not considered relevant here, being concerned only with sensory modulation.)

A consistent feature on functional magnetic resonance imaging (fMRI) scans is a widespread hyperexcitability of cortical areas connected to the lesion site. The hyperexcitability, associated with reduced local activity of inhibitory, smooth stellate (GABA) neurons, appears within days and gradually diminishes over a period of up to a year or even more.

Reorganization within the affected M1

• Cell columns adjacent to those inactivated by the infarct, now liberated from lateral (surround) inhibition, become especially active. Previously silent hand-specific columns within the arm and shoulder representations are likely to become active. Existence of outlying hand-specific columns would be analogous to the cortical representation of the tongue, which in the homunculus is shown entirely below that for the face, although outlying tongue-specific columns extend halfway up the motor cortex.

• *Change of allegiance*. In monkeys, a significant contribution to recovery from paralysis (e.g. of the hand) produced by excising a patch of motor cortex arises from neighboring (e.g. arm-related) cortical cell columns activating hand rather than arm motor neurons in the cord. This phenomenon is easily explained by the extensive overlap of cortical motor territories in the spinal cord, the focusing factor normally being the recurrent (Renshaw cell) inhibitory shield surrounding the zone of maximal activation. Suspended activation of spinal cell columns includes loss of surround inhibition, thereby rendering the silent motor neurons accessible to excitation by collateral branches of nearby corticospinal fibers.

Contributions originating outside the affected M1

• Active secondary motor areas contributing to the contralesional (left) corticospinal tract include premotor cortex, supplementary motor area and anterior cingulate cortex, and arm–shoulder area of the left M1 (Figure CP 35.8.1). All three are active during the recovery period. Functional MRI monitoring indicates that, in those patients with greatest damage to the corticospinal tract, there is greatest reliance on secondary motor areas to generate some motor output. Their recruitment is often bilateral, presumably because of bilateral hand representations.

Opinions differ concerning the contribution of the *hand* area of the left M1 to motor recovery, although some contribution would be expected in view of its 10% contribution to the left lateral corticospinal tract.

• The cerebellum and motor thalamus (ventrolateral nucleus) are also active bilaterally throughout, and share the progressive reduction of activity during later stages. Cerebellar activity during *motor learning* was touched on in Chapter 23, whereby an 'efference copy' of pyramidal tract activity is sent to the cerebellar cortex via red nucleus and inferior olivary nucleus, representing intended movements. Sensory feedback during movement enables the cerebellum to detect any discrepancy between intention and execution, leading to adjustment of the cerebellar discharge via thalamus to motor cortex. As accuracy improves, cerebellar corrective activity declines.

Sensory system contributions

Visual and tactile areas of the cortex show above-normal activation during the recovery period; so too does the dorsolateral prefrontal cortex. These activities suggest a heightened level of sensory attention, with the objective of optimizing task performance.

Collectively, functional MRI observations reveal the recruitment of alternative pathways capable of activating the anterior horn cells that have been functionally deprived by the stroke.

Clinical Panel 35.8 *Continued*

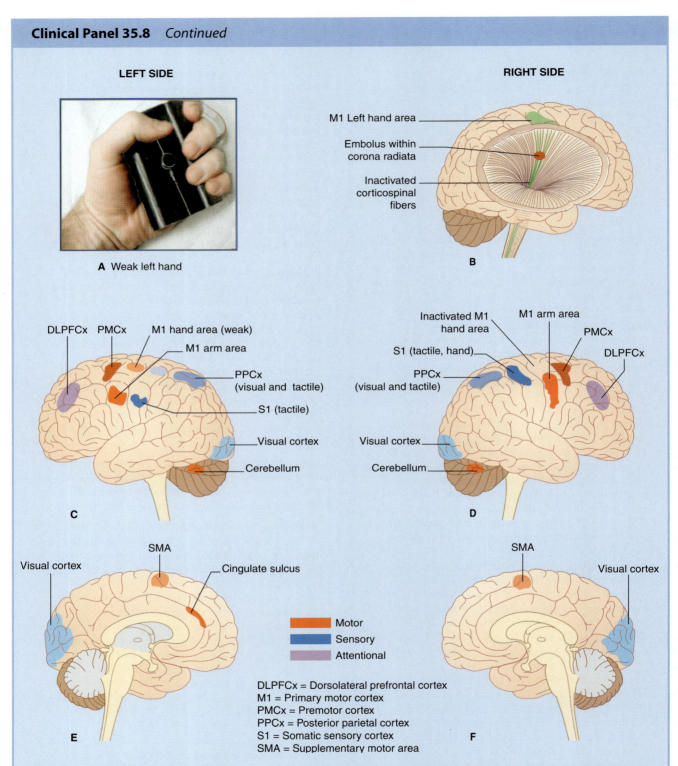

LEFT SIDE

RIGHT SIDE

A Weak left hand

M1 Left hand area

Embolus within
corona radiata

Inactivated
corticospinal
fibers

B

DLPFCx PMCx M1 hand area (weak)

M1 arm area

PPCx
(visual and tactile)

S1 (tactile)

Visual cortex

Cerebellum

C

Inactivated M1
hand area

M1 arm area

PMCx

S1 (tactile, hand)

DLPFCx

PPCx
(visual and tactile)

Visual cortex

Cerebellum

D

Visual cortex

SMA

Cingulate sulcus

E

SMA

Visual cortex

F

Motor

Sensory

Attentional

DLPFCx = Dorsolateral prefrontal cortex
M1 = Primary motor cortex
PMCx = Premotor cortex
PPCx = Posterior parietal cortex
S1 = Somatic sensory cortex
SMA = Supplementary motor area

Figure CP 35.8.1 Areas of cortical activity revealed by functional magnetic resonance imaging during recovery from stroke caused by an embolus within the white matter containing the right corticospinal tract. **(A)** The 'manipulandum' used by Ward et al. to register the squeezing power of the affected hand. **(B)** Depiction of embolic compromise of the right corticospinal tract. **(C)** Lateral view of areas of increased cortical activity in the left (contralesional) cortex of cerebrum and cerebellum. **(D)** Corresponding view of the right (lesional) side. **(E)** Medial view of the left side. **(F)** Medial view of the right side. (The assistance of Dr. Nick Ward, Institute of Neurology, University of London, UK, is gratefully appreciated.)

Clinical Panel 35.8   *Continued*

REFERENCES

Arboix A, García-Plata C, García-Eroles L, et al. Clinical study of 99 patients with pure sensory stroke. J Neurol 2005; 252:156–162.

Brozici M, van der Zwan A, Hillen B. Anatomy and functionality of leptomeningeal anastomoses: a review. Stroke 2003; 34:2750–2758.

Bütefisch CN, Netz J, Wessling M, et al. Remote changes in cortical excitability after stroke. Brain 2003; 126:470–481.

Schaechter JD. Motor rehabilitation and brain plasticity after hemiparetic stroke. Prog Neurobiol 2004; 73:61–72.

Ward NS, Brown MM, Thompson AJ, et al. Neural correlates of motor recovery after stroke. Brain 2003; 126:2476–2496.

Ward NS, Cohen LG. Mechanisms underlying recovery of motor function after stroke. Arch Neurol 2004; 61:1844–1848.

The assistance of Dr. Nick Ward, Honorary Consultant Neurologist, National Hospital for Neurology and Neurosurgery, Queen Square, London, UK, is gratefully acknowledged.

Core Information

Etiology of cerebrovascular accidents

The three chief underlying disorders are atherosclerosis of the internal carotid artery or vertebrobasilar system, thrombi issuing from the left side of the heart, and hypertension. Hypertension may lead either to sudden hemorrhage into the white matter or to production of small lacunae there. Hemorrhage into a tumor may mimic the effects of a vascular stroke. Some 10% of vascular strokes are caused by rupture of a 'berry' aneurysm.

Arterial supply of internal capsule

The anterior choroidal artery supplies the posterior limb and the retrolentiform portion. The medial striate artery supplies the anterior limb and genu. Lateral striate branches supply anterior limb, genu, and posterior limb.

Transient ischemic attacks

Transient ischemic attacks (TIAs) are episodes of vascular insufficiency causing temporary loss of brain function with complete recovery within 24 h. Anterior circulation TIAs may cause motor and/or sensory deficit and/or dysphasia, sometimes monocular blindness. Posterior circulation TIAs may cause vertigo, diplopia, ataxia, or amnesia.

Arterial occlusion within the anterior circulation

Anterior choroidal artery syndrome results from occlusion of the anterior choroidal artery. The complete syndrome comprises contralateral hemiparesis with upper limb ataxia ('ataxic hemiparesis'), hemihypesthesia, and hemianopia.

Clinical effects of anterior, middle, and posterior cerebral artery branch occlusion are summarized in Tables 35.1–35.4.

The effects of middle cerebral artery occlusion are shown in Table 35.2.

Small lacunar infarcts are commonly associated with chronic hypertension. Typical examples are in Table 35.3.

Cerebral hemorrhage most often spreads from the putamen into the internal capsule. Contralateral severe, flaccid hemiplegia results. Sufficient recovery may eventually permit stick-supported spastic ambulation.

Clinical effects of vertebrobasilar arterial occlusion have been summarized in the main text.

Aneurysms

Subarachnoid hemorrhage follows spontaneous rupture of a berry aneurysm at the base of the brain. A typical sequence of clinical effects, in those who survive, is sudden, blinding headache followed by collapse into unconsciousness and development of neck rigidity. About a quarter of patients develop a neurologic deficit within 2 weeks.

Table 35.1 Clinical effects of anterior cerebral artery branch occlusion

Branch	Clinical effects
Orbital-frontopolar	Apathy with some memory loss
Medial striate	Paresis of face and arm
Callosomarginal	Paresis and hypesthesia of face and arm ± abulia ± mutism ± inability to reach across
Pericallosal	Ideomotor apraxia (anterior lesion), tactile anomia (posterior lesion)

Table 35.3 Clinical effects of three common lacunar infarcts

Location	Clinical effect(s)
Genu of internal capsule	Dysarthria-clumsy hand syndrome ± dysphagia
Posterior limb of internal capsule	Pure motor hemiparesis
Ventral posterior nucleus of thalamus	Pure sensory syndrome ± sensory ataxia

Table 35.2 Clinical effects of middle cerebral artery occlusion

Segment	Clinical effect(s)
Left temporal	Wernicke aphasia
Either stem	Hemiplegia, hemihypesthesia, hemianopia
Left stem	Same + global aphasia
Right stem	Same + sensory neglect
Either upper division	Paresis and hypesthesia of face and arm, dysarthria
Left upper division	Same + Broca aphasia
Right upper division	Same + hemineglect or expressive aprosodia
Either lower division	Hemianopia ± agitated state
Left lower division	Same ± Wernicke aphasia, alexia, ideomotor apraxia
Branches	
Orbitofrontal	Prefrontal syndrome
Left precentral	Broca aphasia
Right precentral	Motor aprosodia
Central	Loss of motor ± sensory function in face and arm
Inferior parietal	Hemineglect
Either angular	Hemianopia
Left angular	Alexia
Right temporal	Receptive aprosodia

Table 35.4 Clinical effects of posterior cerebral artery occlusion

Stem	Clinical effects
Either	Homonymous hemianopia
Left	Alexia in visible field
Both	Cortical blindness ± amnesia
Branch	
Midbrain	Ipsilateral third nerve palsy + contralateral hemiplegia
Thalamus	Contralateral numbness ± hemianopia ± thalamic syndrome
Subthalamic nucleus	Contralateral ballism
Corpus callosum	Alexia in contralateral visual field

REFERENCES

Adams RD, Victor M. Principles of neurology. 4th edn. New York: McGraw-Hill; 1993.

Amarenco P, Caplan LR, Pessin MS. Vertebrobasilar occlusive disease. In: Barnett HJM, Mohr JP, Stein BM, et al, eds. Stroke: pathophysiology, diagnosis and management. 3rd edn. New York: Churchill Livingstone; 1998:513–598.

Brust JCM. Anterior cerebral artery disease. In: Barnett HJM, Mohr JP, Stein BM, et al, eds. Stroke: pathophysiology, diagnosis and management. 3rd edn. New York: Churchill Livingstone; 1998:401–426.

Lee JS, Han M-K, Kim SH, et al. Fiber tracking by diffusion tensor imaging in corticospinal tract stroke: topographical correlation with clinical symptoms. Neuroimage 2005; 26:771–776.

Mohr JP, Lazar RM, Marshall RS, et al. Middle cerebral artery disease. In: Barnett HJM, Mohr JP, Stein BM, et al, eds. Stroke: pathophysiology, diagnosis and management. 3rd edn. New York: Churchill Livingstone; 1998:427–480.

Mohr JP, Pessin MS. Middle cerebral artery disease. In: Barnett HJM, Mohr JP, Stein BM, et al, eds. Stroke: pathophysiology, diagnosis and management. 3rd edn. New York: Churchill Livingstone; 1998:481–502.

Sato A, et al. Cerebral infarction: early detection by means of contrast-enhanced cerebral arteries at MR imaging. Radiology 1991; 178:433–439

Savitz I, Caplan LR. Vertebrobasilar disease. New Engl J Med 2005; 352:2618–2626.

Wardlaw JM. What causes lacunar stroke? J Neurol Neurosurg Psychiatry 2005; 76:617–619.

Glossary

ABBREVIATIONS: Chapter containing main reference. *Fr.* Signifies French origin; *Gr.* signifies Greek origin; *L.* signifies Latin origin.

Abducens *L.* 'leading away'. Abducens nerve stimulates lateral rectus muscle to abduct the direction of gaze (Ch. 23).

Absence seizures Childhood seizures associated with frequent brief spells of unconsciousness ('absence') (Ch. 30).

Absolute refractory period Time interval following generation of an action potential, during which the membrane is refractory to a second stimulus (Ch. 7).

Abulia *Gr.* 'lack of will'. Loss of willpower associated with prefrontal cortical disorders (Ch. 32).

Accommodation Focusing light by allowing the lens to become more convex (Ch. 23).

Action potential A brief fluctuation in membrane potential caused by rapid opening and closure of voltage-gated ion channels (Ch. 7).

Action tremor See *Intention tremor.*

Active zone Site of release of neurotransmitter through the presynaptic membrane (Ch. 6).

Adaptation Attenuation of response to a sustained sensory stimulus (Ch. 11).

Adrenaline Synonym for epinephrine (Ch. 13).

Adrenoceptor Sympathetic junctional receptor (Ch. 13).

Affective disorder A disorder of mood, e.g. major depression (Ch. 26).

Afferent *L.* 'carrying toward'. Strictly, applies to nerve impulses traveling toward CNS along sensory fibers; is loosely applied within CNS, e.g. afferent connections of the cerebellum (Ch. 2). See also *Centripetal.*

Agnosia *Gr.* 'without knowledge'. Inability to interpret sensory information (Ch. 29).

Agraphia *Gr.* 'without writing'. Inability to express oneself in writing, owing to a central lesion (Ch. 32).

Akinesia *Gr.* 'without movement'. Refers to immobility often seen in Parkinson's disease (Ch. 33).

Alexia *Gr.* 'without reading'. Inability to read (Ch. 32).

Allocortex *Gr.* 'other cortex'. Phylogenically old, three-layered cortex in the temporal lobe (Ch. 29).

Allodynia Pain produced by normally innocuous stimulation (Ch. 34).

Alpha (α) motor neuron The motor neuron that innervates extrafusal fibers of skeletal muscle (Ch. 10).

Alveus *Gr.* 'trough'. Refers to the thin layer of white matter on the surface of the hippocampus (Ch. 29).

Alzheimer's disease A form of dementia (Ch. 34).

Amino acid transmitters Glutamate, γ-aminobutyric acid, glycine (Ch. 8).

Amnesia Loss of memory (Ch. 34).

Amygdala *Gr.* 'almond'. Nucleus at the tip of the inferior horn of the lateral ventricle (Ch. 2).

Amyotrophic lateral sclerosis See *motor neuron disease* (Ch. 16).

Analgesia *Gr.* 'without pain'. Absence of perception of a noxious stimulus (Ch. 15).

Aneurysm *Gr.* 'widening'. Localized dilation of an artery (Ch. 35).

Angiogram Image of blood vessels obtained by intraarterial injection of radiopaque fluid (Ch. 5).

Anomia *Gr.* 'without names'. Inability to name common objects (Ch. 32).

Anopsia *Gr.* 'without vision' (Ch. 28).

Anterior circulation The territory of the internal carotid artery and its branches (Chs 5, 35).

Anterograde amnesia Inability to lay down new memories (Ch. 34).

Anterograde transport Axonal transport from soma to nerve terminals (Ch. 6).

Antidromic *Gr.* 'running against'. Usually refers to nerve impulses that, traveling proximally along one branch of a Y-shaped sensory nerve fiber, arrive at the junction and travel distally along the other branch (Ch. 11).

Aphasia *Gr.* 'without speech', e.g. motor aphasia, sensory aphasia (Ch. 32).

Apperceptive tactile agnosia Right parietal lobe form of tactile agnosia (Ch. 32).

Apraxia *Gr.* 'without movement'. Inability to carry out voluntary movements in the absence of paralysis (Ch. 32).

Aprosodia Disturbance of the melodic interpretation of speech (Ch. 32).

Arachnoid *Gr.* 'spider-like'. Refers to the web-like delicacy of the arachnoid mater (Ch. 4).

Archi- (arche-) *Gr.* 'beginning'. Refers to oldest areas, e.g. archicerebellum (Ch. 25).

Area postrema *L.* 'back end area'. Refers to the posterior tip of the fourth ventricle (Ch. 17).

Arnold–Chiari malformation Maldevelopment of the posterior cranial fossa (Ch. 14).

Ascending reticular activating system (ARAS) A polysynaptic chain of reticular formation neurons involved in maintaining the conscious state (Ch. 24).

Assistance reflex During voluntary movement, positive feedback from actively stretched neuromuscular spindles assists the movement (Ch. 16). Cf. *Resistance reflex.*

Association cortex Area of cortex receiving afferents from one or more primary sensory areas (Ch. 29).

Associative agnosia Left parietal lobe style of tactile agnosia (Ch. 32).

Astereognosis *Gr.* 'without knowledge of solid'. Refers to inability to identify common objects by touch alone (Ch. 29).

Astrocyte The 'star-like' neuroglial cell (Ch. 5).

Asymbolia for pain Abolition of the aversive quality of painful stimuli while preserving the location and intensity aspects (Ch. 34).

Ataxia *Gr.* 'without order'. Describes the uncoordinated movements associated with posterior column (Ch. 15) or cerebellar (Ch. 25) disease.

Atherosclerosis Arterial degenerative disorder associated with fatty subintimal plaques capable of detachment with consequent embolism within the arterial territory (Ch. 35).

Athetosis *Gr.* 'without stability'. Describes the continuous writhing movements sometimes associated with damage to the basal ganglia (Ch. 34).

Autogenetic inhibition Negative feedback from Golgi tendon organs causing a muscle to relax (Ch. 10).

Autonomic Self-regulating (Ch. 13).

Autoreceptor A presynaptic receptor acted on by the transmitter released at the same nerve ending (Chs 8, 13).

Autoregulation The capacity of a tissue to regulate its own blood supply (Ch. 5).

Axolemma *Gr.* 'husk' covering the central part or 'axis' of a nerve fiber (Ch. 6).

Axon reflex Activation of the reflex arc producing the triple response in the skin (Ch. 11).

Axoplasm *Gr.* 'substance' (i.e. cytoplasm) of the axon (Ch. 6).

Axoplasmic transport Orthograde or retrograde transport of materials within an axon (Ch. 6).

Babinski sign Reflex fanning of the toes with extension of the great toe, following a scraping stimulus to the lateral part of the sole; sign of corticospinal tract disorder (Ch. 16).

Ballism *Gr.* 'throwing', with reference to the flailing movements that follow damage to the subthalamic nucleus (Ch. 33).

Baroreceptor *Gr.* 'weight' receptor. Refers to the blood pressure receptors of the carotid sinus and aortic arch (Ch. 24).

Baroreceptor reflex (Baroreflex) Reflex increase of sympathetic vascular tone in response to a fall in intracranial blood pressure, e.g. on assuming the upright position (Ch. 24).

Barosympathetic reflex Reflex reduction of sympathetic tone in response to a rise of arterial blood pressure (Ch. 24).

Barovagal reflex Reflex reduction of heart rate in response to a rise of arterial blood pressure (Ch. 24).

Bell's palsy Peripheral facial nerve paralysis caused by swelling of the nerve followed by its compression against the wall of the bony facial nerve canal (Ch. 22).

Benign essential tremor A tremulous disorder commonly erroneously attributed to Parkinson's disease (Ch. 33).

Benign rolandic epilepsy Childhood disorder having an ictal focus of origin in front of or behind the fissure of Rolando (central sulcus) (Ch. 30).

Berry aneurysm Blister-like aneurysm on or near the circle of Willis (Ch. 35).

Binocular visual field The visual field common to both eyes (Ch. 28).

Bipolar recording EEG montage using varying pairs of electrodes (Ch. 30). Cf. *Referential recording*.

Bitemporal hemianopia Lateral visual field defects resulting from pressure on crossover optic nerve fibers in the midregion of the optic chiasm, usually by a pituitary adenoma (Ch. 28).

Blind sight Patients with cortical blindness may perceive movement in the peripheral field without being able to see anything there (Ch. 29).

Blinking-to-light reflex Reflex blinking in response to a flash of bright light (Ch. 22).

Blinking-to-noise reflex Acousticofacial reflex causing the orbicularis oculi to twitch in response to a loud sound (Ch. 22).

Blood trauma phobia Fainting at the sight of blood at the scene of an accident (Ch. 34).

Body schema Consciousness of the relative position of body parts (Ch. 32).

Border zone Crescentic arteriolar anastomotic zone on the surface of the cerebral cortex (Ch. 35).

Bradykinesia *Gr.* 'slow movement' characteristic of Parkinson's disease (Ch. 33).

Brain attack Stroke (Ch. 35).

Brain Intracranial CNS.

Brain-derived neurotrophic factor (BDNF) A 'nourishing' factor that promotes survival and normal functioning of cortical neurons (Ch. 6).

Brainstem auditory evoked potentials Potentials evoked in the temporal cortex following click excitation of the cochlea (Ch. 31).

Brainstem Comprises midbrain, pons and medulla oblongata (Ch. 3). In the embryo, also includes the diencephalon (Ch. 1).

Broca aphasia Aphasia caused by damage to the motor speech area (Ch. 32).

Broca's area Pars triangularis (area 44) and pars anterior (area 45) of the frontal operculum, involved in generating speech (Ch. 32).

Brown–Sequard syndrome The constellation of signs that follows hemisection of the spinal cord (Ch. 16).

Bulbar *L.* 'bulb' of the brain. A discredited term usually meaning medulla oblongata.

Calcar avis *L.* 'spur of a bird'. Refers to the elevation produced by the calcarine sulcus in the medial wall of the atrium of the lateral ventricle (Ch. 2).

Cataplexy *Gr.* 'struck down'. Term used to signify sudden, brief episodes of muscle paralysis triggered by emotion, notably by surprise of any kind (Ch. 30).

Catecholamines The neurotransmitters dopamine, norepinephrine, and epinephrine, comprising amines attached to catechol rings (Ch. 8).

Cauda equina *L.* 'horse's tail'. Refers to the leash of spinal nerve roots below the level of the spinal cord (Ch. 14).

Caudal anesthesia Pelvic–perineal anesthesia produced by injection of local anesthetic through the sacral hiatus into the epidural space (Ch. 14).

Caudate *L.* having a tail (caudate nucleus, Ch. 2).

Center-surround receptive field A visual receptive field having a center surrounded by a ring of opposite sign (Ch. 28).

Central motor conduction time Measurement of corticospinal tract conduction time using a combination of transcranial magnetic stimulation and surface electromyography.

Central pain state A state of chronic pain, resistant to therapy, sustained by hypersensitivity of peripheral and/or central neural pathways (Ch. 34).

Central pattern generator A neural circuit giving rise to rhythmic motor activity.

Centrifugal *L.* 'fleeing the center'. See *Efferent*.

Centripetal *L.* 'seeking the center'. See *Afferent*.

Cerebellar ataxia Ataxia of cerebellar origin (Ch. 25).

Cerebellar cognitive affective syndrome Constellation of symptoms associated with cerebellar pathology (Ch. 25).

Cerebellar signs Ataxia or intention tremor (in particular) associated with cerebellar pathology (Ch. 25).

Cerebellum *L.* 'little brain'.

Cerebrovascular accident Thrombosis, embolism or hemorrhage in or around the brain (Ch. 35).

Cerebrum *L.* 'brain', comprising cerebral hemispheres and diencephalon (Ch. 2).

Cervical spondylosis A form of vertebral arthritis accompanied by bony outgrowths around the margins of cervical facet joints, resulting in compression of cervical nerve roots (Ch. 12, 14).

Cervicogenic headache Headache caused by pressure on cervical nerves (Ch. 14).

Change of allegiance Plastic response to motor cortical injury, whereby neighboring neurons are recruited to serve a lost motor function (Ch. 35).

Chemical synapse Distinguished from electrical synapses by release of neurotransmitter (Ch. 8).

Chemoreceptor A sensory receptor (e.g. carotid body) selective for a chemical substance (Ch. 24).

Chiasma *Gr.* 'crossing'. Refers mainly to the optic chiasma (chiasm) (Ch. 2).

Chorea *Gr.* 'dance'. Involuntary movements representing fragments of motor programs (Ch. 33).

Choroid plexus *Gr.* 'membranous network' of capillaries invested with choroidal epithelium, within the ventricles of the brain (Ch. 2).

Chromatolysis *Gr.* 'dissolution of color' in the perikaryon, following axotomy (Ch. 6).

Cingulotomy Section of anterior cingulate cortex, usually for relief of pain (Ch. 34).

Cingulum *L.* the 'girdle' of white matter within the cingulate gyrus (Ch. 2).

Clasp knife rigidity Initial resistance to passive movement followed by collapse of resistance; a sign of upper motor neuron disease (Ch. 16).

Claustrum *L.* the 'barrier' of gray matter between insula and lentiform nucleus (Ch. 2).

Clonus *L.* 'turmoil'. Rapid beating movement at ankle or wrist produced by sudden passive extension; associated with upper motor neuron disease (Ch. 16).

Coactivation (a) Simultaneous activation of α and γ motor neurons (Ch. 10).

Cocontraction Simultaneous contraction of prime movers and antagonists (Ch. 16).

Cognition Information-processing associated with thinking (Ch. 32).

Cognitive style Subtle differences in left vs right cerebral function (Ch. 32).

Cogwheel rigidity Ratchety response of muscles to passive movement of joints; associated with Parkinson's disease (Ch. 33).

Colliculus *L.* 'little hill'. Four colliculi comprise the tectum of the midbrain (Ch. 3).

Column A term used interchangeably for the posterior funiculus of the spinal cord (Ch. 15).

Combined processing Simultaneous engagement of more than one sensory modality in a particular sensory task (Ch. 27).

Commissure *L.* 'link' between the two sides of the nervous system, e.g. white commissure of the spinal cord (Ch. 3), anterior commissure of the brain, corpus callosum (Ch. 2).

Compensation reflex Vestibuloocular reflex compensating for movement of the head; associated with fixation (foveation) (Ch. 23).

Complex focal (partial) seizure See *Temporal lobe epilepsy* (Ch. 34).

Complex spikes Complex response of Purkinje cells to stimulation of olivocerebellar fibers (Ch. 25).

Compound MUAPs Compound motor unit action potentials (Ch. 12).

Concussion Brief cortical dysfunction resulting from a blow to the head (Ch. 4).

Conduction aphasia Aphasia caused by damage to the arcuate fasciculus (Ch. 32).

Conductive deafness Deafness caused by disease in the outer ear canal or middle ear (Ch. 20).

Conjugate movement Movement in parallel, e.g. ocular saccades (Ch. 23).

Conscious proprioception Perceived sensations arising within the body, notably from muscle spindles (Ch. 15).

Consolidation The process of storing information in long-term memory (Ch. 34).

Continuous conduction Mode of impulse conduction along unmyelinated nerve fibers (Ch. 6).

Contralateral *L.* Refers to opposite side of the body (cf. *Ipsilateral*, 'same side').

Convolution *L.* a gyrus (Ch. 2).

Cordotomy Incision of the spinal cord for pain relief (Ch. 15).

Corneal reflex Blinking in response to corneal contact (Ch. 22).

Corona radiata *L.* 'radiating crown' of white matter extending from cerebral cortex to internal capsule and vice versa (Ch. 2).

Corpus callosum *L.* 'hard body'. Refers to the great transverse commissure of white matter interconnecting like areas of the cerebral hemispheres (Ch. 2).

Corpus striatum *L.* 'striated body' comprising caudate and lentiform nuclei (Ch. 2).

Cortex *L.* the 'bark' of gray matter at the surface of the cerebrum and cerebellum (Ch. 2).

Cortical blindness Blindness owing to damage to the primary visual cortex (Ch. 29).

Cortical mosaic Mosaic arrangement created by interdigitation of cortical modules of different kinds (Ch. 29).

Cortical-type sensory loss Diminution of tactile perception associated with damage to the primary sensory cortex (Ch. 29).

Crossed hemiplegia Follows unilateral brainstem lesion affecting one or more motor cranial nerve nuclei (resulting in ipsilateral paralysis) together with corticospinal fibers (resulting in contralateral hemiplegia or hemiparesis) (Ch. 35).

Crus *L.* 'leg', e.g. crus of fornix (Ch. 2), crus of midbrain (Ch. 2).

Cuneate *L.* 'wedge-like', e.g. cuneate fasciculus (Ch. 3).

Cuneus *L.* 'wedge', e.g. the gyrus of that shape in the occipital lobe (Ch. 2).

Declarative memory Memory for facts and events (Ch. 35).

Decussation *L.* from Roman numeral X. Refers to X-shaped crossing of nerve bundles at junctional regions, e.g. pyramidal decussation (brain-spinal cord, Ch. 3), decussation of superior cerebellar peduncles (pons–midbrain, Ch. 3).

Déjà vu *Fr.* Epileptic aura where a novel visual field seems familiar (Ch. 34).

Delta waves EEG waveforms characteristic of slow-wave sleep (Ch. 30).

Dementia Loss of cognitive abilities in the presence of intact motor and sensory systems (Ch. 34).

Dendrite(s) *Gr.* 'tree(s)'. Refers to the neuronal processes receiving the axons of other neurons (Ch. 6).

Dendritic sheaves: Within the thalamic reticular nucleus, bundles of dendrites belonging to different neurons, linked to one another by dendrodendritic synapses.

Dentate *L.* 'toothed', e.g. dentate nucleus of cerebellum (Ch. 2), dentate gyrus in the temporal lobe (Ch. 34).

Denticulate *L.* 'little-toothed', e.g. denticulate ligament of pia mater anchoring the spinal cord (Ch. 4).

Depolarization block Conduction block produced by high frequency stimulation (Ch. 34).

Depolarize Make the membrane potential of the neuron less negative (Ch. 6).

Dermatome The strip of skin supplied by an individual spinal nerve (Ch. 11).

Detrusor instability 'Unstable bladder' characterized by spontaneous expulsion of urine despite conscious attempts at restraint (Ch. 13).

Developmental dyslexia Pronounced reading difficulty in children who are the match of their peers in other respects (Ch. 32).

Diabetes insipidus Hypothalamic disorder characterized by polyuria and polydypsia (Ch. 26).

Diencephalic amnesia Amnesia associated with shrinkage of the mammillary bodies (Ch. 34).

Diencephalon *Gr.* 'between-brain', comprising epithalamus (Ch. 25), thalamus (Ch. 25), and hypothalamus (Ch. 24).

Diffuse noxious inhibitory controls Appropriate neural connections whereby painful stimulation of one part of the body may produce pain relief in all other parts (Ch. 24).

Diplopia *Gr.* 'double vision' (Ch. 23).

Discriminative touch Fine touch sensibility (Chs 15, 29).

Disinhibition Release of excitatory neurons by inhibition of inhibitory neurons (Chs 6, 33).

Dissociated sensory loss Loss of one sensory modality with preservation of others (Ch. 15).

Dopa Dihydroxyphenylalanine, a precursor of dopamine, norepinephrine, and epinephrine (Ch. 8).

Dopamine Catecholamine neurotransmitter synthesized from dopa (Ch. 8).

Dura mater *L.* 'hard cover'. Outermost meninx (Ch. 4).

Dynamic (kinetic) labyrinth The semicircular canals (Ch. 19).

Dynamic posturography Testing postural reflex responses to sudden tilting (Ch. 25).

Dys- *Gr.* 'difficult'.

Dysarthria *Gr.* 'difficult articulation' (Ch. 35).

Dysarthria–clumsy hand syndrome Syndrome associated with a vascular lesion of the genu of the internal capsule (Ch. 35).

Dysdiadochokinesia *Gr.* 'difficult successive movements'. Refers to difficulty in performing rapid pronation–supination sequences (Ch. 25).

Dyskinetic cerebral palsy Movement problems associated with damage to the basal ganglia, associated with cerebral palsy (Ch. 33).

Dyslexia *Gr.* 'difficult reading'. Reading difficulty (Ch. 32).

Dysmetria *Gr.* 'difficult measurement'. Refers to reduced motor control in cerebellar disease (Ch. 25).

Dysphagia *Gr.* 'difficult eating or swallowing', e.g. following paralysis of pharyngeal constrictors (Ch. 18).

Dysphasia Mild or moderate expressive aphasia (Ch. 32).

Dysphonia Hoarseness of speech (Ch. 18).

ECT (electroconvulsive therapy) A treatment for depression (Ch. 26).

Ectoderm *Gr.* 'outer skin'. Refers to the outer germ layer giving rise to the nervous system and to the epidermis of the skin (Ch. 1).

EEG (electroencephalography) Taking a record of brain waves (Ch. 30).

Efferent *L.* 'carrying away'. Strictly, applies to nerve impulses traveling away from the CNS; also used regionally, e.g. cerebellar efferents. See also *Centrifugal*.

Ejaculation *L.* 'throwing'. Expelling semen into the vagina (Ch. 13).

Electrical potential Voltage.

Electrical synapse Synapse where current can flow from one neuron to another via gap junctions (Chs 6, 7).

Electrodiagnostic examination The combination of nerve conduction studies and electromyography (Ch. 12).

Electromyogram The recorded tracings of muscle fiber waveforms (Ch. 12).

Electromyography (EMG) Recording the waveforms generated by selected muscles during voluntary contraction.

Electrotonic potentials Decremental positive or negative membrane voltage changes in target neurons induced by transmitter activation (Ch. 7).

Electrotonus A decremental voltage change passing over the neuronal soma following receptor activation (Ch. 7).

Emboliform *Gr.* 'plug-like', e.g. emboliform nucleus of cerebellum (Ch. 25).

Embolus *Gr.* 'plug', e.g. cerebral embolus formed by a blood clot breaking away from the internal carotid artery (Ch. 35).

EMG Electromyography (Ch. 12).

Emission Expulsion of semen into the urethra (Ch. 13).

Endocytosis Vesicular uptake of material from the extracellular space (Ch. 6).

Endogenous opioids Brain-derived neuropeptides (Ch. 24).

Endoneurium *Gr.* 'within nerve'. Refers to the connective tissue sheath surrounding individual nerve fibers (Ch. 9).

Engram The chemical intraneuronal representation of a memory.

Enteroception *L.* 'reception from inside'. Refers to stimuli transduced within the alimentary tract (Ch. 13).

Entorhinal *Gr.* 'in nose'. Refers to entorhinal cortex of the temporal lobe (Ch. 34).

Entrapment neuropathy Peripheral neuropathy caused by nerve compression beneath ligamentous bridges by stretching at bony angulations (Ch. 12).

Ependyma *Gr.* 'upper garment'. Refers to the epithelium lining the ventricular system of brain (Ch. 2) and the central canal of the spinal cord (Ch. 3).

Epidural anesthesia Anesthesia procured by injecting analgesic solution into the epidural space, usually in the lumbar region (Ch. 14).

Epinephrine Catecholamine hormone synthesized in adrenal medulla (Ch. 13); also a neurotransmitter synthesized in the brainstem (Ch. 24). Also known as adrenaline.

Epineurium *Gr.* 'on nerve'. Refers to loose connective tissue investment of peripheral nerves (Ch. 9).

Episodic memory Ability to recollect episodes of one's own earlier life (Ch. 34).

Epithalamus *Gr.* 'above thalamus', includes pineal gland (Ch. 27).

Evoked potentials EEG potentials evoked by motor or sensory stimulation (Ch. 31).

Excitotoxicity Toxic effects on target neurons, of excessive glutamatergic activity (Chs 8, 34).

Expressive aphasia Loss of motor speech (Ch. 32).

Extensor plantar response See *Babinski sign*.

Exteroception *L.* 'reception from outside'. Refers to stimuli transduced at the body surface, rather than within the body wall or limbs (proprioception) or alimentary tract (enteroception).

Extrapyramidal *L.* 'non-pyramidal'. Refers to pathways involving the basal ganglia (Ch. 34).

Eye-righting reflex Reflex torsion of the eyeballs to maintain horizontal gaze when the head is tilted to the side (Ch. 19).

Facilitation Increasing the likelihood of depolarization (Ch. 7).

Falx *L.* 'sickle'. Refers to shape, e.g. falx cerebri, falx cerebelli (Ch. 4).

Familial tremor Benign essential tremor (Ch. 34).

Far response Sympathetic activity causing flattening of the lens for far vision (Ch. 23).

Fascicle *L.* 'small bundle' of nerve or muscle fibers (Ch. 7).

Fasciculation Involuntary twitching of muscle fascicles; often associated with lower motor neuron disease (Ch. 16).

Fasciculation potentials On the EMG record, misshapen MUAPs that appear infrequently and are not under voluntary control (Ch. 12).

Fasciculus *L.* 'small bundle' of nerve fibers within CNS, e.g. fasciculus gracilis, fasciculus cuneatus (Ch. 15).

Fast pain Stabbing pain perceived following activation of Aδ nociceptors (Ch. 34).

Fast receptor See *Ionotropic receptor* (Ch. 8).

Fastigial *L.* 'apex of a roof', e.g. fastigial nucleus in roof of fourth ventricle (Ch. 25).

Feature extraction Separate simultaneous analysis of individual features of the scene by the visual association cortex (Ch. 29).

Fibrillation Minute skeletal muscle contractions resulting from denervation supersensitivity (Ch. 16).

Fibrillation potentials Abnormally small spontaneous motor unit potentials characteristic of relaxed muscles in the early stages of atrophy. They are not clinically visible and the patient is not aware of them (Ch. 12).

Filopodia Antenna-like growth cone processes anchoring growth cones to cell surface adhesion molecules on Schwann cells (Ch. 9).

Fimbria *L.* 'fringe'. Refers to the fringe of white fibers along the edge of the hippocampus (Ch. 29).

Finger-to-nose-test A test of cerebellar function (Ch. 25).

Fixation Visual targeting of an object; cf. *Foveation* (Ch. 23).

Flatness of affect Dearth of emotional expression (Ch. 32).

Focal seizure Seizure originating in a particular lobe (focus) of the brain (Ch. 30).

Foramen *L.* 'opening'.

Forceps *L.* 'pair of tongs', e.g. the forceps minor and forceps major of the corpus callosum (Ch. 2).

Fornix *L.* 'arch'. Refers to the efferent projection of the hippocampal formation (Chs 2, 34).

Fovea *L.* 'pit', e.g. fovea centralis of the retina (Ch. 23).

Foveation Visual targeting so that the center of the scene is aligned with the fovea (Ch. 23).

Functional electrical stimulation Muscle rehabilitation by electrical stimulation at the motor point (Ch. 10).

Functional MRI Functional magnetic resonance imaging (Ch. 29).

Functional sympathectomy Preganglionic sympathetic nerve section (Ch. 13).

Funiculus *L.* 'small cord' (Ch. 3).

Fusimotor Motor to intrafusal muscle fibers (Ch. 10). See *Gamma motor neuron.*

Fusion pore Pore through which transmitter substance passes from terminal bouton into synaptic cleft (Ch. 8).

G protein Cell membrane protein subunits that bind with guanine nucleotides (Ch. 8).

GABA Gamma aminobutyric acid; the main inhibitory neurotransmitter (Ch. 8).

Gag reflex Reflex contraction of the pharyngeal constrictors in response to stroking the oropharynx (Ch. 18).

Gait ataxia Staggering gait associated with posterior column disease (Ch. 15) and cerebellar disease (Ch. 25).

Gamma (γ) motor neuron Neuron with an A diameter axon supplying the intrafusal muscle fibers of a muscle spindle (Ch. 10).

Ganglion *Gr.* 'knot'. Refers (a) to spinal and peripheral autonomic ganglia (Chs 9, 13); (b) to the basal ganglia (Ch. 34).

Gasserian ganglion The trigeminal ganglion (Ch. 21).

Gating Controlling ease of passage of impulses from one set of neurons to another, e.g. from primary to secondary sensory neurons (Ch. 24).

Gene transcription The process of copying sequences from DNA on to messenger RNA (Ch. 8).

Generator potential The membrane voltage (−60 to (−70 mV) at which action potentials are generated (Ch. 7).

Geniculate *L.* 'knee-form', meaning bent.

Genu *L.* 'knee', meaning a bend.

Giant motor unit potential Abnormally large motor unit potential seen in EMG records of muscle reinnervation (Ch. 12).

Glia *Gr.* 'glue'. Refers to the supporting neuroglial cells of the CNS (Ch. 6).

Gliosis Scarring produced by astrocytes (Ch. 6).

Globose *L.* 'ball-like'. Refers to the globose nucleus of the cerebellum (Ch. 25).

Globus *L.* 'ball'. Usually refers to the globus pallidus in the cerebral white matter (Ch. 2).

Glomerulus *L.* 'little ball of yarn', e.g. synaptic glomeruli in the cerebellum (Ch. 25) and olfactory bulb (Ch. 34).

Glossopharyngeal *Gr.* 'lingual-pharyngeal', with reference to the sensory distribution of the glossopharyngeal nerve (and to its motor supply to stylopharyngeus) (Ch. 18).

Glutamate The most common excitatory neurotransmitter in the CNS (Chs 6, 8).

Glutamate toxicity Extensive loss of neurons caused by excessively high rates of discharge of pyramidal cells in any part of the cerebral cortex (Ch. 34).

Gracilis *L.* 'slender', e.g. gracile fasciculus in the spinal cord (Ch. 15).

Graded potentials See electrotonic potentials.

Grand mal *F.* old term for seizure (Ch. 30).

Grapheme A written syllable (Ch. 32).

Growth cone The conical tip of a growing axon (Ch. 7).

Guillain–Barré syndrome An acute autoimmune inflammatory peripheral neuropathy (Ch. 12).

Gustation/gustatory Having to do with taste sensation (Ch. 18).

Gyrus *L.* convolution of the cerebral cortex (Ch. 2).

H response The Hoffman reflex, a test using the tibial nerve to assess radicular nerve conduction velocity (Ch. 12).

Head-righting reflex A reflex, engaging the static labyrinth and medial vestibulospinal tract, designed to keep the head upright (Ch. 19).

Heel-to-knee-test A test of cerebellar function (Ch. 25).

Hemi- *Gr.* 'half'.

Hemianopia *Gr.* 'half-blindness', caused by a lesion of the geniculocalcarine pathway or primary visual cortex (Ch. 28).

Hemiballism *Gr.* 'half-throwing', refers to involuntary, 'throwing' movements of arm and/or leg on one side, associated with damage to the subthalamic nucleus (Ch. 33).

Hemineglect Lack of awareness of the contralateral side of the body and/or the contralateral visual space (Ch. 32).

Hemiparesis *Gr.* 'half-weakness'. Weakness of one side of the body associated with upper motor neuron disease (Chs 16, 35).

Hemiplegia *Gr.* 'half-struck'. Refers to paralysis of one half of the body following a major stroke (Ch. 16, 35).

Hering–Breuer reflex Reflex inhibition of the dorsal respiratory center by pulmonary stretch receptors (Ch. 18).

Hertz (Hz) Cycles per second.

Heteroreceptor Receptor in the membrane of a neuron that does not liberate the corresponding transmitter (Chs 8, 13).

Hippocampus *Gr.* 'sea horse'. Part of the limbic system (Chs 2, 34).

Histaminergic system Widespread histaminergic innervation of cerebral cortex by the tuberoinfundibular nucleus of hypothalamus (Ch. 26).

Homonymous *Gr.* 'matching'. (a) Matching motor neurons (Ch. 10); (b) matching parts of the binocular visual field (Ch. 28).

Horner syndrome Signs of cervical sympathetic paralysis, the complete syndrome comprising miosis, ptosis, and anhidrosis (Chs 13, 23).

Huntington disease A hereditary hyperkinetic disorder (Ch. 33).

Hydrocephalus *Gr.* 'water head', meaning increased volume of cerebrospinal fluid (Ch. 4).

Hyper- *Gr.* 'excessive'.

Hyperacusis Excessive perception of sound (Ch. 22).

Hyperalgesia Hypersensitivity to stimulation of injured tissue and of surrounding uninjured tissue (Ch. 34).

Hypercapnia Excess plasma P_{CO_2} (Ch. 5)

Hyperhidrosis *Gr.* 'too much watering'. Excessive sweating (Ch. 13).

Hyperkinetic states Disorders associated with involuntary movements (Ch. 33).

Hyperpolarize Make the membrane potential more negative (Ch. 7).

Hyperreflexia Exaggerated tendon reflexes; associated with upper motor neuron disease (Ch. 16).

Hypnogogic hallucinations *Gr.* 'accompanying sleep' In narcolepsy, sleep onset characterized by striking visual imagery (Ch. 30).

Hypo- *Gr.* below.

Ictal focus Locus of origin of seizures (Ch. 30).

Ictus Seizure (Ch. 30).

Ideomotor apraxia Failure to perform a learned movement on request (Ch. 32).

Idiopathic Pathology of unknown origin.

Impotence Inability to maintain an erection (Ch. 13).

Incontinence Involuntary voiding of urine or feces (Chs 13, 24).

Infarction *L.* 'stuffed into'. Refers to blood-stuffed necrotic tissue resulting from vascular occlusion (Ch. 35).

Infranuclear lesion Lesion of the trunk of a cranial nerve.

Infundibulum *L.* 'funnel', leading down to the hypophysis (Ch. 26).

Injury potentials, insertional injury Either term signifies potentials elicited by initial contact of an EMG needle with muscle fibers (Ch. 12).

Insula *L.* 'island' of cerebral cortex covered by the opercula (Ch. 2).

Intention tremor Tremor appearing during performance of purposive movements; associated with cerebellar disease (Ch. 25).

Internuncial *L.* 'messenger between'. Refers to small connecting neurons (Ch. 6).

Intra- *L.* 'within'.

Intrafusal muscle fibers Muscle fibers within the fusiform, spindle-shaped neuromuscular spindle.

Ionotropic receptor Receptor containing transmitter-gated ion channels (Ch. 8).

Ipsilateral *L.* 'on same side'.

Irritable bowel syndrome A disorder of the brain–gut axis (Ch. 13).

Iso- *Gr.* 'equal', e.g. isocortex, uniformly containing six layers of neurons (Ch. 29).

Isofrequency stripes Bands of primary auditory cortex responding to particular tonal frequencies (Ch. 29).

Jacksonian seizure Seizure involving sequential activation of adjacent areas of the motor cortex (Ch. 30).

Jamais vu *Fr.* 'never seen'. Epileptic aura where a familiar scene appears novel (Ch. 34).

Jargon aphasia Jumbled speech associated with a lesion of Wernicke's area (Ch. 32).

Jaw jerk Reflex elevation of the mandible in response to a tap on the chin (Ch. 21).

Joint sense The sense of direction of passive movement of a joint; affected in posterior column disease (Ch. 15).

Joint stiffness 'Active' joint stiffness is the element of resistance to movement introduced by autogenetic inhibition and designed to prevent oscillation (Ch. 10). 'Passive' joint stiffness is a progressive resistance to passive movement caused by intramuscular collagen accumulation following an upper motor neuron lesion (Ch. 16).

Jugular foramen syndrome Constellation of symptoms and signs associated with injury to the glossopharyngeal, vagus, and accessory nerves related to the jugular foramen (Ch. 18).

Kindling ('lighting a fire'). Progressively increasing group response of hippocampal neurons to a repetitive stimulus of uniform strength (Ch. 34).

Kinesthesia *Gr.* 'perception of movement' (Ch. 15).

Kinetic or dynamic labyrinth The semicircular canals (Ch. 19).

Kluver–Bucy syndrome Constellation of personality changes following anterior temporal lobectomy (Ch. 34).

Korsakoff psychosis State of anterograde amnesia associated with shrinkage of the mammillary bodies (Ch. 34).

Lacunar infarct Brain infarct in the form of a lacuna ('little lake'); associated with hypertension (Ch. 35).

Latency Stimulus–response interval (Chs 10, 12).

Lateral medullary syndrome Characteristic symptoms and signs following thrombosis of the vertebral or posterior inferior cerebellar artery (Ch. 19).

Lead pipe rigidity Uniform resistance to passive joint movement; characteristic of Parkinson disease (Ch. 33).

Lemniscus *Gr.* 'ribbon'. Used with reference to several afferent pathways (tracts) in the brainstem (Ch. 17).

Lentiform *L.* Lens-shaped nucleus, part of the corpus striatum (caudate and lentiform nuclei) (Ch. 2).

Leptomeninges *Gr.* 'thin membranes' comprising the arachnoid and pia mater (Ch. 4).

Lesion *L.* 'wound'. Refers to tissue damage of any kind.

Ligand-gated channel (*L.* 'binding'). See *Transmitter-gated channel*.

Limb apraxia See *Ideomotor apraxia*.

Limbic *L.* 'marginal'. Refers to limbic structures at the inner margin of the cerebral hemisphere (Ch. 26).

Lissauer's tract The posterolateral tract in the spinal cord (Ch. 15).

Locomotor center Pedunculopontine nucleus (Ch. 24).

Locus ceruleus *L.* 'dark blue place' in the (fresh) floor of fourth ventricle. See *Cerulean nucleus*.

Long-tract signs Signs indicative of damage to major sensory and motor pathways, especially in relation to brainstem vascular lesions (Ch. 35).

Lower motor neuron signs Signs of lower motor neuron disease (Ch. 16).

LTD Long-term (prolonged) depression of neuronal responsivity following conditioning stimuli (Chs 8, 25).

LTP Long-term potentiation of neuronal responsivity following conditioning stimuli (Chs 8, 34).

Lumbar puncture (spinal tap) Needle puncture of the lumbar cistern to obtain a sample of cerebrospinal fluid (Ch. 4).

Macula *L.* 'spot', e.g. macula of utricle (Ch. 19), macula lutea (yellow) of retina (Ch. 28).

Major depression Prevalent psychiatric disorder associated with depressed mood (Ch. 26).

Mammillary *L.* 'nipple-like'. Refers to the mammillary bodies.

Marche à petit pas *Fr.* 'walk with small steps'. Hesitant gait associated with frontal lobe disease (Ch. 32).

Mechanoreceptor Sensory receptor sensitive to mechanical stimuli, e.g. muscle spindles and tendon organs (Ch. 10), some cutaneous receptors (Ch. 11), carotid sinus receptors (Ch. 24) and inner ear hair cells (Chs 19, 20).

Medulla *L.* 'marrow'. Refers to the marrow-like appearance of the fresh brain and spinal cord within their bony shells.

Medulla oblongata *L.* 'oblong' (elongate) part of the hindbrain (Ch. 3).

Membrane potential The voltage across a cell membrane (Ch. 6).

Meningism Neck retraction characteristic of meningitis (Ch. 4) and subarachnoid hemorrhage (Ch. 35).

Meningomyelocele A form of spina bifida accompanied by a cyst protruding from the vertebral canal (Ch. 14).

Meralgia paresthetica ('Thigh-pain with pins-and-needles') Pinching of the lateral cutaneous nerve of the thigh where it pierces the inguinal ligament (Ch. 12).

Mesencephalon *Gr.* 'midbrain' (Ch. 1).

Mesocortical fibers Dopaminergic fibers projecting to the prefrontal cortex from the ventral tegmental nuclei of the midbrain (Ch. 34).

Mesoderm *Gr.* 'middle skin'. Refers to the middle germ layer (Ch. 1).

Mesolimbic fibers Dopaminergic fibers projecting to the limbic system from the ventral tegmental nuclei (Ch. 34).

Metabotropic receptor Membrane receptor capable of generating multiple metabolic effects within the cytoplasm of the neuron (Ch. 8).

Metencephalon *Gr.* 'after the brain'. Comprises embryonic pons and cerebellum (Ch. 1).

Microglia *Gr.* small glial cells (Ch. 6).

Microzone A 'beam' of excitation of Purkinje cells by granule cells (Ch. 25).

Miosis *Gr.* 'constriction' of the pupil (Ch. 23).

Mnemonic *Gr.* 'memory related'. Having to do with memory (Ch. 34).

Modality See *Sensory modality*.

Modulation Effect of neurotransmitters, notably the catecholamines and serotonin, in altering the response of neurons to classic transmitters (Ch. 6).

Module A cell column in the cerebral cortex (Ch. 29).

Monoamines Catecholamines, serotonin, histamine (Ch. 8).

Monocular blindness Blindness in one eye (Ch. 28).

Monocular crescent The C-shaped monocular visual field (Ch. 28).

Monoplegia Paralysis of one limb resulting from upper motor neuron disease (Chs 16, 35).

Monosynaptic reflex Term applied to the tendon reflex (Ch. 10).

Montage EEG printout (Ch. 30).

Motor end plate Myoneural junction (Ch. 10).

Motor evoked potentials Motor unit action potentials detected in surface EMG recordings following controlled excitation of the corticospinal tract (Ch. 31).

Motor learning Learning of a motor skill (Ch. 25).

Motor neuron disease Disease of spinal cord or brainstem motor neurons (Ch. 16).

Motor point The point of entry of a motor nerve into muscle (Ch. 10).

Motor set Posture adopted prior to movement (Ch. 29).

Motor unit action potentials (MUAPs) Individual waveforms representing activation of muscle fibers of an individual motor unit (Ch. 12).

Motor unit An α motor neuron together with the squad of muscle fibers it supplies (Ch. 10).

Movement synergy Manner of operation of the primary motor cortex (Ch. 29).

MRI Magnetic resonance imaging (Ch. 2).

MUAPs See *Motor unit action potentials*.

Multimodal association cortex See *Polymodal association cortex*.

Multiple sclerosis Demyelinating disease associated with multiple areas of sclerosis (hardening) caused by glial scarring (Ch. 6).

Multisystem atrophy A neurologic disorder frequently misdiagnosed as Parkinson's disease (Ch. 33).

Muscarinic receptor Acetylcholinergic receptor historically activated by muscarine (Chs 8, 13).

Myasthenia gravis *Gr./L.* 'serious muscle weakness'. An autoimmune disease of motor end plates (Ch. 12).

Myelencephalon *Gr.* 'marrow brain'. Refers to the embryonic medulla oblongata (Ch. 1).

Myelin *Gr.* 'marrow'. Refers to the myelin sheath of axons (Ch. 6).

Myelocele A form of spina bifida associated with an open neural tube (Ch. 14).

Myelogram Image of the spinal subarachnoid space produced by injection of a radiopaque substance into the lumbar cistern (Ch. 4).

Myoneural junction *Gr.* Motor end plate (Ch. 10).

Myotatic reflex *Gr.* 'muscle touching'. Tendon reflex (Ch. 10).

Narcolepsy *Gr.* 'sleep seizure'. Sleep disorder characterized by daytime distorted REM sleep attacks (Ch. 30).

Near response Passive thickening of the lens, miosis, and convergence, combined for close-up viewing (Ch. 23).

Needle electrode An EMG recording electrode in the lumen of a fine needle (Ch. 12).

Neglect Neglect of contralateral personal or extrapersonal space, associated with inferior parietal lobe disease (Ch. 32).

Neo- *Gr.* 'new', e.g. neocerebellum (Ch. 25), neocortex (Ch. 27).

Nervi erigentes *L.* 'erectile nerves' Parasympathetic nerve fibers responsible for penile erection (Ch. 13).

Neurite *Gr.* process of a neuron, whether axon or dendrite (Ch. 6).

Neuroblast *Gr.* 'nerve germ'. Refers to embryonic neuron (Ch. 1).

Neuroeffector junctions Autonomic nerve endings within target tissues (Ch. 13).

Neurofibril *L.* Refers to matted neurofilaments seen by light microscopy (Ch. 6).

Neurofibrillary tangles Tangles of neurofibrils, most numerous in Alzheimer's disease (Ch. 34).

Neurofilament *L.* The fine filaments seen in neurons by electron microscopy (Ch. 6).

Neurogenic inflammation Inflammation caused by liberation of substance P (in particular) following antidromic depolarization of fine peripheral nerve fibers (Chs 11, 34).

Neurohumoral reflex Reflex with neural afferent limb and hormonal efferent limb, e.g. the milk ejection reflex (Ch. 26).

Neurolemma *Gr.* 'nerve sheath', comprising chains of Schwann cells (Ch. 9).

Neuron *Gr.* 'nerve'. Refers to the complete nerve cell (Ch. 6).

Neuropathic pain Chronic stabbing or burning pain resulting from injury to peripheral nerves (Ch. 34).

Neuropathy Pathology in one or more peripheral nerves (Ch. 12).

Neurotoxicity Refers to damage inflicted on target neurons by excess glutamate (Ch. 29).

Neurotransmitter A chemical liberated at a nerve terminal that activates postsynaptic and/or presynaptic receptors (Chs 6, 8).

Nicotinic receptor Acetylcholinergic receptor historically activated by nicotine (Chs 8, 13).

Nigrostriatal pathway Dopaminergic projection from substantia nigra to the striatum; especially relevant to Parkinson's disease (Ch. 33).

Nociceptive *L.* 'taking injury'. Responsive to noxious stimulation (Ch. 11).

Nociceptors Peripheral receptors whose activation generates a sense of pain (Ch. 34).

Non-conscious proprioception Sensory signals arising within the body that are not perceived, e.g. spinocerebellar tracts (Ch. 15).

Noradrenergic Neurons using norepinephrine (noradrenaline) as transmitter, e.g. postganglionic sympathetic (Ch. 13), cerulean nucleus (Ch. 17).

Norepinephrine (Also called noradrenaline) Hormone released by the adrenal medulla (Ch. 13); also a brainstem neurotransmitter (Ch. 17).

NREM sleep *Non-rapid eye movement* sleep (Stages 1–4); also called *slow-wave sleep* (Ch. 30).

Nuclear lesion Lesion of a motor nucleus in the brainstem.

Nucleus *L.* 'nut'. Refers either to the trophic center of a cell, or to a group of neurons within the CNS (Ch. 6).

Nystagmus *Gr.* 'nodding'. Refers to involuntary oscillation of the eyes (Ch. 19).

Occupational deafness Deafness brought on by noise in the workplace (Ch. 20).

Ocular dominance columns Cell columns in the primary visual cortex activated by geniculostriate neurons (Ch. 29).

Oculomotor hypokinesia Inadequate saccades associated with Parkinson's disease (Ch. 33).

Oculomotor *L.* 'eye-moving'. Refers to the oculomotor nerve (Ch. 23).

Olfactory aura Epileptic aura accompanied by illusion of an odor (Ch. 34).

Oligodendrocyte *Gr.* 'few tree cell'. A myelin-forming neuroglial cell with few processes (Ch. 6).

Operculum *L.* 'cover'. Refers to any of the three opercula covering the insula (Ch. 2).

Ophthalmoplegia *Gr.* 'eye stroke'. Signifies paralysis of extrinsic ocular muscles (Ch. 23).

Organ of Corti The spiral organ of hearing (Ch. 20).

Orthodromic Impulse conduction in a centrifugal direction.

Orthograde transneuronal degeneration Neuronal degeneration in a proximodistal direction (Ch. 9).

Orthograde transport Proximodistal axoplasmic transport (Ch. 6).

Orthography Processing of letter shapes by the visual cortex (Ch. 32).

Oscillation Spontaneous burst-firing of large groups of thalamic reticular neuron; thought to be responsible for occurrence of sleep spindles (Ch. 27).

Osteophytes Bony excrescences associated with spondylosis (Ch. 14).

Otitis media Middle ear disease (Ch. 20).

Otosclerosis Ear disease associated with ankylosis of the footplate of stapes (Ch. 20).

Ototoxic Term referring to drugs causing sensorineural deafness (Ch. 20).

Pachymeninx *Gr.* 'thick membrane'. The dura mater (Ch. 4).

Paleo- *Gr.* 'old', e.g. paleocerebellum (mainly the anterior lobe, Ch. 25), paleocortex (olfactory, Ch. 29), paleostriatum (globus pallidus, Ch. 33).

Pallidotomy Surgical lesion of globus pallidus, a treatment for Parkinson's disease (Ch. 33).

Pallidum *L.* 'pale'. Refers to globus pallidus (Chs 2, 33).

Pallium *Gr.* 'cloak'. The cerebral cortex.

Papez circuit An 'emotional' circuit of the limbic system first described by Papez (Ch. 34).

Papilledema Swelling of the optic papilla, usually in association with raised intracranial pressure (Ch. 4).

Para- *Gr.* 'beside'.

Paradoxical sleep Dreamy light sleep accompanied by EEG voltage patterns resembling those of the awake state (Chs 24, 30). See *REM sleep*.

Paralysis agitans See *Parkinson's disease*.

Paralysis *Gr.* 'disablement'. Loss of voluntary movement.

Paraplegia *Gr.* 'paralysis'. Refers to paralysis of both lower limbs (Ch. 16).

Paresis *Gr.* 'weakness'. Incomplete paralysis.

Paresthesia *Gr.* 'side feeling'. A sense of numbness or tingling.

Parkinson's disease Disorder of the basal ganglia associated with one or more characteristic features (Ch. 33).

Partial seizure See *Focal seizure*.

Passive stiffness Resistance of ankle dorsiflexors to slow passive plantar flexion in patients with longstanding hemiparesis; probably caused by collagen accumulation within these muscles (Ch. 16).

Pattern generators Patterned activities involving cranial or spinal nerves (Chs 16, 24).

Peduncle *L.* 'little foot'. Refers to stem, e.g. cerebral peduncle (Ch. 2).

Peri- *Gr.* 'around'.

Peroneal nerve entrapment Compression of the common peroneal nerve at the neck of the fibula (Ch. 12).

PET Positron emission tomography (Ch. 29).

PGO pathway Pathway from pontine reticular formation via lateral geniculate body to occipital cortex, responsible for rapid eye movements during light sleep (Ch. 30).

Phase cancellation A recording event (not a physiologic one) during compound sensory nerve action potential measurements, where positive and negative phases of adjacent waveforms tend to cancel each other out (Ch. 12).

Phasic motor neurons Motor neurons innervating squads of fast, glycolytic muscle fibers (Ch. 10).

Phoneme A syllable of sound corresponding to a grapheme (Ch. 32).

Phonemic paraphrasia In Wernicke aphasia, the use of incorrect but similar-sounding words (Ch. 32).

Phonology The sounds of words (Ch. 32).

Pia mater *Gr.* 'soft cover'. The innermost layer of the meninges (Ch. 4).

Pineal *L.* 'pine cone'. Pineal gland is part of the epithalamus (Ch. 27).

Plasticity Capacity of neurons to adapt to a changed environment (Ch. 29).

Plexopathy Neuropathy within one or more of the limb plexuses (Ch. 12).

Plexus *L.* 'interwoven'. Interwoven nerves or blood vessels.

Polymodal afferent Afferent neuron receptive to more than one sensory modality, e.g. touch and pain (Ch. 11).

Polymodal association cortex Association cortex processing signals in more than one sensory modality (Ch. 29).

Polymodal nociceptors Peripheral nociceptors (notably in skin) responsive to noxious thermal, mechanical, or chemical stimulation (Ch. 34).

Polyphasic MUAPs Compound motor unit action potentials having an abnormally large number of positive and negative phases, signifying *reinnervation* of motor end plates vacated by earlier degeneration of their nerve supply, followed by takeover by neighboring healthy axons (Ch. 12).

Pons *L.* 'bridge'. Part of brainstem in the interval between midbrain and medulla oblongata (Ch. 2).

Position sense Perception of the position of body parts (Ch. 15).

Positron emission tomography (PET) Imaging technique identifying radioactive atoms that emit positrons (Ch. 29).

Posterior circulation The vertebrobasal arterial system including the posterior cerebral arteries (Chs 5, 35).

Posterior columns The gracile and cuneate fasciculi (Ch. 3)

Postherpetic neuralgia Neuropathic pain sequel to *herpes zoster* inflammation (Ch. 34).

Postjunctional receptors Receptors in the plasma membranes of autonomic target tissues (Ch. 13).

Postsynaptic inhibition Inhibition of a target neuron (Ch. 6).

Posttraumatic stress disorder A disorder where a seemingly innocent sight may precipitate vivid recollection of a horrific experience (Ch. 34).

Postural fixation Fixation of axial musculature prior to voluntary movement of a limb (Ch. 25).

Postural hypotension Fall of arterial blood pressure on assuming the upright posture (Ch. 16).

Posture The position adopted between movements (Ch. 16).

Prejunctional receptors Receptors in the plasma membrane of postganglionic nerve terminals (Ch. 13).

Pressure coning Displacement of the cerebellar tonsils into the foramen magnum with compression of the medulla oblongata (Ch. 6).

Presynaptic inhibition Inhibition of conduction along the axon terminal of a target neuron (Ch. 6).

Priapism Persistent involuntary erection of the penis (Ch. 13).

Primary myopathy Degenerative change originating in muscle fibers, e.g. in muscular dystrophies (Ch. 12).

Primary neuropathy Neuropathy originating in peripheral nerve fibers; cf. *Secondary neuropathy* (Ch. 12).

Primary tonic–clonic seizures Seizures that are generalized from the start (Ch. 30).

Procedural memory 'How to do' memory (Ch. 33).

Progressive bulbar palsy Rapidly lethal medullary variant of progressive muscular atrophy (Ch. 16).

Projection *L.* 'forward throw'. Target of a neuronal pathway, e.g. spinothalamic tract projecting to the thalamus (Ch. 15).

Prolapsed intervertebral disk Herniation of the nucleus pulposus into the vertebral canal with consequent pressure on spinal nerve roots (Ch. 14).

Proprioception *Gr.* 'self perception'. Conscious or non-conscious reception, by the brain, of information from muscles, tendons, and joints (Ch. 15).

Propriospinal tract Fibers of internuncial neurons passing from one segment of the spinal cord to another (Ch. 15).

Pros- *Gr.* 'before', e.g. prosencephalon, the embryonic forebrain (Ch. 1).

Prosody Tonal variations during speech having emotional attributes (Ch. 32).

Prosopagnosia *Gr.* 'face-no-knowledge'. Inability to recognize faces (Ch. 29).

Pseudobulbar palsy Term given to the clinical picture arising from compromise of the corticonuclear supply to motor cranial nuclei in pons and medulla oblongata (Ch. 18).

Psychogenic impotence Impotence probably caused by excessive sympathetic activity (Ch. 13).

Psychosis Mental disorder involving a distorted sense of reality, e.g. schizophrenia (Ch. 32).

Ptosis *Gr.* 'falling'. Drooping of the upper eyelid (Ch. 23).

Pulvinar *L.* 'cushion'. Refers to posterior bulge of thalamus above the midbrain (Ch. 3).

Pure motor syndrome Syndrome associated with a lacuna in the posterior limb of the internal capsule (Ch. 35).

Pure sensory syndrome Syndrome associated with a lesion of the ventral posterior nucleus of the thalamus (Ch. 35).

Putamen *L.* 'shell'. Refers to outer part of lentiform nucleus (Ch. 2).

Quadrantic hemianopia Blindness in one quadrant of the visual field (Ch. 28).

Quadriplegia *L., Gr.* 'four-stroke'. Paralysis of all four limbs (Ch. 16).

Radiculopathy (*L. radix*, 'root'). Nerve root pathology (Ch. 12).

Rage attack Irritability increasing to a state of dangerous aggressiveness, attributed to hyperactivity of the amygdala (Ch. 34).

Raphe *Gr.* 'seam'. Refers to the midline. Greek genitive case is used in nucleus raphes magnus (Ch. 17).

Raynaud phenomenon Painful blanching of the fingers in cold weather (Ch. 13).

Receptor A term with two distinct meanings: (a) sensory receptors, e.g. photoreceptors, neuromuscular spindles, transduce sensory stimuli; (b) molecular receptors on or within cells are protein molecules acted on by messenger molecules, e.g. hormones, neurotransmitters.

Receptor blocker A drug that can occupy a membrane receptor without activating it (Ch. 13).

Receptor potential The membrane potential at peripheral sensory nerve endings at which action potentials are generated (Ch. 7).

Referential recording Montage following EEG recording where electrode pairs have one recording site in common, e.g. the auricle or mastoid area, (Ch. 30). Cf. *Bipolar recording*.

Referred pain Perception of pain at a site distant from the locus of nociceptive activity (Ch. 13).

Reflex reversal Term used to signify the 'change of sign' accompanying a switch from reflex resistance to reflex assistance (Ch. 16).

Reflexive impotence Impotence caused by damage to reflex arcs (Ch. 13).

Relative refractory period Time interval after firing one action potential, during which a greater-than-normal negative current is required to generating another (Ch. 7).

REM sleep Rapid eye movement sleep (Chs 24, 30).

Renshaw cells Neurons in the medial part of the anterior horn exerting tonic inhibition on alpha motor neurons (Ch. 16).

Resistance reflex Resistance to passive movement, notably at the knee joint, produced by passive stretching of neuromuscular spindles (Ch. 16). Cf. *Assistance reflex*.

Resting tremor Tremor at rest; characteristic of Parkinson's disease (Ch. 33).

Reticular *L.* 'net-like', e.g. reticular formation (Ch. 24).

Retrieval Calling up items from memory stores (Chs 32, 34)

Retrograde amnesia Amnesia for past events (Ch. 34).

Retrograde transport Transport of materials from axon terminals to soma (Ch. 6).

Rhinencephalon *Gr.* 'nose brain'. Refers to olfactory areas (Ch. 29).

Rhombencephalon *Gr.* 'rhomboid-brain' containing the rhomboid fourth ventricle. Embryologically, the hindbrain vesicle (Ch. 1).

Rolandic epilepsy See *Benign rolandic epilepsy*.

Rostral limbic system The anterior cingulate cortex and related areas (Ch. 34).

Rostrum *L.* 'beak'. Rostrum of corpus callosum extends from genu to lamina terminalis (Ch. 2).

Rubro- *L.* 'red'. Refers to projections from red nucleus.

Saccade Scanning movement of the eyes (Ch. 23).

Saltatory conduction *Gr.* 'jumping'. Mode of impulse conduction along myelinated fibers (Ch. 6).

Satellite *L.* 'attendant', e.g. the satellite cells in spinal ganglia.

Schizophrenia *Gr.* 'split mind'. Psychiatric state with variable manifestations (Ch. 32).

Scotoma Blind spot (Ch. 28).

Second messenger In the context of the transmitter activating a G-protein–coupled membrane receptor is the first messenger, the second messenger (e.g. cyclic AMP) is the responding molecule that initiates biochemical events in the cytoplasm and/or nucleus. (Ch. 8).

Secondary neuropathy Neuropathy of motor nerve fibers brought about by collapse of myelin sheaths secondary to anterior horn disease (Ch. 12).

Secondary tonic–clonic convulsion Convulsion characteristic of temporal lobe epilepsy (Ch. 34).

Segmental antinociception Relief of pain at segmental level, e.g. by 'rubbing the sore spot' (Ch. 24).

Seizure Epileptic attack (Ch. 30).

Semantic Having to do with meaning (Ch. 32).

Semantic retrieval Retrieving the meaning of a word from memory stores (Ch. 32).

Senile tremor Benign essential tremor, often mistaken for a sign of Parkinson's disease (Ch. 33).

Sensitization Lowering the threshold of peripheral nociceptors (by histamine in particular) following peripheral release of peptides via the axon reflex (Ch. 34).

Sensorineural deafness Deafness caused by disease of the cochlea or of the cochlear nerve (Ch. 20).

Sensory competition An explanation for cortical sensory plasticity (Ch. 29).

Sensory modality A distinct mode of sensation, e.g. touch vs pain (Ch. 11).

Sensory unit A cutaneous nerve fiber together with all of its sensory terminals (Ch. 11).

Septal rage A rage attack thought to originate in the septal nuclei (Ch. 34).

Septum pellucidum *L.* 'transparent partition' separating the frontal horns of the lateral ventricles (Ch. 2).

Sign An objective indicator of a disorder, e.g. Babinski sign (Ch. 16).

Sleep paralysis A state of complete awareness accompanied by complete paralysis, at beginning and/or end of an attack of narcolepsy.

Slow pain Aching pain perceived following activation of C-fiber nociceptors (Ch. 34).

Slow receptor See *Metabotropic receptor*.

Slow-wave sleep Stages 3 and 4 of sleep, characterized by slow, delta waves (Ch. 30).

Sodium pump Membrane ion channel capable of simultaneously extruding Na^+ ions while importing K^+ ions (Ch. 7).

Sole plate An accumulation of nuclei, mitochondria, and ribosomes in the sarcoplasm at the myoneural junction (Ch. 10).

Soma *Gr.* The cell body of a neuron (Ch. 6).

Somatic *Gr.* 'body-related'. Implies body wall as distinct from viscera.

Somatosensory evoked potentials Waveforms recorded at surface landmarks en route from the point of stimulation of a peripheral nerve to the contralateral somatic sensory cortex (Ch. 31).

Somatotopic Containing a body map.

Somesthetic Somatic sensory, e.g. somesthetic cortex (Ch. 29).

Sound attenuation Dampening of sound by contraction of stapedius or tensor tympani (Ch. 20).

Sound field The peripheral field of sound perception by one ear (Ch. 20).

Spastic diplegia A form of cerebral palsy involving corticospinal fibers destined for lower-limb motor neurons (Ch. 33).

Spastic State of increased muscle tone induced by an upper motor neuron lesion (Ch. 16).

Spatial sense Perception of the position of objects in relation to one another (Ch. 29).

Speech metanalysis Post hoc analysis if one's own speech (Ch. 32).

Sphincter–detrusor dyssynergia Failure of the urethral sphincter to relax at the onset of micturition; a feature of spinal cord injury (Ch. 16).

Spike Action potential (Ch. 7).

Spina bifida Varying degrees of failure of the embryonic neural arches to unite in the posterior midline (Ch. 14).

Spinal anesthesia Anesthesia induced by injection of local anesthetic into the lumbar cistern (Ch. 4).

Spinal lemniscus Conjoint anterior and lateral spinothalamic tracts (Ch. 15).

Spinal tap (lumbar puncture) Needle puncture of the lumbar cistern to obtain a sample of cerebrospinal fluid (Ch. 4).

Splanchnic *Gr.* 'visceral'.

Splenium *L.* 'pad'. Refers to posterior end of corpus callosum (Ch. 2).

Startle response Generalized muscle twitch in response to sudden loud noise (Ch. 20).

Static labyrinth The utricle and saccule (Ch. 19).

Static posturography Study of postural responses on an unstable platform (Ch. 25). Cf. *Dynamic posturography*.

Stellate *L.* 'star-like', e.g. stellate ganglion (Ch. 13).

Stereoanesthesia Inability to identify an unseen object held in the hand, as a consequence of cortical-type sensory loss (Ch. 29).

Stereopsis Three-dimensional vision (Ch. 29).

Stimulus-induced analgesia Analgesia induced by electrical stimulation of the periaqueductal gray matter (Ch. 24).

Strabismus *Gr.* 'squinting' (Ch. 23).

Stress incontinence Incontinence brought about by weakness of the pelvic floor (Ch. 13).

Stria *L.* 'narrow band', e.g. stria terminalis originating in the amygdala (Ch. 2).

Striatum *L.* 'furrowed'. Refers to caudate nucleus and putamen taken together (Ch. 2).

Stroke A disorder of brain function following a cerebrovascular accident that does not disappear within 24 h (Ch. 35).

Subfalcal herniation Displacement of brain tissue beneath the falx cerebri (Ch. 6).

Subiculum *L.* 'little layer'. Refers to transitional zone between six-layered parahippocampal gyrus and three-layered hippocampus (Ch. 34).

Substantia *L.* 'substance', e.g. substantia gelatinosa of the spinal gray matter (Ch. 15), and substantia nigra of the midbrain (Ch. 17).

Sulcus *L.* 'groove'.

Supervisory attentional system Attention center within the dorsolateral prefrontal cortex (Ch. 32).

Supraspinal antinociception Pain suppression by pathways descending to the posterior horn gray matter from the brainstem (Ch. 24).

Sympathetic *Gr.* 'with feeling', i.e. responsive to emotional state.

Symptom Clues about a disorder derived from the patient's account.

Synapse *Gr.* 'contact'. Refers to sites of contact between neurons (Ch. 6).

Syndrome *Gr.* 'running together'. A characteristic group of symptoms and signs.

Syringomyelia *Gr.* 'marrow tube'. Refers to central cavitation of the spinal cord (Ch. 15).

T channels Voltage-gated calcium channels causing *transient* (momentary) bursts of thalamic reticular neuronal activity during slow-wave sleep (Chs 27, 30).

Tactile agnosia Inability to identify common objects by touch alone (Ch. 32).

Tactile menisci Merkel cell–neurite complexes.

Tandem Romberg sign A toe-the-line test for presence of ataxia (Ch. 15).

Tapetum *L.* 'carpet'. Refers to sheet of callosal fibers above the lateral ventricles (Ch. 2).

Tarsal tunnel syndrome Compression of the tibial nerve within the tarsal tunnel (Ch. 12).

Tectum *L.* 'roof' of midbrain, comprising the four colliculi.

Tegmentum *L.* 'covering'. Refers to intermediate region of midbrain and pons (Ch. 17).

Tela choroidea *L.* 'membranous web', consisting of vascular pia-ependyma (Ch. 2).

Telencephalon *Gr.* 'endbrain', comprising the embryonic cerebral hemispheres (Ch. 1).

Teloglia *Gr.* 'distant'. Modified Schwann cells surrounding axons in encapsulated nerve endings (Ch. 11).

Temporal dispersion Elongation of the compound sensory nerve action potential over time (Ch. 12).

Temporal lobe epilepsy Temporal plane. Upper surface of temporal area (Ch. 32).

TENS Transcutaneous electrical nerve stimulation (Ch. 24).

Tentorium *L.* 'tent', e.g. tentorium cerebelli (Ch. 4).

Tetanus *Gr.* 'taut'. Spasmodic state of musculature, especially those of the face and lower jaw following infection with *Clostridium tetani* (Ch. 6).

Tetraplegia *Gr.* 'four-paralysis'. Synonymous with quadriplegia.

Thalamic syndrome Syndrome including contralateral intractable pain, following occlusion of the thalamogeniculate artery (Ch. 35).

Thalamus *Gr./L.* 'meeting place'.

Thermosensitive neurons Peripheral sensory neurons having sensory receptors activated by heat (Ch. 11).

Theta rhythm Slow-wave temporal lobe rhythm associated with the septohippocampal cholinergic pathway (Ch. 34).

Threshold The membrane voltage level at which action potentials are generated.

Thrombus *L.* 'clot' (of blood, Ch. 35).

TIA Transient ischemic attack (Ch. 35).

Tinnitus A ringing, booming, or buzzing sound heard in one or both ears (Ch. 20).

Todd paralysis A jacksonian seizure followed by weakness or paralysis of the affected limb(s) for a period of hours or days.

Tonic motor neuron Motor neuron innervating a squad of slow, oxidative–glycolytic muscle fibers (Ch. 10).

Tourette syndrome Involuntary exclamations associated with anterior cingulate cortex dysfunction (Ch. 34).

Tracking Smooth visual pursuit of a moving object (Ch. 23).

Tract *L.* tractus, 'district'. A group of CNS axons having the same origin and destination.

Transcranial electrical stimulation Recording central motor conduction time by means of transcranial magnetic stimulation of the motor cortex (Ch. 31).

Transduction *L.* 'leading across'. Refers to conversion of sensory stimuli into trains of nerve impulses.

Transient ischemic attack (TIA) Episode of vascular insufficiency causing focal loss of brain function, with total recovery within 24 h (Ch. 35).

Transneuronal atrophy Atrophic neuronal degeneration passing from one neuron to another, in either an orthograde or a retrograde manner (Ch. 9).

Trapezoid *L.* 'diamond-shaped'. Collection of second-order auditory fibers in the pons (Ch. 20).

Tremor Involuntary trembling of one or more body parts; may accompany cerebellar (Ch. 25) or basal ganglionic disease (Ch. 33).

Trigeminal *L.* 'triplet'. Refers to the ophthalmic, maxillary, and mandibular divisions of the trigeminal nerve (Ch. 21).

Trigger point A point or zone in the membrane of the initial segment of the axon where the generator potential arises (Ch. 7).

Triple response The line—flare—wheal response to a sharp stroke to the skin, involving an axon reflex (Ch. 11).

Trochlear *L.* 'pulley'. Transferred epithet: tendon of superior oblique muscle passes through a fascial pulley, and the name is applied to the nerve of supply (trochlear) (Ch. 23).

Trophic *Gr.* 'nourishing' (Ch. 9).

Tropic *Gr.* 'turning'. Chemotropic substances attract ('turn') axonal growth cones (Ch. 9).

Truncal ataxia Inability to maintain the upright position; a feature of midline cerebellar disease (Ch. 25).

Two-point discrimination A test used to assess discriminative touch (Ch. 15).

Ulnar nerve entrapment Compression of the ulnar nerve at the elbow (Ch. 12).

Uncal herniation Displacement of the uncus through the tentorial notch (Ch. 6).

Unconscious proprioception See *Non-conscious proprioception.*

Uncus *L.* 'hook'. In the anterior temporal lobe (Chs 2, 34).

Unimodal sensory cortex Primary somatic sensory, visual, or auditory cortex (Ch. 29).

Unstable bladder See *Detrusor instability.*

Upper motor neuron Corticonuclear–corticospinal neuron.

Upper motor neuron signs Physical signs indicating presence of upper motor neuron disorder (Ch. 16).

Up-regulation Increase in the number of membrane receptors (Chs 33, 34).

Urinary retention Inability to void urine (Ch. 13).

Uvula *L.* 'little grape'. A lobule of the cerebellum (Ch. 25).

Vagus *L.* 'wandering', with reference to the tenth cranial nerve (Ch. 18).

Vallecula *L.* 'little valley' between the cerebellar hemispheres (Ch. 25).

Ventricle *L.* 'little belly'.

Verbal paraphrasia In Wernicke aphasia, the use of words of allied meaning (Ch. 32).

Vergence Convergence of gaze for close-up viewing (Ch. 23).

Vermis *L.* 'worm'. Refers to the segmented appearance of the vermis of the cerebellum (Ch. 25).

Vestibular nystagmus Nystagmus resulting from vestibular disorder (Ch. 19).

Vestibuloocular reflexes Reflexes that maintain the gaze on a selected target despite movements of the head (Ch. 19).

Vibration sense A test of posterior column function, using a tuning fork placed on a bone, e.g. shaft of tibia (Ch. 15).

Visceral referred pain Visceral pain referred to somatic structures innervated from the same segmental levels of the spinal cord (Ch. 13).

Viscerosomatic pain Pain caused by spread of disease from visceral to somatic structures (Ch. 13).

Visual evoked potentials Potentials evoked in the visual cortex when the retina is stimulated by a checkerboard pattern (Ch. 31).

Volume transmission Non-synaptic release of neuropeptides into the extracellular space (Ch. 6).

Wallerian degeneration Orthograde degeneration of a peripheral nerve (Ch. 7).

Warm caloric test A test of kinetic labyrinthine function (Ch. 19).

Wernicke's area Area of the temporal lobe concerned with understanding the spoken word (Ch. 32).

Wind-up phenomenon Sustained state of excitation of central pain-projecting neurons induced by glutamate activation of NMDA receptors (Ch. 34).

Withdrawal reflex Reflex retraction of a limb in response to a noxious stimulus; includes the flexor reflex (Ch. 14).

Working memory Holding relevant memories briefly in mind while performing a task (Ch. 34).

Index

Individual arteries, veins, muscles and nerves are listed under the headings Arteries, Veins, Muscles and Nerves, respectively. Numbers in italic refer to figures.